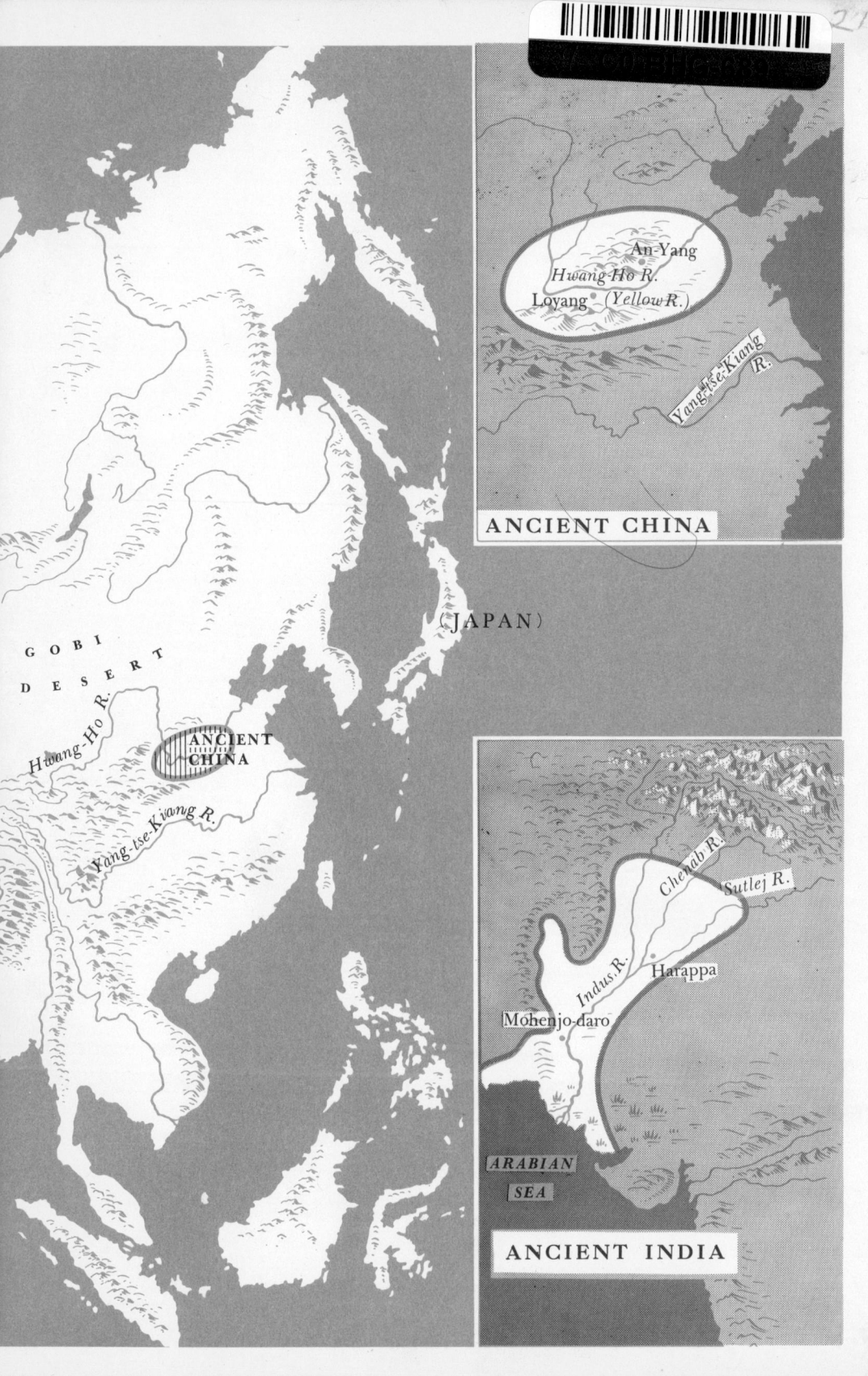

THE COVER

In 1956-57 a French archaeological expedition made an exciting find in the central Sahara desert when over 10,000 New Stone Age pictures, among them this realistic painting of a herd of domesticated oxen, were photographed to scale in their natural colours. This ten-foot mural, frescoed more than 5000 years ago on an exposed sandstone surface, provides concrete evidence that the Sahara once had a climate suitable for cattle-raising. The stylized copper-coloured human figures in the picture represent nomadic herdsmen who originally came from the Upper Nile region, while the negroes show that slaves already formed an accepted part of this early pastoral society. It is also interesting to note the sacrificial ox being dismembered at the far right of the painting.

THE ENDURING PAST

THE

REVISED EDITION

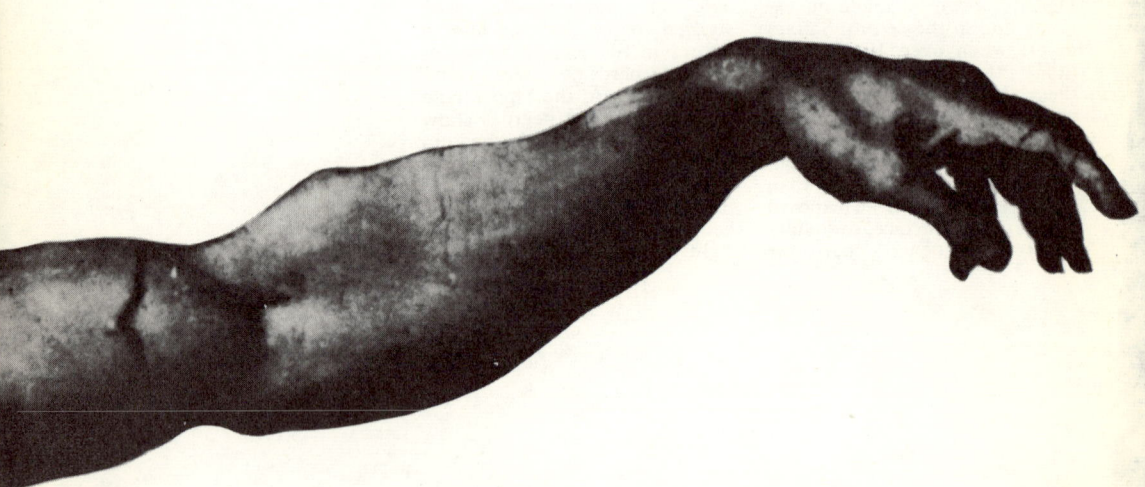

Toronto, The Ryerson Press

ENDURING PAST

EARLIEST TIMES TO THE SIXTEENTH CENTURY

JOHN TRUEMAN, M.A., Ph.D.,
Associate Professor, Department of History, McMaster University

Illustrated by
VERNON MOULD, A.O.C.A.
Head of the Art Department, Upper Canada College

©THE RYERSON PRESS, 1964
PRINTED AND BOUND IN CANADA
BY THE RYERSON PRESS, TORONTO

Published, 1959
Second Printing, 1960, Third Printing, 1962

Revised Edition, 1964, Fifth Printing, 1964

MAPS BY ROBERT KUNZ

The illustration used on the title page is a detail from Michelangelo's Conception of The Creation of Adam *on the ceiling of the Sistine Chapel, Rome. The Hand of God (right) is about to touch and give life to the hand of the first man (left). The ceiling of the Sistine Chapel is one of the truly great masterpieces of Western painting.*

Author's Acknowledgment

Among my scholarly guides I am indebted to the following, who generously read various parts of the manuscript or galley proofs and gave me the benefit of their comments and corrections: three sympathetic colleagues in the Department of History, McMaster University, Hamilton—Professors E. T. Salmon, G. S. French and W. M. Kilbourn; Mr. Heinz Warschauer, Director of Education, Holy Blossom Temple, Toronto; members of the History staff of Saint Michael's College School, Toronto; Professor Thomas R. Millman, Wycliffe College, University of Toronto; Mr. M. H. Baker, Harbord Collegiate Institute, Toronto; and especially Ryerson's indefatigable Mr. James K. Smith.

I have not, then, lacked for good advice, although I have not always taken it. For any errors that may remain, and for the overall selection, emphasis, and interpretation of factual material, I alone am responsible.

Merely to have lived with one dragon while he slew another rates a medal. Yet the manuscript was completely typed and revised—more times than either of us cares to remember—by my wife, who also devised many of the diagrams. For Dawn, the well-worn words "without whom this book would not have been written" are much more than a traditional husbandly garland.

Hamilton, February, 1959   J.T.

Preface to the Revised Edition

In the summer of 1963 President John F. Kennedy, speaking in Frankfurt to representatives of West Germany's government and army, quoted Thucydides' description of the Spartans and their allies. Thucydides shrewdly noted that the Peloponnesians, while potentially mighty in battle, were crippled by their policymaking debates. "Each presses its own end which generally results in no action at all . . . each supposes that no harm will come of its own neglect . . . and so, as each separately entertains the same illusion, the common cause imperceptibly decays." Why should a President of the United States refer twentieth-century Europeans to the history of a little war fought more than 2300 years ago?

This book is an attempt to suggest how we of the twentieth century can view Western man's enduring past. The core is political—though due weight is given to the economic, philosophical, and cultural aspects of man's life—and the treatment of any single person or event must be limited. Yet it is hoped that by concentrating on certain civilizations and their problems, students will derive at least two benefits. First, they may discover in the past the roots of some of the problems faced by their own modern civilization. Second, they may thereby receive some guidance in learning how to face up to these same persistent problems today.

I have bent every effort to make this a thoroughgoing revision. Each page has been carefully scrutinized. Because the narrative now concludes about the year 1600, and because the graded questions and extensive reference bibliography of the first edition are to be published in a separate manual, I have been able to devote considerably more space to the entire ancient world. New sections have also been added on India and China, and the material on the medieval monarchies, the Italian and Northern Renaissance, and the Catholic or Counter-Reformation has been amplified. Students will derive particular pleasure from the additional biographical information now included. Other changes have been made throughout, either to bring the account up to date or to increase the clarity of presentation. All the maps have been redrawn and some new ones added; a number of carefully selected photographs and additional illustrations have been incorporated into the text as teaching aids; new time charts provide chronological reviews for all the civilizations studied; and the pronouncing guide to the index has been revamped and expanded. The popularity of the Source Readings in the first edition has justified the inclusion of more of them. And finally, the revised bibliographies should prove particularly useful in that they now indicate which titles are available in paperback editions as well as which are especially recommended for students.

The suggestions made by teachers who have used the first edition have proved most valuable. I am indebted, too, to Professor Thomas R. Millman of Wycliffe College, University of Toronto, and to Professor Emeritus D. J. McDougall of the Department of History, University of Toronto, for a careful reading of the galley proofs dealing with the history of early Christianity. I also wish to acknowledge the interest of Dr. A. D. Tushingham of the Royal Ontario Museum, whose kindness in providing both material and ideas for the illustration of the Jerusalem dig is greatly appreciated.

What I wrote in 1959 in acknowledging my wife's collaboration is even more apt in 1963. She alone knows how much of midnight—and of dawn—has gone into this revision.

Hamilton, November, 1963 J.T.

Contents

Author's Acknowledgment — v
Preface to the Revised Edition — vi

UNIT I   THE ANCIENT EAST TO 500 B.C.   2
1. The First Civilizations — 3
2. The Middle East Battleground — 48

UNIT II   THE GREEKS   76
3. The Greek City-States — 77
4. The Wars of the Greeks — 103
5. The Triumph of the Greek Spirit — 131

UNIT III   THE ROMANS   160
6. The Foundations of the Roman Republic — 161
7. The Fall of the Republic — 182
8. The Roman Empire and Roman Civilization — 225

UNIT IV   MAKING THE MIDDLE AGES   272
9. The Birth of European Civilization — 273
10. The Feudal Organization of Society — 302
11. The Church and Learning — 335

UNIT V   THE MEDIEVAL TRANSITION   360
12. Trade and Towns — 361
13. The Thirteenth-Century Synthesis — 381
14. Popes, Kings, and Parliaments — 411

UNIT VI   MAKING THE MODERN WORLD   430
15. Europe and Her Wider World — 431
16. The Renaissance — 449
17. The Reformation — 473

Index
Publisher's Acknowledgments

Maps

The First Civilizations: *Front Endpapers*
Europe about 1560: *Back Endpapers*

The Fertile Crescent	14
Ancient Egypt and her Neighbours	20
Bronze Age Invasions (2000-1500 B.C.)	49
Comparative Empires: Egyptian, Assyrian, Babylonian, and Persian	54-55
Ancient Palestine and her Neighbours	62
The Modern Middle East	63
Iron Age Invasions (1200-1000 B.C.)	83
The Colonies of the Greeks	88-89
The City-States of Greece and Asia Minor	110-111
The Hellenistic World and its Kingdoms	148-149
Rome's Conquest of Italy to 218 B.C.	167
The Punic Wars	185
The Roman Republic's Conquest of the Mediterranean (265-44 B.C.)	196-197
The Roman Empire's Expansion (44 B.C.-180 A.D.)	228-229
Products of the Mediterranean World in the 2nd Century A.D.	239
The Barbarian Invasions of the Roman Empire in the 4th and 5th Centuries	246-247
St. Paul's Missionary Journeys	258-259
The Germanic Kingdoms and the Byzantine Empire in 526 A.D.	280-281
Comparative Empires: Alexandrian, Roman, Byzantine, and Moslem	290-291
The Carolingian Empire	293
Vikings, Magyars, and Moslems	308-309
England and France in the 12th Century	323
Medieval Commerce: Principal Sea Routes by the 14th Century	370
The Mongol Empire (about 1250)	374-375
The German Empire in 1250	390
Cities of Renaissance Italy	435
The Widening of the Europeans' World	441
The Religious Divisions of Europe in 1600	486-487

UNIT I
THE ANCIENT EAST TO 500 B.C.

1. The First Civilizations

*The generations pass away,
While others remain.*
ANCIENT EGYPTIAN SONG

Did your grandfather know Julius Caesar? The question, you say, is absurd—yet when you were in Grade 2 you might have taken it quite seriously.

Since that time you have learned a great deal about the measurement of time and can clearly distinguish between the present and the near and distant past. It is this knowledge of the passage of time on which you must draw for an intelligent study of history. If your view of human history is to consist of more than a series of isolated snapshots, if it is to become a living moving picture, then you must have some time scale by which to measure past events.

1/A DAY IS A THOUSAND YEARS

Here is a Canadian scientist's way of looking at the passage of time. He fits thousands of years into a week, beginning on Sunday, January 1st.

Let us imagine that our human time scale can be compressed so that one day represents 1000 years. Thus our scientific history is written over a period of one week beginning on Sunday morning, representing 5000 B.C. Throughout all day Sunday, Monday and part of Tuesday no progress at all is evident. . . . By late Tuesday evening some stone and timber buildings are evident and the Pyramids have been built using human power. Bronze tools are being used on Wednesday and oxen are pulling wheeled carts. Iron tools, hardened by tempering, are ready Thursday. Some canals and reservoirs have been constructed and by nightfall an alphabet has been invented to simplify communication.

Friday morning geometry has been exploited to make accurate land surveys. The screw and pulley have been invented and the Romans are building aqueducts and highways. . . . On this same Friday the stone arch construction is becoming evident in large bridges, palaces and cathedrals. . . .

Saturday the acceleration is apparent. Algebra and trigonometry are available as scientific tools. Magnetism has been discovered; alchemy is flourishing and cast iron is being produced in quantity; sizeable boats are being built and the invention of the magnetic compass is of great importance to their navigation. But it is only on Sunday that things really start to happen. Just before midnight Saturday, an Italian sea captain had his western trip to Asia interrupted by America and by 3 a.m. mass migration was under way. The steam engine was invented at 7 a.m. and immediately applied to land and sea transport. The Bessemer converter

---

Lion-fighting for the Assyrian kings, like bull-fighting for the Cretans (see page 80), was a ritual. This limestone bas-relief from Nineveh shows Ashurbanipal tackling a lion released for the royal sport. The beast has already been wounded with arrows, and the king is finishing him off with a strong sword thrust through the heart. Though it was the custom for the arm to be protected by heavy wrapping, the artist, not wishing to detract from the nobility of the king's figure (or perhaps from his valour), has shown Ashurbanipal meeting the lion's last lunge with bare arms. "Assyrian art," it has been said, "reached its highest expression in scenes of bloodshed"—particularly, it might be added, when the blood shed was that of a wounded animal.

announced at 9 a.m. made large quantities of steel available, followed quickly by Portland cement, rubber, and petroleum. Cheap aluminum was available by 9:30, man flew in an aeroplane before 10, Lindbergh landed in Paris at 10:25, supersonic flights were common by 11 and the first artificial satellites were in the sky before lunch.[1]

This book covers an even longer scale of time than that of our imaginary week. In fact, using the same scale of one day equalling a thousand years, the story of man begins a full year and a third earlier than the week of time that began in 5000 B.C. In other words, man has lived on earth for approximately 500,000 years.

We may represent this long development by a time line. You will notice that the word "Civilization" appears opposite the date 3500 B.C. But what does it mean to be civilized? What can historians tell us of the hundreds of thousands of years before man painfully evolved into a civilized human being?

Civilization Civilization is not an easy word to define. It is derived from the Latin *civilis* meaning "relating to a citizen," that is, relating to a member of an ancient city-state. To be civilized and not to have been under the influence of a town or city, then, is a contradiction in terms. For this reason we date civilization from the establishment of man's first cities, about 3500 B.C.

2/MAN THE HUNTER

About 700,000 years ago[2] the climate of the world changed radically. Cold air flowed southward from the north pole and glaciers descended to cover all of Europe down to the British Isles, while in North America Canada was completely covered by the ice cap. Four times the ice moved south, and four times the climate warmed and the ice retreated. Today we are living in the fourth interglacial period which followed the retreat of the ice between 18,000 and 9000 B.C.

Up until 1959 the oldest predecessor of man was presumed to be an apelike being who once inhabited South Africa. Then discoveries were made which may eventually have the effect of drastically altering our picture of earliest man. For in July, 1959, a British paleontologist, Dr. L. S. B. Leakey, unearthed in the Olduvai Gorge of Tanganyika fragments of a

---

[1] J. W. Hodgins, "Modern Engineering at McMaster," *McMaster Alumni News*, Vol. 28, No. 1 (March, 1958), p. 8.

[2] All dates before 2000 B.C. are approximate since they are constantly being revised. Because of the lack of conclusive evidence or conflicting calendar interpretations, scholars even debate some dates in Greek, Roman and medieval history.

Recently scientists have come to the aid of historians in devising methods of determining chronology. They have found that all living matter, be it plant or animal, absorbs radioactive carbon (C^{14}). When living matter dies it gives up its C^{14} at the fixed rate of 15.3 atoms per minute per gram of carbon. Thus by measuring the amount of C^{14} remaining in any sample, the number of years since the living matter died can be calculated within a margin of error of 200 or 300 years either way. Besides this radio-carbon method of dating, which has a range of about 50,000 years, the rate of decay of potassium in volcanic rocks can also be measured by scientists to fix geological dates much farther back.

In this book regnal dates are given for rulers, life dates for others.

THE FIRST CIVILIZATIONS

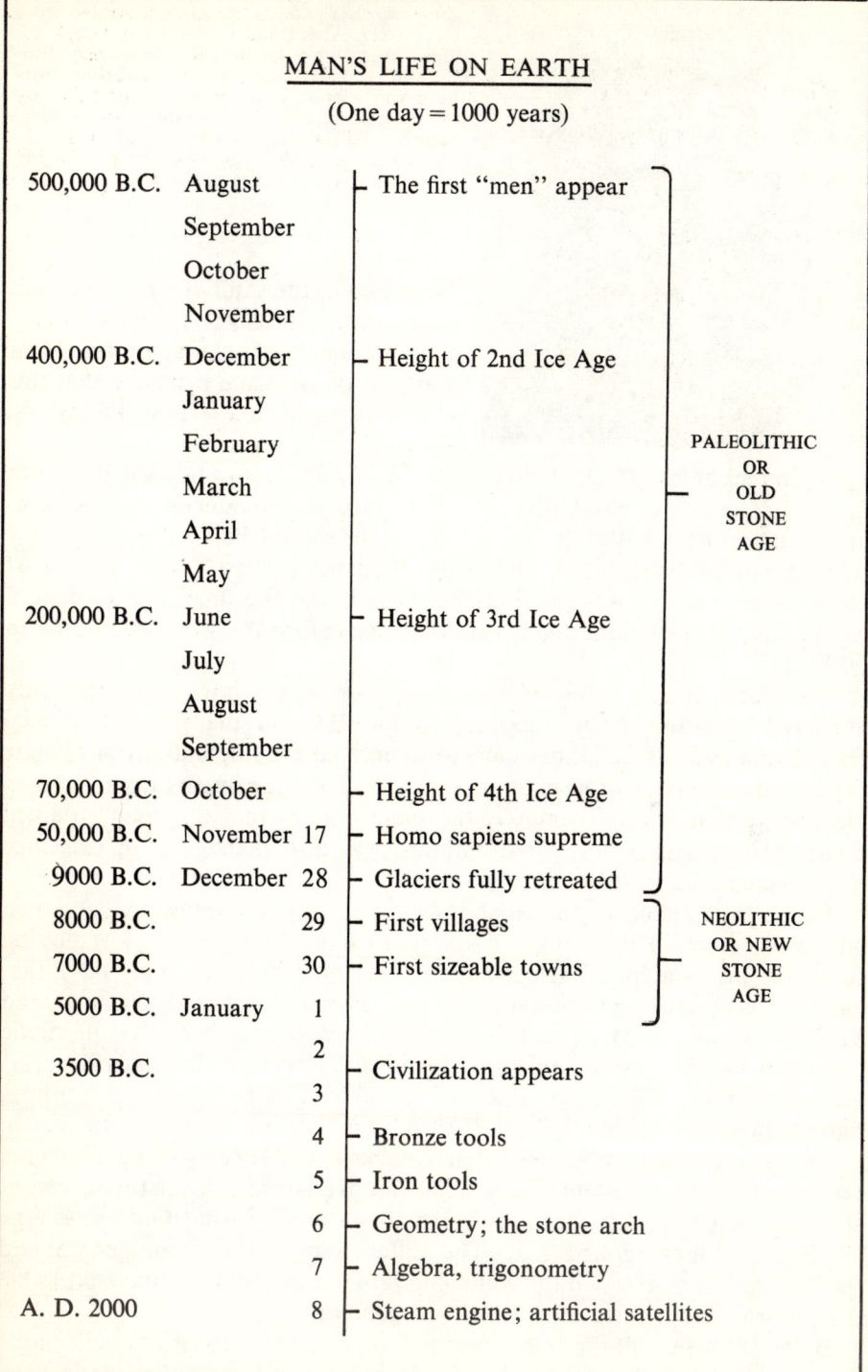

MAN'S LIFE ON EARTH
(One day = 1000 years)

500,000 B.C.	August	— The first "men" appear	
	September		
	October		
	November		
400,000 B.C.	December	— Height of 2nd Ice Age	
	January		
	February		PALEOLITHIC
	March		OR
	April		OLD STONE AGE
	May		
200,000 B.C.	June	— Height of 3rd Ice Age	
	July		
	August		
	September		
70,000 B.C.	October	— Height of 4th Ice Age	
50,000 B.C.	November 17	— Homo sapiens supreme	
9000 B.C.	December 28	— Glaciers fully retreated	
8000 B.C.	29	— First villages	NEOLITHIC
7000 B.C.	30	— First sizeable towns	OR NEW STONE AGE
5000 B.C.	January 1		
	2		
3500 B.C.	3	— Civilization appears	
	4	— Bronze tools	
	5	— Iron tools	
	6	— Geometry; the stone arch	
	7	— Algebra, trigonometry	
A. D. 2000	8	— Steam engine; artificial satellites	

Zinjanthropus (Zinj is the ancient Arabic word for eastern Africa and anthropus is Greek for man) may have looked like this—very low-browed yet long-faced. Behind his shallow forehead was a brain little more than half the size of modern man's. On the shore of a long-vanished lake this early ancestor of ours fashioned his stone tools, possibly some 1,750,000 years ago.

large part of the skull of what he claims is a more advanced type of "man" than the South African ape-man. Most startling of all is the evidence that this Olduvai "man" knew how to fashion his own crude tools.

When had he lived? Dr. Leakey first dated him about 600,000 B.C.; but in 1961 scientists of the University of California announced a potassium-argon dating of 1,750,000 B.C. If this date should be confirmed (many scholars still withhold final judgment), then the earliest "man" is at least three times as old as was previously believed, and the time line on page 5 would have to start some *four years* or more before the week beginning in 5000 B.C.

The first "men"

The oldest "men" about whom scholars do agree, however, apparently appeared in South Africa about 500,000 B.C. in the first interglacial period, and their descendants somehow survived the fluctuations of climate through hundreds of thousands of years. Yet fossil remains discovered in Java prove that at approximately the same time as the "ape-man" roamed South Africa, a more man-like creature was also managing an existence somewhere in the East Indies.

The scientific name of Java man is *Pithecanthropus erectus,* "the ape-man who walks erect." He was like an ape in his shambling gait, his low sloping forehead, his prominently ridged brows and his lack of a chin. But Pithecanthropus erectus had a combination of special features which he shared to some extent with apes and monkeys. The difference was that he alone possessed five in a fully developed form: erect posture, free-moving arms and hands, sharp-focusing eyes, a brain capable of judgment and perception, and the power of speech. With these five, man became master of the earth.

The first modern man

The bones of a number of different genera and species of earliest man (usually parts of the skull) have been discovered, and from these, reconstructions have been made to give us a rough idea of what Old Stone Age (Paleolithic) men looked like. The differences in the main genera and species are summarized in the following table in the order of their probable disappearance up to the time of *Homo sapiens*.

By 50,000 B.C. all the other men had died out and Homo sapiens alone survived. From the neck down he was little different from either other fossil

GENERA AND SPECIES OF EARLY MAN

1. South African Ape-Man—skull of a child and remains of six adults, discovered from 1924 on; adult weighed about 80 pounds and stood 4 feet tall; brain a little larger than ape's.

2. Java Man —skullcap, jawbone and thighbone first discovered in Java in 1890; larger than S. African Ape-Man and having a larger brain, though only half as large as modern man's.

3. Peking Man —remains discovered in a cave near Peking, China, from 1927 on; about the same size as Java Man but with legs more bowed.

4. Rhodesian Man —skull, shinbone, thighbone, etc. discovered at Broken Hill, Rhodesia, in 1921; heavy-boned adult standing 5 feet 10 inches and weighing about 160 pounds; brain almost as large as modern man's; another specimen found in Cape Province in 1953.

5. Neanderthal Man —first specimen found in Neander valley in Germany in 1856, and many others in various locations since; stocky adults about 5 feet 4 inches tall with heavy bones and a brain slightly larger, though less developed, than modern man's.

6. Homo sapiens ("Thinking Man") —part of a jaw discovered in Kenya Colony in 1932; other discoveries since near London, England, in 1935, 1955, and at Fontéchevade, France, in 1947; a large brain set behind a high forehead, and a chin like modern man's.

men or modern man. But from the neck up both Homo sapiens and present-day man proved to be greatly different from the others, for both had high foreheads encasing a large brain, and a lower jaw with a chin. It was probably this larger and more fully developed brain which preserved Homo sapiens for his role of the first modern man.

Thousands of Generations The evidence for prehistoric men would be paltry if it were based only on fossil remains. Fortunately we have also been able to recover millions of tools that the earliest men used. These implements have been discovered in the trash piled up from generation to generation on the floors of caves, some of which collected as much as fifty feet of debris to record the passing of thousands of generations of human occupants.

Prehistoric man did not pass his days in the dark depths at the back of some gloomy cave. Instead, he lived just within the cave mouth or on some adjoining terrace. He might even prefer the shelter of an overhanging cliff, much as the overnight camper does in the Canadian woods. Old Stone Age man hunted game, and, dragging it back to his habitation, littered his floor with the bones of the kill—woolly mammoth, bison, reindeer, rhinoceros, antelope, wild horse, or cave bear.

Partly because the largest social group was that of the family, the inventive level of savage society was extremely low. Calculating one man-power as equivalent to a tenth of one horsepower, it has been estimated that at any one time all of Old Stone Age Europe could produce only as much energy as one modern aeroplane. Anything that might be called progress, anything like a political state or formal laws or even writing, would have to wait for some improvements in methods of obtaining food; for to find and eat raw food took up most of each day.

Paleolithic Accomplishments The products of stone and bone in the early and middle Old Stone Age were mainly the hand-axes and broken bones used to kill or dismember beasts. But toward the end of the Old Stone Age weapons became many and varied as men gradually learned how to improve on their earliest tools. They began to flake off flints to produce scrapers and crude knives, and eventually evolved an elementary chisel with which to rough out other tools of bone, antler or ivory.

Fire

Fully as important as the manufacture of better tools was the use of fire. Probably Paleolithic men "discovered" fire when they came upon an erupting volcano or a roaring forest fire. There is no evidence that the earliest men who had fire knew how to make it themselves: if it died out in one camp it would have to be obtained from another. Fire kept men warm; but it did even more. It allowed them to cook their food so that bones could readily be broken and the marrow sucked out, while the fibres in both meat and roots could be broken down. Consequently, the time taken in eating could be reduced from most of the day to about two hours.

Paleolithic art

In the homes of some of these first men—the caves of northern Spain and southern France—archaeologists have also discovered lifelike sculptures of bone and ivory, including one representation of the head of a woman, who even wears an elaborate head-dress.

On the walls of the caves themselves the engravings and paintings of

This Old Stone Age carving of a girl's head, found in southern France, has a quality of pleasing naturalism and is well executed technically. It portrays either a head-dress or an attempt at a "hairdo." Which do you think it is? Can you suggest why, when Paleolithic Age artists could portray animals so realistically, their representations of human beings were not more lifelike?

mammoths and reindeer may be seen in all their wonderful profusion. Nor were the artists who made these outlines by the light of small stone lamps any crude craftsmen. Whether it was a herd of passing reindeer, a stag pausing to drink, or a horse galloping by, the realism of their work is amazingly successful. For a long time modern scholars were critical of the paintings of the woolly mammoth, critical, that is, until a mammoth found frozen in Siberia proved the scholars wrong and the Stone Age artists right.

The most impressive product of Paleolithic artists is the series of multi-coloured murals to be found in such caves as those at Font-de-Gaume or Lascaux in France, or Altamira in Spain. The paintings were first etched on the limestone walls; then a strong outline of black was applied, followed perhaps by another of red, and finally rich browns deftly represented the hairy portions of the bison. The pigments were so skilfully mixed that today in the darkness of Altamira the colours are still almost as brilliant as when they were applied.

It is inconceivable that the average rude hunter would be capable of such fine work. A few trained artists might, however, be spared from the hunt while their fellows roamed abroad in a pack; only if the menfolk of a number of families went out together could big game be pursued and killed. The reason the artists were allowed to work in the smoky recesses of some cave was that they were "guaranteeing" the success of the hunt by the pictures they produced. This explains why the representations on the walls are so lifelike, for it was believed that only if the artists portrayed the mammoth accurately would it be caught. How did they achieve such realism? Perhaps they used dead models or merely drew from keen observation; perhaps they took rough sketches back into the depths of the caves. We shall never know.

Dim Gropings   Man has long considered himself different from all other animals in that he conceives himself to be a combination of body and spirit. Even Paleolithic man, we find, pondered death, and we can guess that he hoped to ease its finality from the manner in which he buried his dead. The burial usually took place right in the home, beneath the cave floor and often near the hearth, with the dead hunter dressed in his finery and smeared with red ochre (perhaps in the hope of restoring blood to the pale corpse). Alongside the body were placed life's necessities—food and weapons. Whatever the exact form primitive man's belief may have taken we cannot now tell, but he seems to have had the hope (or fear) of a life beyond death. Perhaps in his own childlike way Paleolithic man was dimly groping towards a "religion".

Finally, man alone of all animals has the power of speech. Though the first men could not write, they must have been able to speak, to exchange ideas by means of language. Hunters must have told each other how to trap mammoths, artists must have compared sketches, and the survivors within a family must somehow have communicated with one another about their ceremonies for the dead—all of which presupposes some kind of language.

Here are skilled artists at work over 15,000 years ago in a cave in western Europe. For pigments these Old Stone Age craftsmen probably used natural ochres consisting of iron oxides, mixed with animal fat that formed a binder. The man at the extreme left is blowing powdered colour through a bone tube, while his companion draws with a "crayon" moulded of raw pigments. At the right, others grind the pigments in stone containers. The cave is illuminated by light from fat-oil lamps. The hunter in the foreground holds a primitive spear with a flint head, such a weapon as was probably used to kill the bison being sketched on the wall. The spear-head was first shaped by flaking, and at a later period by pressure chipping. What other sharp weapons and implements did Paleolithic men produce by these methods?

No one knows, of course, what the languages were or exactly how they developed; but a number of ingenious theories have been propounded to account for man's earliest speech. The most colourful of these appear in the following table.

HOW SPEECH MAY HAVE ORIGINATED[3]	
1. The *bow-wow* theory	—imitation of sounds, such as the barking of dogs.
2. The *oh-oh* theory	—instinctive expressions evoked by pain or other intense feelings.
3. The *dingdong* theory	—harmony between sound and sense; for every inner feeling there is an outward expression.
4. The *yo-he-ho* theory	—relief gained by letting out one's breath, and so making expressions, following heavy muscular exercise.
5. The *gestural* theory	—imitation with the tongue of gestures already made with the body.
6. The *tarara-boom-de-ay* theory	—humming or singing as the first and easiest way to give vent to one's feelings.

[3]Adapted from A. Montagu, *Man: His First Million Years* (Signet Books: The New American Library, revised edition, 1962), p. 103.

The Ice Retreats About 20,000 years ago the sun at last began to melt away the glaciers, causing them to retire slowly northwards for a fourth time. As the ice melted, the sea level rose. There was a retreat of tundra and a spreading of forest, and while the reindeer followed the tundra north, other mammals became extinct. Bands of hunters equipped with bows and arrows and helped by dogs (all this we know from cave drawings) now marked the scattered human settlements, and tiny flints would seem to indicate that the game included small animals and birds. Yet at the same time, naturalistic art died out and was replaced by geometrical representations. Was it because the huge mammals had migrated or become extinct? Had man turned to abstract drawings when his vivid pictures were no longer able to place a spell over the great beasts?

The time was now coming when man would have to adjust to a new environment. Beginning about 9000 B.C. there was a permanent climatic change in southern Europe, brought on by the retreat of the ice. Because the northward movement of the Atlantic storm-track left huge areas of southern Europe and north Africa without rain, these areas simply dried up. This geographical change must have reduced both animal and human populations, and, in drawing them to the remaining rivers and oases, began a process destined to convert man from a hunter into a farmer. It is a development that historians have not hesitated to label a "revolution."

3/MAN THE FARMER

The first fundamental change in man's relationship to nature was the domestication of plants and of animals other than the dog. At the beginning of the Neolithic or New Stone Age (before 8000 B.C.) man was developing an aggressive attitude towards nature. He observed animals gathering at the water-hole and determined to corral them, or attempted to cut and cultivate the wild grasses about him. With this revolution came the possibility of increased population.

The food-producing revolution

Presumably this first "farming" took place not in Europe but in the arc of Palestine, northern Syria, and the Tigris-Euphrates valley. Polished tools used to reap grass have been discovered where huts were built at oases. In time, as the old huts collapsed and new ones were built on top of them, a mound or *tell* was formed, with the result that the age of the settlement can now be estimated by measuring the depth of the debris. Such tells are common in modern Israel and Iran; and in the Fayum, a great depression in the Egyptian desert west of the Nile, remnants of Neolithic silos have been excavated along with numerous earthenware and stone vessels.

The first villages

These tells supply conclusive evidence that villages of 200 or 300 people carried on some sort of trade or barter, for shells have been found that could only have come from the Mediterranean or Red Seas. Further proof of settled habits comes from the burial grounds where the villagers buried their dead individually or in groups, the bodies decked out with shell jewellery, stone ornaments, and figurines. The discovery of an undergarment of linen shows that there was even weaving in some of these villages. Pottery was crude—usually simple open bowls, made, of course, without the potter's wheel.

In recent years two sites in particular have yielded rich evidence for the archaeologist and the historian: Jarmo in northern Iraq and Jericho in Jordan. Here are extensive records of man's revolutionary transition from the nomadic existence of Old Stone Age times to a settled mode of life.

Perhaps before 6500 B.C., the 150 inhabitants of Jarmo kept herds of sheep and goats, reaped grain, and supplemented their diet by hunting. But at Jericho a thriving town existed as early as 7000 B.C. In fact the earliest settlement there, which grew up beside a spring, goes back to 8000 B.C. Nor was this community of 7000 B.C. any mean town. Surrounded by fertile fields, its well-built clay-brick houses had plastered floors and walls; and because these dwellings covered some eight acres we can assume that there were probably two or three thousand people living within the massive walls of this "oldest known town in the world." Strangely enough, however, these first town-dwellers did not have pottery at all. Instead they used dishes and bowls made of polished stone.

For various reasons the development of Neolithic culture in Europe did not proceed at the same pace as that in the Middle East. In the Middle East it ended about 4500 B.C., but in Europe at a much later date. It is startling to realize that while New Stone Age farmers in Europe and Britain were slashing the forests and learning to keep cattle and pigs, haughty citizens stalked the streets of great Middle Eastern cities such as Memphis and Ur.

THE EARLIEST HISTORY OF MAN

Old Stone Age: Man the Hunter	500,000 years ago	Man appears
	50,000 years ago	Homo sapiens supreme
	30-15,000 years ago	Great period of Old Stone Age art
	before 8000 B.C.	Old Stone Age ends
New Stone Age: Man the Farmer	8000 B.C.	Food-Producing Revolution; first villages
	7000 B.C.	First sizeable towns
	4500 B.C.	New Stone Age ends
Civilization: Man the City-Dweller	after 3500 B.C.	Urban Revolution Civilization begins
		Bronze Age
	before 1100 B.C.	Bronze Age ends Iron Age begins

The Urban Revolution   The first villages and towns had grown up in an arc that curved from Israel and the Jordan valley across northern Syria and the foothill zone of Turkey and northern Iraq to Iran and the shores of the Caspian Sea. It was no coincidence that farming first occurred there. These territories contained oases (as at Jericho) or grassy uplands (as at Jarmo), and for this reason were more favourable to early food production than the great southern river valleys of the Tigris-Euphrates and Nile where the climate was drier.

Despite Jericho's early date, it seems probable that the grassy uplands were more likely places for agriculture to begin than the areas around isolated oases. These uplands had enough winter and spring rainfall to support wild grains, which in turn encouraged wild goats and sheep to pasture there. Some of these animals were domesticated—partly out of the human desire for pets, a desire easily fulfilled where they roamed plentifully. Finally, the foothill zones also held metal ores, vital resources for any sizeable community. What more natural location, then, for man's first farms than the grassy uplands?

As time went on, overpopulation spilled the farmers out of the hills and onto the plains. Peasant villages multiplied until, sometime before 4000 B.C., the men of the plains began to move into the swamps at the head of the Persian Gulf or along the fringes of the Nile Delta. On little islands of dry land they pursued agriculture, raising crops and domesticating animals. The population grew, and once again new land was required, land gained either by draining sections of the swamps or by moving farther upriver. These earliest of pioneers discovered that the land upriver was extraordi-

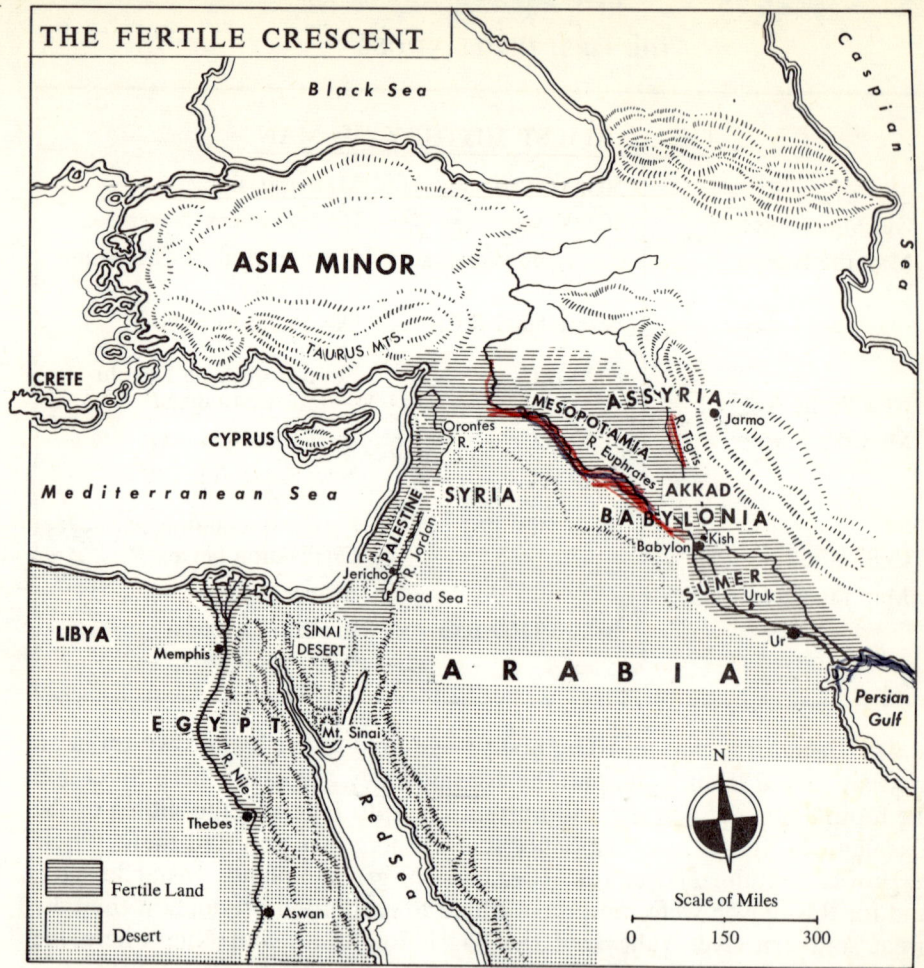

narily fertile. But there was one major drawback: rainfall was insufficient, and more water would have to be obtained somehow. Irrigation was obviously the answer, and irrigation required some central directing authority to mobilize the manpower necessary for digging ditches and building canals and dikes. Only in this way could settled communities produce enough surplus food to allow some of their number to become full-time specialists —craftsmen, miners, or merchants, as well as men involved in directing the work of the community as a whole—rulers, priests and administrators. With these specialists came the birth of cities. So it was that by about 3500 B.C. mature cities developed.

The first cities

In one of the earliest of these cities, Uruk (modern Warka) on the Euphrates River, great temples were built, protected behind a wall 5½ miles long boasting over 900 towers. But from ancient Kish, north of Uruk, comes an even more fundamental feature of city life: our earliest known example of "writing", dated at 3500 B.C.

It took a thousand years—more than twice the length of time that separates us from Columbus—for the Persian Gulf and Nile communities to develop into full-fledged cities in Mesopotamia ("the land between the rivers") and Egypt. Taken together, the amazing changes of the Food-Producing and Urban Revolutions between 8000 B.C. and 3500 B.C. were more far-reaching than all those in the preceding half million years of man's life on earth.

Man was at last civilized.

4/ THE LAND BETWEEN THE RIVERS

Between Mesopotamia and Egypt arches a semi-circle of arable land known as the Fertile Crescent. On one end of this arc, in the rich flood plain between the Tigris and Euphrates Rivers, man built his first cities.

During the Bronze Age, that is, the 2400 years after 3500 B.C. when tools and weapons were made of bronze, the Persian Gulf extended inland at least 60 miles farther than it does today. Various deposits of clay from nine to twelve feet deep bear testimony to the devastating flood that once overwhelmed the city of Ur, a catastrophe which must have seemed like the end of the world to the people in that valley. It is therefore not surprising to find in Mesopotamian mythology a flood epic similar to the Biblical story of Noah.

Agriculture

The various city-states of Mesopotamia based their economy on agriculture, and farmers raised truly amazing crops of barley, wheat, millet, and at least sixty varieties of vegetables. So rich were the harvests that they received this glowing tribute from a 5th century B.C. Greek historian:

> Of all the countries that we know there is none which is so fruitful in grain. It makes no pretension indeed of growing the fig, the olive, the vine or any other tree of the kind; but in grain it is so fruitful as to yield commonly two-hundred-fold, and when production is the greatest even three-hundred-fold. The blade of the wheat-plant and barley-plant is often four fingers in breadth. As for the millet and the sesame, I shall not say to what height they grow, though within my own knowledge; for I am not ignorant that what I have already written . . . must seem incredible to those who have never visited the country.

Naturally, the Mesopotamians became adept at irrigation, and by 3000 B.C. there was a general system of dikes and canals in Sumer, the lower section of the Tigris-Euphrates valley. Their system of canals called for careful legal regulation.

Industry

Industry was organized by crafts of metal-workers in gold, silver, or copper, or weavers of woollens and linens, and wages were fixed for the various workmen. Thousands of clay tablets have survived to show us the careful accounts kept of Mesopotamian business transactions.

Nevertheless, chronology is very obscure for this early period. Ancient lists of kings merely show a series of civil wars between cities in the early part of the third millennium B.C., and about 2350 B.C. the people of Akkad, the northern section of the great valley, rose under Sargon I to take

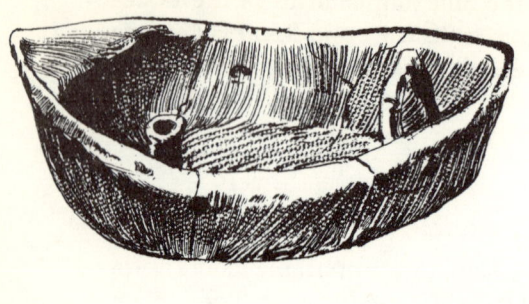 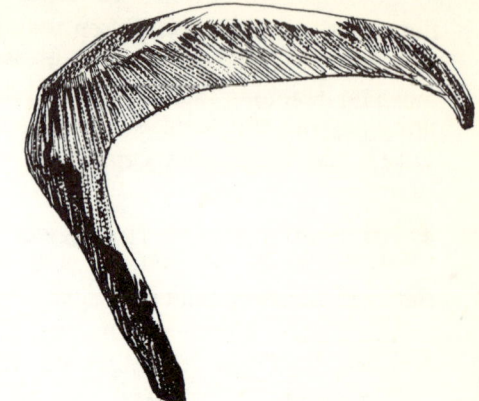

This Mesopotamian clay model of a sailboat (left) shows that by 4000 B.C. man had learned to harness the wind for water transport. It is approximately 10 inches in length and may have been a child's toy. A primitive clay sickle (right) of the same period shows early man's ingenious use of the materials at hand. Without flint, he has used a pottery technique to produce a moderately tough, efficient agricultural implement.

over Sumer as well. Sargon thereby created the first empire in history. His grandson, Naramsin, extended his conquests and covered the country with temples and palaces. About 2050 B.C. the Sumerian city of Ur conquered the Akkadians. Ur in its turn fell to other invaders, who were finally expelled by a line of new kings at the city of Babylon. The sixth and most famous, Hammurabi (1728-1686 B.C.), united the land as far north as Assyria. Later, however, a succession of weak rulers left Babylonia prey to two Indo-European peoples,[4] the Hittites from the north and the Kassites from the east.

Unequal Justice Long before the strong government of Hammurabi, society revolved around the city and its cult. The god of the city supposedly entrusted the administration to an earthly vice-regent, who was at once the overseer of the irrigation system, the commander-in-chief, and the high priest. In the beginning the political power of each city probably rested in the hands of a "general assembly of adult freemen." This body, which was really a form of primitive democracy, could hand over political power to one man in time of crisis.

In any city the temple was the centre of government, administering edu-

[4]Indo-European is a linguistic, not a racial, classification. All Indo-Europeans spoke languages derived from a common stock, though they might be peoples of entirely different nationalities. From their homeland in the western Russian plains they surged out in two great waves, one (2000-1500 B.C.) during the Bronze Age, and another (1200-1000 B.C.) during the Iron Age. In the course of these migrations the original Indo-European language broke down into a number of dialects which proved to be the ancestors of the Celtic, Germanic, Italic (Latin), Greek, Slavic, Persian, and Sanskrit (Indian) tongues.

As far as is known, the Indo-Europeans were the first people to train horses and to use the two-wheeled chariot—factors which account for the wide sweep of their migrations. For maps illustrating the two waves of Indo-European invaders, see pages 49 and 83.

THE FIRST CIVILIZATIONS

cation and landed estates. The god controlled the temple through his human servants. The physical aspect of the city could be both magnificent and squalid: the buildings were of adobe and the streets were narrow and winding, but the temple was an impressive structure. The one at Ur was 200 x 150 feet, and 70 feet high, an artificial mountain or *ziggurat* with platforms or terraces on which flowers or trees might be growing. Perhaps it was on such a ziggurat that the hanging gardens of Babylon flourished, for it is almost certain such a temple inspired the story of the tower of Babel and that of Jacob's ladder.

Prominent in Mesopotamian society were the well-to-do members of the priestly class and the military hierarchy, both of whom possessed hereditary land holdings. The agricultural workers might be slaves or tenant farmers. Justice was available to every man, but, as we know from the famous Code of Hammurabi (selections from which are in the Source Reading) was not equal to all, for a legal distinction was made between the nobles, common people and slaves. An offence committed against a noble was punished more severely than one against a freeman. On the other hand, if a noble committed a crime he was punished more severely than a middle-class transgressor. Perhaps the class distinction in punishment was a result of Mesopotamia's military organization. The freemen were the backbone of the army, and since discipline was all-important their offences were more serious than those of other men.

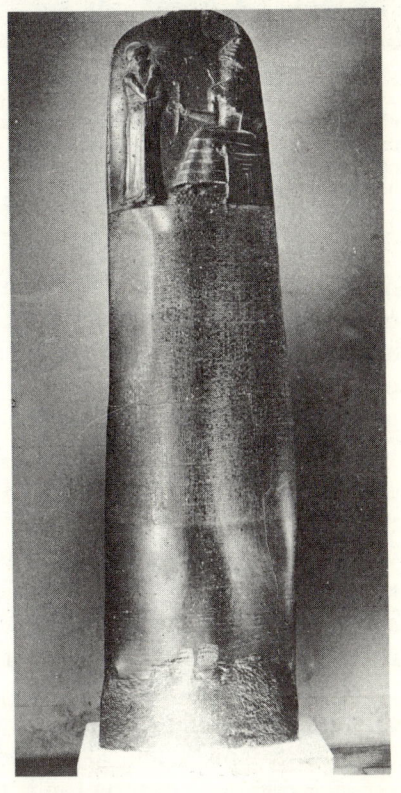

Hammurabi ordered his law code cut in this shaft of black diorite, which was discovered at Susa by French archaeologists in 1901. The pillar is nearly eight feet high. At the top Shamash, the sun god, extends to the worshipping Hammurabi the rod and ring—symbols of his kingly authority—while below, in parallel columns, are inscribed the laws of the code (see the Source Reading on page 44). An Elamite conquerer of Babylon carried this shaft off to Susa in the 12th century B.C.

A Worldly Religion The Mesopotamian religion was primarily concerned with this world. The notions of a life after death were very vague, and in such a shadowy existence the gods had no place or part. The living buried the dead properly only to ensure themselves against being haunted,

Sumerian writing began as a series of crude pictures of common objects carved on stone which in time became stereotyped into a number of lines. The tablet on the left is semi-pictorial. The one on the right shows what happened when the use of stone was abandoned in favour of clay. It was discovered that a transaction could be recorded much more quickly and easily by pressing and turning the end of a square reed, which had been cut obliquely, into wet clay. Since, however, the reed produced only short straight lines, all curved lines had to go in favour of a set of conventionalized wedge-shaped characters—cuneiform.

not to open the way to hell or heaven. The great religious poems such as the *Epic of Gilgamesh,* the story of a legendary king who ruled for 126 years, are hardly moving moral tales. Rather they served to explain man's sad plight, how the world came to be what it is, and the manner in which the gods might be placated. Alongside the refined Hebrew versions of some of these stories, the Mesopotamian epics seem almost disgustingly crude.

The Cradle of Civilization Mesopotamian architects knew how to build the arch and the dome out of brick, and the long lines of the mighty ziggurat were curved to avoid the optical illusion of sag—a principle used later in the Athenian Parthenon of the 5th century B.C. The Mesopotamians were accomplished sculptors, being especially skilled in stone carving. The development of Sumerian art from geometrical designs to well-executed scenes from the Gilgamesh epic has been traced from the thousands of cylinder seals used to authenticate documents. Metal workings in gold, often inlaid with lapis lazuli, also produced brilliant creations. In fact the later art of Babylonia, Assyria, and in some measure of Egypt, is traceable to the great Third Dynasty of Ur three centuries before Hammurabi.

In a civilization so dependent on a central temple administration, a system of writing and notation was, of course, essential. From crude pictographs a system of writing known as *cuneiform,* with wedge-shaped characters, was developed. A square-tipped reed was used to make the pictographs in the wet clay, but these eventually turned into a series of wedge-shaped strokes associated not with the thing pictured but rather with a particular sound.

The Mesopotamians devised the most complete of the early systems of measurement. Their numerical system had a basic unit of 60, and their mathematics was based on both a decimal and a sexagesimal system. Division and multiplication could be easily handled. There was also an elementary form of algebra, although their system of geometry was very crude.

Since medical, astronomical, and geographical theories were rooted in mythology and religion, these subjects were hardly what we would call scientific. However the year was divided into 354 days, with 12 months of 29 or 30 days based on the phases of the moon and an extra month added every three or four years to make the solar and lunar calendars coincide.

Almost 3500 years have passed since the civilization of Mesopotamia died out. Yet the Hittites adopted cuneiform writing, the Middle Eastern diplomatic language was Babylonian, Alexander the Great brought back the arch to Europe, the Hebrews adapted the epics of the Creation and the Flood, and the Mosaic Law borrowed from Hammurabi. Perhaps for these debts, even though archaeologists may dispute her role as the first urban society, Mesopotamia rightly deserves the reputation of the "cradle of civilization."

5 / THE GIFT OF THE NILE

The Nile is Egypt's fount of life. Each year between June and October the great river rushes north out of the highlands of Ethiopia, surging over its banks in a great flood and leaving behind it water and fertile silt. The green gash of the Nile is the life-line of Egypt, a valley some 14 miles wide slicing through the desert sands which stretch away to the east and west. In ancient times Egyptian life revolved about the Nile; in modern times it still does—as may be seen from the construction of the Aswan High Dam.

It is not easy to exaggerate the influence of Egypt's environment on her life. "To go north" in the Egyptian language is the same as "to go downstream"; "to go south" the same as "to go upstream". It was so unusual to be without a boat that there was a special word to convey this meaning. The hieroglyphic sign for a "foreign country" was the same as that used for "highland" or "desert" because the deserts that fringed the Nile were both mountainous and foreign. Thus the influence of her geography was indelibly marked on both Egyptian life and thought.

The long sealed tube of civilization that is Egypt was probably the first large territory to come (in 3200 B.C.) under a single ruler in the ancient

ANCIENT EGYPT AND HER NEIGHBOURS

world. There were two reasons for this. The prosperity of the country depended to such an extent on the annual flooding of the river that the harvesting of crops could not be left to the whims of individual farmers. State control was needed to make sure that water would be hoarded and food harvested, so that famine might be averted. Consequently only a strongly centralized administration could manage the economy properly. Moreover, Egypt was fortunate in her early freedom from foreign invasion, which the insulation of the desert provided, granting her a relative isolation in which to develop government.

There is still much uncertainty about Egypt's chronology. Early Egyptian records name the years not for the rulers but rather for those events which the particular chronicler considered most noteworthy—"the year of fighting and smiting the northerners", or "the year of the second enumeration of all large and small cattle of the north and south". This inconvenient system of reckoning was eventually replaced by a set of tables dividing the rulers of Egypt into 30 dynasties.

> **CHIEF DIVISIONS OF EGYPTIAN HISTORY**
> 1. 2700-2200 B.C. — The "Old Kingdom" (Dynasties 3-6)
> 2. 2050-1800 B.C. — The "Middle Kingdom" (Dynasty 12)
> 3. 1570-1090 B.C. — The "New Kingdom" (Dynasties 18-20)

The Double Crown About 3400 B.C. a race of newcomers began to penetrate Egypt along the eastern edge of the Nile Delta. Over the course of the next two centuries these conquerors set up two monarchies, one symbolized by the red crown of the Delta kingdom and one by the white crown of Upper Egypt. Yet by 3200 B.C. a single king, the legendary Menes, came to wear the double crown of all Egypt as an absolute monarch and a living god.

For five centuries the first two of the dynasties ruled Egypt, and it was a time of peace and prosperity. Egyptians traded their stone pottery along the eastern Mediterranean coasts, mined copper in the Sinai peninsula, travelled to Byblos in Syria for timber and to Crete for wine and oil. From Mesopotamia they borrowed brick buildings, artistic decorations, the potter's wheel, and a theory of writing—borrowed and refined them until they far surpassed their Mesopotamian inspiration. By the time of the third dynasty Egypt had passed through adolescence and entered upon the first of her three great periods of history, the "Old Kingdom".

Houses for the Soul From the beginning the Egyptians were obsessed with the necessity of preserving the body after death. For they believed that a man's soul did not die. Instead it took the form of a bird, often a falcon, flying freely in the world but able to return to the dead body. Hence the body had to be preserved so that the roaming soul could recognize it. A rather crude description of the embalming process by which Egyptians preserved their dead comes down to us from a Greek historian:

First of all with a crooked piece of iron they draw out the brains through the nostrils. . . . Then with a sharp Ethiopian stone they make a cut along the flank and extract all the intestines, and after purifying and washing them with palm-wine, wash them once more with pounded spices. Then after filling the belly cavity with pure finely powdered myrrh, cinnamon, and other spices, except frankincense, they sew it up again. After this they put the body in salt, covering it with natron [salt] for seventy days; they must not leave it in the salt any longer. When the seventy days are over, they wash the body and wrap it around from head to foot with strips cut from a sheet of fine linen, smeared with gum. . . . Then the relatives take it and have a wooden coffin made shaped like a man, and shutting the body up in it they place it in the sepulchral chamber, setting it upright against the wall.

The preserved body required a home. It was placed in a coffin which, along with the dead man's personal possessions, was housed in a series of underground rooms excavated for that purpose. Above ground was erected a plain brickwork structure, gaily painted and adorned to suggest the palace

This cut-away drawing depicts the construction of the first and largest of the pyramids of Gizeh, the Great Pyramid of Cheops or Khufu. The blocks of limestone were brought by boat from quarries across the Nile (1)—a method of transportation available only when the river was in flood. From the landing area the blocks were pulled on sledges by gangs of men along a stone causeway (2), and were finally hauled up a long brick and earthen ramp (3) which was removed when the pyramid was completed. From the entrance to the pyramid (4) a long Ascending Corridor leads to the Grand Gallery (5), and thence to the King's Chamber (6). These rooms were even provided with air vents, as you can see. There is also a Queen's Chamber (7), and an underground chamber (8) with a blind passage at its end. The pyramid originally rose 481 feet (as high as a forty-storey skyscraper), and the base covers 12½ acres. The smooth white limestone casing, which was filched over the years for building elsewhere, is shown along one side (9), while at the top of the structure (10) workmen are fitting the huge blocks into place.

or home of its late owner. From such humble beginnings sprang the immense stone memorials that symbolize Egypt even today. "In sublime arrogance," writes a modern historian, "the royal pyramids dominated the Old Kingdom and sent their shadows down the ages."

The most famous Egyptian monuments are the elaborate tombs to preserve the dead pharaohs' bodies, the pyramids of Gizeh, built in the Fourth Dynasty. The Great Pyramid comprises six and a quarter million tons of stone, with exterior casing blocks averaging two and a half tons each. These casing blocks were fitted with a joint of only one-fiftieth of an inch, while

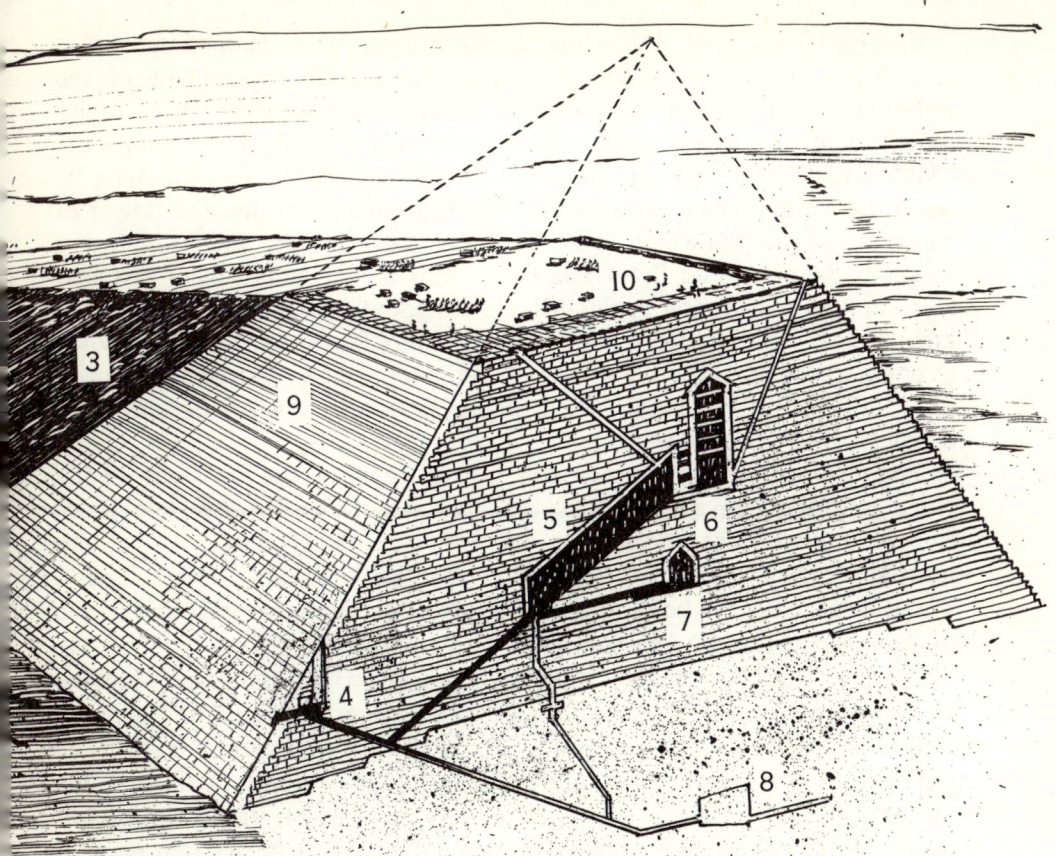

the whole tremendous mass of the pyramid was set on a rock pavement that had a deviation from a true plane of a mere .004%. Not only that, but the Great Pyramid was only partly built from local stone. The granite for the inner chamber was quarried more than 700 miles away.

How could these marvellous structures be built without modern machinery, and especially without the wheel for carriages, pulleys, or cranes? Apparently the immense blocks were hauled and levered up sloping brick and earth ramps greased with sand and gypsum. When certain Greek tourists came to Egypt to gaze at the Pyramids 2000 years later, they were told that it had taken 100,000 men working in three-month relays for 30 years to build the Great Pyramid.

As a pharaoh's reign stretched out he might replan his pyramid, adding chambers and passages until its proportions were even more magnificent than its planner had at the beginning dared to envisage. But magnificent monuments though they may be, it must not be forgotten that they are the symbols of ruthless absolutism and an unlimited supply of labour. "The groans of the slaves have been silenced, the Nile wind has swallowed up the whistle of the whiplash and blown away the harsh odour of human sweat. Nothing but the huge structures themselves remain."

Captive Conquerors For a time the Fifth Dynasty maintained the vigour and originality of its predecessors, but the central government of the Pyramid Age gradually collapsed as different areas broke away under independent princes and lords. The country was not united again until the "Middle Kingdom" period in the reigns of the rulers of the Twelfth Dynasty, when the capital was moved from Thebes to the Fayum and the power of Egypt was again shown in such foreign ventures as the raids north and south to Palestine and Syria and down into Nubia and Libya. Art and literature flourished, and in this Middle Kingdom the middle class—craftsmen, artists, scribes, government officials—were encouraged and promoted to counteract the disruptive power of the local princes. For two centuries peace reigned in the land. Then, before a strange swarm of barbarian invaders, the state collapsed.

The Hyksos

The new peoples who surged into Lower Egypt across the Delta of the Nile, doubtless shoved south by the pressure of the Bronze Age Indo-European migrations, are known to history as the "Hyksos", sometimes translated "shepherd kings" but really meaning "rulers of foreign lands". The Hyksos may have been the first (they were certainly not the last) people to be taken captive by those whom they conquered. They retained the Egyptian system of administration, adopted the royal style, and accepted hieroglyphic writing. They also introduced a superior bow into Egypt, as well as the use of the horse chariot. The Egyptians were quick to learn the use of chariotry, and the Theban princes of the Seventeenth Dynasty drove the foreigner out of Egypt and themselves embarked on imperialistic conquests. With the "New Kingdom", Egypt entered upon her golden age.

The Napoleon of Egypt Thutmose III (1490-1436 B.C.) probably came to the throne by murdering his stepmother. The reign thus begun in violence was to continue ruthlessly. Thutmose was determined to create a mighty empire, and to this end he waged seventeen campaigns in Asia, twice crossing the Euphrates. Everywhere his bowmen, axmen, spearmen and charioteers trampled lesser peoples into the dust, then swept on, leaving governors and garrisons to keep the conquered in subjection. Here is an account of one of Thutmose's victories, won after a seven-month siege.

... they came out, pleading to My Majesty and saying: "Give us breath, Our Lord! The lands of Syria will never again rebel against thee!"

Then that enemy and the princes who were with him sent me great tribute, borne by their children: gold and silver, all the horses they possessed, their great chariots of gold and silver and painted, all their coats of mail, their bows and their arrows, all their weapons. With these had they come from afar to give battle to My Majesty, and now they were sending them to me as tribute, while they stood upon their walls, praising My Majesty and begging that they might be granted the breath of life.

Thutmose was, however, more than a mighty imperialist. He was also a painstaking administrator. No sooner was he back from a conquest than he was off on tours of inspection up and down the Nile. Before his campaigns were over, the boundaries of Egypt stretched to the Taurus Mountains and the Euphrates River. Tribute came from both Babylon and Assyria. Egypt ruled the East. To celebrate the glories of his reign, Thutmose had certain obelisks set up to commemorate his accession to the throne. Four have survived to this day, and the tourist may marvel at them in Istanbul, Rome, London, or even in Central Park in New York City, whither they have been transported to become mute testimonials to the greatness of Thutmose III.

An Attempted Reformation Egypt reached her zenith under the grandson of Thutmose, Amenhotep III, who, unfortunately, could not or would not give his son any direction in wielding the great administrative machinery of the empire. Thus it was that when Amenhotep IV became co-regent with his invalid father, the young ruler was preoccupied with building a new capital on the site of modern Tell el-Amarna, halfway between Thebes and Memphis, and paid little or no heed to the calls for help from the weakly garrisoned provinces.

Amenhotep IV (Akhenaton)

In some ways Amenhotep IV (1369-1353 B.C.) is one of the most remarkable of the pharaohs. The statue of him which is now in the Louvre in Paris shows us an effeminate, flabby young man who probably came to the throne when he was about twenty-five years of age. As a boy Amenhotep may have been a high priest of the sun-god Re, and in Heliopolis, his first capital, he may well have been under the influence of the priesthood who taught that Re should be worshipped above all other gods.

At any rate, the young pharaoh eventually began to openly favour the purest form of the sun-god, the sun itself, which was called "Aton". Hence his new capital was called Akhetaton ("Place of the Glory of Aton"), and was designated as the sacred city of Aton, while the king himself changed his name to Akhenaton ("He Who is Beneficial to Aton"). The new city was rushed to completion in two years, while the temples of the other gods were closed and their property confiscated. For this reason the new religion has sometimes been called *monotheistic*, that is, believing in the existence of only one god, although actually the pharaoh himself became a god along with Aton. There was, apparently, no revolt on the part of the priesthood and high officials at the passing of the old gods. Indeed it is doubtful how much attention the mass of the people paid to the new cult.

We know from preserved royal correspondence that pleas for military aid from the governors of the Egyptian provinces went unheeded. The mystic pharaoh's interest in religion and art left little time or inclination for statecraft. Akhenaton's last years must have been overshadowed by the defection of his beautiful and talented wife, Nefertiti, and the breakup of his empire. As the pharaoh's deputy wrote from Jerusalem: "All the lands of the king have broken away. . . . If no troops come in this very year, then all the lands of the king are lost."

Tutan-khamon

An Unrobbed Grave With Akhenaton's death his religious reformation collapsed. His name was obliterated on monuments and in records, while Tutankhamon (1352-1344 B.C.), the new pharaoh, boasted of how he restored the ruined shrines and temples of the gods and goddesses.

Strangely enough, we have come to know a great deal about this young pharaoh who undid Akhenaton's work. By a happy chance a tomb was discovered in 1922 behind a sealed doorway. This momentous find could mean only one thing to its English discoverer, Howard Carter: that unlike almost all the other tombs, this one had not been rifled by grave robbers.

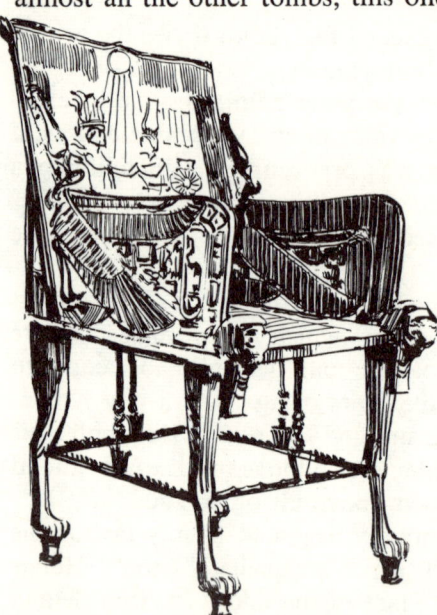

Actually, as Carter found on more careful investigation, robbers had skilfully opened and re-sealed the doors; but most fortunately they had not penetrated to the burial chamber. Best of all, the pharaoh's body lay undisturbed.

One of the richest finds from the tomb of Tutankhamon was this handsome throne. The feline-shaped legs are surmounted by lions' heads, and the arms are wings of crowned serpents. Carved in high relief on the back panel is a scene portraying King Tutankhamon sitting in an unconventional attitude, his arm casually draped across his chair while the queen anoints his shoulder. The figures are done in red glass, silver, and precious stones, all set against a golden background which highlights the Aton disk. Although Tutankhamon overthrew Aton-worship and proclaimed it heresy, what evidence does this throne provide of the persistence of such ideas?

Who was he? The eighteen-year-old Tutankhamon, Akhenaton's successor.

There gleamed the gold inlaid furniture and jewellery, the nesting gold coffins, the untouched body. Yet none of these impressed Carter as much as something he found nestled on the outermost coffin. Here, he writes,

[lay a] tiny wreath of flowers . . . the last farewell offering of the widowed girl queen to her husband. . . . Among all that regal splendour, that royal magnificence—everywhere the glint of gold—there was nothing so beautiful as those few withered flowers, still retaining their tinge of colour. They told us what a short period three thousand three hundred years really was—but Yesterday and the Morrow. In fact, that little touch of nature made that ancient and our modern civilization kin.

With Ramses I (1319-1318 B.C.) the mighty Egyptian dominions of Thutmose III shrank to only Palestine. It was left for Ramses II (1290-1224 B.C.) to conclude a treaty of friendship with the fierce Hittites and marry a Hittite princess. But not even the transfer of the capital of Egypt from Thebes to Avaris-Tanis in the Delta could steady the tottering empire.

And although Ramses III (1195-1164 B.C.) was able to beat off attacks by groups of Bronze Age Indo-Europeans (the so-called "Peoples of the Sea"), Egypt's Asiatic possessions did not outlive him. As the power of the pharaohs declined, that of the priesthood of Amon (god of the Empire, who was united with Re) increased, and during the Twenty-First Dynasty the kings were the high priests of Amon. Then followed Libyan and Nubian kings, until finally Egypt went down before the Assyrians from the northeast. Her influence ended in 605 B.C. with her defeat by the Chaldeans, the conquerors of the Assyrians.

The Grandeur that was Egypt Egyptian religion was a conglomeration of cults. As well as the local city gods and goddesses there were many lesser gods and spirits, often conceived in animal form. Nature deities, as one would expect in an environment where sky and sun and water were so important, were everywhere. The sun-god Re might travel in his barque or fly through the sky as a falcon; or in Upper Egypt he might be portrayed as a sun disk with outspread wings.

Ideas about life after death were equally confused. Trinkets and various types of food and drink were put in tombs to provide for the wants of the deceased, while inscriptions besought the passerby to utter a prayer that would help supply the dead man's wants. The so-called *Book of the Dead* was a compilation of magical spells which could be put inside the coffin so that the dead man, having studied them at length, could recite to Osiris, the king of the dead, all the sins that he had *not* committed. Some passages from the *Book of the Dead* may be read on pages 44 and 45.

One might be tempted to think of the Egyptians as a gloomy people oppressed by the thought of dying, but such a view is far from the truth. While literature was mostly given over to religious themes, there were exceptions, sheer adventure stories such as "The Shipwrecked Sailor," about an Egyptian Sinbad, or "The Tale of Sinuhe," which tells of a roving political exile. There are even cartoons depicting the battles of pharaohs and their enemies as fights between cats and mice, or lampooning the vast Egyptian civil service.

Perhaps because the pyramids turned out not to be proof against grave robbers, Egyptian kings came to favour rock-cut tombs in the Valley of the Kings—though these proved no safer. With the Pyramid Age over, immense temples such as those at Karnak or Luxor became the style. For sheer mass they rank with the pyramids: the Temple of Amon at Karnak was planned on a scale large enough to contain not only the Maple Leaf Gardens, but the Montreal Forum, Madison Square Garden, and seven football fields as well.

Egypt's architecture is her most enduring monument.

The sculpture of the Egyptians was dominated by a concept of "frontality". Because the artist sought to portray the most characteristic aspect of what he drew or sculptured, he usually chose a front view, though with a human figure he might have to alternate between a head in profile, eyes

This hypostyle hall—part of the massive Temple of Amon which stands at Karnak on the outskirts of Thebes in Egypt—was built during the New Kingdom. Its columns are 66 feet tall and are crowned with immense capitals on each of which there is enough space for 100 men to stand. Note that the capitals in the centre of the hall are modelled after the bell-shaped papyrus plant. No cement was used in the building, the tremendous weight of the stones alone holding them in place. Once brilliantly coloured and decorated with gold and silver, this distant ancestor of the Roman basilica and the Gothic cathedral is now a forlorn ruin lapped at by the Nile during the flood season.

and shoulders as in full front view, trunk in profile, and arms, legs and feet in profile. This was the way the Egyptian *wanted* to paint and carve. The exaggeration and distortion was deliberate, not the result of faulty technique, as can be proved by an examination of the perfectly proportioned clay figurines put inside tombs to help the dead hunt or fish or do other tasks.

During the reign of Akhenaton Egyptian art did become more naturalistic, so that for a time the dignified and aloof statues began to look almost human. The pharaoh's sculptor certainly did not flatter him, and the magnificent painted sculpture of Nefertiti, while perhaps idealized, is amazingly lifelike. But this trend represents only the influence of one pharaoh (who apparently admired naturalism in art) over a group of sculptors. It was hardly the Egyptian ideal.

In science, too, Egypt demonstrated her genius. The most outstanding scientific accomplishment was a yearly calendar of 365 days: the solar year began when the star Sirius rose above the horizon, and this in turn coincided closely with the annual Nile flood. Both water clocks and sundials were used, and the day was divided into 24 hours. Aside from their calendar, however, the Egyptian achievement in mathematics and astronomy is mainly important as an introduction to a subject that the Greeks were later to perfect. Mathematics was practical and primitive. They had a numeration system that could deal successfully with the mathematical problems they might encounter in erecting the pyramids, or in everyday living. Their astronomical observations, however, were little more than crude schemes bound up with religion.

The Egyptians gained a high reputation in medicine. Many of their medical records have survived in written form on *papyri*, very durable scrolls of paper which they made by pressing together strips of the papyrus, a plant growing in swampy regions. Medical papyri from the Middle and New Kingdoms set forth a large collection of prescriptions for ailments, including the surgical treatment of wounds and fractures. But anatomical and surgical knowledge was elementary, and magic was intertwined with medicine, illness and disease being blamed on demon possession. The process of mummification, however, has had two notable consequences: first, it taught the Egyptians dissection and anatomy; second, it has bequeathed to us some corpses that show the actual ravages of ancient diseases.

The Structure of Society The Egyptians were well governed. Since the pharaoh was both god and man, he was high priest as well as chief justice and commander-in-chief. Naturally much of his work had to be delegated to a vizier, a sort of prime minister. Next in line came the ministry of finance, administered by various overseers who collected taxes and disbursed salaries. The bureaucracy must have been extensive by the time of the New Kingdom. After the expulsion of the Hyksos there was a large standing army—fed by the government and paid off in booty, with the result that it posed a grave threat to the security of the state.

Local government was managed by nomarchs, independent princes who ruled the various provinces or *nomes*. Ability seems to have been recognized, for it was possible for a talented man of lower class to rise to a position of authority. Such opportunity may explain the durability of Egyptian society. "Put writing in thy heart," the young man was urged, "so that thou mayest protect thine own person from any kind of labour and be a respected official."

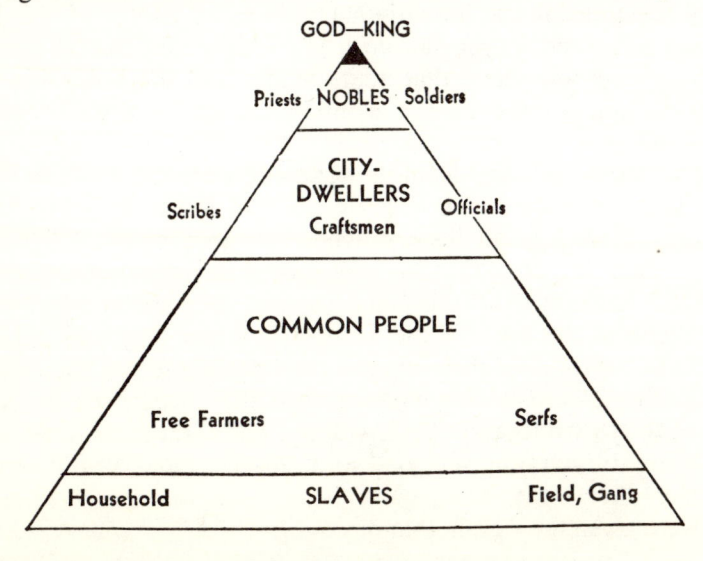

The structure of Egyptian society, and of ancient Middle Eastern societies in general, is shown in the accompanying diagram. In Egypt as in Mesopotamia, agriculture was the basis of economic life. And though many of the slaves had an unenviable lot, the small free farmers do not seem to have fared too badly, despite the fact that they were called upon for forced labour in mines and quarries and in irrigation and construction projects. Nevertheless, all land belonged to Pharaoh. It could be used only with his permission, and he collected an annual tax on it. The following description by an Egyptian scribe was probably not the rule: tax collectors are never popular. Yet it does show how dismal conditions sometimes were under corrupt administrators.

Dost thou not recall the picture of the farmer when the tenth of his grain is levied? Worms have destroyed half the wheat, and the hippopotami have eaten the rest; there are swarms of rats in the fields, the grasshoppers alight there, the cattle devour, the little birds pilfer; and if the farmer loses sight for an instant of what remains on the ground, it is carried off by robbers; moreover, the thongs which bind the iron and the hoe are worn out, and the team has died at the plough. It is then that the scribe steps out of the boat at the landing-place to levy the tithe, and there come the Keepers of the Doors of the [King's] Granary with cudgels, and Negroes with ribs of palm-leaves, crying ,"Hand over the wheat!" There is none, and they throw the cultivator full length upon the ground, bind him, drag him to the canal, and fling him in head first; his wife is bound with him, his children are put into chains. The neighbours in the meantime leave him and fly to save their grain.

In the towns lived the labourers and merchants. Egypt imported raw materials—minerals from her mines in Arabia and Nubia and timber from Syria—and exported such finished products as furniture and linen. The early Egyptians also discovered how to fuse copper and tin to make bronze, from which weapons and tools were manufactured. It is interesting to note that ancient Egyptian exports have been found even as far west as Italy.

History's First Strike   Egypt's decline under the New Kingdom was due in part to the pressing demands of war, in part to the passing of the Bronze Age. By the time of the Twentieth Dynasty the effects of the Iron Age were being felt—and Egypt had no iron. This meant that she had to buy from abroad the new metal that made harder and more durable weapons, and the consequent drain on her economy caused the price of grain to go up alarmingly.

Those who suffered most from the price rise were the workers on government projects, who were paid in grain. In fact it was they who staged the first strike recorded in all history, when the government fell behind two months in their wages. "On this day [occurred] the crossing of the five walls of the necropolis by the gang, saying: 'We are hungry!' . . . And they sat down at the back of the Temple." When they still did not have their grain two days later they invaded the temple and put their case to the officials in these words: "We have reached this place because of hunger, because of thirst, without clothing, without oil, without fish, without vegetables! Write to Pharaoh, our good lord, about it, and write to the Vizier, our superior. Act so that we may live!"

They were given their grain that day.

Line 1 shows the Egyptian *hieroglyph* or word symbol of 4000 B.C. Through centuries of use the symbol became simplified into the *hieratic* script of 2500 B.C. (line 2). A further stylizing is illustrated in line 3 in the *demotic* script, which appeared in the 7th century B.C. As the Egyptian economy became more complex, these new and simpler forms of writing grew up because of the need for a rapid recording of daily business and events.

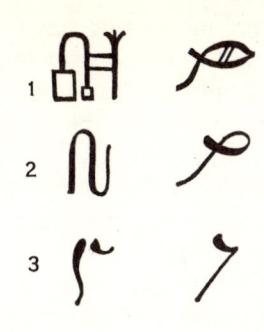

Glory in the Sand Combined with the increasing power of a nobility and priesthood that stamped down the masses, the coming of the Iron Age spelled disaster for Egypt. At home the individual was sacrificed to the group, while abroad Egyptian armies could not stand up to new and powerful states. Foreign troops had to be hired to protect a country that had once united to expel the Hyksos.

Egypt averted her eyes from the whole unpleasant prospect and looked back on her long past. In doing so she ceased to be creative. On that note the glory of Egypt seems to have sunk slowly down into the sand.

What has she bequeathed to us? Later architects adapted her stone columns—but for new types of buildings. Greek mathematicians took over her clumsy arithmetical system—but had to refine it. Later religious thought received little or no inspiration from Egypt. All told, then, Egypt passed on no very significant spiritual or intellectual heritage. Yet because her civilization lasted so long, later civilizations continued to regard her with worshipful awe. The fact remains that more than 1000 years of cultural stability is in itself no mean accomplishment.

6/A VAST SUBCONTINENT

While the Egyptians were building the pyramids and while Sargon I ruled Sumer and Akkad, a brilliant culture was flourishing far to the east of them. In the valley of the Indus River there had grown up a civilization that covered an area larger than either Egypt or Mesopotamia, an area extending 1000 miles from the Arabian Sea to the headwaters of the Indus.

White-capped Sentinels and Rolling Seas The geography of the Indian peninsula has played a major role in making that vast subcontinent, half as large as the United States, a cradle of civilization. Stretching some 1500 miles across northern India soar the Himalayas, whose peaks—over 100 of which reach 20,000 feet—have protected India from two invaders: man, and the dry winds that sweep across Asia. These same Himalayas also nurture the Indian plains. When the monsoons from the south rise over the mountains they release their rain and snow, feeding India's rivers and irrigating her fields.

Yet no mountains have ever succeeded in permanently barring men determined to climb over them; and when invading peoples did force their way through the tortuous Himalayan passes and the deep, jungle-clad foothills, they were able to fan out into the wide river plains of the Indus to the west and the Ganges to the east. In these vast valleys, so fertile that in most places they support two crops a year, many peoples over many centuries have met and fused together in a veritable crucible of nations.

Farther south lie hills, forests, and a great central plateau called the Deccan, while along the east and west coasts the hills give way to narrow tropical plains. All around this ancient peninsula roll her oceans, too wide to permit large-scale immigration from across the seas, but no lasting barrier to the merchants who flocked to Indian shores to sample the mysteries of the East.

From Stone Tools to City Sewers Flaked tools possibly 300,000 years old have been found in India's Sohan valley, and archaeologists tell us that man has been making a home here since at least 400,000 B.C. By about 10,000 B.C. bone and flint implements appeared, and, after several millennia, village life grew up. Sometime after 4000 B.C. farmers from the Iranian (Persian) plateau moved eastward, and sometime after 3000 B.C. they probably entered the Indus valley. These farmers were the forerunners of an advanced civilization, a civilization that flourished from about 2500 to 1500 B.C. in part of what is today West Pakistan and north-west India.

So far, archaeologists have discovered over 70 ancient cities and towns in the Indus valley. In fact widespread excavations in the last fifteen years have yielded evidence that this exciting civilization stretched 1000 miles from east to west and 500 miles from north to south. Here, then, once matured what has been called "the vastest political experiment before the advent of the Roman empire."

The Indus Empire Apparently this Indus empire had two capital cities: Harappa in the Punjab, and Mohenjo-daro (the hill of the dead), nearly 400 miles down-river to the south-west. Each city was laid out in blocks with streets intersecting at right angles, and each centred around a citadel which looked down on the city from an artificial hill. The houses, mostly of baked brick, were built around courtyards. There were even drainage systems which ran off into brick-lined street sewers—something modern India had to wait for until the 19th century.

The Harappans Little is known about the inhabitants of these cities. Few human remains have been found, but judging by some of the skeletons so far recovered the Harappans may have been small in stature. One man was 5 feet 4½ inches tall, and two women were 4 feet 9 inches and 4 feet 4½ inches. There is, however, considerable evidence concerning both the implements the people used and the animals they domesticated. The Harappans fashioned wheel-made pottery, copper, bronze, and silver utensils, knives, spears, arrows, axes and razors. They also domesticated camels, buffalo

Grain was a valuable commodity for the cities of the Indus Valley: not only was it used for food, but it was the universal medium of exchange. Hence some kind of grain storage that would afford protection against both marauders and floods was vital, and the State Granary at Mohenjo-daro, pictured here, was a national bank and revenue department in one. Both human and natural menaces were warded off by the massive mud-brick platform which rose some 40 feet above the surrounding plain, making it necessary for the wheat and barley to be hauled up on ropes to the granary floor above. The holes visible in the brick walls are the ends of ventilating shafts that allowed air to circulate beneath the grain. Incidentally the carts, which are models found in the ruins of Mohenjo-daro, look precisely the same as ox-carts used in West Pakistan today—some forty centuries later.

sheep, and possibly horses and elephants—as well as those two traditional enemies, the cat and the dog. One day, in fact, they must have been the despair of a certain brickmaker; for a clay brick has been recovered on which are still to be seen the footprints of a cat, followed at hair-raising proximity by those of a dog.

As far as we know now, the inhabitants of the Indus valley cities were not a warlike people. Their cities were unwalled, and the citadels were probably primarily centres of local administration (presided over by priest-kings) rather than safeguards against attacks. Weapons have been found it is true; but these are just as likely to have been used for hunting as for war.

And so the Harappans cultivated the arts of peace. They produced some very fine sculptures in stone and bronze, as well as clay figurines, and left behind them an abundance of jewellery—bracelets, rings, and beads fashioned of gold, silver, copper, bronze, and stone. Moreover, rouge, lipstick, tweezers, and ear-piercing implements have been found. Apparently the ladies of the Indus valley were at a sufficiently advanced stage of civilization to spend considerable time over their make-up.

The Indus script is, as yet, undeciphered, though many examples of it have survived on seals and pottery-stamps. The animal and human figures engraved on the seals may have been religious symbols, but we still do not know whether some small clay animals were religious objects or children's toys. Archaeologists are reasonably certain, however, that the Harappans worshipped a male and a female deity, and they have unearthed a three-faced male figure in a sitting position which is probably the original Shiva, a primitive Hindu god. Certain animals, such as the humped-bull, were also revered; the fig tree was regarded as sacred; and bathing seems to have been a sort of religious ritual.

We presume that, by 2500 B.C., there was trade between India and Mesopotamia—probably in incense and cotton, possibly in slaves and timber—and undoubtedly this contact stimulated the Indus civilization. Yet in art, in writing, and in plumbing the Harappans were unique. It was not, however, these physical accomplishments of their civilization that were to be passed on to later ages. It was their gift of the spirit, their religion.

Fair-haired Invaders Doubtless the Indus valley cities had begun to decline before 1500 B.C. At any rate, after that date waves of Bronze Age Indo-European invaders washed down over this peaceful, if somewhat puzzling, people, scattering through the country and slaughtering the Harappans. The tall, fair-haired, fair-skinned Aryans swept across the plains in swift battle-chariots. Their war-god, they said, "rends forts as age consumes a garment", and the trail of death they left behind them can still be seen by the archaeologist. "The attackers left the dead lying where they fell. In one of the houses sprawled thirteen skeletons—men, women,

The Aryans

and children—some wearing bracelets, rings and beads, and two of them with sword cuts upon their skulls."

The Aryans settled down along the banks of the Indus and set up separate kingdoms, each with an elected king who presided over a council of nobles and a body of freemen. They were an agricultural people (their word for "war" meant "a desire for more cows"), and with them they brought the plough, adapting it now to the wet, heavy soils of the river valleys. Widespread cultivation of rice now began, and iron-working was introduced. Cities eventually flourished, and by 600 B.C. there were some sixteen separate states. Nevertheless many of the older inhabitants must have continued their former life in much the same vein as before the invaders usurped their land.

Proud as they were, the conquerors were not above learning from their captives. Some of the ritual and religion of the Indus peoples was destined to be adopted, and with the coming of the Aryans we can clearly see the early stages of Hinduism—one of the world's major religions, and hence our chief interest in early Indian life.

The Wheel of Life As the northern nature religion of the Aryans merged with the beliefs of the earlier peoples it found expression in a series of hymns to the gods. These hymns, known as the *Vedas* (which go back to 1400 B.C.), and a later collection, the *Upanishads* (written in the 8th and 7th centuries B.C.), contain the basic Hindu beliefs, including an emphasis on sharp class distinctions and a caste system. At the head of these classes stood the powerful priests, followed in descending order by kings and warriors, artisans, labourers, and finally the outcastes or "untouchables". According to Hinduism, it was the will of the gods that each man stay strictly within his caste.

The Vedas and the Upanishads

These early Hindu beliefs also incorporated a strong conviction that all living things were part of a world soul, Brahma, from which they came and to which, ultimately, they returned. If, therefore, a man died before his own soul was worthy of final release, the soul would be reincarnated in another body before being reunited with Brahma. In fact successive rebirths, known as transmigrations of the soul, might be required before a final reunion with the world soul.

Thus death was not something to be feared. Life was a kind of wheel.

Never was there a time when I was not, nor thou . . . and never will there be a time when we shall cease to be. . . . Just as a person casts off worn-out garments and puts on new, so does the soul cast off worn-out bodies and put on others . . .

Achieving Nirvana By the 6th century B.C. Hinduism had become extremely formal and ritualistic, and was being examined critically by certain members of the noble class. Amongst those who wished reform was the son of a petty Indian king, a young man named Siddhartha Gautama (567-487 B.C.), better known to history as Buddha, "the Enlightened One". Oppressed by the suffering in the world, Gautama left his wife and

Buddha

Shiva, the Indian god of destruction and creation, is portrayed as Lord of the Dance in this delicate 11th-century bronze figure. With one foot poised on a dwarf (which represents evil), Shiva holds, in two of his four hands, a drum and fire. A ring of fire rising from the lotus pedestal surrounds him. When Shiva dances with the drum and the fire he awakens the powers of nature to life; but fire in turn destroys these powers. This perpetual cycle of creation and destruction eventually releases the soul.

infant son when he was twenty-nine and for six years lived as a wandering beggar, practising all manner of austerities. But he failed to find the truth. Then one day as he was sitting under a fig tree enlightenment came to him. Thenceforth for the next forty-five years he was to preach his vision.

The first sermon that the enlightened Buddha is reputed to have preached to his followers may be read in the Source Reading on page 45. Buddha believed that it was man's desire for material things that chained him to the ever-revolving wheel of life. To escape this repeated rebirth in successive existences man must conquer desire and thereby be absorbed into oneness with the Supreme Spirit, an exalted state called Nirvana. Buddha was even more revolutionary in that his message was for all and for all equally. Not only was the caste system rejected, but monks were to replace the aristocratic Brahmins, and the many gods of Hinduism were renounced.

The Eightfold Path

An eightfold path was set down whereby Nirvana could be achieved:

1. Right views, which meant a knowledge of Buddha's teaching, especially about misery.
2. Right resolve, which meant the resolution to resist desires, to bear no malice, and to do no harm.
3. Right speech, which meant to abstain from lying, slandering, and talking aimlessly.
4. Right conduct, which meant not to kill or steal or live immorally.
5. Right livelihood, which meant not to follow occupations such as slave-dealing, caravan-trading, butchering, dispensing liquors, and selling poisons.
6. Right effort, which meant the suppression of evil states of mind and the stimulation of good intentions.
7. Right mindfulness, which meant the achievement of self-mastery by self-knowledge.
8. Right concentration, which meant the ordering of thought so that, purified of desire, temper, sloth, fretfulness, and perplexity, the sense of duty done gave a feeling of peace and freedom.

Underlying these rules was the spirit of nonviolence. Since all creatures housed living souls the Buddhist refused to take life and practised vegetarianism. Compassion was also a prime virtue. "Never shall hatred cease by hating; by not hating shall it cease."

It is important to note that Buddha did not consider himself to be anything more than a teacher. It was not until some centuries had passed that one branch of his followers began to revere him as a god and saviour of mankind. Nor did Buddhism spread widely until two centuries after its founder's death, while in India it eventually gave way to Hinduism. Only half a million of the world's 155 million Buddhists today live where the Buddha himself once taught. Nevertheless later Hinduism, with its insistence on nonviolence and the sacredness of all life, owes much to Buddhist teachings.

7/THE SLEEPING GIANT

Napoleon Bonaparte called China "the Sleeping Giant", and right up to modern times she has seemed to be dormant—a far-off, mysterious, and isolated phenomenon as far as Westerners were concerned. Early China was almost cut off from the outside world, the only contacts being by caravan across 2500 miles of steppe and desert to Mesopotamia, or by ship around the southward thrust of the Malay peninsula. The few traders who did penetrate the interior found an exotic and brilliant civilization.

The Floods that Feed   Along the Yellow River (Hwang Ho) lies a flood plain much like those through which the Nile, Tigris-Euphrates, and Indus Rivers flow. Here in this North China plain yet another early civilization developed.

The Indian symbol for time is a wheel. The Chinese symbol is a pool—an apt one in a land constantly in danger of erosion. For the mighty Yellow River, bearing heavy yellow silt, builds up its own bed and floods viciously over the northern plains. Through the centuries, this river has posed an apparently insoluble problem. All of China's rulers have had to grapple with its ravages; none have tamed it. Yet if any life was to be supported by the river some attempt at a system of water control was an absolute necessity.

North China is normally dry; hence the floods, when not disastrous, supply much needed moisture. South China, on the other hand, has an ideal climate for agriculture, but its soils have been exhausted by thousands of years of intensive farming. And although most of the area can still produce two crops a year, China has to support a population four times as large as that of the United States on a cultivated area one-half the size. In the last 1800 years there have been over 1800 famines in China.

One of man's earliest ancestors came from China, as the table on page 7 indicates. Peking Man's skeletal remains, which are perhaps 400,000

Along with this Shang noble were buried his most precious possessions—his horses and his chariot. Though the chariot itself has completely decayed, it has left a clear impression of its two wheels and of the heavy pole projecting behind them to support the chariot car in which the dead man was laid. The bead traces and tall bronze decorations that ornamented the horses are also plainly visible.

years old, were discovered in the Chou-k'ou-tien cave thirty miles from Peking. Then, in 1933, a second cave yielded skeletons dating from about 20,000 to 10,000 B.C.

We know next to nothing about the life of these earliest Chinese. There is evidence, however, that during the New Stone Age people lived in North China in sunken pit-dwellings and raised pigs and dogs for food. At Yang-shao an entire village has been excavated which goes back to 2000 B.C. Its inhabitants painted their large red earthenware pots with black geometric designs, and used bone sewing needles and other tools of bone and horn.

The Shang Dynasty

It is only with the coming of the so-called Shang dynasty about 1400 B.C. that anything in the way of a clear historical picture emerges. The Shang dynasty ruled for some three centuries from the "City of Shang" (An-yang). Modern historians had for a long time known of the Chinese literary traditions concerning a Shang dynasty, but many doubted that these accounts were historical. Then in the late 1920's came dramatic proof of their genuineness.

Bones and Bronze Strange bones with writing on them had for some time been turning up on the Peking market, usually to be ground up for medicinal purposes. When they were finally traced back to their point of origin, it proved to be none other than An-yang—the City of Shang. Excavators proceeded to unearth thousands of these bones, with exciting consequences. They were found to date from 1400 to 1100 B.C., and bore the names of the traditional Shang rulers. Also on these under-shells of tortoises and shoulder-bones of cattle were discovered inscriptions in early Chinese characters. Why had this writing been put there?

What the archaeologists had discovered were "oracle bones", devices used to predict the future. The oracle worked like this: a question was scratched on a bone; then heat was applied to it to produce cracks, and by interpreting the size and direction of these cracks the questioner received his answer. All sorts of inquiries were made—and received answers. For example, one bone asks, "Will it rain tonight?" A later disgusted notation reads, "It really didn't rain."

The Shang people had 2000 different symbols for words. These symbols were the ancestors of the later Chinese script with its 50,000 or more characters, a complex system of writing as can be seen from the sample reproduced here.

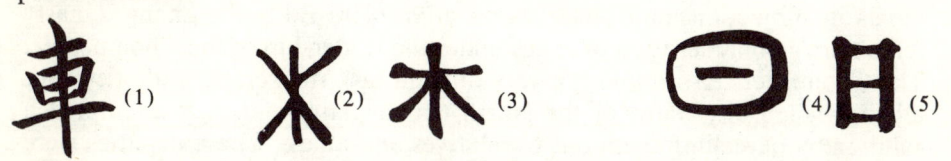

車 (1)　　　米 (2)　木 (3)　　　　　⊙ (4) 日 (5)

The great complexity of the written Chinese language derives considerably from its use of individual symbols to represent objects, ideas, and other words. This is in contrast to the largely phonetic character of Western writing. Example (1) is the character for vehicle and represents a common two-wheeled Shang chariot drawn by two horses as seen from above (compare with illustration on page 38). Examples (2) and (3) are the pictographs for "tree" used in Shang and modern times respectively. Pictographs (4) and (5) are the characters for "sun" used in Shang and in modern times.

Next to the discovery of such an advanced system of writing, the most exciting find at An-yang was the collection of bronze weapons and of vessels used in religious ceremonies. These bronzes have never been surpassed. Not even the craftsmen of the Italian Renaissance could match the Shang technique of casting bronze, with its patterns and lines formed by square—not rounded—grooves, perfect miniature trenches with perpendicular sides and flat bottoms. The glory of any museum with a good Chinese collection is its Shang bronzes.

Both in their writing and their bronze-working the Chinese probably drew from the already existing civilizations to the west of them. But it was only the *idea* that came from the West. The execution of the idea was characteristically and exquisitely Chinese.

The Shang rulers probably exercised firm authority over much of the North China plain, although they were frequently at war with neighbouring peoples. In their fighting the kings and nobles used a two-wheeled chariot which could be manoeuvred with great ease over the flat, open country. So highly esteemed was this vehicle that it was entombed, along with the charioteer and horses, with the An-yang rulers when they died. Even yet, a drawing of one of these chariots (as seen from above) forms the Chinese character for "vehicle" or "wheel".

The royal palace was large and imposing, built in a style which remains characteristic of modern China, and facing south, as is the custom in North China today. Yet the common people still lived in the crude pit dwellings of Neolithic times. And life was precarious. We know, for example, that human sacrifice was practised—probably another sign of the absolute supremacy of a ruling military class over the surrounding nomadic tribes.

The one unifying element in this motley state, the glue that held their society together, was the Shang religion. The world for the Chinese was inhabited by all sorts of spirits, especially the spirits of one's ancestors. "Pray to grandmother Yi for rain," reads one of the oracle bones. Even after death the spirits of the ancestors hovered near to affect the living.

Rule by Underlings In the last years of the Shang dynasty there may have been as many as 1700 villages scattered across the Yellow River flood plain. It is little wonder that it proved impossible to hold all these settlements in subjection, and finally, soon after 1100 B.C., the Shang dynasty fell before a confederation of tribes under the leadership of the Chou people. The Chou rulers also found it a very difficult task to exercise authority over all the walled city-states of the North China plain, and they attempted a solution by parcelling them out to relatives and allies. These deputies were expected to support the king with money and soldiers in return, but were otherwise left free to govern their territories pretty much as they pleased.

As always, below the Chou aristocracy as below any ruling class, toiled the masses of peasantry, bound to the land for life. An old Chinese adage sums up their lot: "The superior man uses his mind; the commoner uses his body. He who uses his mind rules; he who uses his body is ruled. He who rules feeds on others; he who is ruled provides food for others." Little wonder that when the peasant died a hoe was buried with him.

The Chou dynasty, too, fell on evil days. A traditional story tells us that a Chou king of the 8th century B.C. had, by repeatedly lighting beacon fires, summoned his subject armies merely to amuse a certain court lady. Then in 771 B.C. disaster struck. Rude invaders swept in from the west. The fires were once more lighted—but this time no armies came. The king was killed, his capital destroyed, and never again did the Chou rulers assert real control over their wide lands. Despite the fact that the royal line was re-established at Loyang in Honan, for the next five centuries the Chou were weaker than many of their underlings. Indeed, their subjects' fears of each other were about all that allowed the kingship to continue.

Nevertheless the later Chou period was an exciting and important one in Chinese history. Although in many ways 8th century B.C. China had trailed behind Western Asia and the Mediterranean lands in her development, by the 3rd century B.C. she had caught up. Iron weapons now replaced bronze, the iron-tipped plough came into use, large-scale irrigation works were constructed, wealthy traders and merchants plied a lucrative business, copper coinage was developed, and chopsticks were in use. Mediterranean ideas and inventions now flowed into China, speeded up by the Chinese adoption of horseback riding. Moreover, the hodge-podge of tiny semi-independent states now gave way to larger units as the border areas of the west and south were incorporated into the Chou domains.

Shang bronze vessels provide a peculiar expression of An-yang culture: their use in the sacrificial rites of ancestor worship stems from the deep-rooted Chinese reverence for the family unit; their frequent casting in the shape of an animal (such as the elephant or the deer) marks them as the product of an agricultural people. Here we see an even more common style—the tripod, a shape originally used for cooking pots. It is particularly graceful, the legs in this instance taking the form of flat-sided dragons. Note also the conventionalized dragon motif on the body of the vessel. So meticulously were these bronzes cast that little, if any, final finishing was necessary. Still more amazing was the Shang ability to cast body, handles, and base in one piece, a skill which has not been surpassed even by modern mechanical methods. One of the Shang bronzes' greatest charms today is, however, accidental. Their beautiful shades of blues and greens, often iridescent, come from the accretion of a patina created by the interaction of the metal alloys and the earth as the bronzes lay buried for centuries, usually in the grave of some important personage.

This age, which coincided with the period of the Hebrew prophets and Buddha, was also a time of vigorous intellectual activity. It produced the first, and the greatest, Chinese professional teacher on record—K'ung-fu-tzu (551-479 B.C.), or as we know him in the latinized form, Confucius.

Confucius

More Cruel than the Tiger Confucius was apparently a tall, homely youth with large ears, a flat nose, and buck-teeth. He was a member of the lower aristocracy, and his great ambition was to achieve high political office. But the ambition was never realized. Confucius failed at politics and took up teaching as a sideline. His hobby made him an outstanding success.

Confucius taught respect for authority, tradition, law, and custom. Yet, like the prophets of the Old Testament, he saw the basic weakness of his society—the turbulent rule of petty monarchs who felt themselves above reproach. He taught that Chinese society must build on its past, and that the rulers must be an example to those they controlled. "The oppressive rule," he said, "is more cruel than the tiger." It was, then, as a reforming but conservative force that Confucius' teachings were important, and many of his sayings were addressed to men who would have tortured him to death as readily as they would have crushed an insect.

Confucius never claimed divine inspiration: it was his followers who eventually deified him. Nor was he outstandingly original in his thought, although we must recognize his courageous independence in the harsh age in which it was asserted. But he expressed his ideas in short pithy sayings that were easily remembered. These were collected after his death under the title *Analects,* some examples of which are given in the Source Reading on page 46.

The Analects

Confucius' teachings had little effect on Chinese society during his lifetime. Yet through his disciples, men of the pen rather than the sword, he was destined to have a profound and abiding influence. The conservative nature of Confucianism made it popular with the ruling classes, to whom it became a code of gentlemanly behaviour. Thus, in a vastly overpopulated country where great masses of people were forced to live close together, Confucianism buttressed both the state and the family. So enduring was the need for some rules to make life in China at all bearable that only in our own century, under the impact of Chinese communism, has the Confucian system come to be seriously questioned.

8 / SUMMARY AND CONCLUSION

Man appeared on earth at least 500,000 years ago, and by 50,000 B.C. a human being who may readily be recognized as the direct ancestor of modern man had superseded the others. All these Old Stone Age men were hunters, and it was only with the climatic changes completed by 9000 B.C. that both men and animals were drawn to water-holes and grassy uplands, where farming was invented.

By 8000 B.C. these first farmers lived in villages of 200 or 300 people, and by 3500 B.C. the first communities that raised enough surplus food to support full-time specialists had arisen. Man was at last a city-dweller: man was civilized.

The first civilizations grew up in two great river valleys—that of the Tigris and Euphrates Rivers east of the Mediterranean, and that of the Nile River in Egypt. These civilizations were based mainly on agriculture, but developed trade and industry and distinct social classes. Both also created a rich Bronze Age culture in which a method of writing, legal and religious systems, primitive scientific learning, and some progress in the arts were accomplished.

Mighty empires grew from the individual city-states. When an empire's ruler prospered he was usually supported by the officials and priests of his kingdom; but when his power faltered they were apt to strike out for themselves. The land itself was owned by the king, or the temple, or the priests, or all three, and the mass of the populace took no part in the political life of the state. Slavery was a universal institution.

It is true that in the ancient Middle Eastern empires we witness the first large-scale governments. And while it is also true that the earliest Mesopotamians had a kind of "primitive democracy," the assembly could hand over power to a dictator in times of danger. As soon as cities increased in number and began to dispute among themselves, emergencies became daily occurrences—and kings became permanent. Thus for all practical purposes government in the Middle East tended to become authoritarian or dictatorial. In the Bronze Age, the *people* were of almost no account.

THE TREE OF CIVILIZATION

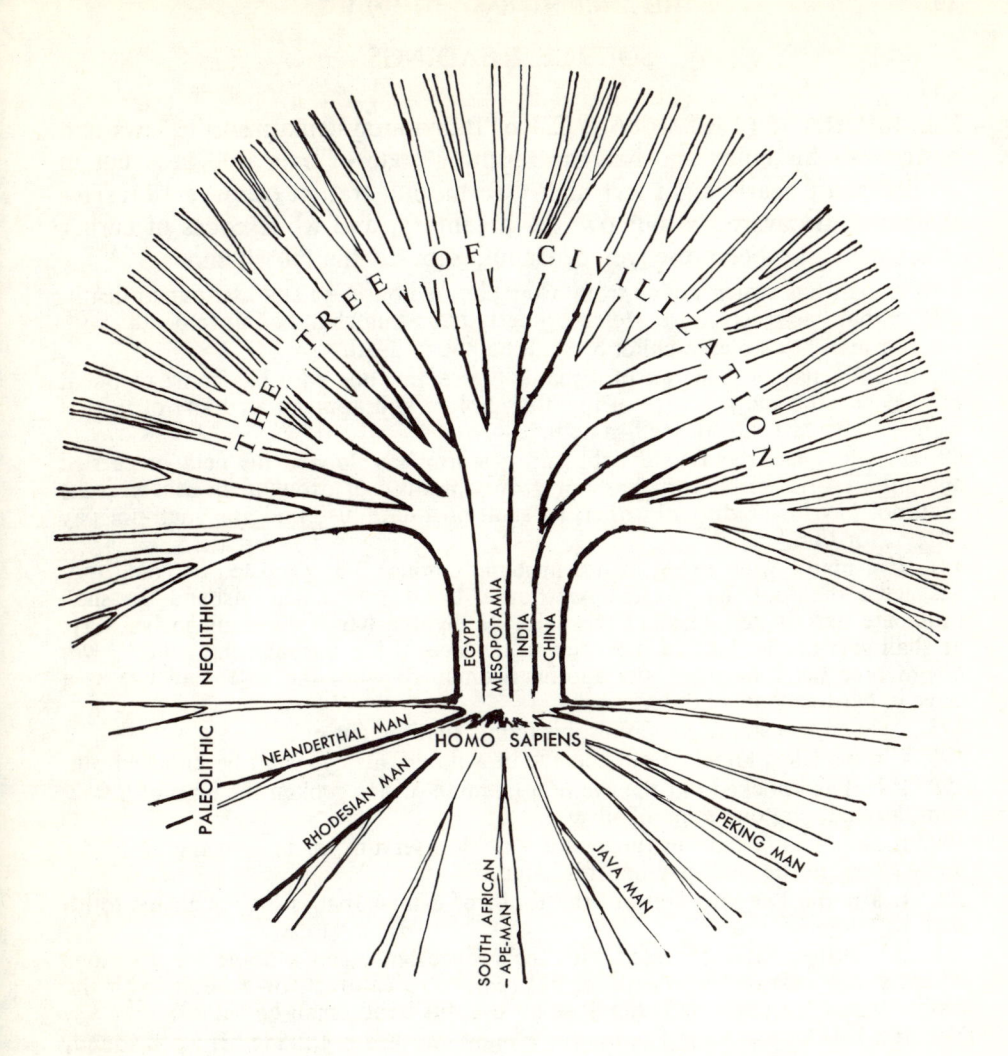

PALEOLITHIC • NEOLITHIC
NEANDERTHAL MAN
RHODESIAN MAN
HOMO SAPIENS
EGYPT • MESOPOTAMIA • INDIA • CHINA
SOUTH AFRICAN APE-MAN
JAVA MAN
PEKING MAN

The people were of even less account in two other early civilizations that developed in river valleys to the east. In the Indus River valley of northern India and the Yellow River valley of northern China, man passed through stages of development comparable to those he had already experienced in Mesopotamia and Egypt. But though undoubtedly the Indus and Chinese people were stimulated to build up their own particular kind of civilizations by contacts with the west, their cultures remained unique.

Of all these first civilizations, those in the Far East have provided the most lasting contributions: the religious teachings of Buddha and Confucius. Would patterns of thought change in the Middle East, too, in the clash of Iron Age empires? Let us examine civilization under the Assyrians, the Persians, and the Hebrews.

SOURCE READINGS

(a)

HAMMURABI (1728-1686 B.C.) of Babylon systematized the laws and customs of his lands and had the resulting code of over 300 laws cut in a stone shaft nearly eight feet high (see the illustration on page 17). This shaft was discovered in our own 20th century, and while codes of earlier rulers have also been discovered, Hammurabi's is the most famous.

6. If a man has stolen goods from a temple, or house, he shall be put to death; and he that has received the stolen property from him shall be put to death.

14. If a man has stolen a child, he shall be put to death.

25. If a fire has broken out in a man's house and one who has come to put it out has coveted the property of the householder and appropriated any of it, that man shall be cast into the self-same fire.

48. If a man has incurred a debt and a storm has flooded his field or carried away the crop, or the corn has not grown because of drought, in that year he shall not pay his creditor. Further, he shall post-date his bond and shall not pay interest for that year.

141. If a man's wife, living in her husband's house, has persisted in going out, has acted the fool, has wasted her house, has belittled her husband, he shall prosecute her. If her husband has said, "I divorce her," she shall go her way; he shall give her nothing as her price of divorce. If her husband has said, "I will not divorce her," he may take another woman to wife; the wife shall live as a slave in her husband's house.

195. If a son has struck his father, his hands shall be cut off.

196. If a man has knocked out the eye of a noble, his eye shall be knocked out.

198. If he has knocked out the eye of a freeman or has broken the limb of a freeman, he shall pay one mina of silver.

199. If he has knocked out the eye of a noble's servant, or broken the limb of a noble's servant, he shall pay half his value.

200. If a noble has knocked out the tooth of a man that is his equal, his tooth shall be knocked out.

218. If a surgeon has operated with the bronze lancet on a noble for a serious injury, and has caused his death, or has removed a cataract for a noble, with the bronze lancet, and has made him lose his eye, his hands shall be cut off.

229. If a builder has built a house for a man, and has not made his work sound, and the house he built has fallen, and caused the death of its owner, that builder shall be put to death.

G. H. Knoles and R. K. Snyder, editors, *Readings in Western Civilization* (J. B. Lippincott, revised edition, 1954), pp. 3-7, slightly adapted.

(b)

THE BOOK OF THE DEAD was a collection of magic spells useful to the soul in making confession before Osiris, the god of the Nile, who presided over the judgment hall of the dead. These "negative confessions" were composed about 1500 B.C.

1. I have not acted sinfully towards men.
2. I have not oppressed the members of my family.
3. I have not done wrong instead of what is right.
4. I have known no worthless folk.
8. I have not domineered over servants.

10. I have not filched the property of the lowly man.
12. I have not vilified a servant to his master.
13. I have not inflicted pain. . . .
14. I have not permitted any man to suffer hunger.
15. I have not made any man to weep.
16. I have not committed murder. . . .
17. I have not given an order to cause murder.
18. I have not made men and women to suffer calamities.
21. I have not carried off the cakes of the dead.
27. I have not added to the weights of the scales [to cheat the seller].
30. I have not driven away the cattle from their pastures.
31. I have not snared the geese in the preserves of the gods.
34. I have not made a cutting in a canal of running water.
46. . . . I have not played the eavesdropper.
49. . . . I have not spoken treasonably about the king.

Knoles and Snyder, *Readings in Western Civilization*, pp. 26-27.

(c)

BUDDHA (567-487 B.C.) is reputed to have preached his first sermon in a park outside the city of Varanasi (modern Benares). It was delivered to five of his former associates who had left him in disgust when he gave up trying to find wisdom by practising self-denial, and outlines to them the enlightenment he had earlier gained while sitting under a fig tree at Gaya. In the following extracts note three basic concepts of Buddhism: the Noble Eightfold Path, the Middle Way, and the Four Noble Truths.

There are two ends not to be served by a wanderer. What are these two? The pursuit of desires and of the pleasure which springs from desire, which is base, common, leading to rebirth, ignoble, and unprofitable; and the pursuit of pain and hardship, which is grievous, ignoble, and unprofitable. The Middle Way of the Buddha avoids both these ends. It is enlightened, it brings clear vision, it makes for wisdom, and leads to peace, insight, enlightenment, and Nirvana. What is the Middle Way? . . . It is the Noble Eightfold Path—Right Views, Right Resolve, Right Speech, Right Conduct, Right Livelihood, Right Effort, Right Mindfulness, and Right Concentration. This is the Middle Way. . . .

And this is the Noble Truth of Sorrow. Birth is sorrow, age is sorrow, disease is sorrow, death is sorrow; contact with the unpleasant is sorrow, separation from the pleasant is sorrow, every wish unfulfilled is sorrow—in short all the five components of individuality are sorrow.

And this is the Noble Truth of the Arising of Sorrow. It arises from craving, which leads to rebirth, which brings delight and passion, and seeks pleasure now here, now there—the craving for sensual pleasure, the craving for continued life, the craving for power.

And this is the Noble Truth of the Stopping of Sorrow. It is the complete stopping of that craving, so that no passion remains, leaving it, being emancipated from it, being released from it, giving no place to it.

And this is the Noble Truth of the Way which Leads to the Stopping of Sorrow. It is the Noble Eightfold Path—Right Views, Right Resolve, Right Speech, Right Conduct, Right Livelihood, Right Effort, Right Mindfulness, and Right Concentration.

W. T. de Bary, editor, *Introduction to Oriental Civilizations: Sources of Indian Tradition* (Columbia University Press, 1958), pp. 101-102, slightly adapted.

(d)

CONFUCIUS (551-479 B.C.) was not accepted as a great teacher during his lifetime, except by a small band of disciples. But slowly his teachings gained influence until, by the 2nd century B.C., Confucianism was declared the state creed, and all scholars and statesmen were expected to study it. The chief source of Confucian philosophy is the record of the sage's activities and conversations written down after his death, the 20 chapters and 497 verses of the *Analects*. Here are a few selections.

Confucius said: "The young are to be respected. How do we know that the next generation will not measure up to the present one? But if a man has reached forty or fifty and nothing has been heard of him, then I grant that he is not worthy of respect."

Someone inquired: "What do you think of 'requiting injury with kindness'?" Confucius said: "How will you then requite kindness? Requite injury with justice, and kindness with kindness."

Confucius said: "He who sins against Heaven has none to whom he can pray."

Confucius said: "Were any prince to employ me, even in a single year a good deal could be done, and in three years everything could be accomplished."

When Confucius was travelling to Wei, Jan Yu drove him. Confucius observed: "What a dense population!" Jan Yu said: "The people having grown so numerous, what next should be done for them?" "Enrich them," was the reply. "And when one has enriched them, what next should be done?" Confucius said: "Educate them."

Tzu Kung asked about government. Confucius said: "The essentials are sufficient food, sufficient troops, and the confidence of the people." Tzu Kung said: "Suppose you were forced to give up one of these three, which would you let go first?" Confucius said: "The troops." Tzu Kung asked again: "If you were forced to give up one of the two remaining, which would you let go?" Confucius said: "Food. For from of old, death has been the lot of all men, but a people without faith cannot survive."

Tzu Kung asked about the gentleman. Confucius said: "The gentleman first practices what he preaches and then preaches what he practices."

Tzu Lu asked about the worship of ghosts and spirits. Confucius said: "We don't know yet how to serve men, how can we know about serving the spirits?" "What about death?" was the next question. Confucius said: "We don't know yet about life, how can we know about death?"

Tzu Yu asked about filial piety. Confucius said: "Nowadays a filial son is just a man who keeps his parents in food. But even dogs or horses are given food. If there is no feeling of reverence, wherein lies the difference?"

Confucius said: "I won't teach a man who is not anxious to learn, and will not explain to one who is not trying to make things clear to himself. If I hold up one corner of a square and a man cannot come back to me with the other three, I won't bother to go over the point again."

Confucius said: "At fifteen, I set my heart on learning. At thirty, I was firmly established. At forty, I had no more doubts. At fifty, I knew the will of Heaven. At sixty, I was ready to listen to it. At seventy, I could follow my heart's desire without transgressing what was right."

W. T. de Bary, editor, *Introduction to Oriental Civilizations: Sources of Chinese Tradition* (Columbia University Press, 1960), pp. 22-35, slightly adapted.

THE FIRST CIVILIZATIONS

FURTHER READING

NOTE: Books marked with an asterisk (*) are available in a paperback edition; those marked with a dagger (†) are especially recommended for students.

1. GENERAL
- † BIBBY, G., *Four Thousand Years Ago* (Knopf, 1961)
- * CLARK, G., *World Prehistory: An Outline* (Cambridge, 1961)
- * COTTRELL, L., *The Anvil of Civilization* (Mentor, 1957)
- DIEZ, E., *The Ancient Worlds of Asia* (Macdonald, 1961)
- † DURANT, W., *Our Oriental Heritage* (Simon and Schuster, 1935)
- HITTI, P. K., *The Near East in History* (Van Nostrand, 1961)
- * MOSCATI, S., *The Face of the Ancient Orient* (Anchor, 1962)
- † PIGGOTT, S., editor, *The Dawn of Civilisation* (Thames and Hudson, 1961)

2. EARLY MAN
- † BIBBY, G., *The Testimony of the Spade* (Knopf, 1956)
- BRODRICK, A. H., *Man and His Ancestry* (Hutchinson, 1960)
- CHILDE, V. G., *What Happened in History* (Penguin, 1954)
- * ———, *Man Makes Himself* (Mentor, 1951)
- † COON, C. S., *The Story of Man* (Knopf, 1962)
- HOWELLS, W., *Mankind in the Making* (Doubleday, 1959)
- † * MONTAGU, A., *Man: His First Million Years* (Signet, 1962)
- † QUENNELL, M. and C. H. B., *Everyday Life in Prehistoric Times* (Batsford, 1960)

3. MESOPOTAMIA
- CHAMPDOR, A., *Babylon* (Elek, 1958)
- † CHIERA, E., *They Wrote on Clay* (Phoenix, 1956)
- * FRANKFORT, H., *The Birth of Civilization in the Near East* (Anchor, 1956)
- † * KRAMER, S. N., *History Begins at Sumer* (Anchor, 1959)
- † SAGGS, H. W. F., *The Greatness that was Babylon* (Hawthorn, 1962)

4. EGYPT
- † * COTTRELL, L., *The Last Pharaohs* (Pan, 1956)
- † * ———, *Life Under the Pharaohs* (Pan, 1957)
- * EDWARDS, I. E. S., *The Pyramids of Egypt* (Penguin, 1961)
- * EMERY, W. B., *Archaic Egypt* (Penguin, 1961)
- * FAIRSERVIS, W. A., *The Ancient Kingdoms of the Nile* (Mentor, 1962)
- HAYES, W. C., *The Scepter of Egypt*, 2 vols. (Harvard, 1959-60)
- † MONTET, P., *Everyday Life in Egypt in the Days of Ramesses the Great* (Arnold, 1958)
- MURRAY, M. A., *The Splendour that Was Egypt* (Philosophical Library, 1949)
- * STEINDORFF, G. and SEELE, K. C., *When Egypt Ruled the East* (Phoenix, 1963)
- * WILSON, J. A., *The Culture of Egypt* (Phoenix, 1956)

5. INDIA
- * BASHAM, A. L., *The Wonder that Was India* (Evergreen, 1959)
- * PIGGOTT, S., *Prehistoric India to 1000 B.C.* (Penguin, 1950)
- WHEELER, M., *Early India and Pakistan to Ashoka* (Thames and Hudson, 1959)

6. CHINA
- CREEL, H. G., *The Birth of China* (Day, 1937)
- * FAIRSERVIS, W. A., *The Origins of Oriental Civilization* (Mentor, 1959)
- † * GOODRICH, L. C., *A Short History of the Chinese People* (Torchbook, 1963)
- NEEDHAM, J., *Science and Civilisation in China*, Vol. I (Cambridge, 1954)
- REISCHAUER, E. O. and FAIRBANK, J. K., *East Asia: The Great Tradition* (Houghton Mifflin, 1960)

2. The Middle East Battleground

Israel is a hunted sheep driven away by lions. First the king of Assyria devoured him, and now at last Nebuchadrezzar king of Babylon has gnawed his bones.
JEREMIAH 50: 17

Thus says the Lord, your Redeemer . . . of Cyrus, "He is my shepherd. . . ."
ISAIAH 44: 24, 28

Anyone who reads the newspapers of the second half of the 20th century must be aware of the importance of the area we call the Middle East. The modern independent states of Israel, Lebanon, Jordan, Iraq, Iran, Egypt, and Syria make the headlines almost daily. If there had been newspapers in the Iron Age, the same areas would have made front-page news under their ancient names of Assyria and Persia. And occasionally on the inner pages you might have found a brief dispatch dealing with another people, the Hebrews.

In this chapter we shall be studying the clash of empires that took place in the Middle East between two and three thousand years ago. How did this clash come about, and what were its results?

1/ASSYRIA

On the upper Tigris River, in the north-east corner of the Fertile Crescent 300 miles north of Babylon, lies a triangular rolling plateau of land which for centuries was a prey to invaders. One of the early invading forces settled down on the banks of the Tigris River to form the city-state of Ashur, which dates back to the third millennium B.C. The strategic position of Ashur as a pivot to hold up caravans or control the great east-west highway along the Crescent was of vital importance. Only constant warfare or constant preparedness for warfare could hold this bridge-head. From it the ambitious Assyrians could strike out against their neighbours such as the Babylonians, and in 1247 B.C. they sacked and razed to the ground the ancient city of Babylon.

Assyria was getting ready to step on to the stage of world politics—and the times were ripe for her. The Kassites, who had been in at the kill of the Babylonian empire in the 16th century B.C., had been absorbed by other peoples; and the Hittites, who had vied with the Egyptians for Middle Eastern supremacy, had been conquered by Iron Age Indo-Europeans called Phrygians. Thus by 1200 B.C. the power situation in the Middle East had altered drastically.

The fall of the Hittites may have triggered the spread of the Iron Age, for by 1100 B.C. the knowledge of smelting and forging iron had spread around

BRONZE AGE INVASIONS (2000-1500 B.C.)

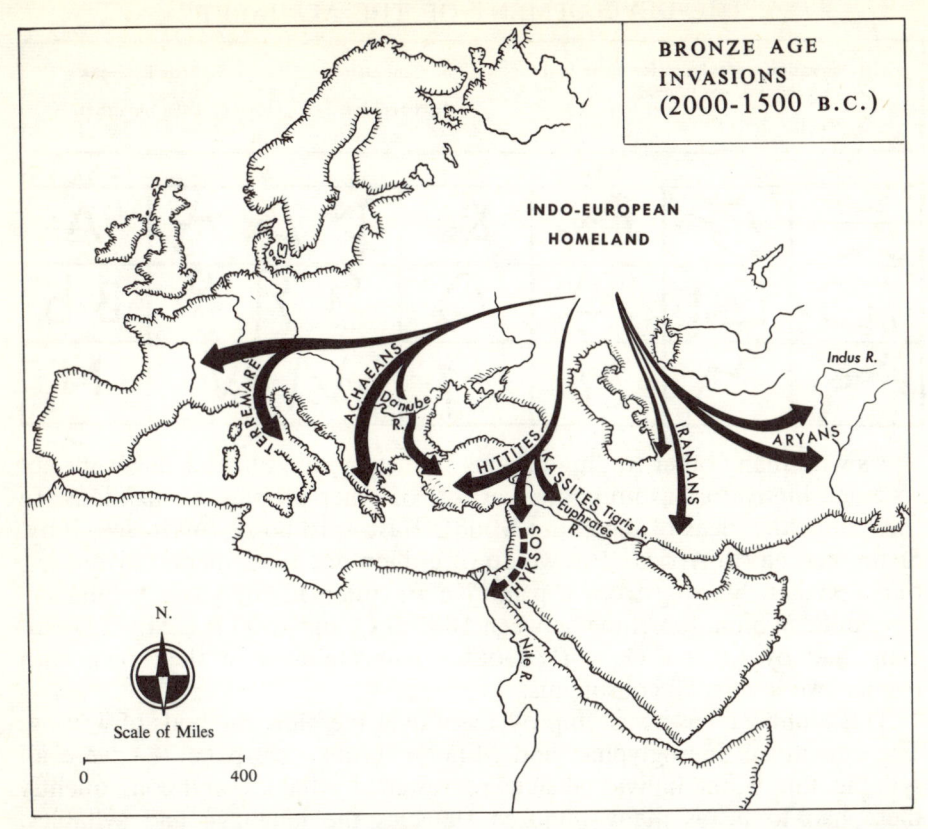

the eastern Mediterranean. Though we do not know who invented this process, the first historical references to it occur in the archives of the Hittite kings. The Hittites valued their iron so highly that one one occasion they suggested to the Egyptians that a fair purchase price for it would be an equal weight of gold! By 1200, however, the Hittites were no longer able to monopolize either the process or the iron ores. The secret was out.

The end of Hittite power coincided with the decline of Egypt, and the eclipse of these two mighty empires provided a lull of three centuries during which a number of smaller independent nations could prosper. Two in particular are worth noting.

The Greek word *phoinix* means "red", and the English version of it, "Phoenician", is the name we apply to the people the Greeks called "red" men. These Phoenicians lived on a narrow strip of coast about 100 miles long and 30 miles wide at the eastern end of the Mediterranean. From this meagre coastline they took to the sea with a will, spreading their sails from such ports as Gebal, Beyrut, Sidon, and Tyre, founding trading stations along their routes, and exploring the Atlantic coasts of Europe and Africa —even getting as far as Britain in their search for tin. Their most celebrated product became the deep red or purple dye of Tyre; hence their peculiar name of "red" men.

THE DEVELOPMENT OF THE ALPHABET

1. Egyptian hieroglyphs for *aleph* (ox), *beth* (house), *nun* (snake)		3. Canaanite 4. Phoenician		5. Archaic Greek 6. Latin alphabet	
2. Sinaitic					
1.	2.	3.	4.	5.	6.

The Phoenician alphabet

As with many other trading nations in history, the Phoenicians were not so much innovators as imitators, carriers of other peoples' civilizations. In their ships the ideas of the entire Middle East were borne to Greece, Italy, Spain, and North Africa. But we do owe them one fundamental invention: our alphabet. Starting from a primitive Egyptian hieroglyphic "alphabet" developed in Sinai sometime between 1800 B.C. and 1500 B.C., the Phoenicians had by 1100 B.C. worked out a true alphabet of their own with twenty-two letters, all consonants.

This alphabet was a vast improvement over the older methods of writing. The cuneiform, hieroglyphic, and Minoan scripts (see page 78) were all syllabic, that is, the individual signs represented syllables, and consequently there had to be as many hundred signs as the language had syllables. But the new Phoenician alphabet could express all the syllables and all the words with a scant twenty-two letters. No longer did men of business need the services of specially trained scribes, nor rulers have to depend on a whole scribal class who jealously guarded their learning and their privileged position within the state. Henceforth almost anyone could, with a little instruction and practice, learn to read and write. By 850 B.C. Greek merchants in Cyprus or Syria had taken over the Phoenician alphabet and added vowels. From the Greeks it passed to the Romans, and so to us. The fact that you can now read these words is your greatest debt to the "nomads of the sea", the Phoenicians.

The Aramaeans

To the east of the Phoenicians dwelt another trading people, the Aramaeans. From their chief city, Damascus, long strings of camels plodded to the bazaars of the Middle East, exporting something infinitely more enduring than bundles of wares. For the Aramaeans adapted the Phoenician alphabet to their language, and passed it on to the Hebrews, Persians, Arabians, and other Asians. Along with these letters they passed on their own language, with the result that Aramaic became a standard tongue in the Middle East and displaced Babylonian as the language of diplomacy.

On these small nations, enjoying a brief Indian summer after the season of great empires, the chilling Assyrian wind was soon to blow.

Building an Empire The first Assyrian king to bring order out of chaos by systematically defeating his neighbours was Tiglath-pileser I (1114-1076 B.C.). Here was a man of action. He chased sharks, hunted lions, and conquered cities—nay, states, "forty-two" of them, he boastfully records. But with his death Assyria's early prominence evaporated. It would take another 350 years for her to reach her peak.

The Assyrian Empire was really founded by Tiglath-pileser III (745-727 B.C.), who made himself king of Babylon and reduced the Hebrews to vassalage. It was Tiglath-pileser III who began the Assyrian practice of deporting rebellious peoples to other parts of his empire: in the next twenty-five years over 200,000 captives were to suffer this fate. Sargon II (722-705 B.C.) went on to crush the petty Palestinian kingdoms, as did his successor Sennacherib (704-681 B.C.). Sennacherib's arrogant account of his siege of Jerusalem in 701 B.C. may be read in the Source Reading on page 72.

Tiglath-pileser III

The kings of Palestine tried to protect their tiny country by playing off Egypt against Assyria, a policy that only egged the Assyrians on to greater conquests. Sennacherib's son, Esarhaddon (680-669 B.C.), invaded Egypt and rebuilt old Babylon, and under Ashurbanipal (669-633 B.C.) the Assyrian Empire achieved its greatest extent when it stretched from Babylon to Thebes.

Assyrian records cease shortly before the end of Ashurbanipal's reign. We know, however, that the New Babylonians, or Chaldeans as they came to be called, succeeded in destroying the Assyrian capital of Nineveh in the summer of 612 B.C. "I turned the hostile land into heaps and ruins," writes the Babylonian king. Archaeologists can still poke in the ash heaps and calcined sculptures which bear evidence as to how completely the great city was destroyed. None of the surviving records mourn its passing.

A Policy of Terror The militaristic aspect of Assyrian civilization is well known. The army was a model of its kind, armed with bows, pikes, and swords, and protected with iron breastplates. Iron-tipped battering-rams and siege towers on wheels pummelled enemy cities, with the terrors of psychological warfare added for good measure. For the Assyrian reputation for ruthlessness preceded them. One inscription thus casually describes the treatment meted out to a conquered city:

I slew one of every two. I built a wall before the great gates of the city; I flayed the chief men of the rebels, and I covered the wall with their skins. Some of them were enclosed alive within the bricks of the wall, some of them were crucified with stakes along the wall; I caused a great multitude of them to be flayed in my presence, and I covered the walls with their skins.

How much of the Assyrian reputation for frightfulness depends on their own boastfulness, a propaganda designed to head off any revolt in the empire? We cannot tell. But some scholars have claimed that if the many competing small states had not been absorbed by the Assyrians, their

The Assyrian armies, probably the first large armies extensively equipped with iron, were made up mainly of archers and spearmen. (A spearman is portrayed here carrying a large shield.) There were also cavalry, at first with three men—driver, warrior and shield-bearer—to a chariot, and later (700 B.C.) with a single mounted warrior armed with spear and bow. The Assyrian bow was composed of a combination of animal sinews, wood, and horn, and while relatively small it was immensely powerful.

anarchy might well have destroyed that Middle Eastern civilization which their conquerors protected. Cruelty and brutality there was in abundance; yet culture did not die.

Assyrian government and society

As might be expected, the Assyrian government was a military despotism; and Ashur, the national god, required reports from his representative, the king, on the progress of his wars. Even so, the inhabitants of the various city-states had self-rule and a status separate from that of agricultural serfs. The middle class of city-dwellers (bankers, merchants, carpenters, weavers, and metal-workers) was usually exempt from military service. Slaves captured in warfare were often employed in handicraft industries and lived a life little worse than the serfs. On the other hand, the slaves employed in government building projects fared no better than those who sweated on the Egyptian pyramids. The lot of serfs was not enviable. Not only were they forced to hand over part of their produce, but also to work on the crown lands or serve in the army.

The majority of traders in the cities were not Assyrians. Yet it was for them that the great trade routes were kept open, so that in a sense the Assyrian kings sold "protection" to the traders of the Middle East.

Preservers of Culture Although the Assyrians modelled their religion on that of Babylon, the fact that they placed their god Ashur above all others showed a tendency to monotheism. Nor did the priests ever attain the power of their Babylonian or Egyptian counterparts.

The Assyrians produced the first literary works of history: their military annals. Great royal libraries were built in Asia, and under Ashurbanipal a huge library was set up at Nineveh—although much of its contents consisted of copies of Babylonian cuneiform tablets rather than original Assyrian literature. From this library alone 22,000 tablets have been recovered.

It is in the palace of this same king, Ashurbanipal, that Assyrian art reached its highest point of development, for the bas relief of the lion-hunt there excelled all earlier works in fidelity of detail (see page 2). Assyrian art was bursting with life. In the beginning it was little more than an imitation of Babylonian art; but its animal sculptures developed a vividness and realism that artists were unable to equal until the days of the Roman Empire. Since the use of brick for walls and arches had also been learned from the Babylonians, Assyrian architects employed stone only for foundations, columns, and decorative facings.

In science, these ingenious people combined fact with superstition. Astronomers made systematic reports of observations to the court—though it was mostly for astrological purposes, while liver divination was employed to foretell the future. As might be expected in a civilization that put great stock in a healthy army, the use of medicines was increased. Over 500 drugs were catalogued.

Thus in many ways Assyrian civilization may be characterized as more than a brutal militarism. Along with militarism there existed a cultural life whose source historians have continued to debate. Was this culture a genuine Assyrian growth on Babylonian foundations, or a purely Babylonian inheritance? One modern historian, although fully aware of the Babylonian influence on Assyria, still asserts, "From 900 to 600 B.C. Assyria was the intellectual centre of the ancient-oriental world."

Marble Magnificence The fall of Assyria destroyed the balance of power in the Middle East. New Babylonia and Egypt now attempted to fill the vacuum, and with the decisive defeat of Egypt at Carchemish in 605 B.C. the Babylonian king Nebuchadrezzar (or Nebuchadnezzar) established New Babylonia—that is, the Chaldean Empire—as Assyria's successor to power. In due course all Syria and Palestine fell to him, as is well known from the Biblical story of the captivity of the Jews in Babylon. The magnificence of that ancient capital must have dazzled its rustic Jewish captives—the streets paved with marble, the three- and four-storey houses, the imposing temples and palaces, and the city walls 85 feet thick. Nebuchadrezzar was surely a masterly builder.

COMPARATIVE EMPIRES:—

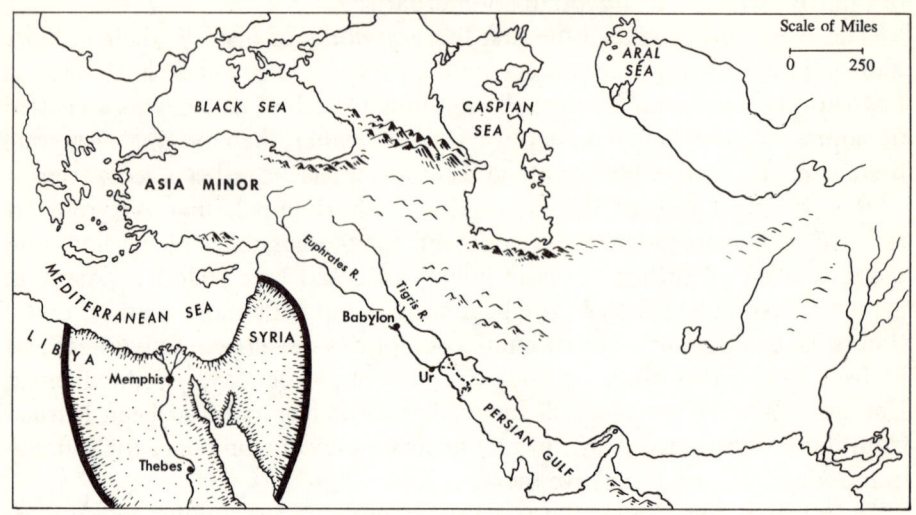

EGYPTIAN NEW KINGDOM (about 1400 B.C.)

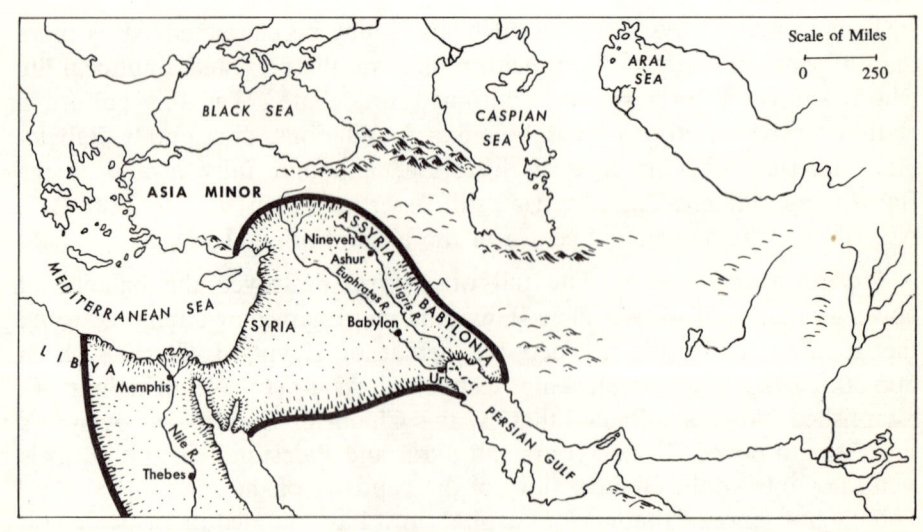

ASSYRIAN (about 700 B.C.)

COMPARATIVE EMPIRES:—

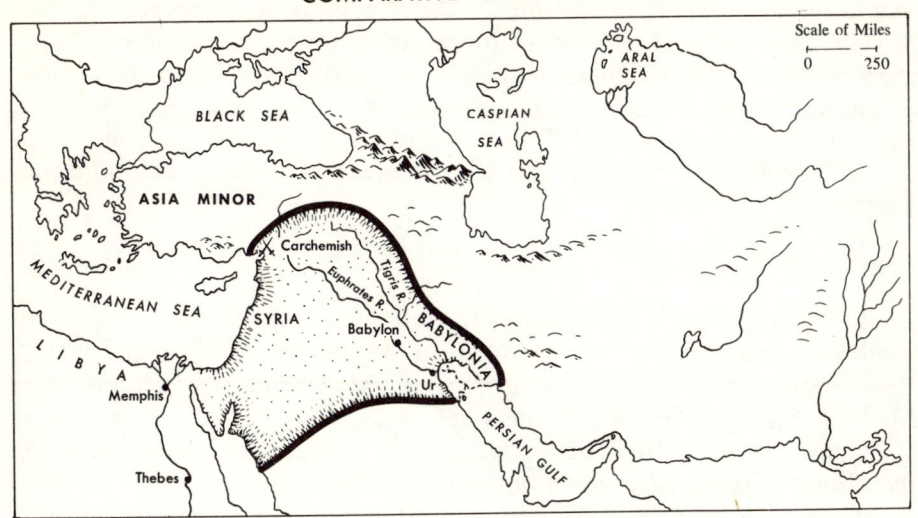

BABYLONIAN (about 550 B.C.)

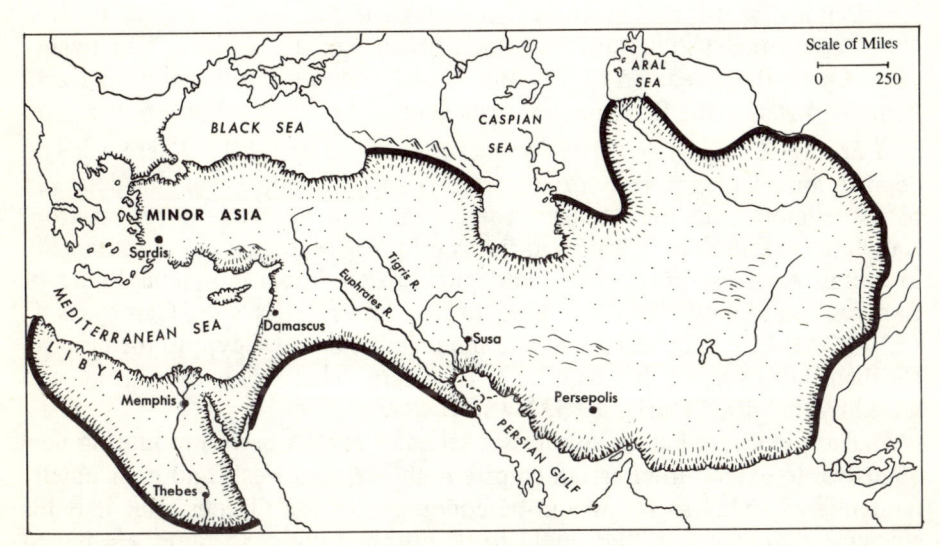

PERSIAN (500 B.C.)

Chaldean mathematics was sufficiently developed to allow more exact astronomical calculations than had been possible under Old Babylonian science. The lunar calendar was brought into agreement with the solar year, and accurate observation and prediction of eclipses was undertaken. Yet astronomy went hand in hand with astrology: when we "thank our lucky stars" we are reverting to a superstition as old as the Babylonian civilization.

The successors of Nebuchadrezzar were weak men, and when Cyrus the Persian entered Babylon in 539 B.C. it was apparently with the connivance of the Babylonians. Another slice had been added to the growing Persian Empire.

2/PERSIA

In the Bronze Age, Indo-Europeans known as Iranians began to move southward and eastward from the Caspian Steppes. By the 9th century B.C. these invaders had split into two groups, the Medes and the Persians. The Medes established an empire in what is today Iran, with a capital at Ecbatana (modern Hamadan), while their subjects the Persians made their home in south Media east of the Persian Gulf.

Cyrus the Great

A local prince, Cyrus of Anshan, united these Persian tribes and led them against Media in 549 B.C. Nor did he stop there. He went on to defeat the fabulously rich Croesus of Lydia in western Asia Minor, a victory which ultimately brought the Persians face to face with the Greeks. Then he looked covetously at the Chaldean Empire.

For the Babylonians the handwriting on the wall was all too plain: "The kingdom is divided, and given to the Medes and Persians." (*Daniel* 6: 28). Cyrus could boast that Babylon fell "without fighting a battle." In twenty years Cyrus the Great (549-529 B.C.) had conquered the civilized world from the Indus to the Dardanelles, from central Asia to the borders of Egypt.

A Strange Conqueror The conqueror behaved strangely. There were no deportations. Cities were not destroyed. A subject people was not deprived of its religion, and the Jewish exiles were even allowed to return from Babylon to Jerusalem. Cyrus had the good sense and shrewd statesmanship to mobilize all his propaganda and put himself forward first as "king of Babylon", and only then as "king of the lands." His son, Cambyses II (529-522 B.C.), succeeded him and managed to annex Egypt to the empire.

Darius I

He died, however, on his way back to Persia, whereupon the throne was seized by the gifted Darius I (521-486 B.C.).

Darius the Great had to suppress a series of revolts before mounting new offensives to extend the Persian empire both east and west, and he is chiefly remembered today as the would-be conqueror of the Greeks. But it is his efficient administration that ought to be his chief claim to fame. He issued a law code, established a system of uniform weights and measures and an imperial currency, completed a canal connecting the Nile and the Red Sea, and even built a fleet that sailed down the Indus to the Indian Ocean and thence on a two-and-a-half-year voyage to Egypt. Under Darius, Aramaic became the international language of diplomacy and commerce.

When Darius decided to build a royal palace at one of his capitals, Susa, he determined to dazzle the world. For this purpose he summoned the varied resources of his vast empire:

> The cedar timber was brought from a mountain called Lebanon. The Assyrians brought it to Babylon; from Babylon the Carians and the Ionians brought it to Susa. The *yaka* wood was brought from Gandara and from Carmania. The gold wrought here was brought from Sardis and from Bactria. The precious stone wrought here, lapis lazuli and carnelian, was brought from Sogdiana. The turquoise wrought here was brought from Chorasmia. The silver and the ebony were brought from Egypt. The ornamentation with which the wall was adorned was brought from Ionia. The ivory wrought here was brought from Ethiopia and from Sind and from Arachosia. The stone columns wrought here were brought from a village called Abiradu, in Elam.
> The stone-cutters who wrought the stone were Ionians and Sardians. The goldsmiths who wrought the gold were Medes and Egyptians. The men who wrought the wood were Sardians and Egyptians. The men who wrought the baked brick were Babylonians. The men who adorned the wall were Medes and Egyptians.

Over this "community of nations" presided Darius, one of the greatest organizers of history. At his death his empire stretched from India to Macedon—about as far from east to west as from Toronto to Vancouver. This "King of Kings" immortalized his achievements in an impressive inscription, some extracts from which are given in the Source Reading on page 73. The ancient Middle East had come to Europe, and the history of Persia from about 500 B.C. to 330 B.C., when it was conquered by Alexander the Great, is so bound up with that of Greece that it will best be considered in the account of Greek history in Unit II.

An Able Administration The Persians made two outstanding contributions to the ancient world: the organization of their empire, and their ethical religion. Both of these contributions have much significance for our western world. The system of imperial administration was inherited by Alexander the Great, adopted by the Roman Empire, and eventually bequeathed to modern Europe. The Zoroastrian religion passed on much to the Hebrews and thence to Christianity.

In their empire, the Persians turned the conquered kingdoms into provinces (*satrapies*), each under a governor (*satrap*). The distinction between a ruling and a conquered people was gradually reduced, and a diversified yet unified empire was built up. Excellent roads enabled the Persian king to keep in close touch with his extensive empire, and an imperial postal system was operated of which a Greek wrote: "Neither snow nor rain nor heat nor gloom of night stays these couriers from the swift completion of their appointed rounds." There could be communication between Sardis and Susa (a distance of 1500 miles) in one to two weeks. Over these roads could also travel the king's agents—"the eyes and ears of the king"—to report lapses of duty on the part of his satraps.

The grand stairway to the immense 72-columned Audience Hall of the royal palace at Persepolis, begun by Darius I in 512 B.C., leads up to the stone platform on which the palace was built. The Persians preferred the horizontal style of Egypt to the Babylonian arch, and with their plentiful supply of limestone made striking use of graceful columns. Sculptural reliefs along the stairway portray a stately procession of representatives from the vast Empire at a New Year's festival, their crisply chiselled details evincing a dignity and courtliness in sharp contrast to the dynamic realism of Assyrian art. Darius's majestic palace graced Persepolis until the city was captured and partly destroyed by Alexander the Great in 330 B.C.

Because the separate satrapies were allowed to retain their own language and religion, the private life of the citizens of the empire was not dislocated by the conquests. Moreover, the Persian kings willingly adopted the royal style of the conquered peoples: in Babylon the god Marduk ordained the Persian king; in Egypt the alien monarch became the son of Re.

In such a vast empire there was a great variety of products. Fruits and vegetables, wheat and barley were cultivated in the fertile valleys and oases. Irrigation of small areas was practised (the Persian "a well-kept park" = the English "paradise"), and in the northern plains horses, cattle, goats, and sheep were raised. Mining and metal-working were both important industries.

Of course the provinces were taxed; but it was not a burdensome tax, and possibilities of local economic development were probably promoted by the steady administration and open communications of the empire as well as by the introduction of a uniform coinage and a uniform system of weights and measures. Eventually such encouragement of local autonomy in economic matters, as well as the dwindling nucleus of native Persians in the army, tended to disrupt the empire. Even so it lasted for two centuries—and then succumbed only to external pressure.

While it flourished, Persian society was held together by its remarkable army. The peasantry were archers; the nobles, cavalry. When war was declared every able-bodied male between the ages of fifteen and fifty was drafted. There was even a story—and it may be no more than that—of a father of three sons who petitioned the king to exempt one of them from military service. All three were executed.

Light versus Darkness   The Persian religion, which had always emphasized a moral code, was further refined by the 6th century prophet Zoroaster. He elevated the conflict between good and evil into a dualistic system, proclaiming it to be a struggle between the only good god, Ahuramazda, and Ahriman, the god of evil and darkness. According to Zoroaster, there would finally be a resurrection of the dead and a Last Judgment in which good would ultimately triumph.   *Zoroastrianism*

It is not hard to see how the mighty drama of a struggle between good and evil, between light and darkness, deeply affected Jewish thought, and how Ahriman became a model for Satan.

3/THE HEBREWS

So far in this survey of the Middle East up to 500 B.C. no mention has been made of a rather insignificant band of nomads who settled in Palestine at one end of the Fertile Crescent. This area, arching between Egypt and Mesopotamia, was a great highway for military conquest, trade, and the transmission of ideas. Often the native rulers of Palestine were able, by balancing off their powerful neighbours, to win an uneasy truce; but never were they able to maintain any prolonged political independence. Thus Egyptian, Babylonian, and Persian elements were blended into their civilization.

The land of Palestine, in ancient times an area 400 miles long and 80 to 100 miles wide, had no fewer than forty different climatic and geographic units. This great variety of environments produced many tribes, and a marked division between the wandering hill people and the settled dwellers of the plains. The tension that from the beginning existed between the ideas and ideals of the tribesmen of the hills and the city merchants and traders goes far toward explaining the internal conflicts in Palestine's history.

A Band of Nomads   From the 20th century B.C. a group of people called *Habiru* or *Hapiru* ("nomad" or "bandit" in Babylonian) are mentioned in Middle East documents. Among these nomads were the Hebrews, who seem to have come out of the general area of Ur and who lived for a time in the north-west section of the Babylonian empire. Eventually a group came to inhabit the land called Canaan or Palestine. What was their society like?

The hill-dwelling patriarch Abraham (1800-1750 B.C.) and his family formed a close-knit tribal unit in a society of roving shepherds. As the Hebrews settled down in Palestine the tribal chiefs lived in fortresses, to which their subjects could come in the winter or when danger threatened. At other times the populace lived in tents or huts.

A vivid portrayal of a Hebrew tribe has been found on the wall of an Egyptian tomb of about 1900 B.C. Both men and women wear woollen tunics, and they carry the tools and baggage of their profession: simple weapons, a lyre, and bellows. Donkeys are their beasts of burden, for the camel was not domesticated for another eight centuries. Although it was a century before the time of Abraham that this eye-catching group of Hebrew tinkers visited Egypt, the description recorded of them could apply just as well to later generations of Palestinian nomads.

The Exodus

Early in the 17th or late 16th century B.C. a group of these nomads seems to have penetrated into Egypt. This was at the time of the Hyksos domination, and perhaps Joseph became an administrator when these foreigners ruled Egypt. With the overthrow of the Hyksos came the enslavement of non-Egyptians and the period of bondage. The escape or *exodus* from Egypt probably occurred in the early 13th century B.C. under the leadership of Moses. The miraculous crossing of the Red Sea, the meeting of Moses and his God on Mount Sinai, and the subsequent wanderings in the Sinai peninsula for "forty years" in search of the "Promised Land" are well known.

Then the Hebrew tribes, some 6000-7000 strong, invaded Canaan a second time. They stormed Jericho under Joshua, and it is possible that the famous Biblical story describing the falling walls refers to the havoc wrought by an earthquake. This second Hebrew conquest may have gone on for from two to four centuries—the confusing years celebrated in the Old Testament books of *Joshua* and *Judges* when "there was no king in Israel; every man did what was right in his own eyes."

A Dynasty and a Temple   In the course of their second conquest the Hebrews swept down from the hills into the plains inhabited by the Canaanites, gaining control of the land by slow infiltration and intermarriage as well as by military conquest. As time went on, they adopted laws and local religious customs from neighbouring peoples. Elders dispensed justice by the town gate, and the local divinities of the towns, called *Baals*, were worshipped to the neglect of the Hebrew God.

Yahweh

In the hills and on the frontier, however, the desert ideals were preserved. In the back country God was still the god of the desert, a male deity and a god of war. The Hebrews called this God *YHWH*, and vowels have been added to make the name *Yahweh*, mistakenly transliterated as "Jehovah" by a Christian scholar of the Middle Ages.

Saul

With the collapse of the Canaanite power in Palestine, the Philistines and other "Peoples of the Sea" who had earlier been turned back at the Nile Delta attacked the Hebrews. This danger forced a national unity which brought to an end the period of the Judges and placed King Saul (1020-1005 B.C.) on the throne. Saul was chosen king over Israel primarily to wage war against the invading Philistines, for the early kings had to be capable military leaders. The nucleus of a standing army was formed under

Saul, and the usual weapons seem to have been javelins, darts, and slings —as who that reads of David's encounter with Goliath can forget.

The brooding, moody King Saul, unsure of himself and jealous of the growing popularity of his military chieftain, David, committed suicide after a Philistine defeat. The throne then passed to David who, succeeding where Saul had failed, managed to weld the tribes into a permanent nation and to found a dynasty which ruled for over four hundred years. Jerusalem was made the national capital of a united Israel.

David

Few men have been more fortunate than David. His rise to power about the year 1000 B.C. coincided with the decline of the Middle Eastern kingdoms. David inherited the advantages of this political power vacuum, and he and his successor, Solomon, were thereby given much-needed time to build up an administration. Power was centralized. A great Temple was planned for Jerusalem, and forced labour was introduced to complete the building programme. Population increased, and about four hundred towns were joined together in a national federation. The old tribal boundaries were also revised, probably to aid tax assessment and to break down local prejudices. The army was made into a professional body under the king's command, and the chariot was introduced as an engine of war.

But the passionate, unpredictable king died before all his projects could be completed. It was Solomon (965-925 B.C.), his son by Bathsheba, who pushed David's plans to their ultimate realization.

Solomon

Solomon's most famous work was the Temple, which made Jerusalem the religious as well as the political heart of Israel. Great imports of wood, craftsmen from Phoenicia, forced labour, crushing taxes—all were necessary to build the Temple. Trade boomed. Solomon's ships and caravans could be seen throughout Egypt or in the land of the Queen of Sheba in southern Arabia. The twelve ancient tribal divisions were continued in the twelve administrative units of the country. Solomon married foreign princesses and established altars for the deities of his alien wives. He was a sort of sultan, surrounded by opulence.

Solomon in all his glory impoverished his country so that he might live in luxury. His very drinking vessels were of gold: "none were of silver; it was not considered as anything in the days of Solomon." Solomon's tragedy was that of a man with a heritage that proved too great for his personality.

A northerner called Jeroboam dared to criticize the king, and was forced to flee to Egypt. But in the reign of Solomon's son, Rehoboam (922-915 B.C.), he returned to lead a rebellion which split the kingdom. The southern tribes remained loyal and kept the old capital at Jerusalem in their new kingdom of *Judah*, while the northerners established a separate capital for their new kingdom of *Israel* at Samaria—learning too late that the rebel Jeroboam (922-901 B.C.) was only another monarch intent on riches who cared not a fig about helping the poor. The prophet Amos protested that the righteous were sold for silver, "and the needy for a pair of shoes."

The Kingdoms of Judah and Israel

ANCIENT PALESTINE AND HER NEIGHBOURS

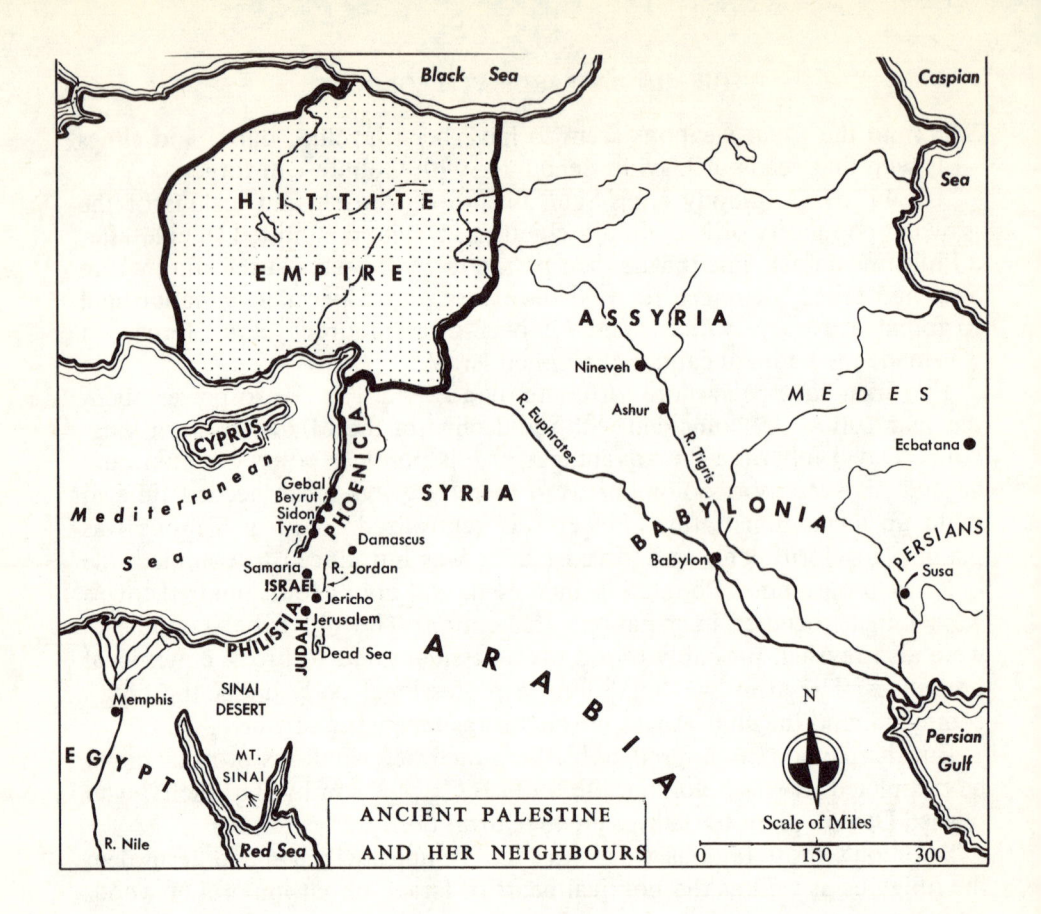

From about the middle of the 9th century B.C. Assyria began to decide the political fate of the two Hebrew kingdoms. When both became vassal states of that empire in the 8th century, Hebrew intrigues began with other Assyrian satellites—now Egypt, now New Babylonia—to throw off the foreign yoke. Finally a coalition of Middle Eastern states destroyed first the Assyrian (612 B.C.), then the Egyptian (605 B.C.) power. The crowning disaster for Palestine came when Nebuchadrezzar the Babylonian shipped off approximately one-third of the population of Judah (mostly the artisans and the ruling class) to exile in Babylon.

The two Hebrew kingdoms were at an end.

A Nation Apart After two generations of exile the Hebrews had settled into an influential role in the Babylonian economy, while back in Judah chaos reigned and the old religion of Moses and the prophets dropped into disuse. There the whole story of the Hebrews might have ended—but it did not. The exiles ought to have been assimilated into the attractive Babylonian culture—but they were not. A small remnant remembered Moses and studied the prophets, and gave expression to their longing for the homeland in the Biblical books of *Ezekiel* and the *Second Isaiah*.

Possibly, too, it was during the exile that the institution of the synagogue (from the Greek *synagogue* meaning a "bringing together") developed.

THE MODERN MIDDLE EAST

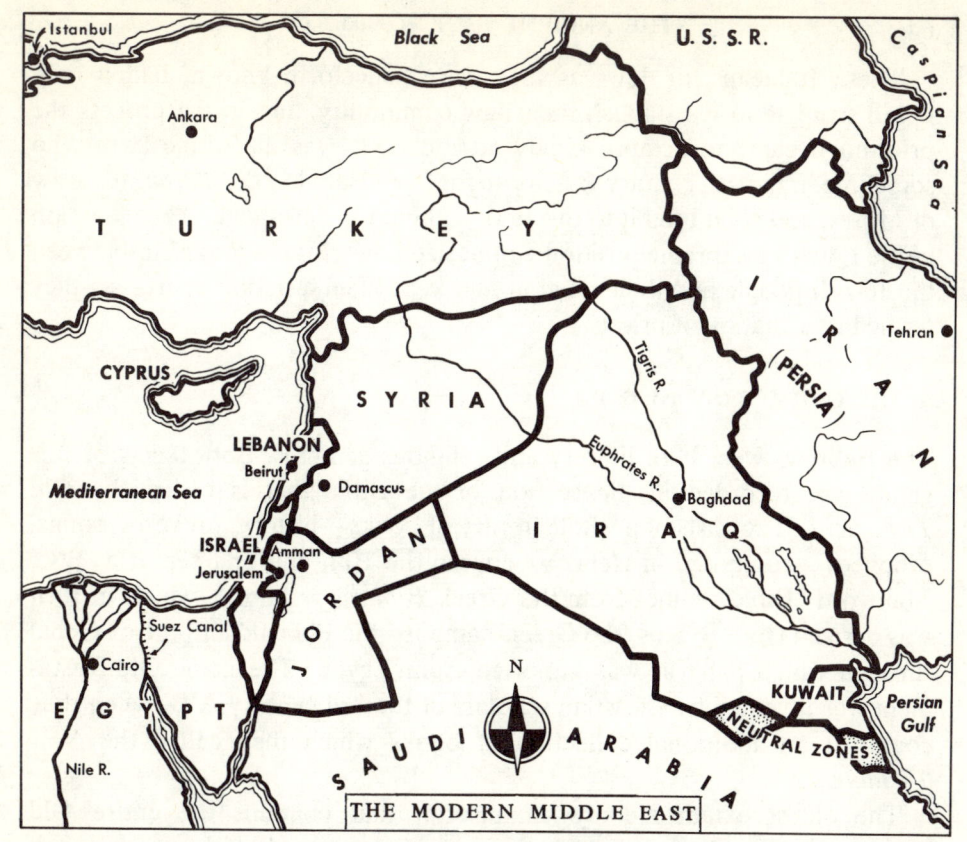

Undoubtedly the faithful in Babylon congregated to worship and be instructed in religious traditions. Stubbornly they kept themselves apart. They did not intermarry with their captors. Above all, they did not forget their homeland, as Psalm 137 poignantly shows.

By the waters of Babylon,
   there we sat down and wept,
   when we remembered Zion.

On the willows there
   we hung up our lyres.
For there our captors
   required of us songs,
and our tormentors, mirth, saying,
   "Sing us one of the songs of Zion!"

How shall we sing Yahweh's song
   in a foreign land?
If I forget you, O Jerusalem,
   let my right hand wither!
Let my tongue cleave to the roof
   of my mouth,
   if I do not remember you,
if I do not set Jerusalem
   above my highest joy!

Thus it came about that when the New Babylonian Empire fell, and Cyrus of Persia issued an edict of liberation about 538 B.C. giving some of the Jews permission to return to Jerusalem to rebuild their Temple, a small group straggled back to the Persian province of Judaea.

The Torah

These Judaeans, or Jews as they were henceforth known, had a long uphill struggle to re-establish their new community, and in the process the priesthood came to acquire kingly attributes. It was the scribe Ezra who sometime in the 4th century B.C. brought from Babylon the *Torah*, the Law of Moses, and submitted it to the Jews gathered in Jerusalem. The elevation of the Law to a supreme position emphasized the close relationship between the Jewish people and their religion and kept them a nation apart—as they termed it, a "nation of priests."

4/THE TRIUMPH OF ONE GOD

The Bible

The Hebrews were both literary and religious geniuses. Both facets of this genius are recorded in the section of the Bible that is called the *Old Testament*. It consists of a whole library of books—history, proverbs, songs, prophecy—composed in Hebrew between the 10th and 2nd centuries B.C. The word "Bible" comes from the Greek *Biblos* (= book), which in turn was derived from Byblos, the Greek name for the Phoenician port of Gebal through which papyrus was imported from Egypt. The name Old Testament was invented by Christian scholars of the 3rd century A.D. when they compiled an additional collection of books which they called the *New Testament*.

The oldest extant Hebrew manuscript that contains the entire Old Testament dates from the 10th century of our era, but there are much earlier manuscripts of individual parts of the Hebrew text. The oldest Greek manuscript dates from the 4th century A.D.

In 1947 there came to light some other manuscripts which scholars have termed the "Dead Sea Scrolls" from their discovery in a cave in Palestine overlooking the Dead Sea. They were unrolled with great care, and from their script and an examination of the cave and the pottery jars in which they were sealed, experts have concluded that these scrolls must be earlier than the middle decades of the 1st century B.C. The most impressive of them is a complete Hebrew text of *Isaiah*, which has been dated in the second half of the 2nd century B.C.

The Covenant

God's Chosen People During the wanderings in the desert, Moses had had an experience which gave the Hebrews the audacity to claim to have had a direct personal encounter with God. Moses met Yahweh on Mount Sinai, and this exalted experience was sealed by the Covenant, a sort of contractual agreement. Just as Abraham had made a pact with his god, a god who would devote himself to the tribe in return for their exclusive obedience and trust, so now the whole of Israel elected to become God's chosen people.

And Moses went up to God, and the Lord called him out of the mountain, saying, "Thus you shall say to the house of Jacob, and tell the people of Israel: You have seen what I did to the Egyptians, and how I bore you on eagles' wings and brought you to myself. Now therefore, if you will obey my voice and keep my covenant, you shall be my own possession among all peoples; for all the earth is mine, and you shall be to me a kingdom of priests and a holy nation. These are the words which you shall speak to the children of Israel."

Exactly what the religion of Moses was it is difficult to say. Certainly he passed on to his followers a new concept of God. Perhaps he even brought with him from Mount Sinai some sacred stones which, enclosed in a wooden coffer called the Ark, represented the presence of the new God on the field of battle, or in the temple or tent that sheltered the Ark. At any rate he brought down from the mountain the Laws. These Laws have already been mentioned in connection with the Code of Hammurabi; but it is the differences rather than the similarities that are striking. The Hebrew laws seem to have had an independent existence, although they did incorporate some Arab customs as well as a number of laws native to the Babylonian period, to which were added the social regulations necessary for a people passing from desert life into that of a settled population.

The Laws

This whole body of Law (found in *Exodus,* chapters 21, 22, and 23) was probably first recorded under Joshua some time about 1200 B.C. in a framework partially derived from Hammurabi's Code. It was in this way that the Hebrews adapted the customs of other peoples and refined them for their own use. The fact that this law was associated with the dictates of a living God gave it a divine as well as a human authority.

The absolute authority of Yahweh, however, was soon to be challenged by false gods. When the Hebrew tribes came to Canaan they encountered a religion based on nature worship. The Baal or local divinity was to the agricultural communities of Canaan what the tribal god was to the Hebrew people, and no doubt many went over to Baal worship, or in some way practised it and associated the various festivals with Yahweh. But the old faith of the desert would brook no compromise with mere nature gods.

The establishment of the Hebrew monarchy under Saul could only bring a strengthened Yahweh into opposition with the Baals in a most spectacular way. Nevertheless, King Ahab of Israel (869-850 B.C.) and his queen, Jezebel, fostered Baal worship. Accordingly, Elijah summoned the Israelites to Mount Carmel and there bested the "four hundred and fifty" prophets of Baal, proving the superiority of his God by slaying all the false prophets.

Yet if Yahweh could speak to Elijah out of the fire, he could also call him with "a still small voice." Gradually the Hebrews were spiritualizing Yahweh through their experiences with him, a process that the earlier loss of the Ark to the Philistines in battle and its eventual return minus its sacred stones had probably encouraged.

Watchmen of Yahweh It was the revolutionary social change of the centuries after Solomon that gave the prophets their greatest stimulus. When the rate of interest rose to 25% in rich and powerful Assyria, how much must money-lenders have charged in the poor country of Israel in the 9th and 8th centuries B.C.? But no matter how the peasants and artisans suffered, the kings seem to have been powerless to effect sweeping social changes. In short, the situation was ripe for the prophets.

IMPORTANT HEBREW PROPHETS

1. Amos	
2. Hosea	8th century B.C.
3. Isaiah	
4. Micah	
5. Jeremiah	7th and 6th century B.C.
6. Ezekiel	6th century B.C.
7. Second Isaiah	

The prophets proclaimed no democratic ideals or social security programmes. Much of what they said had no definite political implication: they were interested mainly in social justice, and lashed out against the rich. Yet they could not rise completely above the partisan politics of their time and often found themselves embroiled with either Assyria or Egypt in their attempt to remain independent from them. They were little heeded by the king, who often, like Ahab, kept his own band of "prophets" to echo his wishes. Nor were the genuine prophets great lovers of the priesthood. Many of them were ecstatic men, men who heard voices and saw visions and babbled in strange tongues.

But these watchmen of Yahweh were no mere psychosomatic cases unable to adjust to their environment, else their message would probably have perished, or, if it survived at all, would be looked upon as a literary curiosity. Instead, they created matchless Old Testament poetry which constitutes the special glory of Israel.

The prophets emphasized two fundamentals: the Covenant and the Law. If Israel did not prosper it was because she broke the Covenant, the compact made with God. God must be worshipped both in the letter and the spirit of the law if the Covenant was not to be violated. Moreover, the Covenant equalized all Hebrews. The Jewish law code had no separate privileges for various classes. Surely then the law ought not to countenance grave economic inequalities. God's laws were to be obeyed, not only in the Temple but in the heart. By doing nothing less could Israel be saved. So reasoned the prophets.

This view of Jerusalem from across the Kidron valley shows the great trench outside the present city walls that has been opened by a combined British, French, and Canadian (Royal Ontario Museum) archaeological expedition. This site was actually within the city from about 2500 B.C. until the rebuilding of Jerusalem following its destruction in A.D. 70 by the Emperor Titus.

It is often thought that the strata of occupation on an ancient site are like the layers of a cake—one need only strip them off layer by layer, and the deeper one goes, the older the deposit that is uncovered. This trench, however, reveals that this idea can be very misleading. Since bedrock is just below the surface of the ground and since the site is on a steep (45 degrees) slope, the strata are not horizontal; the constructions found lower on the slope may therefore be of a later date than those higher up. Several points of special interest are identified to illustrate this point. **A.** Steps leading down to the spring Gihon (I Kings 1: 33 ff.), the main water supply of ancient Jerusalem. From this spring tunnels were cut into the bedrock to bring water inside the city walls. The earliest tunnel seems to have been cut by the Jebusites. Hezekiah dug a much longer tunnel to lead the water to the Pool of Siloam (Isa. 8: 6; II Kings 20: 20; II Chron. 32: 2 ff., 30; Neh. 3: 15; John 9: 7). One can still traverse this tunnel, which continues in use. **B.** The earliest city wall: built about 1800 B.C. and rebuilt and re-used down to 7th Century B.C. This was the wall of the Jebusite city captured by David (II Sam. 5: 6-9; I Chron. 11: 4-7) and it continued to stand through most of the period of the Hebrew monarchy. **C.** City wall; built in 7th century B.C., perhaps by Manasseh (II Chron. 33: 14). **D.** Ruins of houses of the Hebrew city that fell to Nebuchadrezzar in 586 B.C. rest on the remains of terraces here. In them were found inscribed weights. **E.** A small section of Nehemiah's wall (about 440 B.C.) has been discovered here (Neh. 2: 11 ff.). **F.** The present south wall of Jerusalem. Built by the Turkish sultan Suleiman the Magnificent in the 16th century, it follows the line of the wall built by Caliph Aziz in the 10th century. **G.** A natural rock shelter that contained principally the remains of poor squatters of the Hebrew monarchy period. On the floor there is evidence of occupation from the middle of the third millennium B.C.—the earliest signs of occupation yet discovered at Jerusalem.

Foundations of Monotheism   The prophet Amos lived in the time of Jeroboam II of Israel (876-746 B.C.). Amos was a shepherd from the south, an alien in northern Palestine who was shocked by the injustice and dishonesty of the courts, the economic inequalities of the market-place, and the degeneracy and corruption of the court and the holy places. According to his own words, he did not want to prophesy; but the Lord laid hold of him and he was compelled to. In a ringing denunciation which he must have delivered throughout the northern kingdom, Amos told his appreciative audience of the punishment to be meted out to all of Israel's powerful and wicked neighbours.

Amos

But his message reached a terrifying climax. He prophesied the doom awaiting Israel herself for breaking the Covenant, and when Assyria and Egypt were portrayed as the instruments of Yahweh's wrath in chastising His people Israel, the prophet's sedition went too far. Jeroboam ordered Amos to return to Judah. Whether this exile preceded or followed his other prophetic messages is not known. At any rate these political pronouncements seem to have been sufficient to produce deportation.

About a decade after Amos thundered his warnings, a contemporary of his, a man named Hosea from the northern kingdom, began to prophesy. The fortunes of the Israelites were now on the wane, and Hosea's own

Hosea

unhappy marriage seemed to him to be an allegory of the faithlessness of Israel to Yahweh. Instead of turning to Yahweh, Israel had begged help from Egypt to resist Assyria.

When he preached spiritual weapons rather than the use of force, Hosea was going far beyond the old idea in *Judges* of Yahweh as a great warrior whose armed might could save his people. Hosea also had harsh words for Baal worship, bull worship, and the like that were practised in his day. What did Yahweh desire? Simply "steadfast love and not sacrifice, the knowledge of God rather than burnt offerings." Hosea's God is compassionate whereas that of Amos is merely a stern moralist. Both made more of God than a fickle, cruel, and jealous deity. It is hard to see universal monotheism in either Amos or Hosea, but the foundations for it had been surely laid.

Isaiah

The first of the major prophets was Isaiah. He lived and carried out his main work in Jerusalem and seems to have played a prominent role in the political life of Judah during the second half of the 8th century B.C. When King Ahaz of Judah (735-715 B.C.) was faced with a coalition of Israel and Syria, it was in vain that Isaiah put forward a policy of isolation. Ahaz appealed to the Assyrians for help. But things turned out much as Isaiah had predicted they would: Assyria ended by taking over Judah. Later Israel fell to Assyria, and when the armies of Sennacherib besieged Jerusalem in 701 B.C. Isaiah saw the invaders as the punishing arm of Yahweh —but for this reason he considered them to be *also* under the control of Yahweh. The prophet who asserts that Yahweh is the God who controls the armies of the Assyrian is approaching a position very close to monotheism.

Even if the salvation from Assyria were only temporary, and were followed by fresh and greater disasters, Isaiah was convinced that a "remnant" of disciples would survive to form the nucleus of a new nation. To lead this remnant, Isaiah prophesied that God himself would provide a ruler, an anointed one, a *Messiah*. For, said Isaiah,

to us a child is born, to us a son is given; and the government will be upon his shoulder, and his name will be called "Wonderful Counselor, Mighty God, Everlasting Father, Prince of Peace." Of the increase of his government and of peace there will be no end, upon the throne of David, and over his kingdom, to establish it and to uphold it with justice and with righteousness from this time forth and for evermore. The zeal of the lord of hosts will do this.

Micah

The last of the 8th century prophets was Micah. He, like Amos, concentrated on social injustice. And though it is sometimes stated that he contributed little to the development of Hebrew religion, he did express the prophetic position with great simplicity:

> And what does the Lord require of you
> But to do justice, and to love kindness,
> And to walk humbly with your God?

The Climax of Prophecy The prophetic movement reached its climax with Jeremiah (641-586 B.C.), and with Ezekiel and the Second Isaiah of the period of the Babylonian exile. Like Isaiah, Jeremiah found himself caught in the cross-currents of Middle Eastern empires, and like Isaiah he preached that Israel should trust only in Yahweh. Indeed his advice to his countrymen to surrender to the invading Chaldeans and not to resist them was what today might be called high treason.

Jeremiah, Ezekial and Second Isaiah

When the Israelites were carried off to Babylon, Jeremiah counselled them to settle down to their sojourn in exile—for it was to be a long one—but not to forget Yahweh. The national covenant between Yahweh and the Hebrew people was transformed, in Jeremiah's eyes, into an individual agreement between a Jew and his God (as can be seen from the Source Reading on page 74). What was most important to Yahweh was the right mental and moral attitude. Perhaps this was why a later Jew, Jesus of Nazareth, admired Jeremiah most of all the prophets.

Jeremiah's God knew all, saw all, heard all, was omnipresent. There were, for Jeremiah, other gods. But they were little better than "broken cisterns." In the history of Hebrew religion Jeremiah is usually acclaimed as the founder of individualism and personal faith.

The exile transformed prophecy. Ezekiel, a priest at Jerusalem, was taken captive to Babylon in 598 B.C. and so had the unusual experience of worshipping Yahweh both in Israel and in a foreign land. He was to realize that Yahweh was not just a national God who remained in Palestine—for after the destruction of Jerusalem by Nebuchadrezzar there was not even any homeland toward which to look with longing. Ezekiel predicted, in strangely idealistic terms, the regeneration and restoration of Israel, and he revived the old inscrutable Yahweh whose purposes were hid from men and who spoke out of the thundercloud. God could enforce his laws anywhere and at any time. Yet even for Ezekiel He was peculiarly the God of Israel, of an Israel that was really the centre of the world.

About the year 540 B.C.—two years before Cyrus of Persia's edict of liberation—an anonymous prophet whom scholars have called "the Second Isaiah" wrote down his interpretation of God's actions. His message may be found in chapters 40-55 of the book of *Isaiah*, which fortells Cyrus's restoration of Israel and enthusiastically proclaim that Yahweh is responsible for it. For the Second Isaiah, Yahweh had become lord of the Persian as well as the Hebrew, and all gods of wood or stone or metal were nothing, "empty wind."

Let all the nations gather together, and let the peoples assemble. Who among them can declare this, and show us the former things? Let them bring their witnesses to justify them, and let them hear them say, It is true. "You are my witnesses," says the Lord, "and my servant whom I have chosen, that you may know and believe me and understand that I am He. Before me no god was formed, nor shall there be any after me. I, I am the Lord, and besides me there is no saviour."

TIME CHART FOR THE ANCIENT EAST*

DATE (B.C.)	EGYPT	PALESTINE	MESOPOTAMIA	INDIA	CHINA
8000		Earliest Jericho			
7000		Jericho a sizeable town			
5000			Jarmo		
4000	Farming villages in Delta		Farming villages at head of Persian Gulf		
3500			City-states such as Uruk		
3400	Invaders from east				
3200	Unification of Egypt				
3300			City-state of Ashur	Farmers enter Indus valley	
2700	Old Kingdom				
2600	Pyramids of Gizeh			Indus civilization	
2500					
2300			Sargon I creates first empire in history		Yang-shao
2000	Middle Kingdom	Abraham	Ur conquers Akkad		
1700	Hyksos		Hammurabi in Babylon		
1600					
1500	New Kingdom	Hebrews in Egypt	Hittites and Kassites	Aryan invaders	Shang dynasty
1400	Thutmose III			Earliest *Vedas*	
1300	Akhenaton				
1200	"Peoples of the Sea"	Moses and Joshua	Rise of Assyria		
1100		Hebrew conquest of Canaan			
1000		Saul and David			Chou dynasty
900		Solomon			
700		Amos, Hosea, Isaiah, Micah; Assyrian conquest	Tiglath-pileser I		
			Sennacherib	*Upanishads*	Competing states
600	Battle of Carchemish	Jeremiah	Ashurbanipal; fall of Nineveh; Median empire	Competing states	
500	Persian conquest	Babylonian Captivity Ezekiel and Second Isaiah Liberation of Jews	Nebuchadrezzar; Persian conquests of Cyrus and Darius Zoroaster	Buddha	Confucius

*In this chart, the various areas are arranged from left to right in accordance with their geographical location from west to east. The missing centuries were not, of course, devoid of political or intellectual progress, but they are less important for the topics we have studied. If you wish, you can fill the blanks in for yourself by reading other books such as the ones listed in the bibliographies at the end of each chapter.

Here was a far cry from the exclusively Hebrew God of Moses. At last monotheism was an accomplished belief.

The political decline of Egypt, Assyria, and Babylon must have affected the prophets and also increased Israel's awareness of the extent of the world. According to the Second Isaiah, God must be universal. But Israel is to be "a light to the nations." Israel is to be the "servant" of Yahweh, as she alone has true knowledge of God. As the "Suffering Servant" of Yahweh Israel may undergo all manner of humiliation and torture, yet must do so gladly on behalf of others as the instrument of Yahweh's purpose. It is not hard to see the roots of that strength which has enabled Judaism to survive the first and all later dispersions.

A Unique Legacy Why take as much space for the story of a politically insignificant people living in a land half the size of Nova Scotia as for the thousands of years of Egypt's history? Because the legacy of the Hebrews is unique and lies at the heart of Western civilization.

Although the Hebrews lived in a land that was the natural meeting-place of great nations and great ideas, their religion did not fit into the general pattern of other Middle Eastern religions. In fact the various ancient religions of the Middle East are more like each other than any one of them is like Old Testament religion. The Hebrews did, of course, take over many of the forms of other faiths; but they did not stop there. What they took over they refined and elevated. The inspired Hebrew genius really burst the old religious forms asunder. It was a masterly combination of unique environmental contacts and a great capacity for abstract thought that made the Hebrews vehicles for the proclamation of the glory of God.

We are all in their debt.

5/SUMMARY AND CONCLUSION

In Iron Age times and earlier, the Middle East was a battleground of empires: Egyptian, Assyrian, Chaldean and Persian. The accompanying table places in parallel columns a selection of the main political developments of these Middle Eastern empires, as well as those of India and China, down to 500 B.C.

As with the first civilizations, the mass of the people were not consulted in the making of political decisions. Often they were enslaved by their foreign conquerors: the Egyptians, Assyrians, and Chaldeans resorted to garrisons, to a conquering religion, and even to mass deportations to control their far-flung empires. The Persians worked a startling revolution when they tolerated the foreign customs and usages of their defeated peoples.

In Middle Eastern religion there was a tendency to pass from the competing anarchy of many gods to a sort of heavenly monarchy with one national god. But only in Palestine did a genuine monotheism evolve. Often religious sanctions reinforced laws which were usually said to be (as in Hammurabi's Code or the Mosaic Law) the gift of a god from on high.

Since Middle Eastern societies accepted the rule of gods almost without question, literature and art often attempted to explain certain natural phenomena as due to the intervention of some god. Because of this overshadowing belief in the supernatural, science could hardly be more than a compound of superstition and a bit of accurate observation.

It is a strange truth that the small country of Palestine, ground between the upper and nether millstones of Assyrian and Egyptian imperialisms, has had a more lasting influence on us than all the great empires put together. While the mighty powers crashed around them—Assyria (612 B.C.), Egypt (605 B.C.), and New Babylonia (539 B.C.)—the prideful and stubborn Hebrews became the spiritual leaders of the ancient Middle East. Their heritage has come down to us in Judaism and Christianity.

If we are the heirs of Middle Eastern civilizations, spiritually, scientifically, and artistically, are we also their heirs politically? No. In politics we owe them little more than the recognition that they were the first to set up governments over larger areas than cities and to produce written codes of law.

In the political field our important link is rather with a vibrant, energetic people—the Greeks. They invented a new political system called *demokratia*, "control by the people". Let us see how they invented it and what they did with it.

SOURCE READINGS

(a)

SENNACHERIB of Assyria (704-681 B.C.) besieged Jerusalem in 701 B.C. This is his own account of the seige.

As to Hezekiah, the Jew, he did not submit to my yoke, I laid siege to 46 of his strong cities, walled forts and to the countless small villages in their vicinity, and conquered them by means of well-stamped earth-ramps, and battering-rams brought thus near to the walls combined with the attack by foot soldiers, using mines, breeches as well as sapper work. I drove out of them 200,150 people, young and old, male and female, horses, mules, donkeys, camels, big and small cattle beyond counting, and considered them booty. Himself I made a prisoner in Jerusalem, his royal residence, like a bird in a cage. I surrounded him with earth-works in order to molest those who were leaving his city's gate. His towns which I had plundered, I took away from his country. . . . Hezekiah himself, whom the terror-inspiring splendour of my lordship had overwhelmed and whose irregular and elite troops which he had brought into Jerusalem, his royal residence, in order to strengthen it, had deserted him, did send me, later, to Nineveh, my lordly city, together with 30 talents of gold, 800 talents of silver, precious stones, antimony, large cuts of red stone, couches inlaid with ivory, chairs inlaid with ivory, elephant-hides, ebony-wood, box-wood and all kinds of valuable treasures, his own daughters. . . . In order to deliver the tribute and to do obeisance as a slave he sent his personal messenger.

W. C. McDermott and W. E. Caldwell, editors, *Readings in the History of the Ancient World* (Rinehart, 1951), pp. 53-54.

(b)

DARIUS I of Persia (521-486 B.C.) had to put down many revolts during the first two years of his reign. He describes his successful wars and his other accomplishments in "the most important historical monument in Asia," an inscription carved 300 feet high on a rock cliff. This Behistun inscription is near Ecbatana in modern Iran, and was carved in the three main imperial languages. Some translated passages from it follow.

1. I am Darius the Great King, King of Kings, King in Persia, King of countries. . . .

4. Saith Darius the King: VIII of our family there are who were kings afore; I am the ninth; IX in succession we have been kings.

5. Saith Darius the King: By the favour of Ahuramazda I am King; Ahuramazda bestowed the kingdom upon me.

6. Saith Darius the King: These are the countries which came unto me: by the favour of Ahuramazda I was king of them: Persia, Elam, Babylonia, Assyria, Arabia, Egypt, those who are beside the sea, Sardis, Ionia, Media, Armenia, Cappadocia, Parthia, Drangiana, Aria, Chorasmia, Bactria, Sogdiana, Gandara, Scythia, Sattagydia, Arachosia, Maka: in all, XXIII provinces.

8. Saith Darius the King: Within these countries, the man who was excellent, him I rewarded well; him who was evil, him I punished well; by the favour of Ahuramazda these countries showed respect toward my law; as was said to them by me, thus was it done.

52. Saith Darius the King: This is what I did by the favour of Ahuramazda in one and the same year after that I became king. XIX battles I fought; by the favour of Ahuramazda I smote them and took prisoner IX kings. . . .

58. Saith Darius the King: By the favour of Ahuramazda and of me much else was done; that has not been inscribed in this inscription; for this reason it has not been inscribed, lest whoso shall hereafter read this inscription, to him what has been done by me seem excessive, and it not convince him, but he think it false.

67. Saith Darius the King: If thou shalt behold this inscription or these sculptures, and shalt destroy them and shalt not protect them as long as unto thee there is strength, Ahuramazda be a smiter unto thee, and may family not be unto thee, and what thou shalt do, that for thee may Ahuramazda utterly destroy!

68. Saith Darius the King: These are the men who were there at the time when I slew Gaumata the Magian who called himself Smerdis; at that time these men co-operated as my followers. . . .

69. Saith Darius the King: Thou who shalt be king hereafter, protect well the family of these men.

70. Saith Darius the King: By the favour of Ahuramazda this inscription in other ways I made. In addition, it was in Aryan, and has been made on leather. In addition, this inscription as a whole has been confirmed by the impression of a seal. And it was written, and the written document was read off to me. Afterwards this inscription was sent by me everywhere among the provinces; the people universally were pleased.

McDermott and Caldwell, *Readings in the History of the Ancient World*, pp. 58-64.

(c)

The prophet JEREMIAH (641-586 B.C.) was called by God to his mission about 626 B.C. Though he lived most of his life in Jerusalem he died in exile in Egypt, having been taken there against his will by a group of Jews fleeing the vengeance of Nebuchadrezzar. In the selections that follow Jeremiah describes the Judaean violations of the Covenant, but nevertheless looks to the day of a "New Covenant" when Yahweh will write his Law in the hearts of Judaeans and Israelites alike.

The word that came to Jeremiah from the Lord: "Hear the words of this covenant, and speak to the men of Judah and the inhabitants of Jerusalem. You shall say to them, Thus says the Lord, the God of Israel: Cursed be the man who does not heed the words of this covenant which I commanded your fathers when I brought them out of the land of Egypt, from the iron furnace, saying, Listen to my voice, and do all that I command you. So shall you be my people, and I will be your God, that I may perform the oath which I swore to your fathers to give them a land flowing with milk and honey, as at this day." Then I answered, "So be it, Lord."

And the Lord said to me, "Proclaim all these words in the cities of Judah, and in the streets of Jerusalem: Hear the words of this covenant and do them. For I solemnly warned your fathers when I brought them up out of the land of Egypt, warning them persistently, even to this day, saying, Obey my voice. Yet they did not obey or incline their ear, but every one walked in the stubbornness of his evil heart. Therefore I brought upon them all the words of this covenant, which I commanded them to do, but they did not."

Again the Lord said to me, "There is revolt among the men of Judah and the inhabitants of Jerusalem. They have turned back to the iniquities of their forefathers, who refused to hear my words; they have gone after other gods to serve them; the house of Israel and the house of Judah have broken my covenant which I made with their fathers. Therefore, thus says the Lord, Behold, I am bringing evil upon them which they cannot escape; though they cry to me, I will not listen to them. Then the cities of Judah and the inhabitants of Jerusalem will go and cry to the gods to whom they burn incense, but they cannot save them in the time of their trouble. For your gods have become as many as your cities, O Judah; and as many as the streets of Jerusalem are the altars you have set up to shame, altars to burn incense to Baal.

"Therefore do not pray for this people, or lift up a cry or prayer on their behalf, for I will not listen when they call to me in the time of their trouble. . . ."

(11:1-14)

"Behold, the days are coming, says the Lord, when I will make a new covenant with the house of Israel and the house of Judah, not like the covenant which I made with their fathers when I took them by the hand to bring them out of the land of Egypt, my covenant which they broke, though I was their husband, says the Lord. But this is the covenant which I will make with the house of Israel after those days, says the Lord: I will put my law within them, and I will write it upon their hearts; and I will be their God, and they shall be my people. And no longer shall each man teach his neighbour and each his brother, saying, 'Know the Lord,' for they shall all know me, from the least of them to the greatest, says the Lord; for I will forgive their iniquity, and I will remember their sin no more."

(31:31-34)

The Holy Bible: Revised Standard Version (Thomas Nelson & Sons, 1953), pp. 797, 822.

FURTHER READING

1. ASSYRIA
   † CONTENAU, G., *Everyday Life in Babylon and Assyria* (Arnold, 1955)
   OLMSTEAD, A. T., *History of Assyria* (Scribner, 1923)

2. PERSIA
   * GHIRSHMAN, R., *Iran* (Penguin, 1954)
   * OLMSTEAD, A. T., *History of the Persian Empire* (Phoenix, 1948)

3. PALESTINE
   ANDERSON, B. W., *Understanding the Old Testament* (Prentice-Hall, 1957)
   * ALBRIGHT, W. F., *From the Stone Age to Christianity* (Anchor, 1957)
   BRIGHT J., *A History of Israel* (Westminster, 1959)
   * BURROWS, M., *What Mean These Stones?* (Living Age, 1957)
   FINEGAN, J., *Light from the Ancient Past* (Princeton, 1959)
   † HEATON, E. W., *Everyday Life in Old Testament Times* (Batsford, 1956)
   * _____, *The Old Testament Prophets* (Penguin, 1961)
   † HOWSE, E. M., *The Lively Oracles* (Nelson, 1956)
   NOTH, M., *The History of Israel* (Black, 1958)
   † * ORLINSKY, H. M., *Ancient Israel* (Cornell, 1960)
   † SMITH, J. W. D., *God and Man in Early Israel* (Methuen, 1956)

UNIT **II**

THE GREEKS

3. The Greek City-States

. . . a state exists for the sake of a good life, and not for the sake of life only.
ARISTOTLE, *Politics*

"The government of the Athenians . . . I do not approve of," wrote a crusty old Greek of the 5th century B.C., for "in choosing it they chose that scoundrels should become better off than decent citizens." To a present-day North American such words come as a shock. Is not democracy the best possible form of government? And did not the Greeks gain immortality by forming the first democracies?

The ensuing Unit shows us how the Athenians attained their form of government, and why it did not immediately meet with the approval of all the Greeks.

The Greek epic begins on Crete.

1 / THE MINOAN CIVILIZATION

European man did not progress as rapidly as Middle Eastern man, and New Stone Age culture persisted in Europe long after the dawn of civilization in Mesopotamia and Egypt. Eventually, however, the introduction of copper into Europe produced a Bronze Age, which varied in time from area to area. Greece and the lands bordering on the Aegean Sea seem to have been the first European regions to pass from Stone Age culture to a more complex form, at the beginning of the third millennium B.C.

Digging Up the Past We owe much of our knowledge of this more complex civilization that grew up around the cities of Troy and Mycenae to the excavations of Heinrich Schliemann. His work, along with that of Sir Arthur Evans on Crete, has built up a fairly complete picture of the oldest European civilization. Their life-work forms one of the most spectacular chapters in the science of archaeology. After these excavations the most sceptical scholars were forced to concede that the Greek past could, with scientific certainty, be pushed back to dim beginnings almost four thousand years ago.

Schliemann went to Hamburg as a youth in 1841 and signed up as a cabin boy on a ship bound for Venezuela. The ship sank off the Netherlands, and the cabin boy became a clerk in an Amsterdam office. In a cold attic room he managed to teach himself Greek and began reading Homer. As his fascination grew, Schliemann developed a passionate belief that the Troy and Mycenae of Homer's epics were no mere poetic inventions. They

---

This temple to Apollo the Helper was built about 420 B.C. by the city-state of Phigalia to commemorate relief from an epidemic during the Peloponnesian War. Rising high among some of the loftiest mountains of the Peloponnesus, this beautiful shrine on a slope of Mount Cotylium in Arcadia seems to reach half-way to heaven in an attempt to entice its god to come down to earth.

had existed as surely as Amsterdam, and somewhere, somehow he would find proof of that existence.

The dream did not disappear with manhood. Schliemann made a fortune in Russian oil, and retired to use it for his own detective work on Homer. His eventual discovery of the ruins of the two ancient cities provided an exciting vindication of his faith in the historical basis of Homer's writings.

Schliemann did not live to excavate the ruins on Crete. That task was undertaken by Arthur Evans, an English scholar. A professor who had been educated at English and German universities, Evans was the direct opposite of the talented amateur, Schliemann. But both men were great archaeological pioneers, and between them they unearthed a whole civilization. Since it surrounded the Aegean Sea, it is known as "Aegean" civilization. It is usually, however, subdivided into "Minoan", that part centering on the island of Crete, and "Mycenaean", that part centering on the Greek mainland city of Mycenae.

In his excavations on Crete after 1900 Evans unearthed the long-buried civilization which he called Minoan after the legendary king Minos, and which he divided chronologically into three main periods. These divisions are still followed, though the dates originally assigned to them have subsequently been altered.

1. 3000-2200 B.C.—Early Minoan
2. 2200-1600 B.C.—Middle Minoan
3. 1600-1150 B.C.—Late Minoan

It is important to note that the Cretans were *not* Greeks. Rather they came overseas from Asia Minor during the Neolithic Age, as early as 5000 B.C. As time went on and more immigrants brought a knowledge of copper and bronze from the lands in the south and east, the Cretans entered the Bronze Age. There is evidence that in the Early Minoan period a metal industry was established, and commercial relations with Egypt have been traced from Cretan pottery. Great variations in pottery occurred all over Crete, as well as in architecture, religious ritual, and burial customs. Political power seems to have been in the hands of local princes whose palace-temples became the centres of prosperous towns.

It is only in the Middle Minoan period that uniformity can be detected over the entire island. This is the great age of the palaces of Knossos and Phaestos. The Minoan civilization had become mature. Highly artistic creations were achieved in pottery and bronze, and a system of writing emerged—although until recently Minoan records could not be deciphered.

These are the Minoan-Mycenaean signs for the archaic Greek word for tripod. Each sign stands for a syllable rather than, as in English, a single letter.

The crowning achievement of Minoan architecture was the Palace of Knossos. The restored wooden columns, thinner at the bottom than the top, were possibly a legacy from Neolithic times when builders may have found the upper end of a tree trunk easier to drive into the ground. Though the wall frescoes lack true perspective they blaze with vivid colours, and in the murals pictured here large griffins (dating from the 15th century B.C.) stand out against a background of ruby and yellowish brown. This room, which derives its name of the "Throne Room" from the high-backed gypsum chair placed against the frescoes, is the very one that was abandoned so hurriedly one wild spring day in 1400 B.C.

In 1952, however, an exciting discovery was made when a young English scholar succeeded in deciphering one of the types of Minoan script, the so-called "Linear B", as an archaic form of the Greek language. Linear B clay tablets have also been discovered on the Greek mainland at Mycenae and Pylos. Many of these tablets are inventories of household goods, and hence make about as fascinating reading as a telephone directory. But their decipherment is revolutionizing our knowledge of early Greek history, and we now know that a form of Greek was written in Crete and on the mainland a full six centuries before the Greeks adopted the Phoenician alphabet.

The only samples of Linear B to be discovered on Crete have been found at Knossos, and date from 1450 B.C. We know, however, that for three centuries before this date another form of script, "Linear A", was in use. This script, which has so far resisted decipherment, is definitely not Greek, and no examples of it have been found outside Crete. We may presume, then, that the Mycenaean invaders of Crete got Cretan scribes to devise a new Minoan script—Linear B—for the Greek language.

The Cretan house was a two-storeyed dwelling of wood, brick, or stone, with windows that were either shuttered or covered with some sort of transparent panes. After 1700 B.C. some mysterious disaster, possibly an earthquake, destroyed the palaces of Crete. The ensuing restoration ushered in a golden age of Minoan civilization.

The new cities were unfortified, and the rulers of Knossos—a city of 100,000 people—probably dominated the whole island. They kept their surpluses of grain and olive oil in huge pottery jars (over six feet high) in the basements of the palaces, so that the palaces were both commercial and governmental centres.

From wall frescoes and various carvings we can reconstruct the Cretan sport of bull-leaping. As the bull charges, the athlete grabs its horns. The bull then tries to throw the man by tossing its head back, propelling him over its shoulders in somersault fashion. As the man goes over he lets go of the horns and lands feet down on the bull's back, and from this precarious standing position leaps to the ground to complete the feat. Both men and women practised this sport, such a dangerous one that modern acrobats and rodeo performers say it is virtually impossible. The Cretans, however, appear to have mastered it.

Agriculture, industry, and trade united to produce the happy prosperity of the island. The chief crops were barley, flax, olives, grapes, figs, and dates, while goats, sheep, and cattle were pastured on the hillsides. The main industries were pottery manufacturing, metalworking and shipbuilding. In their sleek seventy-foot masted ships the Cretans exported olive oil, wine, bronze utensils, vases, jewellery, and cameos, and imported marble, ivory, tin, and precious metals. Their vessels journeyed so widely that some southern Greek cities were definitely under Cretan influence, and trading posts and colonies were established from Cyprus to Sicily. No wonder the Cretan idea of death was of a last long voyage—an appropriate symbol for a civilization whose life was the sea.

The Minoan civilization at its height reflected the achievements of a prosperous and contented people. They seem almost modern in their love of athletics, the fashions of women's clothes and their many hats, the sewage system for their streets and palaces, and their bath tubs and flush toilets. Moreover, poised as they were outside the orbits of the Middle Eastern empires, they were less autocratic than their Oriental neighbours. The kings mingled with the ordinary people, there were no royal harems, and women took part in athletic contests and worked side by side with men in the "factories". It is an interesting fact of history that great sea powers usually seem to be more democratic than great land powers.

From Myth to History One of the most famous Greek myths tells the story of the youth Theseus. Brought up by his mother far from Athens and in ignorance of his father's identity, Theseus was finally sent to King Aegeus in Athens. Here he was found to be none other than the king's son. One day he saw the Athenians sorrowing as the time approached for their tribute to King Minos of Crete—seven maidens and seven youths to be devoured by the Minotaur, a frightful monster half bull and half man. The Minotaur lived in a labyrinth of halls and passageways built for it on Crete. After insisting on being one of the sacrificial youths, Theseus wooed and won Minos's daughter, who helped him kill the monster, find his way out of the labyrinth, and return to Athens.

Before Evans' excavations this story was regarded as pure fiction. Today we are not so sure. Frescoes of the strange sport of bull-leaping have been discovered in Knossos. And the royal palace has been found to be a veritable maze of small rooms and corridors. Could it have been the labyrinth?

A Wild Spring Day It is certain that the Late Minoan period saw a slow decline, but exactly what was the final fate of the island is a subject of much scholarly debate. There are at least five possibilities. Perhaps the Mycenaeans conquered Crete about 1480 B.C. and then decided, after a couple of generations, to give up and withdraw, pillaging as they went. Or the Cretans themselves may have risen up and expelled their conquerors. A further possibility is that Mycenae first conquered Knossos about 1400 B.C. and made it part of her empire for several centuries. Or piratical sea-raiders may have made hit-and-run raids on the island. It may even have been that the Cretan cities were overwhelmed by natural calamities—earthquake, volcanic eruption, and fire.

Which of these was the precise fate of Crete has not yet been determined. Certain it is that there was some cataclysmic disaster.

... in the last decade of the fifteenth century on a spring day, when a strong South wind was blowing which carried the flames of the burning beams almost horizontally northwards, Knossos fell. ...

The final scene takes place in the most dramatic room ever excavated—the Throne Room. It was found in a state of complete confusion. A great oil jar lay overturned in one corner; ritual vessels were in the act of being used when the disaster came. It looks as if the King had been hurried there to undergo, too late, some last ceremony in the hopes of saving the people. Theseus and the Minotaur! Dare we believe that he wore the mask of the bull?

Crete had fallen and henceforth she was to be a mere satellite of the world centering round Greece, gradually drawing nearer until she was absorbed in the general Hellenic culture which she herself had done so much to found.

For centuries the palace lay deserted except for the ghosts of its departed glory mournfully wandering down the empty mouldering stairways. . . . With that wild spring day at the beginning of the fourteenth century B.C. something went out of the world which the world will never see again: something grotesque, perhaps, something fantastic and cruel, but something also very lovely.

The legend of Theseus' slaying of the Minotaur probably enshrines a folk memory of the control Mycenae won over Crete some time in the 15th century B.C. At any rate, after 1400 B.C. Crete was not mentioned in Egyptian records, and the island likely was tributary to the Greek mainland.

The date 1400, then, may be taken as marking the founding of Europe. Mainland European civilization was to dominate the Mediterranean from that time until the fall of the Roman Empire.

2/THE FIRST CIVILIZATION IN EUROPE

After 1400 B.C. the Greeks of the mainland were supreme for some 250 years. Their chief city, Mycenae, had been settled as early as 2500 B.C., probably by peoples who had come overland from Asia Minor. Then about 1900 B.C. the so-called Achaean Greeks—part of that movement of Indo-European peoples that we have already noted during the Bronze Age—settled on the hilltop at Mycenae.

Mycenae was not like the cities of Crete. It was provided with strong fortifications, and by 1350 B.C. had been rebuilt on a grand scale. Around the Mycenaean *acropolis* (the fortified hilltop) were cemeteries with vaults or shafts for the bodies of the dead, while the princes themselves later came to be buried in beehive-shaped tombs.

These early Greeks were a warlike people. Pottery discovered in the ruins of the seventh layer of the cities built one on top of another at Troy has produced a new version of that famous 13th century B.C. siege. Probably mainland Greeks (led by Mycenae) were merely conducting a "wholesome exercise" in pillaging—not a noble siege to recover a kidnapped queen or even an attack to free the Dardanelles from Trojan toll-levying ships.

Golden Mycenae

Mainland Greek culture owed much to Crete. The athletic contests of the Minoans, their devotion to music and the dance, their artistic creations, even, perhaps, their dependence on a navy, all these were passed on to the later Greeks. But there were also imports from the north: the Greek language, wheel-made pottery, and a northern style of architecture which the Mycenaeans added to their palaces in the *megaron*. The megaron was an oblong building with a hearth in the centre of the floor, a porch supported by pillars, and a sloping roof—a structure better adapted to cold, rainy winters than the open, flat-roofed Minoan house.

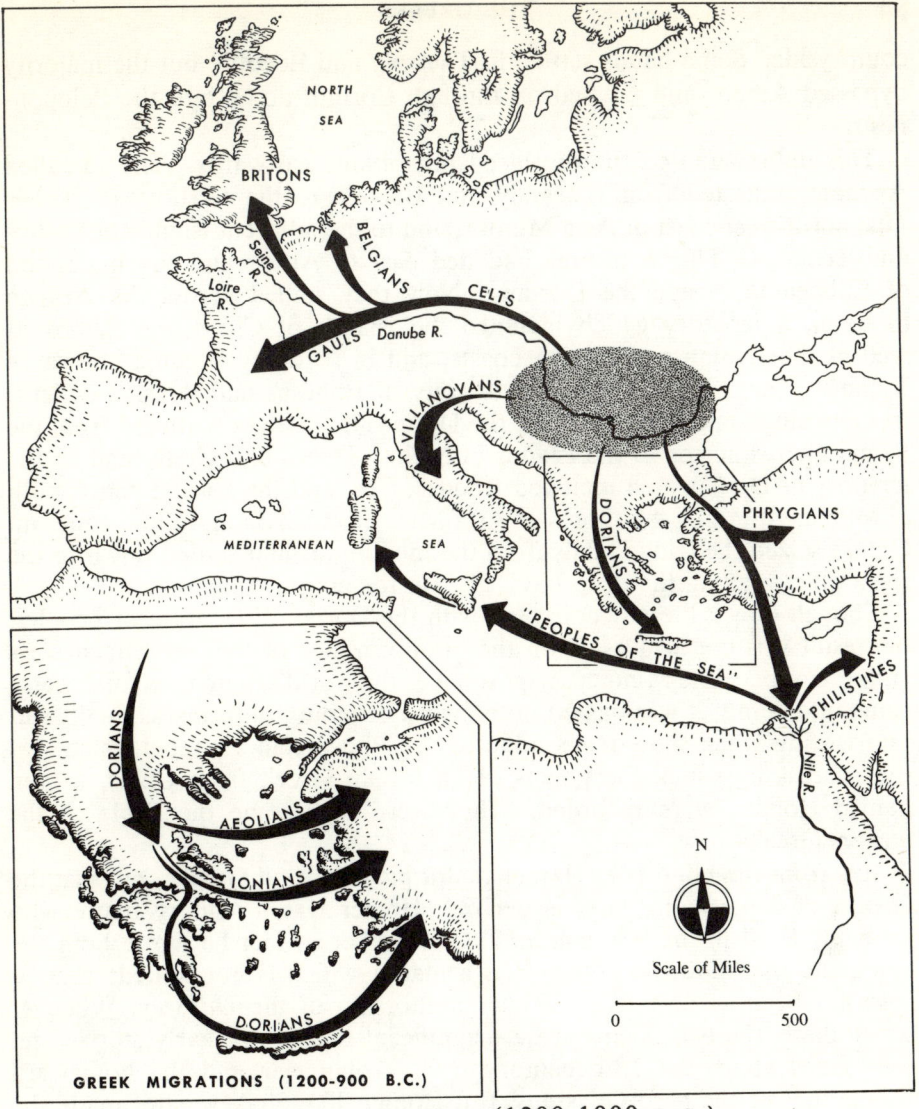

IRON AGE INVASIONS (1200-1000 B.C.)

The wealth of weapons and ornaments found by Schliemann and later excavators in the shaft graves at Mycenae fully justifies the use of the adjective "golden" in speaking of Mycenaean civilization. This glamorous civilization collapsed in the 12th century B.C. before the Dorian Greeks, a rougher people from the north with iron weapons and scant appreciation of the culture they destroyed.

Displaced Persons The Dorian Greeks were a part of the second wave of Indo-European invaders, that shattering expansion of Iron Age peoples in the two centuries following 1200 B.C. As the Dorians filtered down from the north they destroyed palaces, wrecked cities, and ravaged the countryside. Some bands settled in Thessaly and Boeotia, but the majority bypassed Athens and pressed on through Corinth down into the Peloponnesus.

Greek migrations

This influx shunted the displaced inhabitants eastwards. The so-called Aeolians migrated from Thessaly and Boeotia to the territory of Aeolis (the north-west coast of Asia Minor), and to the adjacent islands of Lesbos and Tenedos. The Achaeans had fled east to Attica and the big island of Euboea to escape the Dorians. Now they spilled across the Aegean to Ionia, a territory which included the islands of Chios and Samos as well as the Lydian and Carian coasts, and in so doing became known as Ionians. These Ionians founded the city of Ephesus and wrested Miletus from its inhabitants. And finally, the Dorians themselves scattered from the Peloponnesus across to the islands of Melos, Thera, and Crete, and to the territory of Doris which included Rhodes, Cos, and the Carian coast south of Miletus. Finding the shores of Asia Minor to their liking, these emigrant Greeks seized the fringe lands from the inhabitants and settled down in the river valleys, along the deep bays, and on the coastal islands.

Though Greek literacy collapsed with the Dorian dislocation of peoples, the result was not all loss. With the disappearance of the old official class and its cumbersome Minoan script some of the world's matchless folk poetry came into being. It was passed on by word of mouth until it was finally put into writing when the Greeks adopted the Phoenician alphabet. Such was the process by which the poems of an Ionian Greek from Chios, a man named Homer, were recorded. They were to become the Bible of the ancient Greeks.

Homer

It is from the blind poet Homer that a picture of *all* Greeks, not just the Greeks of a particular city, is derived. Homer was a solitary figure who probably lived in the 9th century B.C. Whether in fact he wrote both the epics, the *Iliad* and the *Odyssey,* is a matter which is not settled: a good case has been made out for a later authorship of the *Odyssey*. Whoever wrote them, the two poems are a magnificent tapestry of early stories and folk songs about the 13th century B.C. Trojan war and the homeward wanderings of the hero Odysseus. But Homer, like Shakespeare, made the settings of his stories (which took place centuries earlier), reflect his own day and age. Thus while it is appropriate to note the Greek poet's stories of Mycenaean civilization, it is even more appropriate to observe his descriptions of what purported to be Mycenaean but was actually 9th century Greek society.

City-States

Homer's Greece lacked Crete's political unity. There was a sort of heroic monarch who ruled each city: Agamemnon of Mycenae was one. But neither the kings of Mycenae nor later Greeks brought political unity to the mainland, and until the days of Alexander the Great the inhabitants of Greece could not manage to create a unified nation. Greek history is really the history of a number of Greek cities. The story of their accomplishments —and failures—is one of the brightest pages of European history.

The Lion Gate was the main entrance to the citadel of Mycenae, an impregnable palace-fortress from which the lords of Mycenae dominated the centre of Greece. The pillar guarded by lions is an oriental symbol of divine protection for the Royal Palace and, dating from the 13th century B.C., constitutes one of the earliest pieces of monumental Greek sculpture. Probably Agamemnon passed through this very gate on his way to Troy.

3/HELLAS AND THE HELLENES

The Greeks called themselves Hellenes and their land Hellas: it was the Romans who first called them "Greeks". The tiny land of Hellas, the southern projection of the Balkan peninsula, was only about the size of the provinces of New Brunswick and Nova Scotia combined. Yet most of Greek history was enacted on an even smaller stage—the southern part of Hellas, roughly the same area, in all, as New Brunswick alone.

Greece is a poor land of low mountain ranges whose limestone absorbs rainfall and will not hold topsoil. The valleys are small and the coastline heavily indented: no one place is further than seventy miles from the sea. Less than a third of the country can be cultivated; the rest is covered with scrub vegetation. The rainy season lasts from October to March, a climatic feature that makes Greek rivers quite unnavigable, torrential in winter, dry gulches in summer. What little arable land there is is sealed in small plains where grain, wine, and oil can be produced. And there are relatively few minerals.

A Land of Poverty

Greece has undoubtedly become poorer than it was in ancient times, yet even then the usual dinner consisted of only one staple, porridge. Aside from the arable plains, the country is fit for little more than goat- and sheep-raising—and that in the summer. Truly, in the words of the 5th century B.C. historian Herodotus (484-425 B.C.), "Hellas and Poverty have always been foster-sisters."

From the beginning, Greece's harsh geography has shaped her history. Her lack of natural resources and the difficulty of overland communications combined to prevent the growth of large states. Moreover, Greece was almost cut in half by the Corinthian and Saronic Gulfs, a fact that made the districts near the Isthmus of Corinth strategic, for here land and sea routes converged. Much of Greek history can be seen in terms of a conflict between north and south.

Yet if the land divided Greeks, the sea united them. All the best harbours were along the eastern coast, and as a whole the country looked east—east to their "Greek lake", the Aegean. Scattered across the sea like stepping-stones were innumerable islands, so many that the sailor was rarely out of sight of land. As long as the sea was open and free, the Greeks could use it as a safety-valve to work off some of the tensions born of a crowded existence in a stark terrain.

Homeric Society With the Dorian invasions and the introduction of iron, a new era of civilization began in Greece. This is the society in which Homer lived, and which is reflected in the pages of the *Iliad* and the *Odyssey*. It was a strangely primitive society, and one with many small isolated political "states". The Greek king, like the oriental monarch, combined in his person the functions of commander-in-chief, supreme judge, and high priest. He was not, however, a despot, but only first among equals. The society was aristocratic, and the nobles formed a council of advisers. One of the chief means of making a living was by warfare and plundering. Wealth was measured in flocks and herds, and commercial transactions were carried on by barter.

The religion had both Mycenaean and Asiatic elements. The gods were all too human—jealous, playing favourites among men, and living in glory on Mount Olympus. Immortality was vaguely believed in. The body was cremated, and the soul somehow survived in Hades, a land of shadows. "I had rather," said Achilles, "be the slave of the meanest landless man on earth than be king in Hades." Ethics were based on tribal custom, hospitality being, as in the desert, the supreme virtue.

As the nobles increased their power in the 9th and 8th centuries B.C. the various kingships disappeared. These nobles, distinguishable from the rest of the population only by wealth, ruled the state; but all citizens were felt to have equal political rights and an equal right to express themselves in the mass meetings.

For protection, the villagers found it wisest to settle permanently around the foot of their acropolis. It was easier than living in isolated country hamlets far from the hill-fortress. They also wanted a place to meet for religious festivals. Most important of all—even more important to the Greeks than the need for defence—was the necessity for a centre where political and public affairs could be discussed and justice dispensed. For these reasons the Greeks came together to form cities, and, in the process, founded principles of constitutional government.

4/THE COLONIES AND CITIES OF THE GREEKS

With the end of the 8th century B.C. Greece was gripped by a severe economic depression. In his *Works and Days,* the Boeotian farmer Hesiod vividly portrays the lean times in his grim picture of the hard life of the peasant.

> Get a house first and a woman and a plowing ox—a slave woman—not a wife—who might also follow the oxen: and get all gear arrayed within the house, lest thou beg of another and he deny thee and thou go lacking, and the season pass by, and thy work be diminished. Neither put off till the morrow nor the day after. The idle man filleth not his barn, neither he that putteth off. Diligence prospereth work, but the man who putteth off ever wrestleth with ruin.

Unable to support their starving and increasing population, the Greek city-states resorted to colonization in the 7th and 6th centuries B.C.

Frogs Round a Marsh Little planning seems to have preceded the founding of the first colonies, but later ones were organized by parent cities which usually consulted the Delphic Oracle for guidance in choosing a site. The colony was politically independent, though bound to its parent city by ties of sentiment reinforced by trade. Greek colonies dotted the shores from the Black Sea and the Crimea to modern Gibraltar; as Plato said, the Greeks came to dwell round the sea as "frogs round a marsh". One of the Ionian Greek cities of Asia Minor, Miletus, was reputed to have founded at least eighty colonies by herself, and even secured a foothold in the Nile Delta.

Colonization spread Greek ways and ideas around the shores of the Mediterranean and Black Seas—but colonization also affected Greece itself. There were great advances in trade and industry. Raw materials from the colonies were exchanged for manufactured goods from the mother cities. The wheat of the Crimea, northern furs and gold, and western timber, hides, and wool flowed to Greece in return for fine textiles, jewellery, metal utensils, arms, and pottery.

Industry developed on only a small scale. Most manufacturing was done in family shops: a shield "factory" employing 120 workmen was considered to be a large enterprise. Often prisoners of war worked as slaves in industry or in the mines.

It was natural that shipbuilding should flourish. No longer, as in the time of Hesiod, did Greeks go to sea only in the summer. By the 6th century B.C. they voyaged from spring to fall, often out of the sight of land, and war galleys over 100 feet in length were built to convoy the large Greek

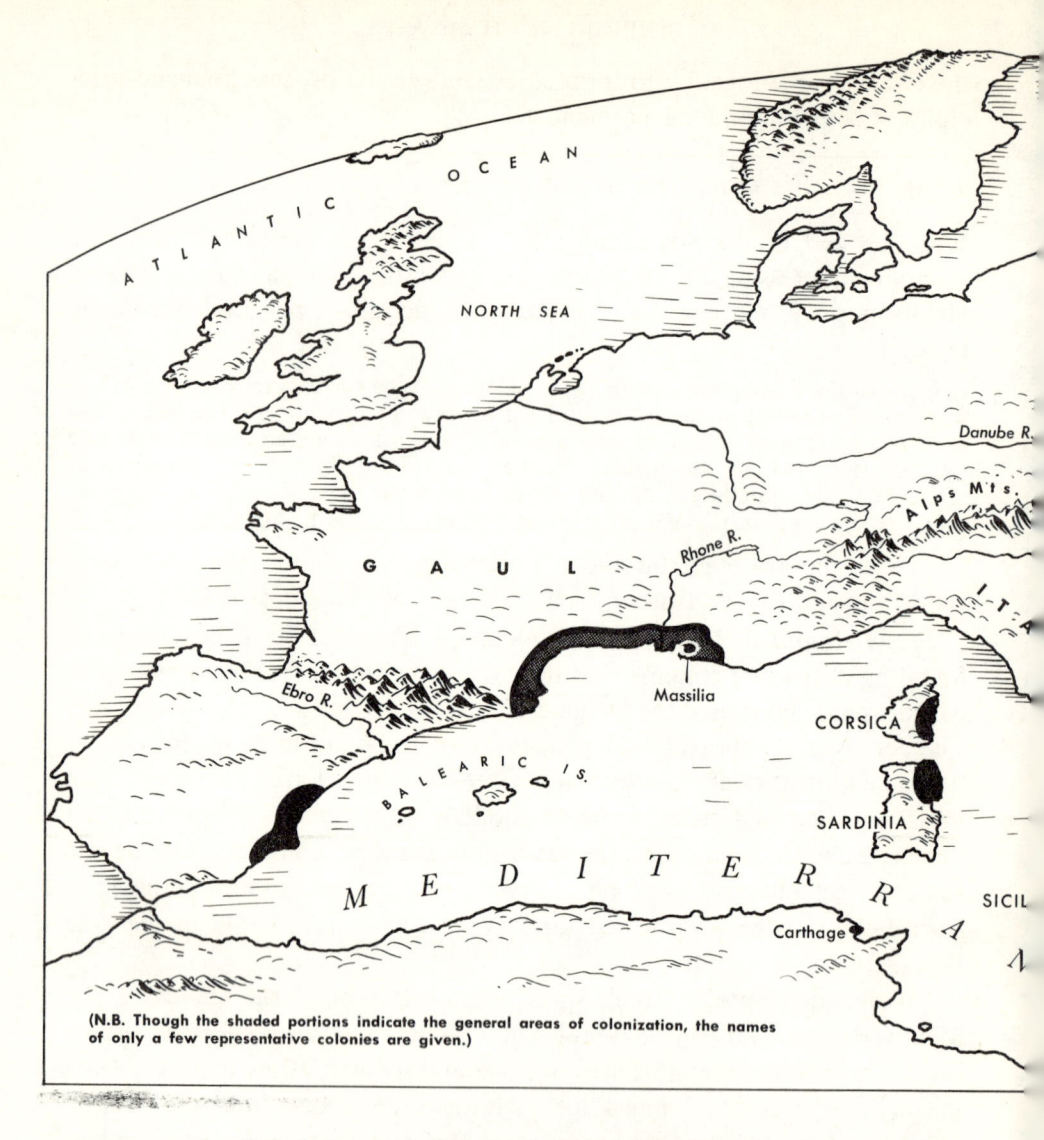

(N.B. Though the shaded portions indicate the general areas of colonization, the names of only a few representative colonies are given.)

cargo vessels. Scholars are still not certain exactly what these galleys, called *triremes,* looked like, but a plausible theory is that the trireme had three banks of rowers, with each man pulling one oar (see pages 96-97). A wood and bronze beak gleamed on the prow, ready to ram enemy vessels.

The city-states from which the colonists set forth were not large, though they included not only the city but its surrounding territory. Athens, with 1000 square miles, was one of the biggest, but four or five cities might share an area of several hundred square miles. In fact, since many Greek states must have had a total population of not more than 10,000-15,000 men, women, and children, we today would call them towns rather than cities.

Each city-state was independent of its neighbours, and the smallness of

THE COLONIES OF THE GREEKS

■ Areas of Colonization

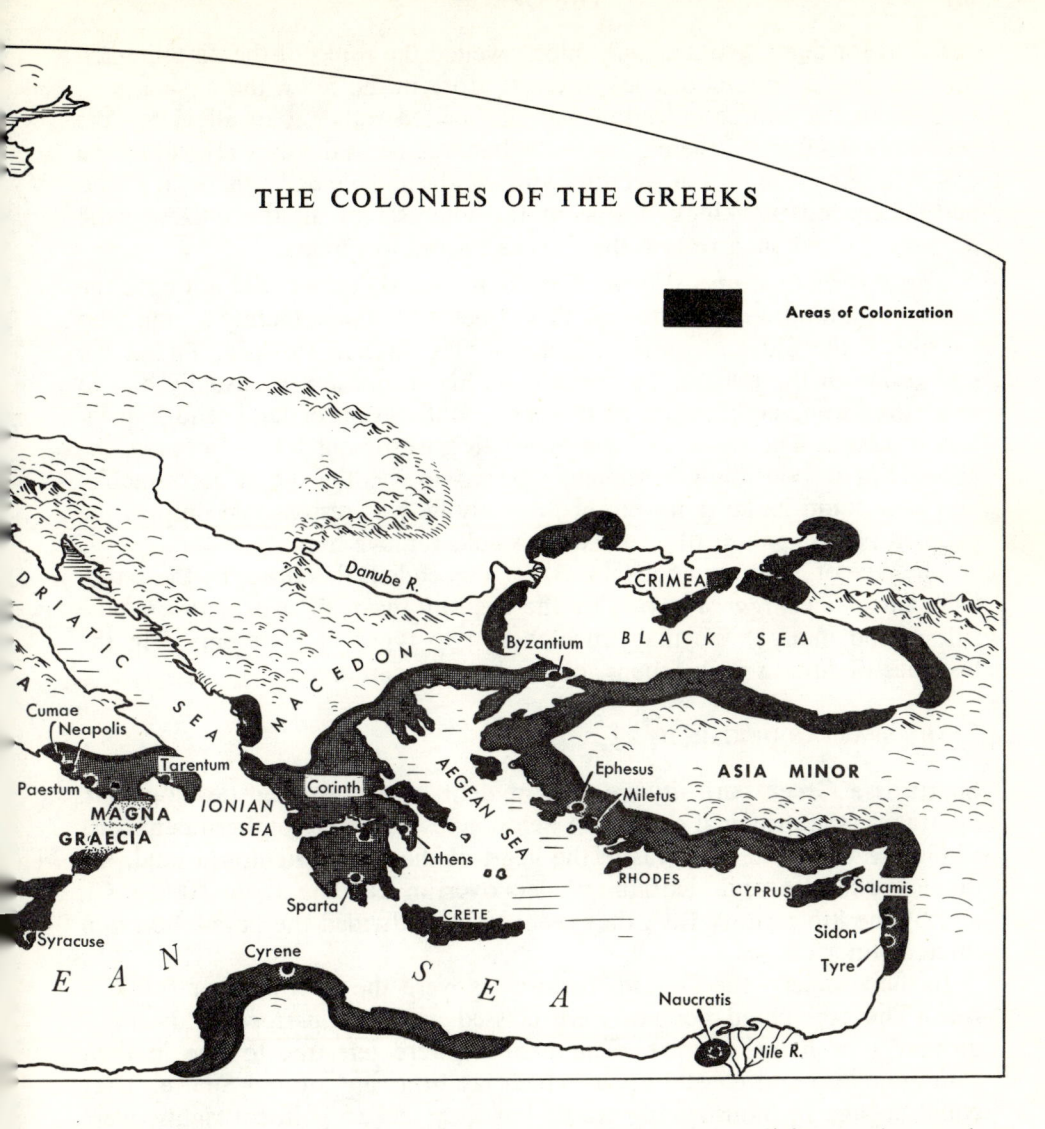

the states enabled all citizens to participate in public affairs so that each city-state developed its own peculiar personality. This Greek separatism was counteracted to a certain extent by the meeting of all Greeks around the council table or at some universal religious or athletic festival, held at such sites as Delos or Delphi or Olympia.

The economic transformation brought about in Greece by colonial expansion soon made the old currency of oxen or metal bars hopelessly cumbersome, and the whole Greek world learned the use of coinage from Lydia and the Ionian cities of Asia Minor. Gradually noble birth and possession of land came to mean less than portable wealth, and in the economic transition the poorer classes often lost their land and were enslaved for debt. Artisans and sailors swelled the ranks of the discontented in the cities. A conflict of class interests arose in the *polis*, the city-state.

Thus in the end colonization had aggravated rather than alleviated the economic troubles of the city-states. What was needed was a reform of the city-state government, an overthrowing of the old landed aristocracy who had succeeded the kings, a rule in the interests of all the citizens (the *demos*). To get such reform the Greeks turned to tyrants.

The Age of Tyranny For the Greeks the word "tyrant" did not have the sinister connotation it has today. The Greek tyrant was merely a man who overthrew the old government of the nobles, and in so doing ended the oppression of the masses. To consolidate his revolution, he usually had to undertake some codification of the laws, confiscation of land, and imposition of taxes. The weakness of a one-man government was, of course, the absence of responsible government. To whom was the tyrant accountable, and how could he be removed? Ultimately he was responsible only to the people. Nothing short of a revolution could remove him.

The Age of Tyranny (650-500 B.C.) was ended, therefore, by the establishment of an *oligarchy* (rule by the wealthy few), or a *democracy* (the rule of the many). We may trace these two types of government in the histories of Sparta and Athens, oligarchy and democracy.

5/THE SPARTAN OLIGARCHY

Sparta is a Greek curiosity—a case of deliberately arrested development. Sparta was the principal city of the Peloponnesus, as the southern peninsula of Greece is named, and was in the heart of the best agricultural land, the district called Laconia. Dorian invaders overran Laconia about 1000 B.C., and by the 8th century B.C. they seem to have divided the population into three distinct classes.

In their society, the Dorians themselves were the *Spartiates* or full citizens. The conquered peoples were classed as either *perioeci* ("dwellers-around") or *helots* (serfs). The perioeci were left free to live in their own mountain and coastal towns which lay in a ring around Sparta. They could engage in industry and trade, but they lacked political rights, were heavily taxed, and were liable to military service. Helots worked the land for their Spartan overlords, giving up part of the produce to them. They were not exactly slaves since they could not be sold or killed by their masters. But they were owned by the state, had no political rights or independence, and were kept in permanent subservience. Thus it was partly the exclusiveness of the Spartan conquerors that was responsible for such a highly stratified, closely knit society.

A second cause for the distinct layers of Sparta's society was the means by which she sought to relieve the economic hardships of the 8th century B.C. Most Greek cities, of course, sent out colonies to remedy the situation at home. Sparta also sent out one, but the main remedy she followed was to annex Messenia, her western neighbour, and by the end of the 7th century B.C. to reduce the Messenians to the status of helots.

Discipline without Culture The Spartiate minority found itself confronted with the grave problem of how to maintain itself as the dominant group. Since a permanent standing army was the only possible safeguard against a Messenian revolt, society had to be organized to one purpose: the production and maintenance of the professional soldier.

The name of Lycurgus is usually associated with the far-reaching reforms designed to establish such a military society, though little is known about this semi-mythical 9th century lawgiver. (Indeed it has been said that Lycurgus was not a man but a god.) We do know, however, that the institutions for which Lycurgus is given the credit were the result of a long evolution covering at least the 7th century B.C.

The problem faced by Lycurgus is clearly evident from these approximate population figures at the end of the 6th century B.C.

Spartiates	—	25,000
Perioeci	—	100,000
Helots	—	250,000

In other words, there was probably about one Spartiate to every 10 or 15 non-citizens. The Spartan overlords must have felt as if they were living on top of a volcano.

Plutarch, a Greek historian of the 2nd century A.D., says that Lycurgus's reforms included the early banishment of "the unnecessary and superfluous arts"; henceforth Spartan artisans were to be "freed from useless tasks." In

This Greek pottery vessel was created in Attica about 535-525 B.C. It is an *amphora,* a general storage jar used for wine, grain, oil or honey. The amphora was made with an opening large enough to admit a ladle, and usually had a cover to protect the contents. The figures were painted on the natural red clay in a velvety jet-black glaze, after which the details were etched with some hard-pointed instrument so as to expose the red beneath. Occasionally white and purple pigments were also used in the decoration. Greek pottery was made on a terracotta wheel such as is still used in some countries today.

earlier times the Spartans had been noted for arts and manufactures. Architecture, pottery, music, poetry, and a busy trade with foreign ports were all known in Sparta in the 8th and 7th centuries B.C. But for the sake of the army all this was now renounced. The Spartan ideal became a military one with no room for cultural pursuits, and no time even for the "laconic" inhabitants to indulge in flowery speech.

At the age of seven, Spartan boys were taken from their mothers and put into barracks to begin their education (well described by Plutarch in the Source Reading on pages 100-101). Until they were twenty these boys were drilled and disciplined to be self-reliant, tough, unthinking soldiers. At twenty, adult military training began, and although the citizen was expected to marry he was to spend most of his time on active service and to visit his wife only by stealth. The troops ate in common messes—a visitor entertained in one of them is reputed to have said that he now understood why Spartans did not fear death!—and from thirty (the age of full citizenship) to sixty the Spartan could be drafted for military service. The army was made up of regiments of heavy infantry. With such rigorous and continuous training, the Spartan heavy-armed hoplite (infantryman) was, as you might expect, about as formidable a soldier as the world has even seen.

Spartan women were expected to be hardy mothers. Girls were given physical training, and were indoctrinated to expect their sons back from battle either carrying their shields or being carried on them. As might be expected, in order to preserve such a rigid system it was felt necessary to exclude foreign ideas and foreign travellers from Sparta, and Spartans ordinarily were forbidden to leave the country—possibly to prevent their desertion from the army. Moreover, a secret police (the *krypteia*) was organized, a kind of commando corps of picked youths who were at the service of the state to murder any helots whom they suspected of plotting rebellion.

It is not necessary to believe all the stories of boys' voluntary submission to flogging and the ritual murder of helots to see in Sparta a state dedicated to physical fitness and obedience to authority. There was no time or room for either culture or individuality.

Spartan government

Spartan government was an adaptation of primitive Greek ideas. Two kings ruled simultaneously, exercising priestly and judicial as well as monarchal functions. After the 7th century B.C., however, the real power rested in the hands of five overseers (*ephors*) elected annually by the Assembly (*Apella*) of free Spartan citizens over the age of thirty. The Apella elected the Council or Senate (*Gerousia*) composed of thirty members—twenty-eight members over sixty years old from noble families, and the two kings. They served for life, prepared legislation for the assembly, and were advisers to the kings. The assembly did not initiate legislation or debate; it only voted to accept or reject. Nor was there any tedious business of counting ballots: the faction that the magistrates decided had shouted the loudest carried the day.

Thus the five annual ephors really controlled kings and senate. It was not, however, this peculiar constitution that marked Sparta off from other Greek states. It was her totalitarian and militaristic organization of society that guaranteed Sparta the military leadership of Greece.

In the 6th century B.C. a league of states was organized in the Peloponnesus with Sparta at the head. Stability and singleness of purpose had made Sparta a first-class military power, and under this state's inspiration the centralized military machine of the Peloponnesian League was, fortunately, created and ready for the Persians when they came in the 5th century B.C. It is little wonder that on the memorial of those who "helped to overthrow the Persian" the name of Sparta leads all the rest.

The Peloponnesian League

6/THE ATHENIAN DEMOCRACY

While Sparta was winning leadership in the Peloponnesus, Athens was a second- or third-class power. Some time during this period (about 900-600 B.C.) all the little cities of the area of Attica were united into the one city-state of Athens. In the development of Athenian democracy four men are outstanding:

Draco	— archon	} 7th century B.C.
Solon	— archon	
Pisistratus	— tyrant	} 6th century B.C.
Cleisthenes	— leader of democrats	

Two Archons and the Law In the beginning, the Athenian government was aristocratic. The executive, which consisted of a supreme judge (*archon*), a commander-in-chief of the army (*polemarch*), and the king (by this time mainly a high priest), governed along with a Council of Elders (*Areopagus*) and an assembly of citizens (*Ecclesia*). This aristocratic city-state government of Athens, like others of its kind, failed to adapt to social and economic changes. It is therefore not surprising that there was a premature attempt at tyranny in the 7th century B.C.

A few years later, in 621 B.C., a man named Draco was elected archon. Draco is famous for the codification and publication of Athenian law, a law that was regarded as unduly harsh ("draconian"), though it was designed, it seems, to end blood-feuds and to make the state responsible for punishing those guilty of homicide.

Draco

But all these legal reforms did not help the small farmer who had been unable to make his farm pay, had lost his land by defaulting on a mortgage to some noble, and had ended by being sold into slavery. What was needed

Solon

to stave off revolution was a radical redistribution of land in Attica. With great foresight, the assembly in 594 B.C. elected a remarkable man as the new archon having supreme power: Solon—poet, merchant, traveller and statesman. He has been called "the first live figure in Athenian history".

Solon was no radical. He was a conservative reformer who, instead of redividing the land, reduced debts and freed all who had gone into slavery on account of them. Moreover, he decreed that in future no one could be enslaved for debt. This meant that the whole population had achieved the status of citizens: there would be no Attic helots.

Oil and pottery

Solon also encouraged trade and industry. He fully realized that if Athens were to feed herself at anything above a bare subsistence level she would always have to import grain. He also knew that Attica was one of the best olive-growing regions in all Greece. Accordingly he forbade the export of any agricultural products except olive oil, with the result that many farmers turned to olive-growing and Attica became a leading producer of olive oil. Henceforth the Athenians had a prime commodity they could export in exchange for grain.

Olive-growing, however, did not require as many labourers as grain-growing, and a policy of industrial expansion would have to be initiated if work were to be found for the unemployed. Once again the answer centred around the olive grove. Oil was transported in pottery containers; hence more of these would now be needed. To speed the manufacture of these jars foreign craftsmen were encouraged to settle in Athens in return for Athenian citizenship, and thanks partly to them and partly to the good beds of reddish-brown clay near the city, the pottery industry boomed. Soon the craftsmen of Attica were excelling their rivals in other city-states. Their beautiful jars, designed simply as containers for olive oil (and later for wine and honey)—the tin cans of the 6th century B.C.—,now grace many a museum.

The virtual monopoly that Attic pottery enjoyed in the latter half of the 6th and the entire 5th century B.C. is an immediate result of Solon's encouragement of manufacture. It is, in fact, Solon's economic rather than his political reforms that seem more important to historians. Yet Solon's political reforms, though they failed, did lay foundations for the later democracy.

Solon's reforms

Four classes of Athenians were recognized on the basis of wealth. Although most of the magistracies went to the two richest classes, the really important distinction was between the members of the first three classes (large and small landowners) and the fourth class, the *thetes* (landless men who worked for hire). The first three classes had the rights and responsibilities of citizens, but the fourth class now won certain political rights which included membership in the assembly. The assembly, too, was changed from a rather aimless meeting of citizens into a body with power to elect magistrates and to decide on vital matters such as war and peace.

Solon recodified and republished the law, as well as creating a popular court of appeal *(Heliaea)*, of which all citizens were entitled to be members. The magistrates or archons now numbered nine, and were chosen annually by the assembly from men of the two upper classes. After their year of office they became members of the Areopagus, which functioned as a court to try those accused of crimes against religion or of plotting to subvert the constitution. In this way the Areopagus became the guardian of the new constitution.

Nevertheless Solon's view was hardly democratic. Since there was no pay for jury service or assembly duty it is probable that few thetes could afford to attend, and that in practice both the juries and the assembly were dominated by the propertied classes. Thus Solon did not end class strife. The dissension was to be further complicated by a war with a rival city-state, Megara.

Pisistratus

The hero of the war with Megara was a young soldier named Pisistratus, who had ambitions to make himself tyrant of Athens. As leader of a coalition of peasants, herdsmen, and miners ("Men of the Hills"), Pisistratus found himself opposed by the nobles with estates in the best farming districts ("Men of the Plains") and the wealthy traders and manufacturers ("Men of the Shore"). The Source Reading at the end of this chapter recounts the masterly trickery Pisistratus used to seize power in 560 B.C. Twice he was overthrown; but finally, backed by armed force, he consolidated his tyranny in 546 B.C., drove many of the wealthy landowners into exile, confiscated their land, and divided it up among the landless.

Pisistratus then proceeded to levy a tax on the income of the first three classes and to furnish employment for the city poor by undertaking extensive public works. He also encouraged greater production of olive oil and wine for export. More far-reaching were the steps he took to increase trade. To this end he not only negotiated treaties but built a navy to protect the grain route from the Crimea, thus making himself at least partly responsible for launching Athens on the road to empire.

Of course all these projects were expensive, and it was fortunate that the silver mines at Laurium could provide him with money. (Probably the famous series of Athenian coins stamped with an owl was issued for the first time during the tyranny of Pisistratus.) Sculpture and vase-painting flourished. Recitals of epic poetry and a stressing of tragic drama were also part of this enlightened ruler's programme, which transformed Athens from a small country town into an international centre.

Pisistratus was an outstandingly able tyrant, yet a tyrant who prepared the way for democracy. He strengthened the whole of the Athenian populace, both high and low, by improving the lot of the small peasants who now became independent farmers, the merchants who were now able to compete with those of Miletus and Corinth, and even the landed aristocracy who found more profit in producing oil and wine for export than in growing grain for home consumption. Above all, Pisistratus destroyed the strong

The Greek *trireme* pictured here was invented about 550 B.C. and superseded the *penteconter*, the earliest type of Greek warship, which had had either 24 oarsmen to a side on a single level or later 12 oarsmen to a side in two banks. The trireme was propelled by a crew of 170 seated at three banks of oars, and could also be assisted by its square sail. In battle the mast was lowered or put on shore, and the oarsmen tried to manoeuvre the ship so as to sink their opponents by means of the bronze ram attached to the prow.

political influence of the noble families. Bitterly one of them lamented its passing:

Shame has perished; pride and insolence have conquered justice and possess the earth. . . .

The city is still a city, but the populace has changed; once they knew nothing of laws, wrapped their flanks in goatskins, and dwelt like deer outside the walls; but now they are the nobles and the one-time nobles base—Oh who can bear the sight? . . .

Grind them hard and let their yoke be heavy: that is the way to make them love their masters. . . .

The mass of the people knows one virtue, wealth; nothing else avails. . . .

Not to be born is best nor look upon the sunshine; or once born to pass as soon as may be through the gates of death and lie beneath a heap of earth.

Hippias and Hipparchus

In 527 B.C. Pisistratus was succeeded by his two sons, Hippias and Hipparchus; but their hold on Athens was soon weakened. For one thing, tyranny was beginning to go out of style as many Athenians thirsted for more freedom. Furthermore, the Persian seizure of Athenian mines in the

Though the seating arrangement of the rowers in a trireme has been much debated, the latest view favours the three-bank theory. The rowers were seated as shown in this cross-section drawing. By means of a projecting outrigger a third bank of oarsmen (whose rowing job was most wearing of all because of the sharp angle) could be added without the construction of a deeper, heavier hull. The result was a ship that still had all the speed and manoeuverability of the earlier two-banked penteconters, but with greatly increased power. It is easy to understand why, in the 5th and 4th centuries B.C., triremes became "the unchallenged queens of the sea."

north Aegean meant a drying up of revenues. And finally, political exiles from Athens began to intrigue with Sparta and Thebes in an attempt to get themselves restored to positions of influence in Attica.

In 514 B.C. Hipparchus was assassinated in the course of an uprising, and though the rebellion was quelled Hippias's severity in putting it down only increased his unpopularity. When the people of Athens blamed him for the encircling alliance of Thebes, Corinth, Sparta, and Aegina which was now building up against them, Hippias turned in desperation to Persia to keep himself in power. In 510 B.C. he was expelled from Athens. The hapless tyrant fled to the court of Darius.

Democracy Triumphs With the tyrants gone, an oligarchy might have been expected. Certain nobles aided by Sparta did attempt it, but another aristocratic group led by Cleisthenes took the side of the people to establish a new constitution.

Cleisthenes was an ambitious noble who had been in the forefront of

Cleisthenes

those opposing Hippias. Although he has been credited by writers of ancient times with more reforms than he actually accomplished, he did base his constitution on the support of a vigorous middle class. He was able to fuse into one party the merchants and manufacturers of the coastal districts and the peasants, herdsmen, and miners of the hill districts, thereby preventing any revival of the political influence of the noble families. Democracy's triumph was assured.

Cleisthenes reorganized the electoral system so that the political influence of the powerful old families was greatly reduced. Ten new electoral districts, each of which included a complete cross-section of Athenian society, replaced the four old groupings. Henceforth the individual counted for more than the family in Athenian politics.

The Boulê

A new Council of Five Hundred (*Boulê*) was instituted. It was chosen annually by lot from citizens over thirty years of age, fifty from each of the ten electoral districts. Since no citizen could be elected more than twice, it was likely that every citizen would get the chance to serve on the Council of Five Hundred at least once during his lifetime. This new body had power to draft legislation, to deal with foreign affairs, and to supervise the entire administration. Each group of fifty acted as a standing committee (*prytany*) of the Boulê for a tenth of a year. The result of this system was that at any given time about a quarter of the citizens who voted in the assembly had gained experience in the conduct of public affairs as members of the Boulê.

By the end of the 6th century B.C. a new board of ten generals (*strategoi*) had been instituted, a board to be elected yearly by direct ballot. It was taking too big a gamble to leave the conduct of military affairs to a chance selection by lot; not even Athenian democracy was prepared to take that risk. Then in 487 B.C. the archons began to be chosen by lot, instead of by ballot as previously. As a result the generals were henceforth the most important elected officials in Athens.

Ostracism

Two other measures that used to be attributed to Cleisthenes were probably not enacted until a generation after his reforms. The first was *ostracism*, a unique practice devised by the Athenians about 488 B.C. to prevent the overthrow of democracy and a return to tyranny. By this process, votes were recorded on broken pieces of pottery (*ostraka*). If at least 6000 people scratched on their ostraka the names of men whom they wished banished, then the citizen with the greatest number of votes cast against him had to leave Attica within ten days and remain in exile for ten years, although he did not lose his citizenship, property, or revenues.

This device was an effective safety-valve: it substituted the ballot for violence. But it could be abused. In a deposit of broken bits of pottery found by excavators in a well in Athens, the name of Themistocles (who we know was ostracized in 470 B.C.) occurs on 190 out of 191 ostraka. The interesting thing is that the 190 names are written in only 14 separate hands, and are all on pieces of vases of good quality. These ostraka were presumably forged by aristocratic enemies of Themistocles for an ostracism of 482 B.C., and though we know that he was not actually ostracized for another twelve years it does not seem likely that his enemies would be any more scrupulous then than they had been before.

Somewhat later, possibly about 462 B.C., the second measure was enacted when the old Heliaea was broken up into ten panels (*dicasteries*) of jurors. The jurors were 6000 citizens chosen by lot each year from among those who volunteered for jury duty. So large were the juries (they numbered 201, 401, 501, or even in some cases 1001 or 1501) that although justice was meted out by amateurs, it was a sufficiently large number of amateurs that it was quite impossible either to browbeat or to bribe them.

Jury duty

Thus were the foundations of Athenian democracy laid down in the four generations that separated Solon (594 B.C.) from the dicasteries of 462 B.C. Solon had placed the administration in the hands of the aristocratic well-to-do, hoping that they would govern the poorer classes with wisdom. But the merchants and manufacturers from these poorer classes rose in the scale to the point where they wanted to govern themselves—something which, under Cleisthenes, they managed to do by means of an alliance with the landed peasants.

Democracy did not, of course, mean to the Greeks what it means to us. Women, slaves, and non-Athenians had no vote. Even so, no other state we have so far studied granted political rights to such a high proportion of its citizens. The word *demokratia* came to mean *demos*-rule, and the demos came to be associated with the landless thetes. These citizens outnumbered all the rest, so that Athenian "democracy" in the 5th century became almost synonymous with control of the state by the poorest class.

This fact explains why most ancient Greek writers use the word democracy as a term of abuse. Such a critic was the so-called "Old Oligarch", whose bitter condemnation of Athenian government opened this chapter.

7/SUMMARY AND CONCLUSION

About the same time that city-states flourished in Mesopotamia, and Egypt was unified, a new civilization grew up on the Mediterranean island of Crete. Crete dominated the Aegean Sea and its vessels ranged far and wide. It developed a more democratic civilization than the Middle Eastern empires, and was almost modern in its comforts.

In the middle of the 15th century B.C. the Cretan power was destroyed from the Greek mainland, and the first civilization in Europe grew up around the rich city of Mycenae. This Bronze Age civilization collapsed in turn before northern invaders who had acquired the use of iron. The resulting 9th century society was made up of a number of small and varied states, each with its own king. Conditions of life were far less comfortable than they had been in the preceding Mycenaean age.

Unable to support an increasing population, by the 7th century B.C. the various Greek city-states sent out colonies, spreading the Greek way of life around the Mediterranean and Black Seas. But colonization did not end the economic difficulties of the city-states, and aristocratic governments were overthrown by tyrants who in time gave way to oligarchies or democracies.

These two forms of government may be symbolized by the military state of Sparta on the one hand and the democracy of Athens on the other. The two cities, about as far apart on the map as Ottawa and Montreal, followed radically different ways of life. Could they continue to exist independently? How would the dozens of neighbouring city-states be affected by their example? For our answer we must examine the wars of the Greeks in the 5th century B.C.

SOURCE READINGS

(a)

PLUTARCH (A.D. 46-127) was a Greek biographer and essayist. He is best known for his *Parallel Lives*, which appeared in the years A.D. 105-115 and now comprises fifty biographies of Greek and Roman statesmen. Plutarch's *Lives* "has been one of the most widely read books of the world," and Shakespeare drew heavily on it for several of his plays. The following selection from the life of Lycurgus describes the education of young Spartans.

[The] offspring was not reared at the will of the father, but was taken and carried by him to a place . . . where the elders of the tribes officially examined the infant, and if it was well-built and sturdy, they ordered the father to rear it . . .; but if it was ill-born and deformed, they sent it to . . . a chasm-like place at the foot of Mount Taÿgetus, in the conviction that the life of that which nature had not well equipped at the very beginning for health and strength, was of no advantage either to itself or the state. On the same principle, the women used to bathe their new-born babes not with water, but with wine, thus making a sort of test of their constitutions. For it is said that epileptic and sickly infants are thrown into convulsions by the strong wine and lose their senses, while the healthy ones are rather tempered by it, like steel, and given a firm habit of body. Their nurses, too, exercised great care and skill; they reared infants without swaddling-bands, and thus left their limbs and figures free to develop; besides, they taught them to be contented and happy, not dainty about their food, nor fearful of the dark, nor afraid to be left alone, nor given to contemptible peevishness and whimpering. . . .

. . . But Lycurgus would not put the sons of Spartans in charge of purchased or hired tutors, nor was it lawful for every father to rear or train his son as he pleased, but as soon as they were seven years old, Lycurgus ordered them all to be taken by the state and enrolled in companies, where they were put under the same discipline and nurture, and so became accustomed to share one another's sports and studies. . . .

Of reading and writing, they learned only enough to serve their turn; all the rest of their training was calculated to make them obey commands well, endure hardships, and conquer in battle. Therefore, as they grew in age, their bodily exercise was increased; their heads were close-clipped, and they were accustomed to going bare-foot, and to playing for the most part without clothes. When they were twelve years old, they no longer had tunics to wear, received one cloak a year, had hard dry flesh, and knew little of baths and ointments; only on certain days of the year, and few at that, did they indulge in such amenities. They slept together, in troops and companies, on pallet-beds which they collected for themselves, breaking off with their hands—no knives allowed—the tops of the rushes which grew along the river Eurotas. . . .

. . . one of the noblest and best men of the city was appointed . . .inspector of the boys, and under his directions the boys, in their several companies, put themselves under the command of the most prudent and warlike . . . [youths of twenty years of age. Each such youth] commands his subordinates in their mimic battles, and in doors makes them serve him at his meals. He commissions the larger ones to fetch wood, and the smaller ones potherbs. And they steal what they fetch, some of them entering the gardens, and others creeping right slyly and cautiously into the public messes of the men; but if a boy is caught stealing, he is soundly flogged, as a careless and unskilful thief. They steal, too, whatever food they can, and learn to be adept in setting upon people when asleep or off their guard. But the boy who is caught gets a flogging and must go hungry. For the meals allowed them are scanty, in order that they may take into their own hands the fight against hunger, and so be forced into boldness and cunning.

. . . The boys make such a serious matter of their stealing, that one of them, as the story goes, who was carrying concealed under his cloak a young fox which he had stolen, suffered the animal to tear out his bowels with its teeth and claws, and died rather than have his theft detected.

Plutarch's Lives, translated by B. Perrin, Loeb Classical Library (William Heinemann, 1914), I, 255-263.

(b)

ARISTOTLE (384-322 B.C.), the Greek philosopher (see pages 135-136), wrote an account of the constitutions of Greek states. The extract which follows is from the *Constitution of the Athenians*. Aristotle wrote it about 328 B.C., although it was only recovered from the sands of Egypt in 1890.

Pisistratus appeared to be most devoted to the popular cause, and had won a brilliant reputation in the war with Megara. Having wounded himself, he persuaded the people, on the supposition that his injuries were inflicted by political enemies, to grant him a guard for his person. Taking the club-bearers, as they were called, he conspired with them against the state, and seized the Acropolis. . . . The story is told that when Pisistratus was asking for a guard, Solon opposed him, saying that he was wiser than some and braver than others—wiser than those who failed to see that Pisistratus was aiming at the tyranny, and braver than those who knew it but kept silent. As he accomplished nothing with words, he brought out his armour and placed it before his door, saying he had aided his country to the best of his ability (for he was at this time a very old man) and asking the rest now to perform this service. But Solon accomplished nothing by his exhortations at that crisis. Pisistratus, however, assuming the government, managed affairs constitutionally rather than despotically. Before his supremacy was firmly rooted, the party of Megacles [the "Men of the Shore"], joining in friendship with that of Lycurgus [who led the "Men of the Plains"], expelled him. . . . But . . . afterward

Megacles, harassed by sedition, again made overtures of peace to Pisistratus on condition that the latter should take the daughter of the former in marriage. Megacles brought him back in an exceedingly old-fashioned and simple way. Spreading a report that Athena was restoring Pisistratus, he found a tall, handsome woman . . . and dressing her up in imitation of the goddess, he brought her in along with Pisistratus, the latter seated in the chariot with the woman at his side, while the people of the city on their knees received them with adoration.

Thus was brought about the first restoration. He went again into exile . . . ; for he did not maintain himself long, but because he was unwilling to treat the daughter of Megacles as his wife, and consequently feared a combination of the two factions, he secretly withdrew from the country. First he colonized a place called Rhaecelus about the Thermaic Gulf; then he crossed over to the neighbourhood of Mount Pangaeus. Making money in that locality and hiring soldiers . . . he attempted to recover his supremacy by force. . . . The people he deprived of their arms in the following manner. Holding a review of the citizens under arms at the Theseum, he attempted to address them, but spoke in a low voice; and when they declared they could not hear him, he bade them come up near the gateway of the Acropolis in order that his voice might sound louder. While he was passing the time making his speech, persons appointed to the task took the arms and locking them in a building near the Theseum, came and made a sign to Pisistratus. He finished his speech and then told them about the arms, bidding them not wonder or be dejected but go and attend to their private affairs, as he would himself manage all public matters.

Such was the origin of the tyranny of Pisistratus and such were its vicissitudes. . . .

G. W. Botsford and E. G. Sihler, editors, *Hellenic Civilization* (Columbia University Press, 1915), pp. 150-152.

FURTHER READING

1. GENERAL

    BOTSFORD, G. W. and ROBINSON, C. A., *Hellenic History* (Macmillan, 1956)
    BOWRA, C. M., *The Greek Experience* (Weidenfeld and Nicolson, 1957)
    BURY, J. B., *A History of Greece to the Death of Alexander the Great* (Macmillan, 1951)
    † DURANT, W., *The Life of Greece* (Simon and Schuster, 1939)
    † FINLEY, M. I., *The Ancient Greeks: An Introduction to Their Life and Thought* (Macmillan, 1963)
    † GRANT, M. and POTTINGER, D., *Greeks* (Nelson, 1958)
    HAMMOND, N. G. L., *A History of Greece to 322 B.C.* (Oxford, 1959)
    † * KITTO, H. D. F., *The Greeks* (Penguin, 1951)
    † * ROBINSON, C. E., *Hellas: A Short History of Ancient Greece* (Beacon, 1955)
    ——————, *A History of Greece* (Methuen, 1957)
    † * SMITH, M., *The Ancient Greeks* (Cornell, 1960)
    TOYNBEE, A, J., *Hellenism* (Oxford, 1959)

2. MINOANS AND MYCENAEANS

    * CHADWICK, J., *The Decipherment of Linear B* (Penguin, 1961)
    † * COTTRELL, L., *The Bull of Minos* (Pan, 1955)
    * HUTCHINSON, R. W., *Prehistoric Crete* (Penguin, 1962)
    MYLONAS, G., *Ancient Mycenae* (Princeton, 1957)
    PALMER, L. R., *Mycenaeans and Minoans* (Faber, 1961)
    † VAUGHAN, A. C., *The House of the Double Axe: The Palace at Knossos* (Doubleday, 1959)

3. THE IRON AGE AND HOMER
    FORSDYKE, J., *Greece before Homer* (Parrish, 1956)
    † MIREAUX, E., *Daily Life in the Time of Homer* (Allen and Unwin, 1959)
    * PAGE, D. L., *History and the Homeric Iliad* (California, 1963)
    STARR, C. G., *The Origins of Greek Civilization, 1100-650 B.C.* (Knopf, 1961)
    † WARNER, R., *Greeks and Trojans* (MacGibbon and Kee, 1951)

4. CITY-STATES
    * ANDREWES, A., *The Greek Tyrants* (Torchbook, 1963)
    BURN, A. R., *The Lyric Age of Greece* (Arnold, 1960)
    EHRENBERG, V., *The Greek State* (Blackwell, 1960)
    * FREEMAN, K., *Greek City-States* (Norton, 1963)
    HIGNETT, C., *A History of the Athenian Constitution to the End of the Fifth Century B.C.* (Oxford, 1952)
    MICHELL, H., *Sparta* (Cambridge, 1952)
    † * ZIMMERN, A. E., *The Greek Commonwealth* (Oxford, 1961)

4. The Wars of the Greeks

They have filled the city with harbours and dockyards and walls and tributes instead of with righteousness and temperance. . . . Themistocles and Cimon and Pericles, who caused all the trouble. . . .
PLATO, *Gorgias*

In the 5th century B.C. democratic Athens faced her most severe test. She faced it and she failed. Within the space of two generations Athens plunged from the heights to the depths. The city that had pioneered in political democracy took liberty from others, and ended by losing it herself.

It was a soul-stirring tragedy, and later generations have never ceased to be fascinated by how it happened and why.

1/THE PERSIAN WARS

So far fortune had smiled on the Hellenes. In the 8th and 7th centuries B.C. they had been given a period of comparative calm during which they could establish a fringe of Greek cities in Ionia along the Asia Minor coast without interference from Assyria or Babylonia. But in the 6th century B.C. Croesus of Lydia managed to gain control of this coast and most of the Greek cities, and when he was conquered by Cyrus of Persia in 546 B.C. the Ionian cities found themselves part of the Persian Empire.

These Asia Minor cities were important to Persia as sources of revenue and as naval bases. Despite the fact that they prospered, their rule by tyrants

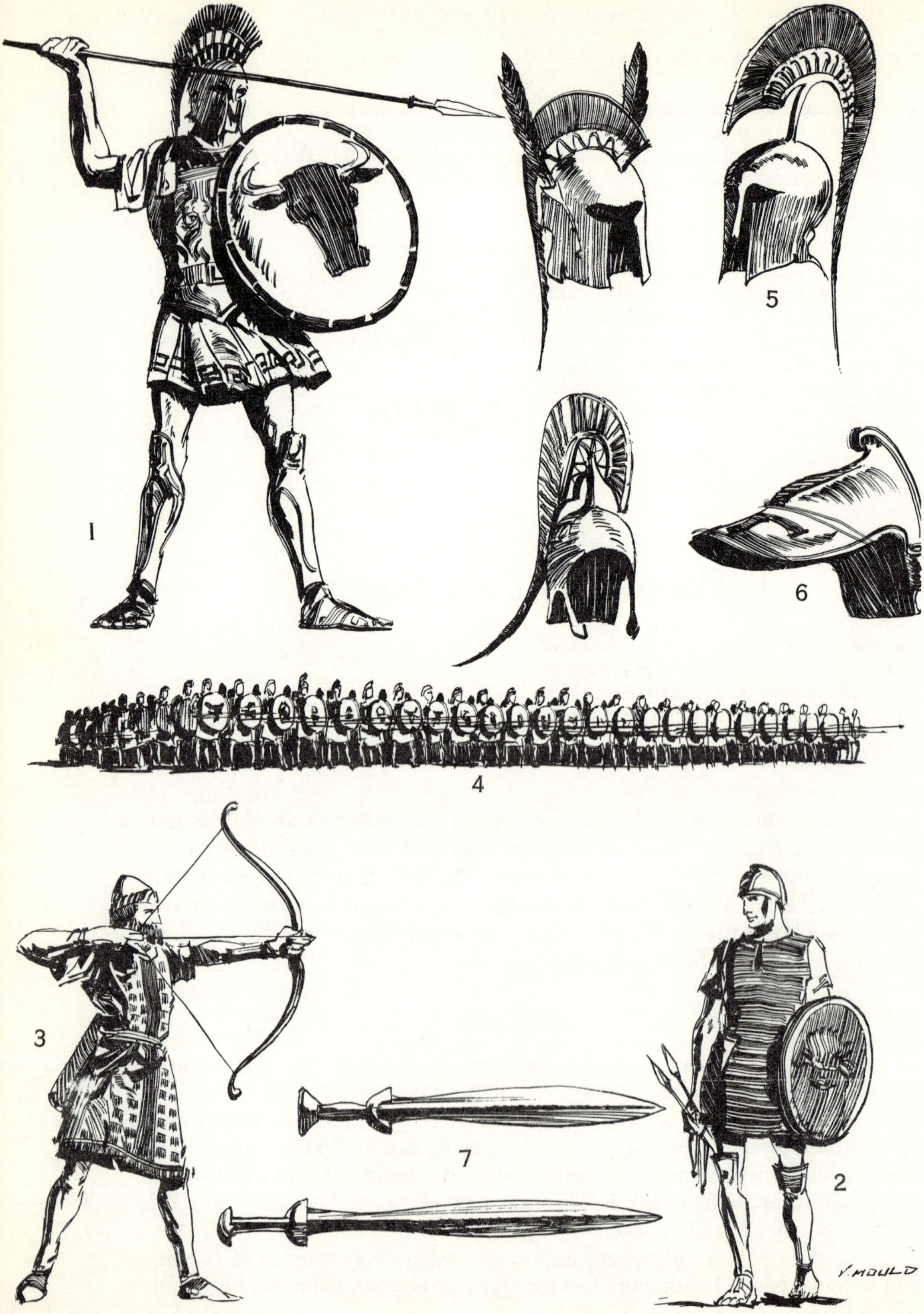
1
5
6
4
3
7
2
V. MOULD

with Persian backing hardly measured up to Greek ideals of government by discussion, and a revolt led by Miletus broke out in 499 B.C. When the immense resources of the empire were mobilized to crush the rebels, Athens and Eretria rallied to their assistance—but in vain. The Greeks suffered a devastating naval defeat in 494 B.C., and the Persians proceeded to reorganize the Ionian cities as democracies—perhaps to ensure their goodwill in the ensuing struggle with Greece—and conquered up to the Hellespont. These Persian Wars broke on Greece in two waves from 490 to 479 B.C.

The Miracle of Marathon The Persians first struck in 490 B.C. to punish Eretia and Athens, and to restore Pisistratus' son Hippias whom they brought with them. The force of approximately 40,000 Persians[1] met with little opposition, took Eretria, and partially disembarked at Marathon on the east coast of Attica, about 25 miles north of Athens.

The First Persian Invasion

Sparta had promised the Athenians aid. But when the famous distance runner Pheidippides covered the 140 miles to Sparta in 48 hours to solicit aid, tradition has it that the Spartans refused because they were in the midst of a religious festival. So 10,000 Athenians were left alone, except for 1000 Plataeans, to face a force of about 20,000 Persians at Marathon.

But a strange quirk of fate had removed one of the Greeks' most formidable obstacles. Thanks to some Ionians in the Persian camp, the Athenian general Miltiades (550-489 B.C.) got the information that the Persians were in the habit of grazing and watering their horses at night by the springs at the far end of the Marathon plain. For the time being, then, the dreaded

---

[1]Contemporary figures for the size of armies in ancient and medieval times are usually unreliable. While it is impossible to be certain, the figures given in the following pages are reasonable modern estimates.

The Greek hoplite of about 500 B.C. (1), who ordinarily came from the third property-owning class, was the elite fighting man of Greece. Here he wears body armour called a *cuirass*, made of two sections shaped to the body which were laced together at the sides and connected over the shoulders by curved metal plates. Beneath the cuirass a short skirt was usually worn. *Greaves* (armour tailored to the legs) needed no straps to be held in place and extended high enough to protect the knees. The *peltasts* or light-armed skirmishers (2) were equipped with slings as well as javelins, and had smaller shields, leather instead of metal armour, and less elaborate helmets. Greek armour was generally made of bronze, although iron had been in use for some time.
It is more difficult to portray a typical Persian soldier (3) because of the great variations in the composition of Persian armies. Scale armour, which was used in the East at this time, was sometimes worn, and the favoured weapon was a short bow, quite unlike the later long-bow of medieval times.
The famous Greek phalanx (4) was a tactical formation of infantry perfected by Philip of Macedon and Alexander the Great. The hoplites stood in solid lines up to 16 ranks deep with shields close or overlapping, a formation virtually invincible when supplemented by cavalry. The Greeks wore a variety of helmets. A common type was a kind of bronze bucket (5) with eyeholes and a nose guard. The helmet had a high crest of widely varying shape, which was probably inconvenient to the wearer but served the double function of terrifying the enemy and providing a rallying point for friends. The Corinthian type (6) had a visor projecting in front which was pulled down to protect the face in battle. Like most Greek objects, Greek swords (7) were extremely beautiful in design, with both edges curved and usable. They were usually bronze. Which of these armours would you consider most battle-proof and why?

Persian cavalry was absent. The Greeks had to act quickly. At dawn the hoplites advanced across the plain, breaking into a run as they came within range of enemy arrows. Then, letting his centre ranks sag inwards to engage the bulk of the Persian host, Miltiades wheeled his heavily reinforced wings to envelop the enemy from the rear. The Persians were thrown into confusion. The superior defensive armour, longer spears, and better discipline of the Greek infantry had won the day.

The brilliant victory at Marathon is forever the glory of Athens. Leaving behind their 6400 dead (the Greeks lost 192), the Persians fled to their ships. The Spartans did not arrive until two days after the battle, and then brought only 2000 men, their tardy arrival with this comparatively small force having more to do with fear of a helot revolt at home than with a religious festival.

The Persian fleet sailed round the coast hoping to catch Athens undefended, but Miltiades marched his Athenians back to the city quickly enough to face the Persians again. They retreated to Asia without giving battle.

The "miracle of Marathon", as it has been called, had a threefold significance. First, it keyed the Athenians up to a hitherto unimaginable patriotic pitch. "They were the first of the Greeks," wrote Herodotus, "who dared to look upon the Median garb, and to face men clad in that fashion. Until this time the very name of the Medes had been a terror to the Greeks to hear." Second, Sparta had been shown that it was possible for Greeks to defeat Persians. Third and most important, other Greek city-states were inspired and emboldened to resist the future Persian invasions that were bound to come.

A Fight to the Death The Greeks were fortunate in having a decade to prepare for the next Persian onslaught. During these ten years the democrat Themistocles rose to power. He was a man of strong and uncompromising opinions who did not hesitate to remove his rivals by the device of ostracism. Above all, Themistocles had an extraordinary ability to analyze a complex situation and then to deal with it by rapid action. It was providential for Athens that she had such a leader at this stage of her history. When a fellow democrat named Aristides proposed that a new rich vein of silver which had been discovered in the mines at Laurium be divided as spoil among the citizens, Themistocles insisted instead that the new wealth be used to build a strong fleet of at least two hundred ships. In the end Aristides was ostracized and the Athenian fleet was made ready to give battle to the Persians. Henceforth it was to be the bulwark of the Athenian empire.

The Second Persian Invasion The Persian expedition of 480 B.C. was organized on a vast scale with 100,000-150,000 men and 500 warships under the new king, Xerxes (485-465 B.C.). The Peloponnesians, headed by Sparta, turned to Athens and her navy. In order to prevent the Greek fleet from being cut off from its land bases it was decided to oppose the Persian army at the narrow pass of Thermopylae. A small force of about 7300 manned the pass, but they were betrayed when a traitor led the Persians by a secret mountain road to attack the Greeks from the rear. The 300 Spartans and 700 Thespians who stayed behind to hold the pass while the main Greek force retreated in the hope of gaining time for a decisive meeting to the fleets were slaughtered to a man.

Athens had to be evacuated, and when the Persians arrived they sacked and burned the acropolis. But though Xerxes had taken a city he had not won a war. There still remained the Greek navy, and once more Themistocles had his way. Let the Greek ships, he argued, be stationed in the bay of the island of Salamis, there to await the Persian fleet in a narrow strait four miles long and one mile wide. Then he took further steps to insure the success of his plan. He saw to it that false information reached the Persians to the effect that the Greeks were about to slip out of Salamis bay. The Persians reacted precisely as Themistocles had intended. Determined to keep the Greek ships bottled up, the Persian fleet rowed all night to meet them.

At daybreak one September day in 480 B.C. the two fleets clashed. The confident Xerxes had his golden throne set up on a headland, from which vantage point he could watch in ease and elegance while his fleet destroyed these Greek upstarts once and for all. But here, in the words of Herodotus, is the scene that met his eyes:

Salamis

Far the greater number of the Persian ships engaged in this battle were disabled.... For as the Greeks fought in order and kept their line, while the barbarians were in confusion and had no plan in anything that they did, the issue of the battle could scarce be other than it was. Yet the Persians fought far more bravely here than at Euboea, and indeed surpassed themselves; each did his utmost through fear of Xerxes, for each thought that the king's eye was upon himself.

There fell in this combat Ariabignes, one of the chief commanders of the fleet, who was son of Darius and brother of Xerxes, and with him perished a vast number of men of high repute, Persians, Medes, and allies. Of the Greeks there died only a few; for, as they were able to swim, all those that were not slain outright by the enemy escaped from the sinking vessels and swam across to Salamis. But on the side of the barbarians more perished by drowning than in any other way, since they did not know how to swim. The great destruction took place when the ships which had been first engaged began to fly; for they who were stationed in the rear, anxious to display their valour before the eyes of the king, made every effort to force their way to the front, and thus became entangled with such of their own vessels as were retreating.

The King of Kings had seen enough. His glorious armada, unable to manoeuvre in the narrow strait, had been fatally crippled. Its shattered remnants were ordered to sail to the Hellespont, where Xerxes met them with the larger part of his army, which had been evacuated after a gruelling forty-five day march from Thessaly.

But the Greeks dared not relax their vigilance. The best cavalry and infantry of Persia had been left in Greece for the winter, and all could yet be lost for the allies. The next August they took a solemn oath: "I shall fight to the death, I shall put freedom before life, I shall not desert colonel or captain alive or dead, I shall carry out the general's commands, and I shall bury my comrades-in-arms where they fall and leave none unburied." For three weeks the Persian cavalry harried the Greeks on the plain outside Plataea; but they were not to be provoked into fighting before they were ready. When at last the battle was joined in September of 479 B.C., 60,000 Greeks defeated 80,000 Persians, and the Spartan hoplites proved themselves to be the finest troops in the civilized world.

The same day, according to tradition, the Greek navy caught and burned the remainder of the Persian fleet, which was beached at Mycale in Asia Minor. Complete control of the Aegean was thereby won. Meanwhile the West Greeks of Sicily had defeated the Carthaginians. Greek trade and Greek ideas were now free to spread in both the eastern and western Mediterranean.

By 479 B.C. the Greeks were supreme. A group of tiny city-states had defeated the mighty Persian empire. Looking back some twenty-three centuries later it still seems miraculous, and all the sympathies of the West are with the Greeks. We seldom take the trouble to see the struggle from the Eastern viewpoint:

> Greek poets, storytellers and historians covered the battlefields of victory with honour and glory; those who laid their lives thereon were raised to the rank of national and immortal heroes. The issue, according to Greek interpretation, was between the preservation of liberty and subjection to Oriental despotism. Persian culture then was no lower than Greek culture, and Persian religion was certainly higher than Greek religion. To Darius and Xerxes the Athenians and Spartans represented a nation of barbarians and sea pirates, a source of constant trouble to the Persian coastal domain of Asia Minor.[2]

Yet it must be remembered that the Greeks were fighting for their very homes.

> ... the real significance of the Greek victories in this great decade is to be found not so much in the field of politics as in the domain of spirit. A tiny people had defeated a great empire. Something spiritual had, by the help of the favouring gods, vanquished wealth, numbers, material strength. Insolence had been curbed; the pride of power had received a fall. The goddess Athena had protected her chosen people in the hour of need. The exaltation which ensued bred great designs and a body of achievement in literature and art so astonishing in its beauty, its variety, and the permanence of its human appeal, that of all the elements which have entered into the education of European man, this perhaps has done most for the liberation of thought and the refinement of taste.[3]

[2]Philip K. Hitti, *The Near East in History* (D. Van Nostrand, 1961), p. 53.
[3]H. A. L. Fisher, *A History of Europe* (Edward Arnold, 1936), p. 29.

The Persians had done for the Greeks what the Greeks could not do for themselves. They had immeasurably strengthened the sense of unity of all Greeks as opposed to all "barbarians" (a term which simply meant non-Greeks). Athens came out of the wars the leading power in Greece, and her navy was henceforth her wall of defence and source of power—two factors which, combined with the increased political importance of the Athenian generals, were to make her the head of a large and unwilling empire.

2/ATHENS BECOMES IMPERIALISTIC

Although the Persians were gone from mainland Greece by 479 B.C., it was only natural for the Greek confederacy to continue. There was the danger of further Persian aggression, and the Greek colonies in Asia Minor had still to be liberated. Yet it did not take long, once outside pressure was removed, for Athens and Sparta to fall out. Sparta had long been the leader of Greece, a leadership which the Athenians acknowledged. But now that the enemy had been driven out, Sparta made a disconcerting proposal: Athens, she said, should not rebuild her ruins; they should be left as a perpetual reminder of what Persia had done. Immediately Athens' suspicions were aroused. If she did not rebuild her city—and this included her walls—she would be at the mercy of any attacker, even of Sparta. There was only one thing to do.

Themistocles went to Sparta to discuss the whole matter. He assured the Spartans that Athens would not refortify; but even as he parleyed the walls of Athens were rising again. Spartan observers were sent to Athens, only to be detained there—on the instructions of Themistocles. Then, when the word came to the wily Athenian leader that the walls were high enough to be defensible, he admitted all to the Spartans. Sparta had been outfoxed. She could only accept what had already been done.

Next, trouble arose involving the commander of the allied expeditionary force in Asia Minor. Throughout the war, military leadership had been given almost as a matter of course to the Spartans. But when the Spartan commander of the Greek force that captured Byzantium began to intrigue with the Persians to become despot there, he was recalled by his home government. The Greek contingents had already been antagonized by the overbearing conduct of the Spartan generals, and now they indicated that they preferred Athenian leadership. Sparta withdrew her troops and commanders from Asia Minor and sulked in the Peloponnesus, while Athens accepted the bid for leadership with alacrity. After all, she was the great sea power among the states, and only a sea power could command the coasts and islands of the Aegean.

The inevitable result of Sparta's withdrawal into isolation was the formation of a formal organization without her, and under Athenian leadership.

The Confederacy of Delos

THE CITY-STATES OF GREECE AND ASIA MINOR

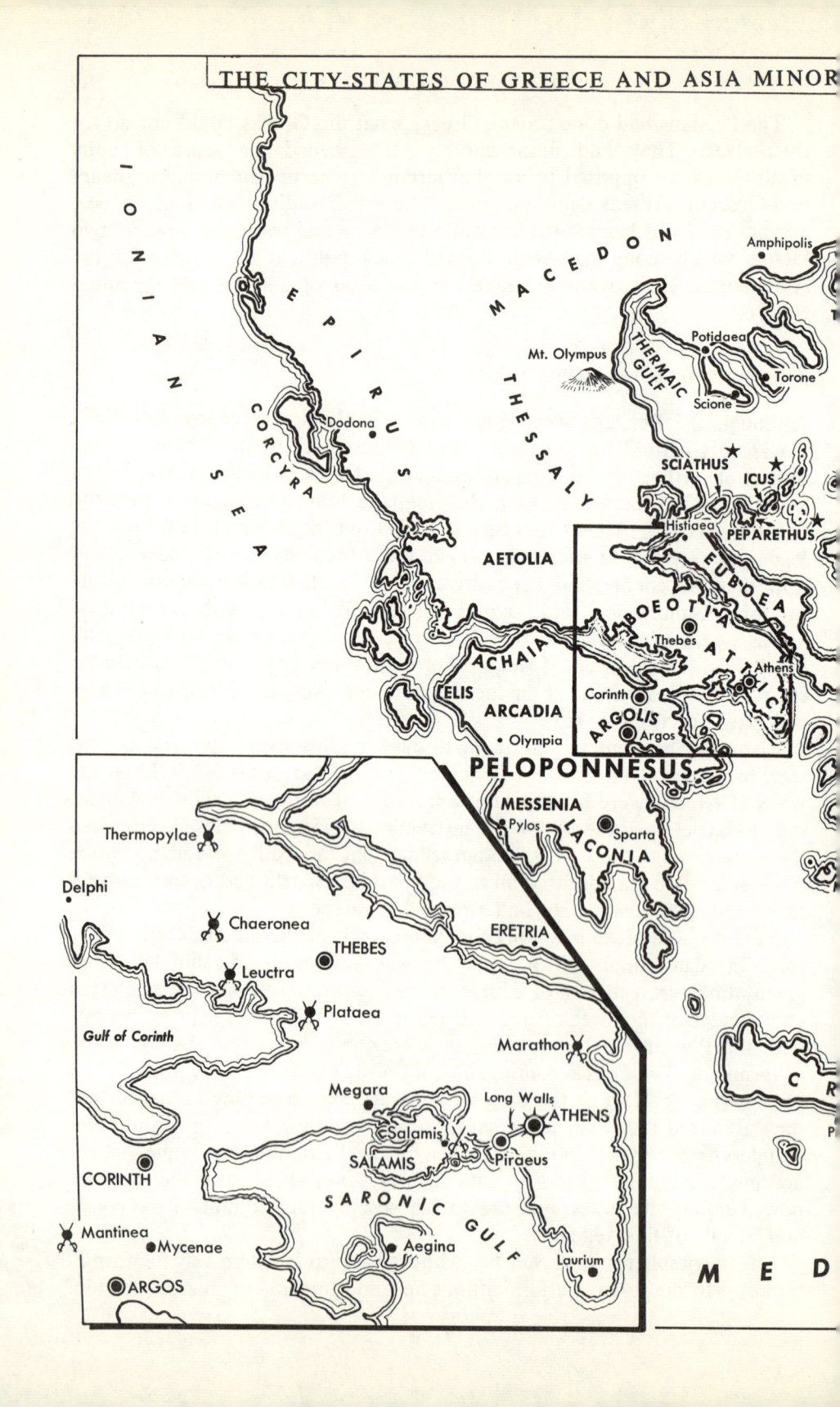

BLACK SEA

Bosporus

Byzantium

BITHYNIA

THRACE

Hellespont

Ilium (Troy)
TENEDOS

AEOLIS

PHRYGIA

MYSIA

Mytilene

LESBOS

Arginusae

ASIA

CHIOS

LYDIA

IONIA

Ephesus

SAMOS

Miletus

MINOR

CARIA

NAXOS

Cos

DORIS

LYCIA

RHODES

★ Principal Members of the Delian Confederacy

N

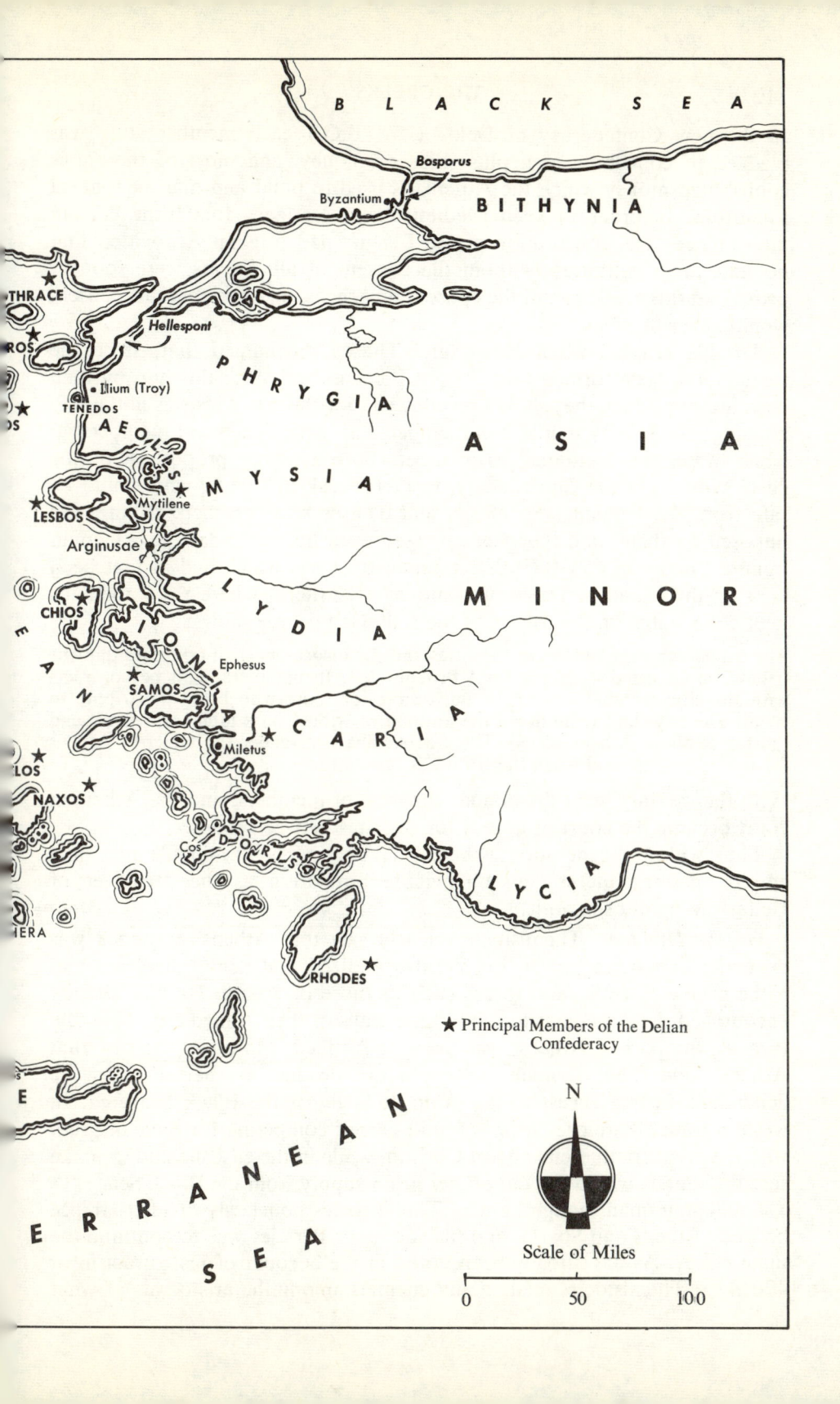

Scale of Miles

0   50   100

MEDITERRANEAN SEA

In the new Confederacy of Delos (478 B.C.) each member state was pledged to contribute men plus ships or money, and most of the states contributed money which the Athenians used to build and man new naval squadrons, in order to keep Mediterranean sea-lanes free from Persian interference. The constitution of the League had originally provided that no state might withdraw without the consent of all. Events were soon to prove that this really meant the consent of Athens, and also that this consent would never be given.

Trouble erupted when Naxos and Thasos proclaimed that, since the League had been formed to drive out Persians and since this had by then been accomplished, they should be allowed to withdraw. This was not agreeable to Athens. Why should non-members enjoy the benefits of membership while those who continued as members bore a higher proportion of the cost? Would not the Confederacy, in fact, break up? Then who would be safe from the Persians? So Naxos and Thasos were crushed, tribute was imposed on them, and they were pressed even harder under the Athenian thumb. Finally in 454 B.C. the common treasury, which had at first been kept on the island of Delos, was moved to Athens, where the Athenians kept close watch on the tribute as the following decree shows:

The council and the officers in the cities and the inspectors shall take care that the tribute be collected each year and brought to Athens; there shall be contracts with the cities so that it shall be impossible for those who bring the tribute to cheat. The city shall write down the amount of tribute on a tablet and shall send the tablet off to Athens sealed. The carriers shall surrender the tablet in Athens for the council to read when the tribute is brought in. . . .

With the treasury went the annual congress of members, and the Athenian court became the court of appeal for disputes between states.

Thus within a generation Athens had built up a formidable maritime empire, gaining thereby an unenviable reputation as the "enslaver of Hellas". Why did she do it?

Greece Divided Themistocles clearly saw that Athens' existence was dependent upon sea power, but he also realized that her naval bases had to be protected from land forces such as those of Sparta. He had already encouraged the Athenians to rebuild the walls of their ruined city. Now the Piraeus, the port of Athens, was likewise fortified. Moreover, he saw that Athens' food supply, which she could not produce by herself, must be dependent on free access to the grain of Sicily or the Black Sea area. In western trade, therefore, Athens found herself competing for Sicilian grain with the great commercial city of Corinth, while in the east she had to make sure that Persia would not cut off her grain supply from the Black Sea.

Great statesman though he was, Themistocles hopelessly divided Greece between Athens and Sparta, and his successor Pericles was to continue the same policy. As has already been noted in the account of his ostracism of 470 B.C., Themistocles made many enemies among the aristocrats of what may be called the conservative party, with the result that Athens was torn between conservatives and democrats for the rest of the century. The democrats usually moved in the direction of liberalizing the government, passing measures that reduced the power of the Areopagus by taking away all its administrative and judicial functions, or establishing regular pay for jurymen in the Athenian popular courts. Conservatives, on the other hand, felt they must support oligarchical factions in other city-states. (In 462 B.C., for example, the Spartans had been aided by Athenian troops after an earthquake and a helot rising. Cimon, the conservative leader in Athens, won this intervention, though the democrats in Athens may well have wished that the Spartans would destroy themselves.)

It is just possible, too, that if the Delian Confederacy had extended Athenian citizenship to all its members it would have been mainly oligarchs who would have travelled to the congress at Athens, and the Athenian democrats would have been swamped. As it was, the unpropertied Athenians controlled the Delian congress.

Athens' Greatest Citizen   After Cimon came Pericles, a democrat who broke openly with Sparta and who was to dominate Athenian politics almost continuously for over thirty years. Since he was the son of a niece of Cleisthenes, he came by his democratic sentiments honestly. In 450 B.C. he succeeded in passing a measure for the payment of jurors—who, it has been calculated, were employed an average of at least 300 days a year—at the rate of two obols (about 6 to 10 cents) a day. By 430 B.C. all the magistrates and members of the Boulê were also to be paid, as well as soldiers and sailors on active service. In addition, Pericles protected the four miles of road between the city of Athens and its seaport, Piraeus, by Long Walls two hundred yards apart. Hence there were no longer two separate cities to defend.

Pericles

Thus it seems likely that under Pericles over a third of the Athenian adult male citizens depended on the empire for a livelihood. Of 43,000 male Athenian citizens perhaps 17,000 found employment in the jury courts and the fleets, both of which stemmed from the empire. Add to this the necessity of maintaining the fleets to guarantee the Athenian food supply and we have reason enough for Pericles to be a convinced imperialist.

Whereas under Cleisthenes a third of the citizen body had been urban dwellers, under Pericles at least a half lived in the city. It was natural for the propertied classes of Athens to resent their loss of political influence to the masses in the city. It was natural for the propertied classes of rebellious city-states to resent the colonists whom Pericles sent to take over their confiscated land. It was natural for allied states to resent their tribute being used for the beautification of the Athenian acropolis. And it was certainly natural for Sparta to look with increasing fear on the ever-growing power of Athens.

Yet when all that is said, it was equally natural for Athens to take the means she did to guarantee her food supply, to support a substantial section of her poorer citizens, and to ensure the political leadership of the democrats. It does not seem likely that Pericles pursued his policy of imperialism merely as the only means he could think of to maintain himself in power. If, as seems probable, he was convinced that only by such a policy could the best interests of Athens be served, the final tragedy is that he led his state into a world war which destroyed Greek democracy.

3/THE GREEK WORLD WAR

The Great Peloponnesian War broke out in 431 B.C., and continued, off and on, until 404 B.C. This twenty-seven years' war has been given a remarkable emphasis in history books—but not because of the scale of military operations, the number of casualties, or the tales of individual heroism. By a strange quirk of fate one of the Athenian admirals was a man named Thucydides (460-400 B.C.), who was alleged during the war to have cared more for the protection of his own property than for the city that he was supposed to be guarding. The charge produced his banishment—and the leisure necessary to compose his *History* of the Peloponnesian War.

It is a remarkable narrative, composed, according to its author, with a set purpose: "The absence of everything mythical from my work perhaps will make it less agreeable to those who hear it read. I shall be contented, however, if it appears useful to those who wish to have a clear idea of the past and hence of the conditions and events which, according to the course of human affairs, will be repeated." How modern this ancient historian sounds!

Even though he did not finish his book (it goes only to 411 B.C.), Thucydides made the war forever famous. But he did something more. He wrote of the breakdown of his own world in such a way that even in the 20th century we feel, on reading his pages, that we have experienced all this before. As the noted historian Arnold Toynbee expressed it, World War I "opened my eyes to the historical and at the same time philosophic truth that my world in my generation was entering upon experiences which Thucydides, in his world in his generation, had already registered and recorded." After World War II the parallel is even more apt.

Might Becomes Right   The events of the war may be passed over briefly. Of far more interest and importance are those elements which the Peloponnesian War holds in common with the two World Wars of our own century.

Embittered by her trade rivalry with Athens, Corinth appealed to Sparta and her allies to join in an attack on her rich and powerful competitor. A series of incidents involving Corinthian colonies produced a "cold war" mentality, until finally Athens threw down the gauntlet by demanding that the old Corinthian colony of Potidaea cease receiving Corinthian magistrates. Forthwith Potidaea secured a promise that Sparta would invade

In recent years the Athenian Agora, the market-place at the foot of the Acropolis, has been excavated and reconstructed by the American School of Classical Studies at Athens. This stoa (porch), built by Attalus II of Pergamum (159-138 B.C.), was a sort of colonnaded shopping plaza and is an excellent example of the way in which Hellenistic princes in the days after Alexander the Great took pride in embellishing the cultural capital of the world. Incidentally, the marble for this modern reconstruction was taken from the same quarry used by King Attalus in the middle of the 2nd century B.C.

Attica if Athens attacked Potidaea. But if the Spartans were to cross the Corinthian isthmus for an invasion of Attica, they would have to pass through the city-state of Megara. Pericles acted decisively. He blockaded Potidaea, and then, without waiting for a Spartan attack, placed an embargo on Megara's trade with all ports and markets of the Athenian empire.

Without such trade Megara faced starvation—as, indeed, would any other state whose grain imports Athens might later decide to cut off. Athens had taken one more step to dominate the Isthmus of Corinth. She had shown the other states that she could close her fist on their trade if they did not please her, and already she was allied with Plataea, who controlled the best pass on the north-south road between Boeotia and the Peloponnesian League. If Athens' enemy Thebes wanted to keep in communication with her allies in the Peloponnesus, she would have to command that pass.

It was almost inevitable that a nervous Thebes, fearing Athens' stranglehold on the isthmus, should inch to the brink of war. One rainy spring night in 431 B.C. three hundred Thebans attacked Plataea. The Greek world was plunged into a generation of war.

RIVALS IN THE GREEK WORLD WAR	
ATHENS	SPARTA
—a sea power —imaginative in strategy —aided by a widely scattered and discontented empire of 300 city-states —500 triremes —30,000 hoplites	—a land power —unimaginative in strategy —aided by a compact, loyal League, and by the Persians —200 triremes —100,000 hoplites

Land warfare for the Greeks often consisted of a series of summer raids. In the winter everyone went home until the next spring. When the Plataeans beat off the Theban attack of 431 B.C., the Spartans invaded Attica and the Athenians retired inside their walls to watch their crops and homes burn—not an easy thing to do.

Their land was being laid waste in front of their very eyes—a thing that the young men had never seen happen and that the old men had seen only at the time of the Persian invasion. Naturally enough, therefore, they felt outraged by this and wanted, especially the young, to march out and stop it. There were constant discussions with violent feelings on both sides, some demanding that they should be led out to battle, and a certain number resisting the demand. . . . Thus the city was in a thoroughly excited state: they were furious with Pericles and paid no attention at all to the advice which he had given them previously; instead they abused him for being a general and not leading them out to battle, and put on him the whole responsibility for what they were suffering themselves.

However, Pericles managed to hold the Athenians in check, and at the close of the campaigning in 430 B.C. inspired them to new military efforts by his famous *Funeral Oration* (selections from which may be read in the Source Reading on pages 126-128). The speech, as reconstructed by Thucydides, is patriotic propaganda of a high order, and with its Churchillian ring it is the best single statement ever made of the Athenian way of life. The next year an epidemic of typhus ravaged the city and killed the great statesman.

The Athenian war strategy was by now clear: it was to stay inside the walls, but to break down Peloponnesian morale by coastal raids and cut off their food supplies by a naval blockade. Also it was necessary for Athens to maintain naval supremacy in the Corinthian Gulf, and in particular to continue her alliance with the city-state of Corcyra, since Corcyra could control the Italian and Sicilian grain trade routes. The Peloponnesian strategy, on the other hand, was to invade and devastate Attica, to keep open the western trade route to Sicily, and to preserve communications with north-west Greece so that Corcyra and Athens might be attacked from the rear. The war soon turned into an endurance contest, with neither side strong enough to deliver the knockout blow.

The conflict dragged on, until after four years all sense of right and wrong had been abandoned. Might was right. When the people of Mytilene inspired a revolt in the island of Lesbos (one of the charter members of the Delian Confederacy), they asked for Spartan aid. But before the aid arrived, Athens blockaded the island and starved Mytilene into submission. The whole incident badly unnerved the Athenians. How, they debated, should they punish these Mytilenian upstarts? In the assembly at Athens a rabble-rousing orator named Cleon was among the candidates aspiring to fill the position left vacant by Pericles' death. Slapping his thigh, disarranging his clothing, running back and forth, Cleon delivered his harangue. With his shrill voice squealing like a "singed sow's", he exhorted Athens to execute all the men of Mytilene and enslave the women and children. His motion was carried in the assembly, and a trireme bearing the order was dispatched to Mytilene.

Cleon

Fortunately sober second thoughts soon took over. The very next day the assembly met and rescinded its brutal decree, despite Cleon's even blunter speech (see page 129) on this occasion. A second trireme sped across the Aegean, the crew eating as they rowed and sleeping in relays. The people of Mytilene were lucky: the trireme was in time. Mytilene was spared.

On the other side of Greece, at Corcyra in the Ionian Sea, the people were not so lucky. When a revolution broke out in Corcyra in 427 B.C., the horrible massacre of oligarchs by democrats was given the moral support of an Athenian fleet standing by. Thucydides follows his graphic description of the slaughter by a comment on the morally degrading effects of warfare:

... What used to be described as a thoughtless act of aggression was now regarded as the courage one would expect to find in a party member; to think of the future and wait was merely another way of saying one was a coward; any idea of moderation was just an attempt to disguise one's unmanly character; ability to understand a question from all sides meant that one was totally unfitted for action. Fanatical enthusiasm was the mark of a real man, and to plot against an enemy behind his back was perfectly legitimate self-defence. Anyone who held violent opinions could always be trusted, and anyone who objected to them became a suspect. To plot successfully was a sign of intelligence, but it was still cleverer to see that a plot was hatching. ... In short, it was equally praiseworthy to get one's blow in first against someone who was going to do wrong, and to denounce someone who had no intention of doing any wrong at all. Family relations were a weaker tie than party membership, since party members were more ready to go to any extreme for any reason whatever. These parties were not formed to enjoy the benefits of the established laws, but to acquire power by overthrowing the existing regime.

... Thus neither side had any use for conscientious motives; more interest was shown in those who could produce attractive arguments to justify some disgraceful action. As for the citizens who held moderate views, they were destroyed by both the extreme parties, either for not taking part in the struggle or in envy at the possibility that they might survive.

About 500 feet above the city of Athens rises the Acropolis, covering two modern city blocks. Originally the Acropolis was the city itself, but later it became the dwelling-place of the goddess Athena and the site of a magnificent group of buildings. A channel or causeway was left through the steps of the *Propylaea* (1), the entrance to the Acropolis, to admit animals being led to sacrifice. In the north-west wing of the Propylaea was the *Picture Gallery* (2), and to the right, with Ionic pillars, stood the charming little *Temple of Athena Nike*, the Winged Victory (3). The *Erechtheum* (4) was designed to replace certain other sanctuaries and was completed in 407 B.C. Because it had to serve several functions, the Erechtheum was somewhat more complex and less symmetrical than most of the other structures. Attached on its left was the north portico, and on the opposite side was the Porch of the Maidens, in which figures of six beautiful maidens served as columns for the roof. Inside the famous *Parthenon* (5) stood the great 40-foot gold and ivory covered figure of Athena by Phidias, while the sculptures of the gods in the pediment over the columns remain a majestic testimonial to the genius of Phidias and his assistants. In the open area behind the Propylaea towers the gigantic bronze statue of Athena, 70 feet high and visible far out to sea. The Parthenon remained intact until A.D. 1687, when a powder magazine located in the building blew up.

Over the next few years Athens pursued the war vigorously. On the home front Cleon increased the pay of jurymen to three obols a day, and at the same time doubled—even tripled—the amount of tribute owed by allied states to Athens. Had he not been killed in an attack on Amphipolis in 422 B.C., the war might have continued even longer under his aggressive direction. As it was, Sparta had for some time been sending out peace feelers, and in 421 B.C., after ten futile years of war, the two sides drew up a fifty-years' peace treaty.

However Sparta's allies, Corinth and Thebes, refused to sign, and by 418 B.C. an Athenian war party led by Alcibiades was thirsting for war once more. This Alcibiades, who had been a ward of Pericles and now filled the gap left by Cleon's death, was an attractive, brilliant, yet thoroughly unreliable commander. The Athenians were to rue the day they ever elected him general.

Neutrality was as difficult to maintain in the Greek world of that time as it is in our world of the 20th century. To Athens, those who were not for her were against her; and it was in this manner that she reasoned in 416 B.C. to the little state of Melos, a small island that had remained outside her empire. Thucydides, who was no democrat, puts the matter in its most cynical light when he reports the Athenians' callous statement to the Melians: "The powerful exact what they may, and the weak grant what they must." His desire to blacken the record of Athenian democracy prevents his also telling us that the Melians had previously subscribed to the Spartan war fund and had sheltered the Spartan fleet ten years earlier. But even reaching between the lines, even taking into account prior Melian sympathies, the awful Athenian punishment exacted still horrifies us. The men were executed, and the women and children sold into slavery.

Doomed to Disunity Still the war dragged on. In 415 B.C., at the instigation of Alcibiades, the Athenian fleet set sail to support weaker Greek states against Corinth's Sicilian colony, Syracuse. In the next two years another expedition came to relieve the first. Both were lost. Thucydides says simply, "Fleet and army perished from the face of the earth; nothing was saved, and of the many who went forth, few returned home."

Although henceforth the Athenian Empire could only go into a long decline, by a tremendous burst of energy Athens managed to construct a new fleet of triremes within a year. By now Persia had allied herself with Sparta and had returned to the Aegean, and the two proved more than Athens could cope with. When their fleet was trapped in the harbour of Mytilene the Athenians desperately built and manned another, their fourth in less than ten years. A brilliant victory was won at Arginusae, but many crews were lost when the fleet was lashed by a storm, and the democratic generals who returned to Athens were executed for negligence—an insane act on the part of the assembly. Still Athens refused to make peace with Sparta. Finally the last Athenian fleet was caught in the Hellespont and annihilated. When the news of this defeat reached Athens

a wail of lamentation arose from the Piraeus, sweeping up the space between the Long Walls to the city, as each passed the news on to his neighbour. That night not a soul slept, as they mourned not alone for the dead, but still more for themselves, imagining that they were about to suffer such ills as they had imposed on the Melians . . . or on the citizens of Histiaea or Scione or Torone or Aegina, and on many other Hellenes.

The end was not far off. After six months' blockade, with all food gone Athens surrendered—twenty-seven years after the war had begun. The shortest-lived (478-404 B.C.) and the smallest of ancient empires had collapsed. The Athenian attempt to unify Greece had failed once and for all.

With this failure came also the failure of the city-state, the polis, and the temporary eclipse of Athenian democracy, "the rule of the poor majority over the rich minority in their own interest," as the Greek philosophers liked to define it. Culturally some of Athens' greatest glories lay ahead of her, but as a political force she was finished. Hellenism was not to be spread by the city-state. The Greek city-states of the 5th century B.C. were doomed to disunity: they could never solve their political problem.

In the story of the Greek failure to conceive of a workable political unit larger than the polis, the historian cannot help wondering what is the modern moral of ancient Greece. Perhaps it is simply this. Unless we solve the political problems of our own nation states we may not have any other problems to solve. We may tear each other apart as surely as (and more quickly than) the Greek city-states of 2400 years ago.

4/DEATH OF THE CITY-STATE

It remained but to bury the corpse. The Spartans had no long-range plans for Greece. The Long Walls of Athens were demolished, though Sparta resisted the demands of Corinth and Thebes to wipe out the great city itself. In Athens, as in many other Greek cities, the Spartans backed the oligarchical parties. Since Sparta knew no means but force to rule men, the reactionary and brutal regime of the "Thirty Tyrants" in Athens was supported by a Spartan garrison. So ferocious, in fact, was the misrule of the Thirty that within months they were overthrown and the democracy restored.

Sparta loses the peace

The reassertion of the old Persian claims to the Ionian cities brought on a war between Sparta and Persia which amounted to a series of Spartan forays against the mainland Persians. Now Persian money, which had been used to back Sparta in the Peloponnesian War, bought Greek cities to oppose Sparta. The so-called Corinthian War lasted from 395 to 387 B.C., until a coalition of Corinth, Argos, Thebes, and Athens succeeded in at last throwing off the Spartan yoke.

Sparta actually profited by the Persian King's Peace of 386 B.C., for by it she gained Persian support. She resumed her old arrogance, and proceeded to attack her Hellenic enemies one by one. In 383 B.C. she seized the citadel at Thebes and established the oligarchic party there, a conquest that was to be undone in one night a few years later when Theban exiles returned from Athens, assassinated the oligarchs, and set up a new democracy. Thebes defeated the Spartan army at Leuctra in 371 B.C. under the brilliant military leader Epaminondas. The massed weight of Theban spears punched right through the Spartan phalanx to end the Peloponnesian League, with the result that for nine years the Peloponnesus was dominated by a barren Theban supremacy bereft of any statesmanship. In the meantime, however, Athens had been building up an alliance to oppose Thebes, and at the Battle of Mantinea (362 B.C.) she reduced both Sparta and Thebes to the status of second-class powers.

It has been said that in the 4th century B.C. "the Greeks devoted their activities to art, science, and mutual extermination, in all of which they were unprecedentedly successful." What was needed was a strong man to end all bickering. That man was Philip of Macedon, who came on the scene to bring order out of chaos.

Philip of Macedon

Athens' Last Chance   Macedon in northern Greece was a wild, hilly country with a rude Homeric society ruled by a king. In 359 B.C., at the age of twenty-three, Philip II came to a throne hedged about with assassination. He was to be an able ruler—soldier, diplomat, and athlete all in one. In the early years of his reign the new Athenian Confederacy fell apart, and Philip was able to win one Greek city after another in the north. He had been a hostage in Thebes in his youth and had learned well the lesson of

Epaminondas: the Macedonian phalanx, armed with spears some 14 feet long, was invincible.

Philip soon succeeded in welding the unruly hill tribes together, and went on to pin down his conquests by building roads and fortresses. Even more important, he created the most magnificent professional army the world had yet seen. To the phalanx of "foot companions" were added archers, slingers, and light and heavy cavalry. His men were in constant training, and his system of rewards and promotions ensured their complete devotion to him.

Fully aware of the advantage he held because of the Greek cities' family squabbles, Philip managed to build a fleet and to replenish his treasury by seizing their northern gold mines. Athens, a prey alike to internal party strife and discontented allies, was in no condition to stand up to Philip while he consolidated his power. By 352 B.C. he was master of Thessaly. And still the Athenians lived on in a state of false security. Finally in 351 B.C. the Athenian orator Demosthenes sought in ringing words to open the eyes of his fellow citizens to the northern menace. Athens in the 340's, Europe in the 1930's, Asia and Europe in the 1960's—the parallels may not be too strained. Listen to Demosthenes:

> If any man supposes this to be peace, when Philip is conquering every one else and will ultimately attack you, he is mad. If we wait for him actually to declare war on us we are naïve indeed, for he would not do that even though he marched right into Attica, if we may judge from what he has done to others.

Athens made peace in 346 B.C., but Demosthenes regarded it as only a breathing spell. Philip was within a few hours' march of the city—protesting friendship, of course—and allied with Thebes. A few years more and Demosthenes' storm warnings were heeded. A hasty alliance was plastered together with Thebes, and in 338 B.C. an allied army faced Philip at Chaeronea. Though the armies each numbered approximately 30,000, the superior tactical skill of Philip won the day. Philip then marched on into the Peloponnesus and called a congress of Greek states at Corinth, in which each state was granted autonomy but was also enrolled in a Hellenic League which soon pledged itself to assist him in marching into Asia Minor against the Persians. Within a year and a half (336 B.C.), however, Philip was dead—cut down, like many a Macedonian before him, by the assassin's hand. His son Alexander, a boy of twenty, found himself king.

The Greek city-state had once abounded in political creativity. Now this power was not only dead. It was well and truly buried.

5/DEMOCRACY AND THE GREEKS

The classic defence of Athenian democracy is that which Thucydides puts into the mouth of Pericles in the *Funeral Oration*. Though the oration is a moving document, it must still be recognized as a wartime speech designed to whip up patriotic enthusiasm. This does not mean that it is untrue, but it may mean that it is a selection of facts designed to emphasize only the better side of Athenian life.

Pericles praises Athens because its government "is in the hands not of the few but of the many." Who were "the many"? The following table lists the main groups of the Athenian population.

ESTIMATED ATHENIAN POPULATION IN 431 B.C.		
Citizens		
	Adult Males	43,000
	Women and Children	129,000
		172,000
Metics		28,000
Slaves		110,000
Total		310,000

Only the adult male citizens, men eighteen years and over born of Athenian citizens, could be members of the assembly, and of these, the ones eighteen and nineteen years of age were usually absent on military service. *Metics* were aliens who had settled in Athens for business or professional reasons. They may have been like many Canadians living in the United States, who, not being citizens, have no vote, but otherwise enjoy all the benefits of their adopted country. In short, less than one-seventh of the total Athenian population was eligible to attend the assembly, and probably no more than about one-seventh of those (approximately 6000) actually attended at any one time.

This meant that on any single occasion at least one in every 50 Athenians was taking an active part in the assembly. When we consider that 265 members of parliament represent 19,000,000 Canadians, then on any single occasion only one in about every 72,000 Canadians ever takes part in our national government. By modern standards of representation, the Athenians certainly took an active part in politics.

What were the opportunities for political service in Athens?

Government by Amateurs There were several official bodies on which to serve. The Boulê was the committee that prepared legislation for the monthly meeting of the assembly. Any male citizen over thirty years of age was eligible for the Boulê, and fifty men from each of ten electoral districts were chosen yearly by lot. Each committee of fifty (*prytany*) remained in

THE MACHINERY OF GOVERNMENT IN ATHENS
(431 B.C.)

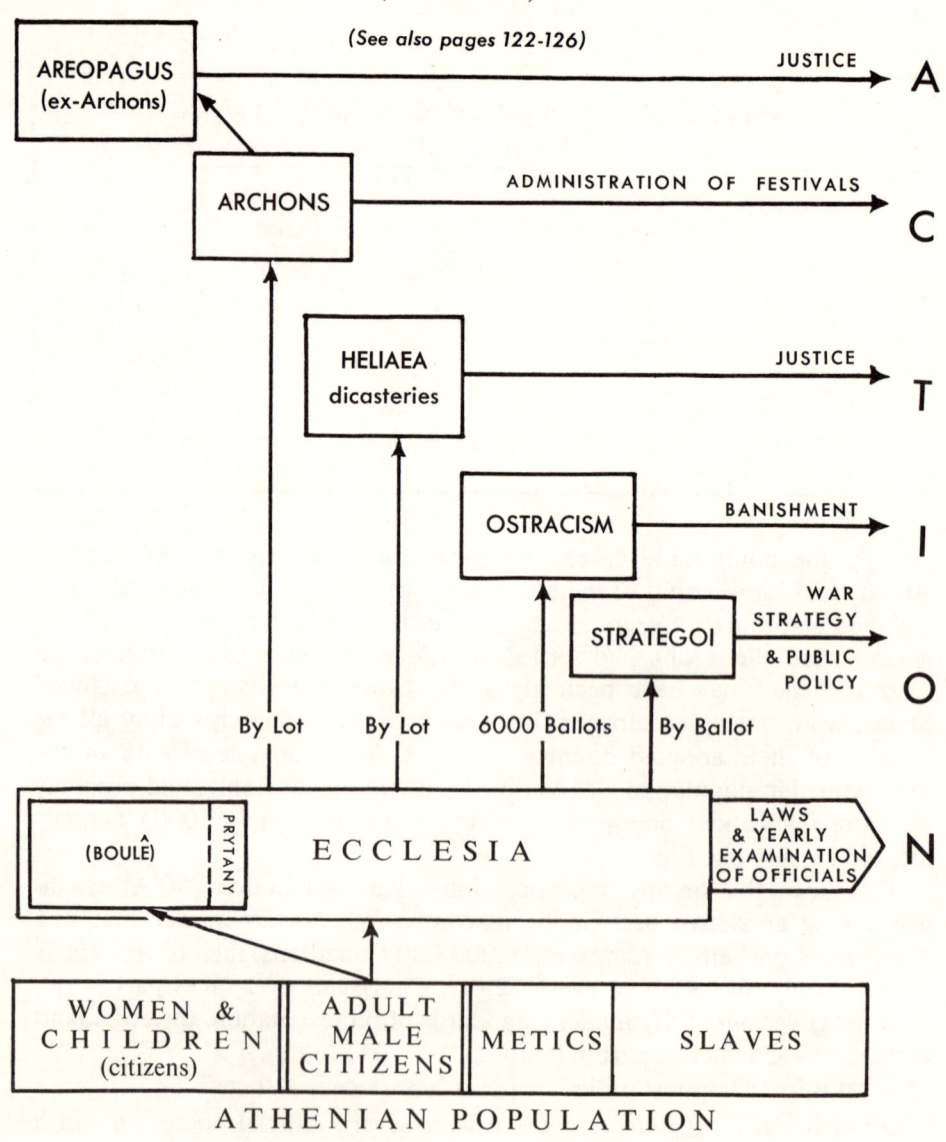

(See also pages 122-126)

AREOPAGUS (ex-Archons)	JUSTICE → A
ARCHONS	ADMINISTRATION OF FESTIVALS → C
HELIAEA dicasteries	JUSTICE → T
OSTRACISM	BANISHMENT → I
STRATEGOI	WAR STRATEGY & PUBLIC POLICY → O

By Lot — By Lot — 6000 Ballots — By Ballot

(BOULÊ) | PRYTANY | ECCLESIA — LAWS & YEARLY EXAMINATION OF OFFICIALS → N

ATHENIAN POPULATION

| WOMEN & CHILDREN (citizens) | ADULT MALE CITIZENS | METICS | SLAVES |

session a tenth of a year, and each day a different member of the prytany was chosen by lot to preside over the assembly and the whole Athenian government for that day.

Thus the assembly, which regularly had forty separate sessions a year, was under a different president each day it met. Every citizen who could command attention could address the assembly. (A leather merchant who criticized a military campaign found himself elected general and told to finish it off!) Moreover, since jurymen also came from the assembly there was ample opportunity to serve in the administration of justice. Amateurs ruled the politics of Athens, and probably every male Athenian citizen had some public service to his credit.

Though the assembly was supreme, any citizen could challenge the legality of any decree passed by it. If the decree was found to be unconstitutional, then the citizen who had originally proposed it might be fined, exiled, or even put to death. Likewise each outgoing magistrate had to submit an account of his official acts to the assembly, and could neither leave Athens nor sell his property before his accounting had been ratified. Demogogues might rant and the mob overrule; the wonder still is that the system worked so well.

The Cause of All Ills When Pericles proudly declared that Athens was an education to all Greece, there were those who thought it was the wrong sort of education. The Old Oligarch's pamphlet *On the Constitution of Athens*, part of which may be read on pages 128-129, provides a striking contrast to the Periclean eulogy. This caustic critic laid all the ills of Athens at the door of democracy. Nor was he alone in his opinion. The philosopher Plato, who in his youth had seen the fall of Athens, also bitterly condemned the "democratic man".

. . . he spends his life, every day indulging the desire that comes along; now he drinks deep and tootles on the pipes, then again he drinks water and goes in for slimming; at times it is bodily exercise, at times idleness and complete carelessness, sometimes he makes a show of studying philosophy. Often he appears in politics, and jumps up to say and do whatever comes into his head. Perhaps the fame of a military man makes him envious, and he tries that; or a lord of finance—there he is again. There is no discipline or necessity in his life; but he calls it delightful and free and full of blessings, and follows it all his days.

Granted that democracy had run to excesses, was it really to blame for all the ills of the Athenians? Thucydides and Plato and the Old Oligarch thought it was. Democrats, they said, had followed economic self-interest, had dabbled in city politics all over the Greek world, had slaughtered oligarchs to promote democrats, and had turned a free confederacy of states into an empire. But the Greek failure was more than the failure of a method of government, be it democracy or oligarchy. It was the failure of the city-state system.

The Greeks refused to accept the idea of a state in which the farthest point from the capital was more than a day's walk away. Any larger state could not be governed by a direct democracy; there would simply be too many citizens for a fair proportion of them to come together in one place. There ought, then, to be no larger political unit than the city-state.

The tragedy of the situation was that the city-states seemed to find it impossible to live at peace with each other. Set up to preserve certain values of civilized living, the city-state ended by destroying those very values. To reconcile the love of liberty with the hard political necessities of survival is never easy. The Greeks attempted it, and their story is at once an inspiration, a failure, and a warning.

6/SUMMARY AND CONCLUSION

For about the first third of the 5th century B.C. some of the Greek city-states were able to unite in the face of an external threat from the Persians. But when the Persian menace had passed, Athens did not allow the defensive Delian Confederacy to disband. Instead she became increasingly imperialistic until, during the second third of the century, the Peloponnesian League and Sparta's allies grew fearful of Athenian domination.

Then came internal upheaval. In the last third of the 5th century B.C. Athens and her allies fought against Sparta and her allies, and the long war ended in the utter defeat of Athens. Sparta, however, was not a good governor of Greece, and was overthrown by Thebes, who could not unite the Greeks either. Finally Philip of Macedon imposed from without the unity the Greeks could not create from within. The city-state had failed as a creative political force.

But this is not the last word on Greece. Although the political failure has been emphasized as a problem particularly relevant to the world of the 20th century, Greece's immortality lies elsewhere than in politics. For 2000 years men have admired the culture of Greece—her philosophy, science, art, and literature. Perhaps men's greatest intellectual accomplishments do not coincide with political superiority. The Greek spirit triumphed even in political adversity.

SOURCE READINGS

(a)

THUCYDIDES (460-400 B.C.) is sometimes called the first scientific historian, and it has been said that "he saw more truly, inquired more responsibly, and reported more faithfully than any other ancient historian." The extract below is from Pericles' *Funeral Oration*, the address delivered by the general at the state funeral held in 430 B.C. for the Athenian soldiers who had been killed in the first year of the Peloponnesian War. Thucydides explains that he and his informants found it difficult to recall the precise words of the speeches that he later recorded in his *History*, and that therefore they "are given here in the form in which I thought the persons concerned would most likely have said what was called for under the circumstances, while keeping as close as posssible to the general gist of what was actually said."

Our constitution is named a democracy, because it is in the hands not of the few but of the many. But our laws secure equal justice for all in their private disputes, and our public opinion welcomes and honours talent in every branch of achievement, not for any sectional reason but on grounds of excellence alone. And as we give free play to all in our public life, so we carry the same spirit into our daily relations with one another. . . .

Yet ours is no work-a-day city only. No other provides so many recreations for the spirit—contests and sacrifices all the year round, and beauty in our public buildings to cheer the heart and delight the eye day by day. Moreover, the city is so large and powerful that all the wealth of all the world flows in to her, so that our own Attic products seem no more homelike to us than the fruits of the labours of other nations.

Our military training too is different from our opponents'. The gates of our city are flung open to the world. We practise no periodical deportations, nor do we prevent our visitors from observing or discovering what an enemy might usefully apply to his own purposes. For our trust is not in the devices of material equipment, but in our own good spirits for battle.

So too with education. They toil from early boyhood in a laborious pursuit after courage, while we, free to live and wander as we please, march out none the less to face the self-same dangers. . . . Indeed, if we choose to face danger with an easy mind rather than after a rigorous training, and to trust rather in native manliness than in state-made courage, the advantage lies with us; for we are spared all the weariness of practising for future hardships, and when we find ourselves amongst them we are as brave as our plodding rivals. Here as elsewhere, then, the city sets an example which is deserving of admiration.

We are lovers of beauty without extravagance, and lovers of wisdom without unmanliness. Wealth to us is not mere material for vainglory but an opportunity for achievement; and poverty we think it no disgrace to acknowledge but a real degradation to make no effort to overcome. Our citizens attend both to public and private duties, and do not allow absorption in their own various affairs to interfere with their knowledge of the city's. We differ from other states in regarding the man who holds aloof from public life not as "quiet" but as useless; we decide or debate, carefully and in person, all matters of policy, holding, not that words and deeds go ill together, but that acts are foredoomed to failure when undertaken undiscussed. For we are noted for being at once most adventurous in action and most reflective beforehand. Other men are bold in ignorance, while reflection will stop their onset. But the bravest are surely those who have the clearest vision of what is before them, glory and danger alike, and yet notwithstanding go out to meet it. . . .

In a word I claim that our city as a whole is an education to Greece, and that her members yield to none, man by man, for independence of spirit, many-sidedness of attainment, and complete self-reliance in limbs and brain.

That this is no vainglorious phrase but actual fact the supremacy which our manners have won us itself bears testimony. No other city of the present day goes out to her ordeal greater than ever men dreamed; no other is so powerful that the invader feels no bitterness when he suffers at her hands, and her subjects no shame at the indignity of their dependence. Great indeed are the symbols and witnesses of our supremacy, at which posterity, as all mankind today, will be astonished. We need no Homer or other man of words to praise us; for such give pleasure for a moment, but the truth will put to shame their imaginings of our deeds. For our pioneers have forced a way into every sea and every land, establishing among all mankind, in punishment or beneficence, eternal memorials of their settlement.

Such then is the city, for whom, lest they should lose her, the men whom we celebrate died a soldier's death: and it is but natural that all of us, who survive them, should wish to spend ourselves in her service.

A. Zimmern, *The Greek Commonwealth* (Clarendon Press, 5th edition, revised, 1931), pp. 203-206, slightly adapted.

(b)

THE "OLD OLIGARCH" is the anonymous author of a pamphlet entitled *On the Constitution of Athens*. We know nothing more about him, though scholars think that he may have been an Athenian who wrote some time between 431 and 415 B.C. At any rate the Old Oligarch, whoever he was, wrote "the earliest known political treatise in any language."

... Everywhere the best people are opposed to democracy, because among the best element there is least excess and injustice, and most self-discipline to useful ends. Among the people, on the other hand, ignorance is at its height, as well as disorder and vulgarity. For poverty, lack of education, and in some cases the ignorance which arises from lack of money lead them more to unseemly conduct. ...

... if the decent people were to do the speaking and the deliberating, for those who were like themselves it would be advantageous, but not so for the popular party. But as it is, any scoundrel who pleases can get up, say his say, and get what is good for him and his like. It might well be asked, "What that is good for himself or for the state would such a man know?" But the people know that this man's ignorance, commonness, and good will profit them more than the virtue, wisdom, and disaffection of the conservative. Perhaps it is not as a result of such practices that a state becomes perfect, but this is the way democracy would be best preserved. For what the people want is not that they should be enslaved in a well-ordered city, but that they should be free and that they should rule; and bad government concerns them but little. ... But if you are looking for good government, first of all you will see to it that the shrewdest men impose the laws on the common people. ...

Speaking of slavery, at Athens the effrontery of slaves and resident aliens is at its height; to strike them is not allowed there, and yet a slave will not stand out of the road to let you pass. And this is why that is the custom there: if it were lawful for a slave to be struck by a free man, or for an alien or a freedman to be beaten, you would often think an Athenian a slave and beat *him*. For in dress the people there are no better than slaves or aliens, and in appearance they are no better either. And if there be any man who marvels at the further fact that they allow the privately owned slaves there to put on airs, and some of them to live downright magnificently, this too you will find they do with good reason. For wherever there is sea power, the slaves have to be *paid* for their slaving, and to be allowed to buy their freedom, in order that *we* may get our percentage of their hire. But in any city where the slaves are rich, it is no longer expedient for my slave to be afraid of you, whereas in Sparta my slave used really to respect you. But in Athens if your slave is afraid of me he will probably even pay blackmail to avoid personal risks.

So we have allowed even slaves free speech in their relations with free men, and resident aliens free speech in their relations with citizens, because the city needs resident aliens on account of the number of trades and on account of sea-faring. This is the reason then why we grant free speech also, reasonably enough, to resident aliens.

The people have driven the practise of gymnastics and music out of fashion in Athens, ostensibly on moral grounds, but really because they know they have not the ability to practise these arts themselves. Moreover, in having the rich pay for choruses, direct gymnasia, and outfit triremes, the people know that they dance while the rich pay the piper, and that in directing the gymnasia and outfitting the triremes the rich have the expense, the poor the entertainment. At any rate, the people take it for granted that they should be paid for singing, running, dancing, and going sailing in the ships, so that they can have money and the rich get poorer. And in the courts they care not so much about justice as about their own advantage.

As for the allies, because the Athenians sail out seemingly to persecute and blackmail those they govern, and hate decent citizens (knowing that the ruler must be hated by the ruled, and that if the rich and the decent gain power in the allied cities short-lived indeed will be the power of the people of Athens), for this reason they disfranchise decent people, confiscate their money, exile them, and execute them, but scoundrels they exalt. The decent people of Athens, on the other hand, come to the rescue of the decent people in the allied cities, knowing that it is to their interest always to help the best element there. One might say that the Athenian strength lies exactly here, in the allies' being able to pay tribute money; but to the popular party the gain seems to be greater if each of the Athenians has the allies' money, while the allies have just enough to live on, and have to work hard, so as to be unable to make plots.

P. MacKendrick and H. M. Howe, editors, *Classics in Translation*, Vol. I, *Greek Literature* (University of Wisconsin Press, 1952), pp. 224-226.

(c)

CLEON (d. 422 B.C.), the son of a rich tanner, was "the first prominent representative of the commercial class in Athenian politics". He is painted in most unflattering colours both by Aristophanes (see page 141) and Thucydides, although since it was Cleon who proposed the decree that exiled Thucydides (see page 114) we may suspect that the latter was not entirely free from bias. Cleon's speech to the Athenian assembly, part of which you are about to read, is recorded in Book III of Thucydides' *History*.

Personally I have had occasion often enough already to observe that a democracy is incapable of governing others, and I am all the more convinced of this when I see how you are now changing your minds about the Mytilenians. Because fear and conspiracy play no part in your daily relations with each other, you imagine that the same thing is true of your allies, and you fail to see that when you allow them to persuade you to make a mistaken decision and when you give way to your own feelings of compassion you are being guilty of a kind of weakness which is dangerous to you and which will not make them love you any more. What you do not realize is that your empire is a dictatorship exercised over subjects who do not like it and who are always plotting against you; you will not make them obey you by injuring your own interests in order to do them a favour; your leadership depends on superior strength and not on any goodwill of theirs. And this is the very worst thing—to pass measures and then not to abide by them. We should realize that a city is better off with bad laws, so long as they remain fixed, than with good laws that are constantly being altered, that lack of learning combined with sound common sense is more helpful than the kind of cleverness that gets out of hand, and that as a general rule states are better governed by the man in the street than by intellectuals. These are the sort of people who want to appear wiser than the laws, who want to get their own way in every general discussion, because they feel that they cannot show off their intelligence in matters of greater importance, and who, as a result, very often bring ruin on their country. But the other kind—the people who are not so confident in their own intelligence—are prepared to admit that the laws are wiser than they are and that they lack the ability to pull to pieces a speech made by a good speaker; they are unbiased judges, and not people taking part in some kind of a competition; so things usually go well when they are in control. We statesmen, too, should try to be like them, instead of being carried away by mere cleverness and a desire to show off our intelligence and so giving you, the people, advice which we do not really believe in ourselves.

. . . .

Let me sum the whole thing up. I say that, if you follow my advice, you will be doing the right thing as far as Mytilene is concerned and at the same time will be acting in your own interests; if you decide differently, you will not win them over, but you will be passing judgement on yourselves. For if they were justified in revolting, you must be wrong in holding power. If, however, whatever the rights or wrongs of it may be, you propose to hold power all the same, then your interest demands that these too, rightly or wrongly, must be punished. The only alternative is to surrender your empire, so that you can afford to go in for philanthropy. Make up your minds, therefore, to pay them back in their own coin, and do not make it look as though you who escaped their machinations are less quick to react than they who started them. Remember how they would have been likely to have treated you, if they had won, especially as they were the aggressors. Those who do wrong to a neighbour when there is no reason to do so are the ones who persevere to the point of destroying him, since they see the danger involved in allowing their enemy to survive. For he who has suffered for no good reason is a more dangerous enemy, if he escapes, than the one who has both done and suffered injury.

I urge you, therefore, not to be traitors to your own selves. Place yourselves in imagination at the moment when you first suffered and remember how then you would have given anything to have them in your power. Now pay them back for it, and do not grow soft just at this present moment, forgetting meanwhile the danger that hung over your heads then. Punish them as they deserve, and make an example of them to your other allies, plainly showing that revolt will be punished by death. Once they realize this, you will not have so often to neglect the war with your enemies because you are fighting with your own allies.

Thucydides: History of the Peloponnesian War, translated by R. Warner (Penguin Books, 1954), pp. 180-181, 184-185.

FURTHER READING

1. WARS AND POLITICS

* ADCOCK, F. E., *The Greek and Macedonian Art of War* (California, 1962)
  BURN, A. R., *Persia and the Greeks: The Defence of the West, c. 546-478 B.C.* (Arnold, 1962)
† ——————, *Pericles and Athens* (Macmillan, 1948)
† CASSON, L., *The Ancient Mariners* (Macmillan, 1959)
  GRUNDY, G. B., *Thucydides and the History of His Age*, 2 vols. (Blackwell, 1948)
  HENDERSON, B. W., *The Great War between Athens and Sparta* (Macmillan, 1927)
  LAISTNER, M. L. W., *A History of the Greek World from 479 to 323 B.C.* (Methuen, 1957)

2. GREEK DEMOCRACY
  † * AGARD, W. R., *What Democracy Meant to the Greeks* (Wisconsin, 1960)
    FERGUSON, W. S., *Greek Imperialism* (Houghton Mifflin, 1913)
    HAMMOND, M., *City-State and World State in Greek and Roman Political Theory until Augustus* (Harvard, 1951)
    JOHNSON, A. C. and others, editors, *The Greek Political Experience* (Princeton, 1941)
    JONES, A. H. M., *Athenian Democracy* (Blackwell, 1957)
    MARSH, F. B., *Modern Problems in the Ancient World* (Texas, 1943)

5. The Triumph of the Greek Spirit

For my part, I assure you, I had rather excel others in the knowledge of what is excellent, than in the extent of my power and dominion.
ALEXANDER THE GREAT,
WRITING TO ARISTOTLE

If you have ever wondered what the universe is made of, what causes disease, what makes one action good and another bad, or what a building ought to look like, then you have been thinking like the Greeks. They too puzzled over such problems, and the answers they gave went far beyond the confines of their little city-states.

In fact so inquisitive a people were the Greeks that they seem to have blazed almost all our intellectual trails. The chances are that if you think you have had an original idea you will find a Greek thought of it first. What problems did the Greeks try to solve, and how did we inherit their learning?

1/GREEK PHILOSOPHY AND SCIENCE

"Philosophy" is a difficult word. It means, literally, "the love of wisdom". The Greeks loved wisdom as no other people have; and this wisdom they believed could be attained only through the exercise of reason. Such a strong belief in reason first showed itself in speculations about the universe.

Thus early Greek philosophy dealt with problems that we today would say belong to science rather than to philosophy. The early Greek philosophers began by attempting to answer two basic questions:

  (1) What exists?
  (2) How does change occur?

The Nature Philosophers Many different answers were given to these questions. Thales, generally conceded to be the first Greek philosopher of note, held that originally everything was water, which may exist as a vapour, or as a fluid, or even as a solid. Anaximenes thought rather that air was the original substance, and that its density as it changed was responsible for fire, clouds, water, or earth. Strange as these ideas may seem to us today, they do show us that the Greeks of the 7th and 6th centuries B.C. believed in an orderly universe in which change was not due to the capricious actions of the gods.

Thales
Anaximenes

Heraclitus was not satisfied with earlier theories, and tried to solve the difficulties of deciding how everything came from water or air by saying that the fundamental fact was the changeability of all matter. "You cannot step into the same river twice," said Heraclitus. It will have flowed on and not be the same by the second step. If, then, any one thing is always changing into something else, what at any instant does exist? Here is the beginning in Greek thought of a distinction between the changing physical world we can see, hear, touch, taste, and smell, and the unchanging world of abstract thought. Parmenides grasped this distinction when he taught that the physical universe was merely an illusion, and that the one true reality was *Thought*.

Heraclitus
Parmenides

If you think the ancient Greeks were merely shadow-boxing with fine shades of meaning, try defining such a simple thing as a tree. What is a tree? It may be waving its branches of red and gold leaves outside your window now. But in a few months it will be bare. Then in the spring it may have blossoms, and later green leaves. Or in the science laboratory in your school you may examine the seed of a tree. Explain how the oak is *in* the acorn. You see, it is much easier to talk about something, even a tree, than it is to define it. The early Greek philosophers were asking fundamental questions to which many of us would be hard put to give answers.

How could the apparent contradictions of the philosophers be reconciled? By the 4th century B.C. a scholar named Democritus reached a high point when he attempted to explain the universe in purely physical terms. Because he believed that the universe could be defined in purely physical or material terms, Democritus is known as a *materialist*. According to him, everything —man, earth, air, stars, even the soul—was composed of infinitely small indivisible particles of matter called atoms. All change was the result of atomic movement.

Democritus

The Quibblers and the Questioner In the 5th century B.C., after the victory over the Persians, the trend in Greek philosophy changed. The questions now centred not on nature but on man:

(1) How can man know anything as a certainty?
(2) What is knowledge?
(3) What are the uses of knowledge?
(4) What, for man, is the good or the just life?

This splendid bronze of a young Greek athlete is the work of an unknown sculptor about the middle of the 4th century B.C. Some rapacious Roman governor—perhaps another Verres—apparently attempted to transfer the statue from Greece to Rome about 50 B.C.; but his shipload of loot was wrecked off Anticythera, an island between the southern Peloponnesus and Crete. The statue was dredged up from the sea, badly corroded, nineteen and a half centuries later, and has been skilfully restored. Particularly naturalistic are the eyes, with their black pupils and brown stone irises set into a white compound. Separate eyelids with short lashes are formed by bronze plates.

Many answers were given to these questions. One group of teachers called the *Sophists* (teachers of *sophia*, wisdom) specialized in grammar and oratory, but focused their attention on the development of good citizens. The greatest sophist, Protagoras, taught that all knowledge is relative to man. "Man," said Protagoras, "is the measure of all things." Clever sophists, however, could easily confuse common men by their involved argumentation. For example, was man's action wicked or did it merely *appear* wicked? If all was relative, that is, if there was no absolute truth and truth differed with each person's interpretation, then what was wrong—or right, for that matter? It is easy to see how the word sophist came to mean "quibbler".

One man was dissatisfied with the emphasis on materialism and the misleading arguments of the Sophists, and he made a great discovery. The man was Socrates (469-399 B.C.), a squat, homely little individual with a razor-sharp intellect; and what he discovered was the "soul", or that quality in any person which makes him an individual. To Socrates, the soul was as real as anything that could be touched, drawn, or measured. It existed independently of physical matter, and was created for the purpose of reasoning.

Socrates taught simply by asking questions. He believed that if he could get men to reason together they would discover the universal soul in man, that they would in other words, find out what was true for all men. Thus by his Socratic method of questioning, Socrates sought to resolve all contradictions and arrive at true knowledge. But he made one naïve assumption: he assumed that if only men knew what was the right thing to do they would go and do it. Unfortunately, not all men had Socrates' will-power.

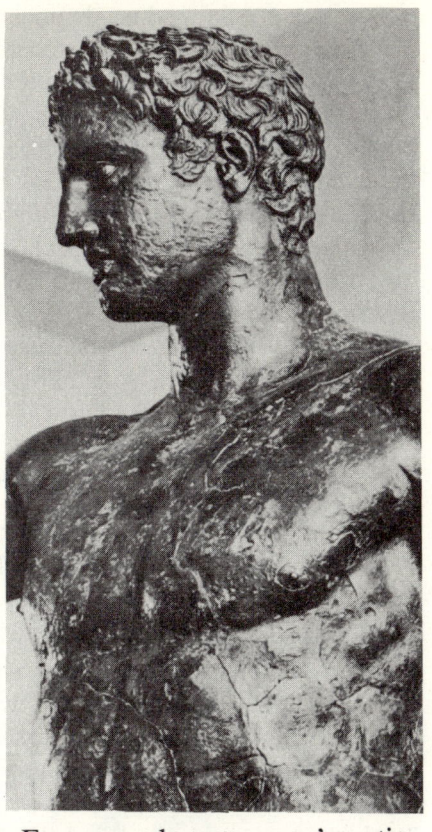

Socrates

Socrates proclaimed no dogmas; he wrote no books; he collected no fees. Yet this great teacher was executed by his fellow Athenians in 399 B.C. Why? Part of the reason was guilt by association. Athens had just lost her Great War, and some of those who had helped Sparta to win had been pupils of Socrates. What was more natural than to blame the pupils' acts on the teacher? Moreover, Socrates' interminable questions ruthlessly exposed the ignorance of those who thought they were wise—something that never makes a man popular. Why did Socrates deliberately antagonize people in his strange quest for wisdom? He gives some of his reasons in the Source Reading on pages 156 and 158.

We will study Socrates because of his unshakable conviction that mankind can know what is true and good. The words that he spoke at his trial are a fair measure of the man:

> I spend all my time going about among you persuading you, old and young alike, not to be so solicitous about your bodies or your possessions, but first of all and most earnestly, to consider how to make your souls as perfect as possible; and telling you that wealth does not bring virtue but that rather virtue brings wealth and every other human good, private or public.

Plato the Idealist   Socrates had a great pupil, Plato (429-347 B.C.), who not only wrote down his master's words but went beyond Socrates' teachings to formulate his own philosophic system. Athens fell when Plato was twenty-four, and in the break-up of his world Plato adopted a more constructive approach than Socrates. Whereas Socrates had concentrated on removing man's preconceived notions, Plato sought to build up positive conclusions.

Platonic Forms

Plato taught that all physical objects were illusions, that is, they were not "real," as we would say. What then *was* real? For Plato, the only real things were the ultimate "Ideas", of which the material world provided only approximate copies; and because of this emphasis he is called an Idealist. Since, however, Plato's Ideas did not exist just in the mind, but were to him realities that actually existed somewhere, a better term for us to use would be "Forms." The Platonic Forms were a natural development from Heraclitus' idea of the permanent reality behind all change.

If Plato were alive today he might say that the things of this world bear somewhat the same relationship to reality as the images on a television screen bear to what is going on in the studio miles away. Let us take one example of a Platonic Form. If you say "tree", you do not necessarily mean a spruce, maple, or oak. No one can tell whether you mean a deciduous or coniferous tree, or whether it is large or small. Yet the word "tree" does conjure up some image for all of us. Plato would say this is because we can dimly perceive—some of us more clearly than others—the Form of a tree, or the Ideal tree of which all trees in nature are only rough copies. In other words, all trees partake of a common quality of *treeness*.

What is true of trees is equally true of abstract ideas such as truth or beauty. When Plato came to plan the ideal state, he wrote his *Republic*, a treatise on how to educate the talented so that eventually they would be able to recognize the permanent Forms of Goodness, Justice, Truth, and Beauty, and thereby be able to rule the state wisely. "Unless either philosophers become rulers," reasoned Plato, "or those who rule become lovers of wisdom, and so political power and philosophy are united, there can be no respite from calamity for states or for mankind."

Whenever men today talk of universal standards of conduct they pay tribute, whether they know it or not, to Plato the Idealist.

Plato Diluted Since Aristotle (384-322 B.C.) studied under Plato for twenty years, he was familiar with Plato's theory of Forms. In fact Aristotle's theories have been described as "Plato diluted by common sense"—a description which does not necessarily mean they are easy to understand. Nevertheless this brilliant pupil, who became one of the greatest philosophers of all time, did not entirely share his master's view of the universe.

Aristotle thought reality *did* exist in particular objects in nature. In other words, every individual object had both Form and Matter. The world that we can see, hear, touch, taste, and smell was real. This specific tree was a real tree. But *this* tree and *that* tree had a universal Form, a quality of "treeness" that we can all see revealed in individual trees. Because of the fact that this Form must find its existence in individual trees, Matter and Form can never exist apart from each other. Thus, in a sense, the oak is the Form that is waiting to be developed in the Matter of the acorn.

Aristotle's theory allows for change, and explains change in a way that Plato's eternal Forms (of which material things are only an imperfect and unreal copy) do not. Although the following diagram greatly simplifies the difference between Plato's and Aristotle's thought, it points out the crucial disagreement in their thinking.

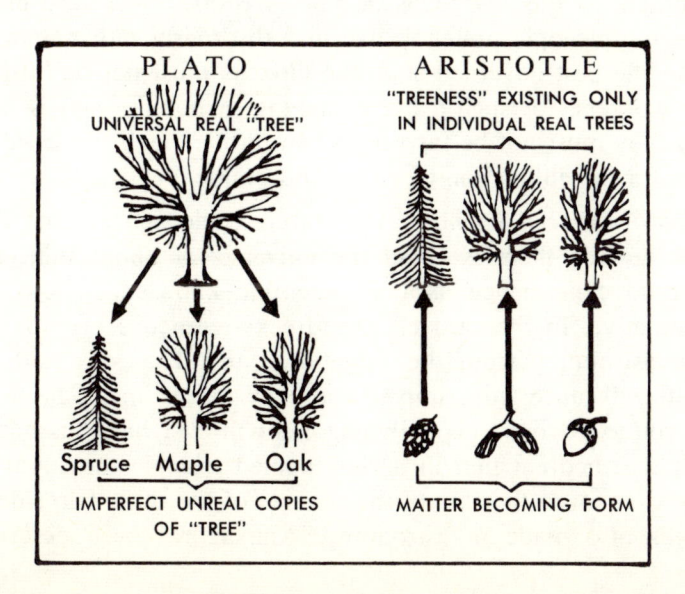

Because he believed individual objects were real, Aristotle found it necessary to describe the universe and man. In his *Ethics* he recognized, as Socrates and Plato had not, that though a man may know what is the right thing to do he may be too lacking in will-power to do it. The common man, says Aristotle, may have to work at being good.

> We acquire virtues by exercising them, as is true of any art. We learn by doing. As men become carpenters by building houses and lyre players by playing the lyre, we become just by acting justly, moderate by doing things in moderation, courageous by showing courage.

And because man lived in a *real* political state Aristotle could describe no abstract ideal state, as Plato had done. Instead, his *Politics* examined the constitutions of 158 Greek city-states in order to find out what made a government good or bad.

Aristotle was also the founder of biology. He studied anatomy and collected specimens, gathering information from farmers and fishermen for *The History of Animals*. In fact he was an encyclopaedic genius whose writings, of which we have mentioned only three, "so combined fact and theory that they fixed the patterns of Western learning for almost two thousand years."

The great weakness in the political ideas of both Plato and Aristotle was that neither could see beyond the narrow, aristocratic city-state. Both philosophers were critics of the democracy that had taken Athens into her Great War and brought about her political downfall. Both emphasized the value of the expert—the true philosopher—to governments, and both condemned the lower classes, who were blinded by their selfishness, as unfit to rule. Yet government by the lower classes, said Aristotle, "will have many supporters, for most people prefer to live in a disorderly rather than a sober way. . . . In the best governed state the citizens must not be businessmen or manual workers, for such a life is lacking in nobility and is hostile to virtue; and they must not be farmers, since farmers lack the leisure which is necessary for the cultivation of virtue and political activity."

The Spirit of Modern Science   The Nature Philosophers from Thales to Democritus devised theories about the universe, or about subject matter that we would call science. Not all scientific knowledge, however, was purely speculative. In their travels the Greeks learned a primitive mathematics and astronomy from the Egyptians, and Thales is said to have calculated the distance from ship to shore by the geometry he is reputed to have introduced from Egypt. Pythagoras will ever have his name associated with the theorem that in a right-angled triangle the square of the hypotenuse is equal to the sum of the squares of the other two sides. Some progress was also made in astronomy. Anaxagoras even described the

This statue of Aristotle—which is probably a Roman copy of the Greek original—shows the realism of which sculptors of the 4th century B.C. were capable. Note, for example, the wrinkled forehead and the veined hand. The Hellenistic emphasis on individualism as contrasted with the classic simplicity of the 5th century is evident in the studied pose of one of the greatest philosophers of all time.

sun as a mass of burning rock *larger than the whole Peloponnesus!* He also explained variations in climate as being the result of the earth's inclination on its axis.

The Greeks made their nearest approach to the spirit of modern Western science in medicine, and some of our Western medical terms, such as "chronic", "dose", "crisis", and "physician" are of Greek origin. The most famous school of medicine was founded by Hippocrates in the 5th century B.C. He bluntly rejected the earlier Greek theory that disease was caused by daemons (evil spirits) entering the body through the mouth:

Hippocrates

The fact is that invoking the gods to explain diseases and other natural events is all nonsense. It doesn't really matter whether you can call things divine or not. In nature all things are alike in this, that they all can be traced to preceding causes.

Greek doctors kept careful case histories of the symptoms and progress of diseases, and Thucydides' report of the Plague that ravaged Athens in the second year of the Peloponnesian War may have been patterned after these medical accounts.

The name of Hippocrates is forever immortal in the medical world. On graduation from medical school, doctors still take the Hippocratic Oath as their promise to observe the proper code of conduct for the medical profession.

I swear . . . that according to my ability I will keep this oath: to regard the man who taught me the art of medicine as dear to me as my own parents; to follow that system of treatment which I believe will help my patients, and to refrain from anything that is harmful to them. I will give no deadly drug if I am asked to do so, nor will I recommend any such thing. . . . In purity and holiness I will practise the art of medicine. Whatever I see or hear which should not be divulged, I will keep secret. While I continue to keep this oath may I enjoy life and the practice of my profession, respected at all times by all men.

2/ART, ARCHITECTURE, AND SCULPTURE

The Greek tendency to emphasize "a sense of the wholeness of things" showed in the harmonious unity of their architecture. Their habit of trying to penetrate below the surface to find true reality was reflected in their idealized sculpture. "When you copy types of beauty," said Socrates, "it is so difficult to find a perfect model that you combine the most beautiful details of several, and thus contrive to make the whole figure look beautiful."

The first Greek art appeared in the 10th century B.C., and consisted of simple bands and geometrical designs on clay pots. Animal and human figures later came to replace these geometrical designs, but the figures were simply done.

In the early days the gods had lived in groves and caves in the country; but as the Greeks became city-dwellers each state had a deity to house. Temples were built, at first simple rooms with porches in front of them supported by a row of columns. The temples can be classified according to the columns used—Doric, Ionic, or Corinthian, the Doric column being the simplest and the Corinthian the most ornate. Both gracefulness and proportion marked Greek buildings.

About the middle of the 5th century B.C. the Athenians began to build a magnificent temple to Athena on their acropolis. This temple is called the Parthenon, and has been described as "the most thrilling building there is." It was financed by the tribute from Delian allies. Thousands of separate contracts were let—one citizen contracted to deliver ten cart-loads of marble, another to flute one column—and the wonder is that the whole building was completed in only ten years.

The simple appearance of the Parthenon is deceptive. The columns lean imperceptibly inward and taper toward the top, and all the prominent lines of the building have a slightly convex curve to correct the optical illusion of sag. Every part of the building blends into the whole—again the Greek sense of wholeness. Nor was the temple as starkly white and solitary as it appears in photographs today. The stainless marble would shock its builders, for originally they reduced its glare in the Mediterranean sunlight by painting the gables blue and tinting the frieze of mythological figures in varied hues. We have lost, too, the huge gilded statues of gods which once crowded Greek temples.

The first Greek statues were stiff, awkward looking creations carved from wooden posts. Stone followed, and, in the 6th century B.C., bronze. By the Age of Pericles sculptors excelled in portraying the human body. Myron, an Athenian, is most famous for his *Discus Thrower*, which has survived only in Roman copies. Yet even in the copy, which shows the athlete at the moment of extreme strain just before releasing the discus, the

The remarkable thing about Greek architecture is that despite its absolute perfection of design it involved no new principles of engineering. Its post-and-lintel construction, which is illustrated in this cut-away drawing of the Parthenon, was used by many earlier civilizations. For example, the Egyptian columns to the right predate the Parthenon by about 2500 years. The Greek columns are of the Doric order, the earliest of Greek architectural styles, and differ from the Egyptian mainly in refinement of line and the use of a bowl-shaped capital between the column and the lintel. The Parthenon, built between 447 and 432 B.C., is about 60 feet in height. The continuous panel of sculpture around the exterior wall inside the portico was designed by the great Phidias.

statue is usually regarded as one of the greatest of all time. Myron is also reputed to have created a bronze cow so lifelike that living cattle mistook it for one of themselves and surrounded it in the field.

Phidias, a contemporary of Pericles, was celebrated for his *Athena* of the Parthenon, a great figure standing about 40 feet high, and for his seated *Zeus,* conceived on an even larger scale for the Temple of Zeus at Delphi. Although both of these works have perished, we know enough of them to state that they were, like the *Discus Thrower*, statues of types rather than individuals. The faces were idealized and almost expressionless. It is only in the 4th century B.C. with Praxiteles that the body and expression became completely free and realistic. His *Hermes* is the only renowned classical Greek statue that has come down to us from the hands of its creator; of the other famous originals only Roman or later Greek copies have survived. The Greeks usually tinted their statues—sun-tan for the body, and red for the hair, lips, eyebrows, and eyes.

There is an anecdote that illustrates the increasing realism of the 4th century B.C. Two of the greatest artists were Zeuxis, who is said to have died laughing at one of his own paintings, and Parrhasius.

The story runs that Parrhasius and Zeuxis entered into competition, Zeuxis exhibiting a picture of some grapes so true to nature that the birds flew up to the wall of the stage. Parrhasius then displayed a picture of a linen curtain realistic to such a degree that Zeuxis, elated by the verdict of the birds, cried out that now at last his rival must draw the curtain and show his picture. On discovering the mistake he surrendered the prize to Parrhasius, admitting candidly that he had deceived the birds, while Parrhasius had deluded himself, a painter.

3/LITERATURE AND RELIGION

Homer was to the Greeks what a combination of the Old Testament and Shakespeare would be to us. The *Iliad* and the *Odyssey* cannot be described: they must be read. In Homer's poems the gods are able to come down from their lofty perch on Mount Olympus to interfere with the fate of mere mortals, or simply to enjoy themselves. But Homer's delight in life is also blended with a shrewd understanding of human character, and he was keenly aware of the final law of mortality which even a being as unique as a Greek must obey. These brooding lines are from the *Iliad*:

As is the life of the leaves, so is that of men. The wind scatters the leaves to the ground: the vigorous forest puts forth others, and they grow in the spring-season. Soon one generation of men comes and another ceases.

In the 7th and 6th centuries B.C. Greek lyric poetry flourished. Let us, however, concentrate on Greek drama and on four dramatists—Aeschylus, Sophocles, Euripides, and Aristophanes. The first three wrote tragedy, the fourth comedy. Their careers span Greek history from the time of the Persian Wars to the downfall of Athens.

Aeschylus   Aeschylus (525-456 B.C.) built up his theme of the relationship of man to the gods in a sweeping series of plays designed to show the folly of human pride. Of the 90 plays he is believed to have written only seven survive in full. His *Agamemnon* is considered by many authorities to be the greatest tragedy produced in Attica. Canadians, however, will be more familiar with the Attic dramatist Sophocles (496-406 B.C.), since one of his most famous plays, *Oedipus Tyrannus (Oedipus Rex)*, was performed at the Stratford Festival in 1954 and 1955. In this tragedy it is prophesied even before Oedipus is born that he will kill his father and marry his mother, and he unwittingly fulfils these prophecies through ignorance of
Sophocles his parents' identities. Sophocles does not mean that Oedipus is the butt

of Fate. He means that there is a design in all that Oedipus does, a design that the gods could see. This great faith in Law as opposed to Fate is a typically Greek concept.

Many of Euripides' plays were composed during the Peloponnesian War, and their speeches reflect the Sophists' rhetorical ideas. It is little wonder that Euripides (480-406 B.C.) was unpopular, for his plays reflect doubt in the gods and the value of human suffering without substituting any answers to man's problems. He seldom won the festival prize, and finally left Athens to live in Macedon.   *Euripides*

The fourth dramatist, Aristophanes (450-385 B.C.), was a writer of outrageous comedies. He ridiculed war, democracy, education, and, of course, Socrates. Actually Aristophanes longed for the good old days before Pericles and his imperialist democracy, and he satirized the politicians of his day mercilessly. In the *Knights*, produced in 424 B.C. after the war had gone well, there are two generals, Demosthenes and Nicias, who want to get rid of Pericles' successor, the democratic leader Cleon. Realizing that the only way to do this is to put up a rival whom the people will prefer to Cleon, they select as their perfect democratic leader a sausage-seller, a simple soul who stands in the market-place selling his wares. Naturally the fellow protests to Demosthenes his unworthiness for the job.   *Aristophanes*

S.S. Tell me this, how can I, just a sausage-seller, be a big man like that?
D. It's the easiest thing in the world. You've got all the qualifications: low birth, market-place training, impudence.
S.S. I don't think I deserve it.
D. Not deserve it? It looks to me as if you've got too good a conscience. Was your father a gentleman?
S.S. By the gods, no! My folks were blackguards.
D. Lucky man! What a good start you've got for public life!
S.S. But I don't know a thing except how to read, and hardly that.
D. The only trouble is that you know anything. To be a leader of the people isn't for learned men or honest men, but for the ignorant and vile. Don't miss this wonderful opportunity!

This is pretty bitter satire.

Gods and Games   The subject of Greek literature and drama cannot be left without a brief discussion of religion and the games. The Greeks were polytheists, that is, they believed in many gods. These gods dwelt on Mount Olympus, but countless local gods and goddesses were worshipped at rustic shrines. Eager to know the future, the Greeks consulted oracles at Delphi and Dodona. At Delphi a priestess wildly proclaimed the will of Apollo; at Dodona, Zeus spoke to men in the rustling of the oak leaves.

Eventually sixty or seventy festivals a year came to be celebrated to honour the gods through athletic games and contests in the arts. Each city had its local festival, but all the Greek world gathered for the festival at

Olympia every four years. We still preserve the Olympic games in our vast international contests; but they began as much more than that. For a month the Greeks suspended warfare in order to stage five days of athletic competitions combined with music, poetry, and the dance.

Greek religion was both confusing and reassuring. There was no priestly class, no organized church, no single sacred book. In fact the Greeks enjoyed more variation in religious beliefs and ritual than any other people before or since. Religion for them was not a matter of beliefs as much as the carrying out of a service of worship. The city-god's statue was kept in his temple, and a bull, sheep, or pig was sacrificed on the altar outside. But the proud Greek did not choose to kneel even before his god. He prayed standing, looking upward, his palms open to receive the blessing that he did not doubt would come. Only the disillusionment of the Peloponnesian War destroyed this hardy presumption.

Much of Greek religious lore seems superstitious to us, if not crude. Nevertheless it was the Greek gift to take these myths and turn them into great dramas, or to celebrate them at national festivals which to some extent counteracted the narrow loyalty of the city-state.

4/THE MINORITY TRANSFORMS THE MAJORITY

This brief survey of Greek cultural achievement from the 7th to the 4th century B.C. shows that Athens' political decline was not matched by a cultural one—at least not in the 4th century. Her defeat in the Peloponnesian War did, of course, affect Greek ideals. While the Athenians tended to concentrate on political Utopias, they at the same time gave their sculpture realism by portraying actual individuals rather than god-like creations. The great tragedians disappeared and literature became more artificial. The Athenians gradually lost the ability to laugh at themselves, as Aristophanes had made them do even in war-time. Yet despite all this, the 4th century B.C. was far from being an age of cultural decline.

It would, of course, be quite wrong to think that *all* the Greeks spent their days in philosophical discussions, breaking off only to create a beautiful statue or attend a tragedy in the theatre. Most Greeks had to work for their living at least three hundred days a year, or considerably more than the average five-day-a-week Canadian. Life had few comforts for the majority of the population, the farmers who produced the necessities of life. Businesses in cities were small; the largest Athenian "factory" employed only over a hundred workers. The ordinary town house was a humble structure of plain stucco or sun-dried brick. Since, however, the climate was much like that of California, the Greeks could spend most of their time out of doors. For relaxation there were sports, festivals, the theatre, even the law courts.

If this, then, was the everyday life of the Greeks, why have we emphasized the culture of the minority? Because, in ways we hardly know, it did affect the majority. How deeply it affected them it is hard to say.

Perhaps the fact that the Greek state was small meant that its culture was not diluted, as is that of a large modern democracy. In his great *Funeral Oration* Pericles addressed not just the intellectual elite but all Athenians. Thousands sat hour after hour in the Athenian theatre (which could seat 14,000), whereas relatively few Canadians, even those in the immediate vicinity, regularly attend the Stratford Festival.

Nor were the Greeks subjected to the bombardment of the too often misleading printed word of our newspapers and magazines, or the low level of much modern television. Most historians seem to agree that the culture of the Greeks was much more than an upper class veneer. And while these same Greeks often sadly failed to measure up to their motto of "nothing in excess," or Aristotle's "Golden Mean" between extremes, we should not be too quick to condemn them.

The Funeral Oration has already given us an Athenian's own opinion of life in his city-state. Another picture of Athenian civilization is given by

This photograph shows a 1962 production of the *Bacchae* of Euripides at Epidaurus in southern Greece. At the centre of the theatre's *orchestra* or circular dancing floor can be seen a small round altar. Behind the orchestra is a long building for properties, a modern reconstruction of the original *skene* from which the actors emerged. Note the absence of scenery, and the members of the *chorus* ranged in a formal circle about the three actors.
The function of the chorus is to articulate the emotions aroused by the tragedy.

some outsiders, Corinthians, whose speech to the Spartans Thucydides has also recorded.

An Athenian is always an innovator, quick to form a resolution and quick at carrying it out. You, on the other hand, are good at keeping things as they are; you never originate an idea, and your action tends to stop short of its aim. Then again, Athenian daring will outrun its own resources; they will take risks against their better judgment, and still, in the midst of danger, remain confident. But your nature is always to do less than you could have done, to mistrust your own judgment, however sound it may be, and to assume that dangers will last for ever. . . . As for their bodies, they regard them as expendable for their city's sake, as though they were not their own; but each man cultivates his own intelligence, again with a view to doing something notable for his city. . . . Of them alone it may be said that they possess a thing almost as soon as they have begun to desire it, so quickly with them does action follow upon decision. And so they go on working away in hardship and danger all the days of their lives, seldom enjoying their possessions because they are always adding to them. Their view of a holiday is to do what needs doing; they prefer hardship and activity to peace and quiet. In a word, they are by nature incapable of either living a quiet life themselves or of allowing anyone else to do so.

5/ALEXANDER THE GREAT AND HIS EMPIRE

What became of Greek civilization after the decline of the polis? It is now time to take up the political narrative where it was left at the death of Philip II of Macedon in 336 B.C.

Philip's son, Alexander, is one of the most romantic figures in all history. He came to the throne at the age of twenty, having had as his teacher the incomparable Aristotle. What kind of boy and man Alexander was is something that historians have debated endlessly. Some have idealized him out of all reality, others have degraded him into a brute. A true appreciation of his character is extremely difficult because scarcely any contemporary estimates of him survive. We are consequently thrown back on evidence that comes from three or four centuries later, but which embodies fragments of contemporary writings.

"Look Thee Out A Kingdom" Alexander was, apparently, fair-skinned and clean-shaven, of average height and athletic build. One of the most engaging pictures of the youthful Alexander tells of the day that a certain Thessalian offered to sell Philip a horse named Bucephalus.

When they went into the field to try him, they found him so very vicious and unmanageable, that he reared up when they endeavoured to mount him, and would not so much as endure the voice of any of Philip's attendants. Upon which, as they were leading him away as wholly useless and untractable, Alexander, who stood by, said, "What an excellent horse do they lose for want of address and boldness to manage him!" Philip at first took no notice of what he said; but when he heard him repeat the same thing several times, and saw he was much vexed to see the horse sent away, "Do you reproach," said he to him, "those who are older than yourself, as if you knew more, and were better able to manage him than they?" "I could manage this horse," replied he, "better than others do." "And if you do not," said Philip, "what will you forfeit for your rashness?" "I will pay," answered Alexander, "the whole price of the horse." At this the whole company fell a-laughing; and as soon as the wager was settled amongst them, he immediately ran to the horse, and taking hold of the bridle, turned him directly towards the sun, having, it seems, observed that he was disturbed at and afraid of the motion of his own shadow; then letting him go forward a little, still keeping the reins in his hands, and stroking him gently when he found him begin to grow eager and fiery, he let fall his upper garment softly, and with one nimble leap securely mounted him, and when he was seated, by little and little drew in the bridle, and curbed him without either striking or spurring him. Presently, when he found him free from all rebelliousness, and only impatient for the course, he let him go at full speed, inciting him now with a commanding voice, and urging him also with his heel. Philip and his friends looked on at first in silence and anxiety for the result, till seeing him turn at the end of his career, and come back rejoicing and triumphing for what he had performed, they all burst out into acclamations of applause; and his father shedding tears, it is said, for joy, kissed him as he came down from his horse, and in his transport said, "O my son, look thee out a kingdom equal to and worthy of thyself, for Macedonia is too little for thee."

This was the lad who at twenty won the allegiance of his father's army and was soon elected by the Hellenic League as commander-in-chief of the war against Persia. By the spring of 334 B.C. he was ready to set forth into Asia—undoubtedly with no idea at this stage of trying to conquer the whole Persian Empire, but rather because his father had, in the last few years of his life, planned a Greek war of revenge. Accordingly, Alexander crossed the Dardanelles to attack Darius III (336-330 B.C.) with an army of over 30,000 infantry and 5000 cavalry.

Alexander made the Macedonian cavalry his striking arm. Small mobile units, each of 250 men, were capable of delivering, collectively, a series of terrific punches against enemy lines. Their success, however, depended upon the prime engagement of the enemy by the phalanx, a formation up to 16 men in depth, that "bristling rampart of outstretched pikes" which struck dumb terror into the hearts of the Persians. Along with the army travelled a series of mobile towers covered with hides to protect them from fire. As high as 150 feet, these structures were immense portable platforms which could be pushed up to the walls of a besieged city. While his Cretan archers rained arrows with deadly accuracy, Alexander's torsion catapults fired huge arrows and fifty-pound stones to a range of 200 yards. These formidable war machines were not rendered obsolete until the advent of cannons and gunpowder—1600 years after Alexander the Great.

Alexander's army

Conquering a Continent Among the countless sieges and minor skirmishes fought by Alexander, four great battles stand out: Granicus (334 B.C.), Issus (333 B.C.), Gaugamela (331 B.C.), and Hydaspes (326 B.C.).

The Persians made their initial stand at the River Granicus in western Asia Minor. Here Alexander's 35,000 men charged across the river, and his heavy cavalry shattered the smaller Persian army. The victory almost cost Alexander his life, but opened the route to the south where he could march along the Asia Minor coast liberating Greek cities.

Granicus

Here we have two fragments of a late Hellenistic mosaic floor—an intricate creation consisting of an estimated million and a half tiny stones, each the size of a grain of rice—from the House of the Faun at Pompeii. Originally laid at Alexandria, the floor was damaged when it was taken up and removed to Italy. The mosaic is a Roman copy of a famous picture of the battle of Issus (333 B.C.) painted about 300 B.C. by the Greek artist Philoxenus of Eretria. Alexander the Great, riding his magnificent charger Bucephalus, presses the attack against Darius who looks back in anguish from his chariot. The highlights on Alexander's face make it glisten with the exertion and tension of the struggle, while his burning gaze seems to pass beyond the battle scene to his vision of world dominion.

Issus

Alexander spent the winter consolidating his position in Asia Minor. Then he resolved to push on into Syria, and late in October 333 B.C. had to fight Darius whose forces threatened the ever-lengthening Macedonian lines of communication and supply. The two armies that faced each other at Issus may have been about equally matched at approximately 30,000 men each (certainly Darius had nothing even approaching the 600,000 that the Greeks attributed to him). At any rate, the Greek cavalry again caused things to go badly with the Persians, and when their king saw his left wing take to its heels he wheeled his chariot and fled. Perhaps it was that night, as he sat down to dinner in Darius's elegantly appointed tent, that Alexander resolved to conquer the entire Persian Empire.

Any conqueror of the Persians, Alexander realized, would first of all have to deal with the Persian fleet. So instead of turning eastward into Mesopotamia in pursuit of Darius, Alexander moved south against Phoenicia, the source of Darius's ships and seamen. Again fortune was with the Greeks. One Phoenician city after another fell, men deserted wholesale, and the fleet simply melted away. The next conquest was Damascus, with the added bonus of the Persian war treasury. Then came a seven-months' siege of Tyre. Before it capitulated, Darius tried to arrange a truce by offering the conqueror a ransom of 10,000 talents for the Persian royal family (captured at Issus), all the land west of the Euphrates, and his daughter's hand in marriage if he would go home. But Alexander had drunk too deeply of the heady wine of conquest. Why stop at the Euphrates? On he went to Egypt, where he was crowned Pharaoh and where he drew up the plans for the most famous of the many cities named after him—Alexandria.

At length, in 331 B.C., having retraced his steps, he was ready for the final reckoning with Darius. It came at Gaugamela across the Tigris. Backed by the largest army he had ever commanded, some 47,000 men, Alexander faced an even greater Persian force. Once again Alexander's cavalry triumphed; once again Darius fled at a critical moment. It was the last time the Persian faced the Macedonian. In 330 B.C. Darius was murdered by one of his own satraps in far-off Bactria (modern northern Afghanistan), just a little while before Alexander caught up with him.

The whole east had opened to Alexander with Gaugamela, and by 327 B.C. a Greek army of 30,000 laboured with superhuman tenacity across the Hindu Kush mountains and through the Khyber Pass into India. The next year the conqueror fought one of his hardest battles when he met

Gaugamela

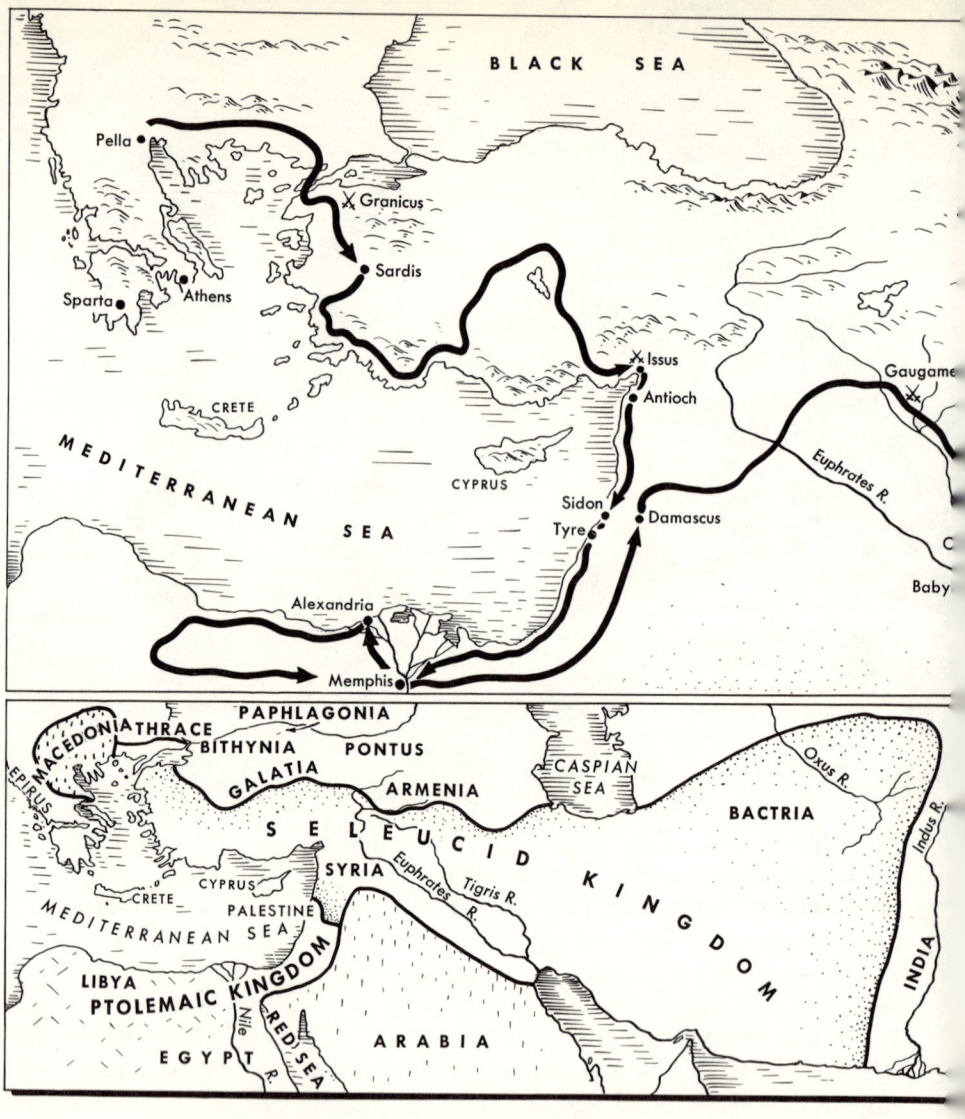

Hydaspes   Porus, king of the Pauravas, at the River Hydaspes, a tributary of the Indus. The Greeks might well have taken to their heels at the fearsome adversary that faced them, for besides outnumbering them the Indian army boasted a mighty advantage: 200 elephants. Since Alexander's horses were not accustomed to these immense beasts, his cavalry was badly handicapped. All he could do was try to draw Porus's cavalry away from the elephants and prevent it from interfering in the battle too much while the infantry fought it out.

The struggle was a desperate one, but the Greeks won. Afterwards, when Alexander met with his regal adversary, he asked him how he wished to be treated. "Like a king," answered Porus. Alexander admired the rajah's proud spirit. Porus was left free to govern his kingdom under Macedonian overlordship.

Alexander was still not satisfied. He had ambitions to push even deeper into the east; but his soldiers at last mutinied. For eight years they had

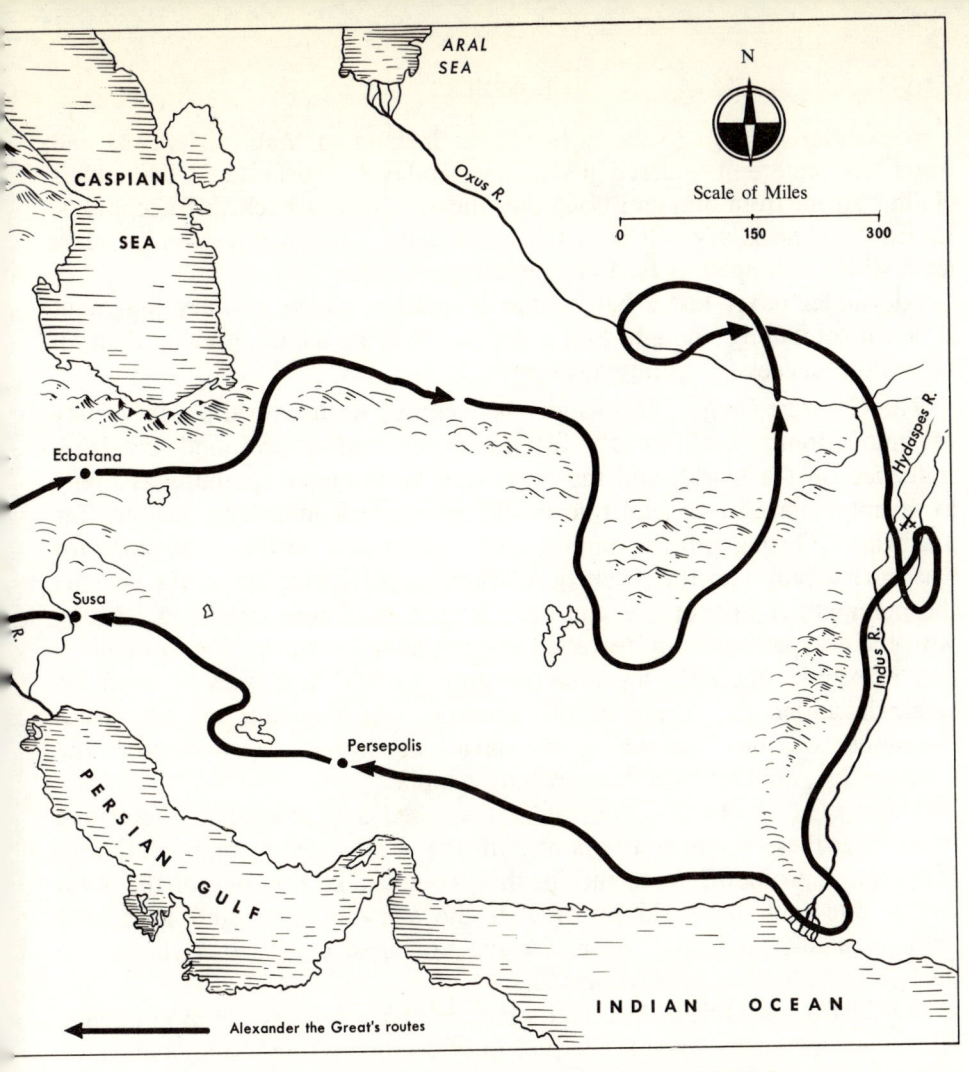

THE HELLENISTIC WORLD AND ITS KINGDOMS

followed him over 11,000 miles to the very ends of the earth. Now they would go no farther. There was nothing for Alexander to do but go home. By 323 B.C. he was back at Babylon, his restless mind still full of the problems of governing his immense empire. But he had exhausted himself both mentally and physically, and when he caught a fever (probably malaria) he was unable to throw it off.

On the sixth day of his fever, he was very seriously ill and was carried into the palace; he could still recognize his officers, but he could not speak. That night he was in high fever, and the day following and the next night, and the next day. His soldiers longed to see him; some that they might behold him while he was yet alive; others because it was announced that he was already dead, and they thought that his death was being hushed up by the bodyguard; most of them, because of their grief and longing for him, forced their way into his presence. They saw that he was speechless, but, as they filed past, he greeted them one by one by just raising his head and signing to them with his eyes.

The next evening he died. He was only thirty-two.

Alexander belongs to the ages. From Iceland to Malaya, legends and romances have embroidered his name. Today in Turkestan tribal chiefs claim descent from him and trace their horses' lineage back to Bucephalus. In Ethiopia he is revered as a Christian saint. Throughout central Asia he is still worshipped as Iskander, founder of cities.

Alexander never lost a battle, and Napoleon, whose military judgment is worth something, thought him a genius. Was he anything more than an incredibly successful young general?

The Greatest Dream   Alexander bequeathed to the world a rich variety of innovations. Undoubtedly his conquests had a profound civilizing influence on the world, and the cities that he founded (perhaps as many as seventy) played a crucial role in diffusing Greek ideas and customs far and wide. Throughout the empire a silver currency on the Attic standard was introduced, with the result that Athens and the Aegean world became the commercial hub of the empire. Science, too, was furthered. Along with the conquering army travelled geographers, botanists, and zoologists, and Alexander set aside an immense grant of 800 talents to be used for research under the direction of Aristotle. At the time of his death Alexander was busily planning two naval expeditions, one to explore the Caspian Sea, another to circumnavigate Arabia.

There is no doubt that Alexander possessed a brilliant and inquisitive mind. But there is one story about him that suggests more than brilliance. It reveals a profound depth in him that would make him one of the most remarkable men in world history. If the story is true—and it is well authenticated—then Alexander was the first man courageous enough to declare that all men could be brothers.

In 324 B.C. at Opis on the Tigris River, after a full decade of conquest, Alexander gave an immense banquet to 9000 men representing every race in his empire; Macedonians, Greeks, Persians, Egyptians, Asians all ate together. The banquet was followed by a libation, a religious act in which all present drew wine from a great bowl on each table and offered it to a universal God. The ceremony culminated in Alexander's prayer.

He prayed for peace, and for partnership in the realm of all the peoples of his empire there assembled, and lastly that all the peoples of the world he knew might be of one mind together and live in unity and concord. Many have dreamt that dream since; but he was the first. Had he lived, he would have tried to make it a reality, and would have failed as men have failed ever since; but that dream was probably the greatest thing about one of the greatest of mankind.[1]

Alexander made vast conquests. But his prayer at Opis was, as some Greeks clearly saw (pages 158-159), his lasting memorial.

We may wonder how all this reconciling of diverse nationalities was to be effected. The answer must be, not by democratic means but by force.

---

[1] W. W. Tarn in E. Barker, G. Clark, and P. Vaucher, editors, *The European Inheritance* (Clarendon Press, 1954), I, 184.

And in a small way some integration of races actually was accomplished—but not by Alexander. Death cut his dreams short.

Yet in his brief career he forever diverted the stream of civilization.

Few men have so changed the world; nothing after him could be as it was before. He did more than enlarge the field of human endeavour; the Hellenistic world was in effect his creation, and it was that world, and not classical Greece, which taught Rome and through Rome has influenced the modern world. And it was due to the impulse he gave to the Hellenization of western Asia that Christianity, when it came, had an easy medium at the start in which to spread. He had set out with Aristotle's belief that Asiatics were only fit for slaves. He soon saw that that was wrong . . .; to him the distinction of Greek and barbarian lost all meaning—a revolution in itself. . . . But his active mind soon went farther: he declared that all men were sons of one Father, the earliest expression known of the brotherhood of man.[2]

The Hellenistic Age   After Alexander's death there followed the Hellenistic (Greek-like) Age, a period spanning the three centuries before Christ.

The mighty empire of Alexander had never been closely knit. Rather it was his own peculiar creation, all its parts bound together solely by personal ties to the conqueror. Alexander's only heirs were a half-witted half-brother, Philip, and a son born posthumously. The half-brother was soon assassinated, and the son imprisoned and murdered. Eventually three of Alexander's generals cut the empire up for themselves: Egypt and some fringe lands went to one, Asia Minor and the old Persian Empire to another, Macedon and Greece to the third. Each kingdom in turn fell to Rome, whose conquest of Egypt in 30 B.C. is often taken as marking the end of the Hellenistic Age.

THE HELLENISTIC KINGDOMS

1. Ptolemy became king of Egypt (306 B.C.). The Ptolemaic Kingdom eventually included Egypt, Cyrene, Cyprus, and southern Syria, and the Ptolemies ruled until 30 B.C.

2. Seleucus founded the dynasty in Babylon (312 B.C.). The Seleucids at one time controlled most of the old Persian Empire, but by 63 B.C., when the dynasty ended, held only Syria.

3. Antigonus and his son Demetrius built a navy to command the Aegean, and dominated Macedon and Greece (316 B.C.). The Antigonid dynasty ruled Macedon until 167 B.C.

It is impossible to trace here the extremely involved political history of the Hellenistic kingdoms. There is one area, however, that is worth singling out: India.

A Brilliant Experiment   In the 6th century B.C. Darius I of Persia had set up the satrapy of Bactria, and thence in 329 B.C. Alexander the Great

---

[2]*European Inheritance*, I, 184.

followed in his footsteps. When Alexander invaded India he found only a number of disconnected states; but soon after his death an Indian adventurer, Chandragupta Maurya (322-298 B.C.), unified all the northern section of the land. His grandson, Asoka (269-232 B.C.), became a convert to Buddhism, renouncing war—but only, it must be noted, after he had consolidated his empire and no further conquests tempted him—and teaching that there was a universal moral law for all men.

After Asoka the Mauryan Empire broke up, and a Greek kingdom was established in the north-west. This kingdom of Bactria and India reached its height under Demetrius (189-167 B.C.), a Greek who modelled himself after the great Alexander. Demetrius proclaimed that Greeks and Indians were to be partners, and issued a bilingual coinage. Indians were admitted to Greek citizenship, the provincial administration became Graeco-Indian under Greek generals, and the two peoples lived harmoniously together.

Although the kingdom of Bactria and India was a half-way house between Europe and the Far East for eastern trade in silk, furs, nickel, and iron, there were no merchants who carried on direct traffic with the Orient. (The first silk caravan to come *direct* from China did not arrive in Bactria until 106 B.C., by which time the country had been conquered by nomads from the north-west.) During Greek rule eastern goods arrived only after passing through the hands of numerous intermediate trading peoples. The Greeks themselves, however, originated the Indian export trade to the West of pepper, ebony, ivory, and peacocks.

The Greeks of Bactria and India finally fell prey to civil wars and foreign attacks, and by the middle of the 1st century A.D. their influence in India was ended. All trace of it vanished. Indeed, it has been said that India would be exactly the same today had the Greeks never existed. Demetrius tried to turn Alexander's dream into reality and failed because he, like Alexander, could not subdue nationalism. Nevertheless it is interesting to look back over 2000 years to a time when, in far-off India, a Greek general tried a brilliant experiment in civilized living.

6/HELLENISTIC CIVILIZATION

It is the cultural consequences of Alexander's conquests that are finally most important. Through him the known world was Hellenized. Foreign philosophy, art, science, literature, and religions flowed into Greek moulds, and a Greek attitude to life became world-wide.

Epicureanism

During the 4th century B.C. conditions of chronic unemployment and a great gap between rich and poor, along with the rise of individualism, found expression in Hellenistic philosophy. Human happiness became the supreme preoccupation. Epicurus (342-270 B.C.) emphasized tranquillity as the solution to human woes. "Be happy with little, for being interested in and needing much, brings unhappiness." Happiness, that is, freedom from pain, ought to be man's goal. Epicurus prescribed a code of social conduct that included honesty, prudence, and justice in dealing with others, not because these attitudes were good in themselves but to save the individual from society's retribution. In later ages a sort of wine, women, and song Epicureanism appeared, but this was a far cry from the original ideas of Epicurus.

The greatest philosophy of Hellenistic times was Stoicism, named after the *stoa* (porch) where Zeno (336-264 B.C.), the founder of this philosophy, taught his followers to repress all physical desires, to live in equality with all other men, and not to complain but to do their duty resolutely. Stoicism was an ennobling, if essentially upper class, philosophy. Its ideals are well illustrated in this short prayer by Zeno's successor, Cleanthes.

Stoicism

> Lead me, O Zeus, and thou, O Destiny,
> Lead thou me on.
> To whatsoever task thou sendest me,
> Lead thou me on.
> I follow fearless, or, if in mistrust
> I lag and will not, follow still I must.

Everything in Excess The Hellenistic Age lost the classic simplicity of Greek architecture and sculpture. More palaces, theatres, and libraries were built than temples. Art was created for the money it would bring, and became the preserve of rich patrons rather than the possession of ordinary citizens. The famous lighthouse at Alexandria would have been something of a monstrosity to the classical Greeks because it rose to an "excessive" 400 feet.

Most of the original Greek sculptures that we have inherited belong to the Hellenistic Age, and one is apt at first to prefer their realism to the idealism of classical Greek art. There is a striving after effect in Hellenistic art, so that the sculpture looks restless and tense. This is nowhere better illustrated than in the churning *Laocoön*, a frighteningly realistic portrayal of a priest and his two sons being strangled by serpents—hardly the Greek ideal of "nothing in excess"!

Sculpture

By far the finest Hellenistic sculpture showing motion is the *Winged Victory* of Samothrace. This majestic portrayal of a winged goddess riding the prow of a ship has about it a flowing movement entirely lacking in the figures of the Parthenon frieze. At the head of a staircase in the Louvre in Paris, where it now stands, it seems literally to surge forward toward the onlooker.

Magnificent as the finest Hellenistic sculptures are, expert opinion rates them below the classical Greek art of several centuries earlier. The Hellenistic passion to provide a thrill, even in stone, too often led to contortion and strain.

The study of science made impressive strides during the Hellenistic Age. Although the work and methods of Aristotle bore few fruits at Athens, the so-called "Museum" established at Alexandria as a home for scholars

Science

whom the government subsidized made that city the scientific capital of the world. Far from being a museum in our sense of the word, this institution was a combined library, research institute, and university where literature, mathematics, astronomy, and medicine were all studied.

Archimedes

Much of the knowledge of Hellenistic scholars remained theoretical. We know that Hero of Alexandria invented an automatic device to open and shut temple doors by steam-power, but in an age of abundant slave labour the steam-engine did not flourish. Archimedes of Syracuse, who established the principle of the lever, constructed a spiral pump for drawing water up out of the Nile. However, he is best remembered for his law of floating bodies. It is said to have been based on his observation of the water displaced in a tub by his own floating body, a discovery that sent him rushing home naked through the streets shouting "Eureka!" ("I have found it!")

Less spectacular was the post-mortem dissection of the human body performed by Hellenistic doctors. Such a thing had been impossible in classical Greece, because the dead were cremated. Of the many consequences of dissection, one was a correction of the earlier notion that the arteries were filled with air.

Euclid

Euclid deserves to be mentioned as the author of a mathematical treatise entitled the *Elements of Geometry*. In 13 books Euclid wrote down 465 propositions, many of which you will be studying in geometry.

The astronomers and geographers of this great age of science came to some startling conclusions. Heraclides believed that the earth rotated on its axis every 24 hours, and Aristarchus claimed that the sun was fixed and the earth and planets revolved around it. Eratosthenes stated that the oceans

The famous Portland Vase, one of the treasures of the British Museum in London, illustrates the superb craftsmanship of the 1st century A.D. This fine piece of glassware was recovered in the 16th century in a tomb beside the Appian Way, doubtless having been produced by a Greek craftsman, perhaps at Alexandria, for export to Rome. The deep blue vase was blown in liquid glass (the handles being moulded separately and added later), and then dipped into opaque white glass. When the white layer was thoroughly hard it was cut away except where a pattern was desired. The result was the precursor of our modern Wedgwood pottery— a cameo-like effect with the raised white figures of Peleus and Thetis, Achilles' parents, in relief against a deep blue background. In 1845 a madman smashed the vase to bits, but it has been skilfully restored.

were all one body of water and calculated the earth's circumference at 24,662 miles (it is 24,857). This estimate was reduced by Posidonius, who also suggested that India might be reached by sailing westward from Cadiz —a theory that fired the imagination of a man named Columbus some fifteen centuries later.

Every kind of literary creation, including works on grammar and philology, abounded in Hellenistic times. Sometimes the term "Silver Age" is applied to this literary period, in contrast to the classical "Golden Age" of the 5th Century B.C. Of the 1100 writers whose names are known, let us mention but two.

Literature

Menander wrote the "New Comedy" ("new" as contrasted with the "Old Comedy" of Aristophanes), which had as its theme the life of the newly-rich of Athens. Although the plots were elaborate the endings were invariably happy, and some of the lines, such as "Whom the gods love die young," have become proverbial. A contrast to Menander's sophisticated writing was provided by Theocritus. Theocritus was the originator of pastoral poetry, verse that deals with happy scenes of rustic life peopled by love-sick shepherds. It was Theocritus who provided inspiration for the Roman Virgil and the English Milton, as well as for hosts of "pastoral" poets smitten by love.

There was also a great deal of pessimism in literature. The civilization seemed tired; life appeared futile and meaningless. Yet among many sour epigrams survives this gentle story.

Along with five others Charmus ran the distance race, and, strange to relate, came in seventh. When there were only six, you will say, how did he come in seventh? Well, a friend of his wearing a heavy mantle ran alongside yelling, "Come on, Charmus!" So Charmus came in seventh. If he had had five more such friends he would have come in twelfth.

Poor Charmus! How the mighty Greeks had fallen by the end of the Hellenistic Age.

Rulers Become Gods The political and economic insecurity of the Hellenistic Age stimulated changes in religion. From the east the Greek world was invaded by countless Oriental cults in which immortality was promised to converts and an elaborate ritual assured communion with a god who had risen from the dead. Superstition also flourished. Cities adopted stars as their guardians, and many a horoscope was cast. Thus it is particularly in religion that the Hellenistic fusion of Greek and Oriental can easily be seen. The appearance of ruler-worship is especially noteworthy. It began by being associated with Alexander the Great, and soon states were deifying their Hellenistic kings. Even the once proud Athenians glorified their ruler in this way: "Thou alone art a true god; the others are asleep or away, or are not."

East Meets West In many ways the Hellenistic Age ought to remind us of our own. More progress was made in science than in any other comparable period of time before the 15th century A.D. Cities flourished—cities built on a checker-board plan like our own. It was a great age of business, perhaps the greatest until today. But poverty was widespread: in Egypt babies left outside to die were collected by the state as saleable articles. There was a great rise in literacy, but also in superstition. And a feeling of unity was developing, traceable in large part to Alexander's wish to reconcile East and West, as well as to a common form of spoken Greek, the *koine*. Yet despite the extraordinary material and scientific achievements of the Hellenistic Age, it is hard to escape the feeling that spiritually Greek creations had suffered a decline and a dilution.

This consideration of Greek and Hellenistic culture concludes with a summary in the form of a "Cultural Calendar."[3] The chart is, of course, highly selective, and it would be a mistake to think that a blank space means nothing of a cultural note happened. Further reading will enable the student to fill in those areas which there has not been enough space to treat in this book.

7/SUMMARY AND CONCLUSION

When historians look back over the whole of Greek history they come to the conclusion that the Greeks were the most remarkable people who ever lived. No other people united such a belief in reason with such a sense of the wholeness and balance of life. But when the city-state that had fostered such ideals collapsed, the ideals faltered. Much more individualism, and much less idealism, came to be evident in all departments of Hellenistic life.

Sparta, with her universal military training, won her Great War; but we remember her as a city without a soul. Athens experimented with democracy and lost her Great War; but she won the hearts and minds of men in every succeeding age. And Alexander the Great, who conquered all, is remembered chiefly for his role in spreading and diffusing the diluted culture of little city-states, states that today we would count political failures.

We have studied the civilization of a number of Greek cities that first experimented with democracy but could never unite to form a nation. Now let us turn to the history of a single city, Rome, who learned how to govern an empire and lost democracy in the learning.

(a) SOURCE READINGS

SOCRATES (469-399 B.C.) was tried and condemned to death on a charge of not believing in the state gods and of teaching his disbelief. Since Socrates himself wrote nothing, our best source of information about him is in the writings of his pupil Plato (429-347 B.C.). Plato's *Apology*, from which the following extract is taken, presents Socrates' speech in his own defence at his trial.

---

[3]Since the chief political events of Greek history overlap those of Rome, they are listed in a parallel time chart on page 220.

A GREEK CULTURAL CALENDAR

DATES B.C.	PHILOSOPHY	ARCHITECTURE	SCULPTURE	SCIENCE	LITERATURE	RELIGION
1200-750 Dorian invasions and Homeric society					Homer, *Iliad* and *Odyssey*	Homeric religion
750-500 Monarchy to democracy at Athens	Thales to Heraclitus	Beginnings of architecture	Beginnings of sculpture	Birth of science	Hesiod, *Works and Days*	Festivals Games Oracles
500-479 Persian Wars		Temples of Greece				
479-461 Post-war age			Myron		Aeschylus	
461-431 Periclean Age	Rise of Sophists	Parthenon	Phidias		Sophocles Herodotus	
431-404 Peloponnesian War	Sophists Socrates			Hippocrates Democritus	Euripides Aristophanes Thucydides	Scepticism in religion
404-336 Political decline of Greece	Plato Aristotle	Temples of Asia Minor	Praxiteles Zeuxis and Parrhasius (painting)			
336-30 Alexander and the Hellenistic Age	Epicurus Zeno	City-planning Lighthouse at Alexandria	Victory of Samothrace Laocoon	"Museum" at Alexandria Hero, Euclid Archimedes Heraclides Aristarchus Posidonius	Menander Theocritus	Oriental influences Mystery religions Astrology

Your witness to my wisdom, if I have any, and to its nature, is the god at Delphi. You certainly knew Chaerephon. He was a friend of mine from our youth, and a friend of your popular party as well; he shared in your late exile, and accompanied you on your return. Now you know the temper of Chaerephon, how impulsive he was in everything he undertook. Well so it was when once he went to Delphi, and made bold to ask the oracle this question—and, Gentlemen, please do not make an uproar over what I say; he asked if there was any one more wise than I. Then the Pythian oracle made response that there was no one who was wiser. To this response his brother here will bear you witness, since Chaerephon himself is dead.

Now bear in mind the reason why I tell you this. It is because I am going on to show you whence this calumny of me has sprung; for when I heard about the oracle, I communed within myself: "What can the god be saying, and what does the riddle mean? Well I know in my own heart that I am without wisdom great or small. What is it that he means, then, in declaring me to be most wise? It cannot be that he is lying; it is not in his nature." For a long time I continued at a loss as to his meaning, then finally decided, much against my will, to seek it in the following way.

I went to one of those who pass for wise men, feeling sure that there if anywhere I could refute the answer, and explain to the oracle: "Here is a man that is wiser than I, but you said I was the wisest." The man I went to was one of our statesmen; his name I need not mention. Him I thoroughly examined, and from him, as I studied him and conversed with him, I gathered, fellow citizens, this impression. This man appeared to me to seem to be wise to others, and above all to himself, but not to be so. And then I tried to show him that he thought that he was wise, but was not. The result was that I gained his enmity and the enmity as well of many of those who were present. So, as I went away, I reasoned with myself: "At all events I am wiser than this man is. It is quite possible that neither one of us knows anything fine and good. But this man fancies that he knows when he does not, while I, whereas I do not know, just so I do not fancy that I know. In this small item, then, at least, I seem to be wiser than he, in that I do not fancy that I know what I do not." Thereafter I went to another man, one of those who passed for wiser than the first, and I got the same impression. Whereupon I gained his enmity as well as that of many more. . . .

Such, fellow citizens, was the quest which brought me so much enmity, hatreds so utterly harsh and hard to bear, whence sprang so many calumnies, and this name that is given me of being "wise"; for every time I caught another person in his ignorance, those present fancied that I knew what he did not. . . . So even now I still go about in my search, and, in keeping with the god's intent, question anybody, citizen or stranger, whom I fancy to be wise. . . .

Plato, On the Trial and Death of Socrates, translated by L. Cooper (Cornell University Press, 1941), pp. 53-56.

(b)

PLUTARCH (see page 100) is most famous for his *Lives*, but he also wrote essays on a wide variety of subjects. His *The Fortune or Virtue of Alexander*, from which the selection below is taken, draws on the geographer Eratosthenes (276-195 B.C.), who preserved a tradition favourable to Alexander the Great. The many legendary histories of Alexander, both ancient and modern, have made it exceedingly difficult for his biographers today to get at the truth. For a good modern evaluation of his complex character see C. A. Robinson, Jr., "The Extraordinary Ideas of Alexander the Great," *American Historical Review*, LXII (1957), 326-344.

And indeed the much admired polity of Zeno, the founder of the Stoic school, is directed to this one point, that we should not live as separate cities or separate peoples, each distinguished by its own rights, but that we should regard all men as fellow countrymen and fellow citizens, and that there should be one manner of life and one order, like that of a herd of cattle feeding together as one, by one common law. Zeno wrote this to shadow forth as it were a dream or image of a philosophic well-ordered State; but it was Alexander who provided the reality which lay behind the principle of Zeno's thought. For Aristotle had advised him to be a leader to Greeks but a master to barbarians, caring for the former as friends and relatives but treating the latter as animals or plants. Had he done this, his leadership would have been full of wars and banishments and festering seditions; but he acted otherwise, for he believed that he had a mission from God to harmonize men generally and to be the Reconciler of the World, compelling those he could not persuade and bringing men from everywhere into unity, mixing their lives and customs and marriages and social ways as if in a loving-cup. . . .

And if the deity who sent Alexander's soul to this earth had not quickly recalled it, one law would have looked down upon all men and they would have ordered their lives by one idea of justice as by one light common to all; but, as it was, those parts of the world which did not see Alexander remained sunless. Therefore the very basis of his expedition exhibits him as a philosopher, since what he had in mind was not to get luxury and extravagance for himself, but to make all men of one mind together and give them peace and partnership with one another.

E. Barker, G. Clark, and P. Vaucher, editors, *The European Inheritance* (Clarendon Press, 1954), I, 255-256.

FURTHER READING

1. GREEK CULTURE
    † * AGARD, W. R., *The Greek Mind* (Anvil, 1957)
        ARMSTRONG, A. H., *An Introduction to Ancient Philosophy* (Methuen, 1949)
        CLAGETT, M., *Greek Science in Antiquity* (Abelard, 1956)
      * DE SANTILLANA, G., *The Origins of Scientific Thought* (Mentor, 1961)
    † * DICKINSON, G. L., *The Greek View of Life* (University, 1962)
      * HADAS, M., *A History of Greek Literature* (Columbia, 1962)
    † * HAMILTON, E., *The Greek Way to Western Civilization* (Mentor, 1948)
      * _____, *The Echo of Greece* (Norton, 1964)
    †   HUXLEY, M., editor, *The Root of Europe: Studies in the Diffusion of Greek Culture* (Clarke, Irwin, 1952)
    †   QUENNELL, M. and C. H. B., *Everyday Things in Ancient Greece* (Batsford, 1957)
      * ROBINSON, C. A., *The Spring of Civilization: Periclean Athens* (Dutton, 1959)
        _____, *Athens in the Age of Pericles*, (Oklahoma, 1959)
      * TAYLOR, A. E., *Socrates, The Man and His Thought* (Anchor, 1956)
      * TOYNBEE, A. J., *Greek Civilization and Character* (Mentor, 1953)
    † * WARNER, R., *The Greek Philosophers* (Mentor, 1958)

2. ALEXANDER AND THE HELLENISTIC AGE
    †   BURN, A. R., *Alexander the Great and the Hellenistic Empire* (Macmillan, 1947)
        CARY, M., *A History of the Greek World from 323 to 146 B.C.* (Methuen, 1951)
        FULLER, J. F. C., *The Generalship of Alexander the Great* (Eyre and Spottiswoode, 1958)
    †   ROBINSON, C. A., *Alexander the Great* (Dutton, 1947)
        TARN, W. W., *The Greeks in Bactria and India* (Cambridge, 1951)
    † * _____, *Alexander the Great* (Beacon, 1956)
      * TARN, W. W. and GRIFFITH, G. T., *Hellenistic Civilization* (Meridian, 1961)

UNIT **III**

THE ROMANS

6. The Foundations of the Roman Republic

Senators, a victory which neither god nor man could begrudge, you and your general have won over us. We surrender to you because we believe (and what could be handsomer for a victor?) that life will be better under your administration than under our own laws.
THE CITIZENS OF FALERII IN THE
SENATE AT ROME, 395 B.C.

Of all the peoples we have so far studied, the Romans are most like ourselves. Why is this? For one thing they are a good deal closer to us in time than the other civilized peoples. Then too, our Western civilization grew most directly out of the Roman experience, which in turn drew on the Orient.

The Romans were a practical people. Their enduring greatness lies principally in the spheres of politics and law. It was when Roman politicians could no longer cope with the problem of administering a vast empire that there began a long process of decline and fall.

This political experiment of the Romans has had a special fascination for modern students of government. How the Romans acquired their empire, why they lost democracy in governing it, how their military strength kept the peace of the world, and, finally, why the whole system collapsed, all are questions that we ponder when we consider the grandeur of Rome.

1/ITALY AND HER CIVILIZERS

Jutting more than 600 miles into the Mediterranean is a long, slender peninsula shaped amazingly like a boot. This peninsula, as you know, is now called Italy. It averages about 100 miles in width, and at its northern end the valley of the river Po, cradled between the Alps and the Apennines, comprises continental Italy. Taken together, these two regions—continental Italy and peninsular Italy—are a little smaller than Newfoundland, but about three times the area of ancient Greece.

Like a rugged backbone, the Apennines run the length of the peninsula. This barrier of mountains has made communication between various parts of Italy difficult, a fact of which the world was reminded during the Italian

---

This stately and powerful eagle, proud symbol of the Roman Empire's supremacy over land and sea, dates from the time of Trajan and is now incorporated into the portico of the Church of the Holy Apostles in Rome. Throughout the long span of history the eagle has been the cherished emblem of peoples as widely separated as the Persians and the North American Indians. During the early Republic, however, Roman armies used to go into battle bearing standards surmounted by a variety of symbols in a set order: eagles foremost, then wolves, minotaurs, horses, and bears. It was the famous general Marius (see page 203) who decreed that the eagle alone should become the special standard of the legions. Here it is framed by the corona civica, *a wreath of oak leaves and acorns bestowed on the soldier who saved the life of a Roman citizen in battle.*

campaigns of 1943 and 1944 in World War II. The mountainous character of much of Italy has another effect: it robs her of any great rivers. With the exception of the Po in the northern plain, the rivers have too swift a current and too variable a volume of water to be navigable.

The mountainous structure of the peninsula and its great length result in a varied climate. While the Po valley has the wide temperature variation of a continental climate (summer rain, winter snow, wet spring and fall), much of the peninsula enjoys the rainy winters and hot, dry summers of a Mediterranean climate. Canadians do not often think of Italian geography in relation to their own, but if we superimpose Italy at its correct latitude on a map of North America, we will find that the northern end will be located over Montreal while the "toe" of the boot is in the vicinity of Washington, D.C. Thus we can see that it is the Mediterranean Sea which moderates the Italian climate rather than Italy's position on the earth's latitude.

Italy's one great geographical advantage is her central position in the Mediterranean basin. Yet her harbours are few, for although the Italian peninsula has a total of 2000 miles of coastline it is too regular to have the countless sheltered inlets of Greece. Moreover, the Mediterranean has no strong tides to scour out the river mouths and they soon silt up. Those good harbours which Italy does possess are found along the west coast. Hence in ancient days Italy faced west, while Greece faced east.

Although Italy's topography hampered her communications, her great economic potentialities contrasted sharply with the poverty of Greece. While Italy did not boast a rich supply of minerals, there was plenty of good building stone and marble, and the wooded Apennine slopes provided ample timber for shipbuilding. The plains, enriched by silt washed down by the rivers and lava dust from numerous volcanoes, yielded profuse harvests of wheat and barley, apples and pears, peas and beans. And as in all Mediterranean countries the cultivation of grapes and olives was important, while sheep, goats, and cattle grew fat on the luxuriant highland pastures.

From Stone to Iron   The earliest stages of history with which we have already dealt can also be detected in Italy. The Old and New Stone Ages, the Bronze Age, and the Iron Age all left remains for modern archaeologists to discover.

The Terramare People

Sometime after 1600 B.C. a Bronze Age Indo-European culture appeared in the Po valley, brought there from the Danube region by the Terramare folk (*terramara* = "black earth", the earth modern Italians have dug from the old settlement mounds). These people, who were superior farmers and livestock breeders, lived in huts of wattle and daub. Many were hunters and fishermen, and some were craftsmen as well, spreading their products—in particular gray and black vases—into central and southern Italy.

The Villanovans

As northern Italy passed from the Bronze to the Iron Age, after 1000 B.C. another people appeared. The origin of these Villanovans, as the

THE FOUNDATIONS OF THE ROMAN REPUBLIC

archaeologists call them (the best known excavations of their culture are at Villanova near Bologna), is uncertain. They may have been caught up in the general dislocation forced by the Dorian invasions and have migrated across the Adriatic Sea. Certain it is that they became expert metalworkers, and by the 8th century B.C. Bologna was the Hamilton of early Italy, the centre of a flourishing export trade in bronze and iron work.

Gradually the Villanovans spread down the east coast of Italy to Rimini and then into Tuscany and Latium, again building their villages of round huts. But the mineral wealth of Italy, in particular the rich copper of Etruria, drew new marauding bands to Italian shores. The Villanovans were eventually to become fused with these latest invaders, the mysterious Etruscans, one of the two peoples who were destined to affect the Romans most deeply.

Bearers of Civilization The Etruscans probably came into Italy between 1000 and 800 B.C. from some point on the shores of Asia Minor. They occupied an area west of the Apennines and north of the Tiber called Etruria (now known as Tuscany). From the cities they established in Etruria the Etruscans began to expand outwards, especially across the Tiber to the south. However, the Etruscan cities remained politically independent of each other and never did succeed in building up a stable political organization.

The Etruscans

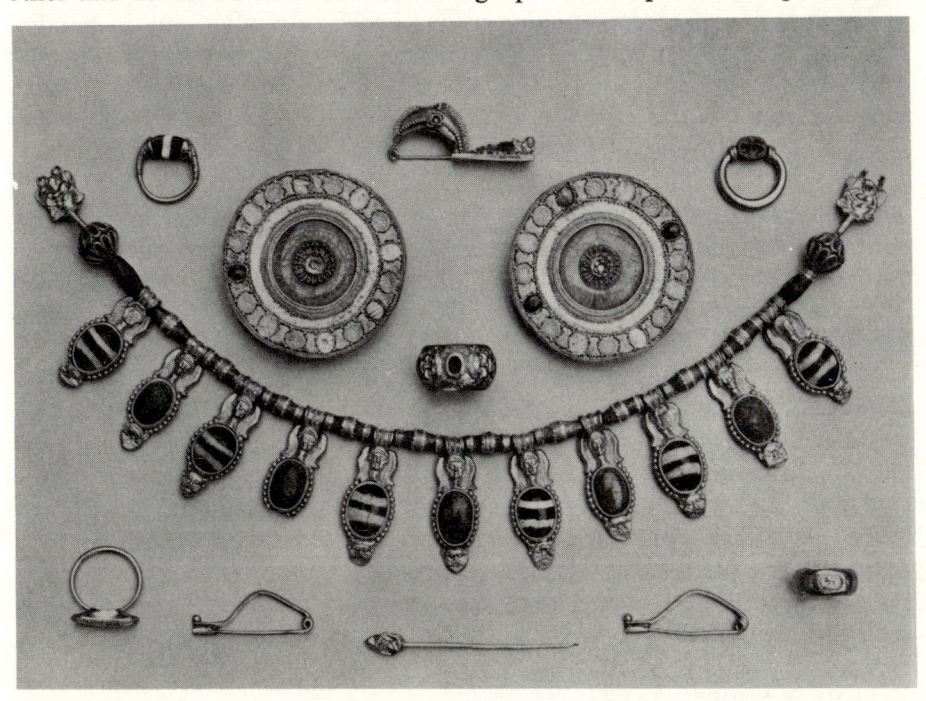

The Etruscan upper classes enjoyed lives of elegance and splendour, as may be seen by these examples of gold, glass, and carnelian jewellery from about 500 B.C. The Greeks probably would have disapproved of the Etruscan women's immodest display of elaborate necklaces, brooches, pins, and rings—to say nothing of their make-up and hair bleach!

At least 10,000 Etruscan inscriptions have survived, and while their language has not been completely deciphered, scholars have been able to partially translate the twenty-seven letter alphabet adapted from the Greeks. Unfortunately these inscriptions do not tell us much because they consist mostly of proper names used for religious dedications or epitaphs.

The Etruscan language reveals little beyond a people whose religion was preoccupied with gloomy thoughts of death. But this would indeed be a misleading impression of their civilization. For the Etruscans loved to bedeck themselves with jewels and live a life of pleasure; they took delight in a good feast, a sprightly dance, and such exciting sports as horse-racing. More surprising is the knowledge that even at this early date women played a prominent role in society, as is known from the fact that often a man's epitaph records the name of his mother.

Etruscan men were first of all warriors: it was Etruscan nobles who introduced the chariot into Italy. They were also pirates and traders; their merchants traded widely with the Phoenicians, Greeks, and even the Egyptians. And they were farmers, who cleared forests, drained land, and began to cultivate grapes and olives. But above all—and here lies their main significance—the Etruscans were great engineers, masters of the arts of building, a skill which they passed on to the later inhabitants of Italy. Even today their genius in constructing walled cities, in mining, and in building aqueducts, harbours, and drainage systems is abundantly evident in Roman ruins. It is a Roman tradition that Rome was first fortified by a wall when she was under Etruscan domination. For these and other reasons the Etruscans have been called the civilizers of Italy.

The other bearers of civilization were the Greeks, who as early as the 8th century B.C. had founded colonies in the western Mediterranean. They settled in Sicily and along the southern and south-western coasts of Italy as far north as modern Naples. It was in the early days on the frontier between Rome and the Greek cities that the Romans coined the name *Graeci*, or Greeks, for their Hellenic neighbours.

The Greek colonies, as we might expect from the history of the Greek homeland, were politically divided; the Greeks had faithfully exported their peculiar city loyalty along with their culture. It was with this culture that the Greeks stimulated the Romans, supplanting Etruscan culture in Rome, and later, in the 3rd century B.C., even more deeply affecting the Romans when the Greek cities of southern Italy, the so-called "Great Greece" *(Magna Graecia)*, were conquered.

Who were these Romans who were caught midway between the Etruscans and the Greeks?

2/ROME CONQUERS ITALY

The People of the Plain

The Latins or "people of the plain" lived south of the Tiber in an area called Latium. For several centuries the Latins were an agricultural people living in rude huts. But from the 7th century B.C. onwards they were influenced by the Etruscans from the north and the Greek traders from the south, and their villages became fortified cities associated in a kind of religious federation which celebrated a common festival.

On the borderland of Etruria and Latium, about fifteen miles up the Tiber, was ancient Rome. Although the Romans later invented the traditional date of 753 B.C. for Rome's founding by Romulus and Remus (distant descendants of the Trojan hero Aeneas), the fact is that villages had grown up on this hilly spot as early as 1000 B.C. Rome provided a good site for settlement because her ring of hills made defence practicable, while at the same time an island facilitated a fording of the river. The story of Rome's founding is, of course, purely mythical, born of the Roman desire to have a glorious history. The Romans had a simple solution for history's gaps: what they did not know they made up.

The tradition that early Rome was ruled by Etruscan kings is, however, undoubtedly accurate. The king *(rex)* was elected from the members of the royal family, and exercised the functions of commander-in-chief of the army, chief priest of the religion, and chief justice in matters affecting public peace. All this authority was summed up by the single word *imperium*. *Early government of Rome*

Associated with the king was a council called the Senate, which was made up of aristocrats. The whole Roman people was divided into 30 groups called *curiae*, and these meeting together comprised the Curiate Assembly *(comitia curiata)*. The assembly was mainly an approving body that sanctioned a new king or a declaration of war.

About 509 B.C. the aristocrats expelled their despotic kings, thereby abolishing the monarchy and ending the Etruscan overlordship. At this time the Roman sphere of influence was restricted to the lower Tiber basin. No one could predict whether it would spread farther.

The Reluctant Imperialist   There was as yet no political unity in Italy—in fact by the 6th century B.C. the name "Italy" did not exist. It was not until the 5th century B.C. that the Greeks adopted the term *Italia*, derived from the Oscan (the chief language of central Italy) word *vitelliu* or "calf-land"; and even then they applied it only to the south-west part of the peninsula adjacent to Sicily.

In the confusion of the 6th century B.C. Rome was just one of many fortified towns in the peninsula, albeit the chief town of Latium controlling about 300 square miles. Soon after the expulsion of the kings, the Etruscans attacked—but did not subjugate—Rome. In the period of readjustment that followed, Rome joined in a defensive league with the other Latin towns, each of which gave the others equal rights of citizenship. Even at this period of her history, however, Rome was beginning to stand out from her allies, for her vote equalled that of all the other members of the League combined. *The Latin League*

The valley of the Tiber was a prey to mountain tribes such as the Volscians, Aequians, and Sabines, northern immigrants who some time after 1500 B.C. settled in Italy and thereafter raided Roman territory year after year. But Rome's most serious enemy in the 5th century B.C. was the Etruscan city of Veii, which lay only a few miles away, north of the Tiber. By about the end of this century Veii was conquered. Thus Rome became supreme in central Italy, not because she was blatantly imperialistic and coveted more territory, but simply because she continually had to capture the bases of enemies who were making sorties into her domains.

The Gauls Rome had barely conquered Veii when a new invader appeared. From north of the Alps swept down an undisciplined horde of fair-haired, blue-eyed giants, the Gauls from *Gallia* or Gaul (modern France). They sacked and burned Rome; but fortunately for the Romans the invaders wanted plunder more than territory, and eventually marched away after ransoming the city. The people quickly rallied to rebuild their ruined homes—and later manufactured legends of how they had beaten off a stubborn foe. Here is one of the most famous of them.

The citadel of Rome and the Capitol were in very great danger. For the Gauls had noticed the tracks of a man, where the messenger from Veii had got through, or perhaps had observed for themselves that the cliff near the shrine of Carmentis afforded an easy ascent. So on a starlit night they first sent forward an unarmed man to try the way; then, handing up their weapons where there was a steep place and supporting themselves by their fellows or affording support in their turn, they pulled one another up, as the ground required, and reached the summit, in such silence that not only the sentries but even the dogs . . . were not aroused. But they could not elude the vigilance of the geese, which being sacred to Juno had, notwithstanding the dearth of provisions, not been killed. This was the salvation of them all; for the geese with their cackling and the flapping of their wings woke Marcus Manlius . . . who, catching up his weapons and at the same time calling the rest to arms, strode past his bewildered comrades to a Gaul who had already got a foothold on the crest and dislodged him with a blow from the boss of his shield. . . . And by now the rest had come together and were assailing the invaders with javelins and stones, and presently the whole company lost their footing and were flung down headlong to destruction. Then after the din was hushed, the rest of the night . . . was given up to sleep.

The Gauls came back on several other occasions, but the second time (in 349 B.C.) the Romans were ready for them and the attackers retreated without a battle. By now Rome was clearly the successful champion of the central Italian peoples.

The Lengthening Shadow Rome's leadership in Latium did not, however, go unchallenged, and eventually the Latin League revolted. Rome defeated the League and dissolved it, incorporating most of the Latin cities into the Roman state.

The Samnites Naturally there were those who feared the lengthening shadow of Rome. These now took their lead from Samnium, the most powerful state in the interior of the peninsula. The Samnites were a brave, hardy people used to mountain fighting, whereas the Romans were supreme fighters in the

ROME'S CONQUEST OF ITALY TO 218 B.C.

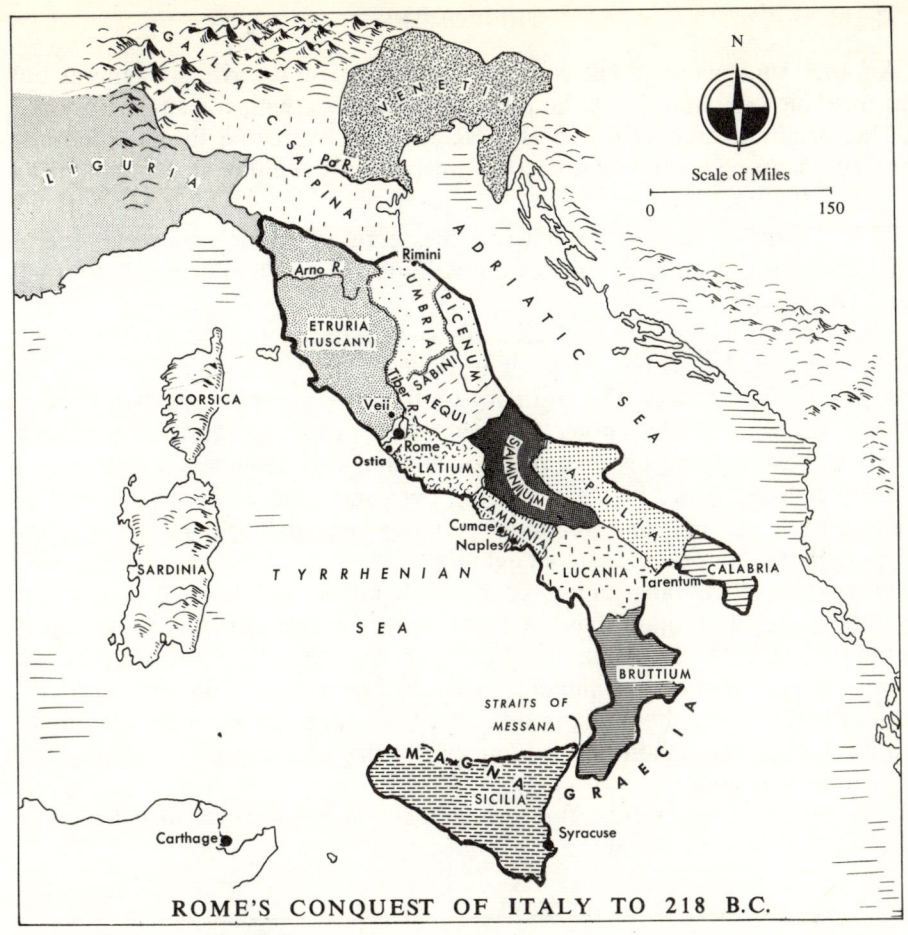

open country. For this reason the Romans planned to defeat the Samnites by hemming them in with encircling alliances.

It took a long time. Off and on during the next generation the Romans warred with the Samnites, who were aided and abetted by Rome's old foes, the Etruscans and the Gauls. By 290 B.C. Italy from Etruria in the north to Lucania at the "instep" of the Italian boot was dominated by Rome. But strange to tell, Rome's many wars did not exhaust her; rather with each victory she seemed to gather strength. This was no accident. In a very real sense, as we shall see, Rome was the architect of her own prosperity.

Rome now came face to face with the Greeks of southern Italy. When some Greek cities became Roman allies, the suspicions of Tarentum, a leading southern city, were aroused, and on the outbreak of war the Tarentines appealed to Pyrrhus, the king of Epirus in north-west Greece. Pyrrhus sent an army of 25,000 men and 20 elephants, but his victory over the Roman forces was so expensive that the term "Pyrrhic victory" has been coined for one in which the victor's losses are extremely high. After a second success, also bought at great cost, Pyrrhus is said to have remarked,

"Another such victory will ruin us." Pyrrhus was a brilliant general but an unstable one, and finally he returned to Greece with the surviving third of his original force. By 265 B.C. Rome had overcome the last Samnite and Etruscan opposition and become mistress of the entire Italian peninsula south of the Po.

The Secret of Success   What was the secret of Rome's success? It was that above all else the Romans were willing to submit to organization and discipline.

The Roman army

Because every soldier provided his own equipment, the landless poor were not enrolled in the army. The richest, those able to afford a horse, composed the cavalry. The ordinary foot-soldier provided his own helmet, shield, armour, and weapons (dagger, short thrusting sword, and spear). Also he had to bring his own pack of auxiliary equipment, consisting of a spade, hatchet, saw, bucket, cooking-pot, and half a month's ration of wheat meal. Armour and weapons probably weighed about 45 pounds, and the auxiliary equipment as much again.

From the Etruscans and Greeks the Romans had learned to use the phalanx, the disciplined mass of foot-soldiers armed with lances and protected by shields and leather or bronze body armour. Some time before the outbreak of the Carthaginian Wars in 264 B.C. Rome developed the *legion* by breaking the phalanx into sections called *maniples* and *centuries*. In addition, each legion had a force of cavalry and a body of light-armed skirmishing troops.

By the 3rd century B.C. the arrangement of the heavy-armed legionary foot-soldiers was as follows:

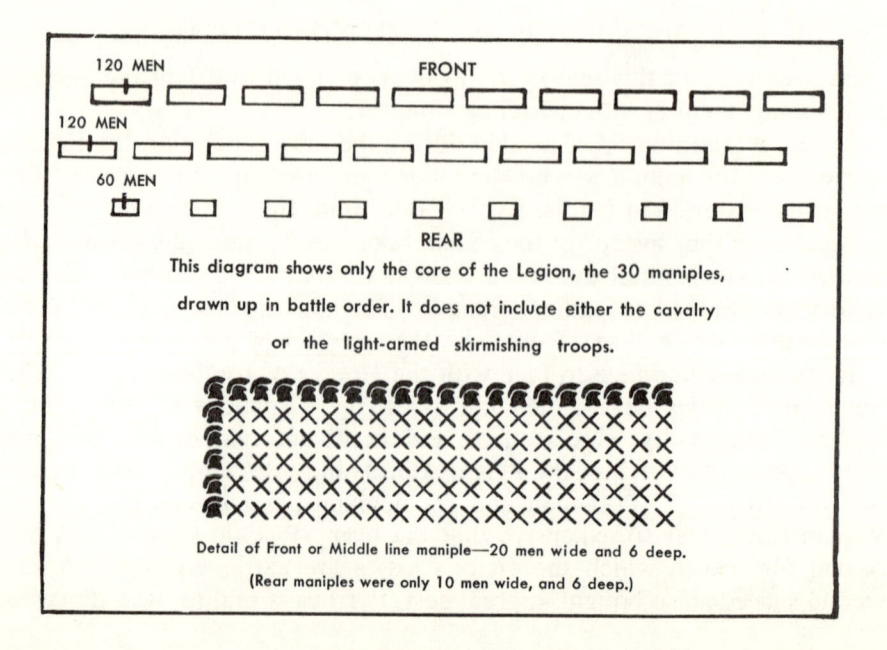

This diagram shows only the core of the Legion, the 30 maniples, drawn up in battle order. It does not include either the cavalry or the light-armed skirmishing troops.

Detail of Front or Middle line maniple—20 men wide and 6 deep.

(Rear maniples were only 10 men wide, and 6 deep.)

The front line was made up of younger men armed with a short sword and one or two heavy throwing spears. In the second line, similarly armed, were the middle-aged legionaries. The seasoned veterans, armed with a sword and a light thrusting spear, formed the rear line. Thus the usual legion consisted of:

10 maniples of 120 = 1200	(front line)
10 maniples of 120 = 1200	(middle line)
10 maniples of 60 = 600	(rear line)
light-armed troops = 1200	
cavalry = 300	
LEGION = 4500 men	

Rome's 3rd century B.C. armies consisted of some 8 legions.

The discipline was magnificent. The best example of their legendary iron will may be found in their camps. At nightfall, no matter what the day had included, no matter how much marching or fighting had been done, the legions would build a fortified camp with a moat, earthen ramparts, and palisades. There was a set plan for the camp, and every soldier was assigned his specific place and task.

With such an army, in less than two and a half centuries Rome had achieved the status of a world power.

The Logical Conclusion It has been said of the British Empire that it was acquired in a fit of absent-mindedness. This implies that there was no long-range plan of conquest, often no desire for foreign territory other than to preserve safety at home—an argument always more convincing to the victor than to the vanquished. At least in the beginning, however, Rome did not seek quarrels with her neighbours. But when they would not leave her alone she absorbed them one by one, thus being led on from one frontier to another. The logical conclusion had to be control of all Italy.

But could Rome stop there? Perhaps she could—if no other power interfered with her plans for Italy. So far, then, a jury pronouncing on the evidence of Rome's imperialism might declare her comparatively innocent. Before we pass final judgment, however, let us examine the way in which Rome had organized Italy.

Rome was now the head of an Italian empire. There were two classes of Roman citizens and two classes of allies, as shown below:

Citizens	1. Full Citizens
	2. Citizens with Private Rights
Allies	1. Latin Allies
	2. Federate Allies

The full citizens lived in the city of Rome, in adjacent towns, or in Roman colonies, and could hold office at Rome and vote in the Roman

Assemblies. The citizens with private rights lived in towns in Etruria, Latium, and Campania, and although independent in trade and local affairs they were denied a vote or office at Rome. They were really half-way along the road to full citizenship.

The Latin allies had full rights of local self-government and a constitution modelled on that of Rome, each ally being bound to Rome by an individual treaty. Colonists in a Latin allied city could trade and intermarry with citizens of Rome and of other Latin allied cities. Federate allies, however, had limited rights of trade (and usually of intermarriage) with Rome and with one another. The extent of privileges granted to a Federate ally depended on whether it had entered the alliance voluntarily or by conquest. In neither class—Federate or Latin—did the allies have any control over their own external affairs. On the other hand, they did not have to pay taxes or tribute to Rome, although they were obliged to furnish troops, which served alongside the Roman legions.

A Sensible Solution   It is, of course, true that the members of the Italian federation had now lost their independence to Rome. Yet it is also true that Rome gave them peace at home and protection from any foreign invader. Because the Romans were clever enough to adapt themselves to varied conditions they did not seek one overall unified system.

Moreover, the Romans appeared as protectors rather than oppressors in the wars of the 5th and 4th centuries B.C. "Rome became the mother of Italy," writes an eminent Roman historian, "training her children by carefully graded stages up to the privilege of full family life. This was an immense stride forward in Rome's history and indeed in the history of mankind. The conquered people were not to be dragged along at Rome's chariot wheels as slaves; they were asked to share in the privileges and responsibilities of their conqueror." The Roman grant of citizenship to subject states was something of which the Greeks could not conceive, but given this sensible Roman solution to a difficult problem, Rome's rule was firmly established in Italy by the 3rd century B.C.

3/THE EVOLUTION OF REPUBLICAN GOVERNMENT

"Our Republic," a stern Roman once said, "was not made by the services of one man, but of many, not in a single lifetime, but through many centuries and generations." The Roman ability to make practical adjustments according to the demands of circumstances, to learn by experience rather than to act according to theory, is shown in the development of their government.

The Men Who Ruled Rome   The Roman constitution evolved from the 6th to the 3rd century B.C., during which time there came to be thirteen elected officials, the so-called "magistrates". These magistrates represented the whole state, and were invested with rights, duties, and executive power *(potestas)* by the authority of the Senate and the people.

This unknown Roman by an anonymous sculptor of the 1st century B.C. is an excellent illustration of the Romans' skill at portraiture, a skill learned from the Etruscans. The hard-headed, practical citizen of the Republic whom we see here is no mere type, but an individual in his own right, a worthy exemplar of the three cardinal Roman virtues: *gravitas* (dignified seriousness), *pietas* (dutiful performance of obligations to the gods and the state), and *virtus* (courage).

When the kings were expelled from Rome in the 6th century B.C. they were replaced by two chief magistrates called *consuls*. These consuls were elected annually by the Centuriate Assembly, and each consul could veto any public act of the other. This check, it was felt, would prevent the office from growing too strong for the good of the state. In peacetime the consuls alternated their rule within Rome monthly, but on the battlefield their command of the army, incredible as it may sound, alternated daily. The consuls also presided over the Senate, although they were there as much to listen as to give advice.

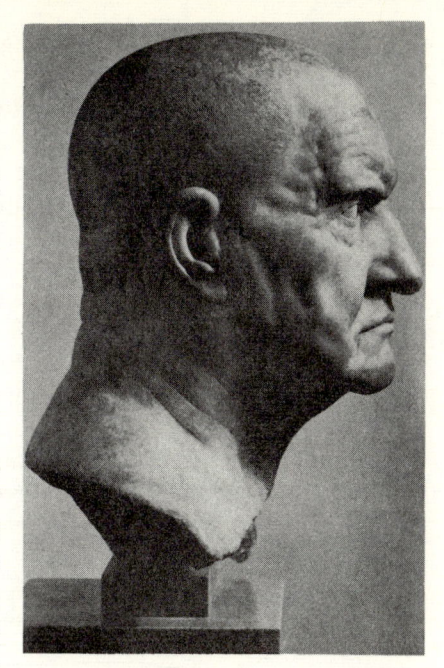

If the two consuls reached an impasse or if a crisis overtook the state, the Senate could endorse one man as *dictator* for a term of six months, a step that amounted to a temporary decree of martial law.

The magistrates ranking next to the consuls were the *praetors*. At first there was only one praetor, who was in charge of the courts; but later eight were elected, one of whom presided over the Senate if the consuls were away from Rome. While all the magistrates had potestas, only the consuls, dictator, and praetors had *imperium*—the right to command an army, preside over an assembly, and try important cases. Next in importance were four *aediles* (later eight), who were responsible for the administration of public works, roads, and the control of weights and measures in Roman markets. Last came the four *quaestors*; two were in charge of the public treasury and two were consular assistants.

Praetors

Aediles and Quaestors

Once in five years two *censors* (usually ex-consuls) were elected to register all Roman citizens and their property, and to revise the list of Senators. So honoured were these magistrates that they were buried in the rich purple toga of royalty.

Tribunes of the People

All these magistrates, excepting the censors, were elected yearly, re-election to the same office being permissible only after a ten-year interval. However, under unusual circumstances (for example a consular military campaign) the Senate could prolong the command. A consul whose office was thus prolonged for more than a single year was said to possess power *pro consule* (in the place of a consul).

Finally, from the mid 5th century B.C. there were ten officers known as *tribunes of the people*. These tribunes were not, technically speaking, magistrates, since their authority did not extend beyond the city of Rome. They attended meetings of the Senate and proposed legislation for the Tribal Assembly, but their chief function was to protect the lower classes *(plebeians)*, whose officers they were, from exploitation at the hands of the aristocrats *(patricians)*. This protector of the commons was not allowed to be absent from the city overnight, and always had to keep his door open.

The regularly established order of progression up the stairway of civic and political fame in Rome was from quaestor to consul thus:

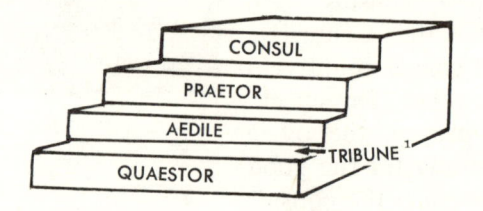

```
            CONSUL
        PRAETOR
     AEDILE
                  ← TRIBUNE [1]
  QUAESTOR
```

The Senate

A Far Cry from Democracy   Sharing the conduct of the day-to-day business of the state with the elected magistrates was the *Senate*. The origins of the Senate can be found in the traditional council of 100 aristocrats, the clan leaders who had advised the king. After the expulsion of the monarchy the Senate came to consist of a body of experienced men chosen for life by the consuls to advise in the governing of the state. By the 5th century B.C. there were 300 Senators, and by the 4th century they were being chosen for life by the censors and consisted of ex-magistrates.

The Senators provided the state with both experience and knowledge, and at first their sanction was necessary before assembly measures could become law. Even when this sanction was no longer required, after 287 B.C., the Senate still exercised considerable influence. Its members not only advised magistrates and assigned them their duties, but they also drew up the state budget and supervised foreign affairs. Above all, in times of emergency the Senate could, by what was called a *Senatus consultum ultimum* (final decree), suspend constitutional liberties and declare martial law. The first known use of this decree occurred in 121 B.C.

It is no accident that in those four famous initials that the Romans were to carve on thousands of their monuments—*S.P.Q.R. (Senatus Populusque Romanus)*—the Senate precedes the people.

[1] Only in the late Republic did the tribunate become a regular "step" after the quaestorship.

The people of Rome were represented by two Assemblies *(comitia)*, the Centuriate Assembly *(comitia centuriata)* and the Tribal Assembly *(comitia tributa)*. Both had been developed in the 5th century B.C., and their membership overlapped.

The Centuriate Assembly

The Centuriate Assembly gradually replaced the older Curiate Assembly (see page 165, and was organized into groups of voters called *centuries*. The centuries originated in the grouping of citizens for military service according to the arms they could afford. This army classification was adapted to the Centuriate Assembly because it was felt that a citizen's voting power ought to correspond to his worth as a soldier. If, for instance, he could afford a horse, he was more valuable to the army (and should have more power in the Assembly) than the foot-soldier. Accordingly the richest citizens formed 18 centuries of Equestrians (cavalry), and were supported by 170 centuries of foot-soldiers (who were divided again into 5 classes depending on the cost of their arms and armour). The least power went to the 5 centuries of the great mass who could not afford to arm themselves and hence had scant military value.

Because the particular century to which any voter belonged depended on his wealth, the centuries were of unequal size. There are always more poor than rich; hence there might be 100 to 300 voters per century in the wealthiest group, whereas in the poorest there might be 25,000 or more. Yet each century had only one vote. Consequently, the proportion of votes to population was greatly distorted, as the diagram on page 174 shows.

The power of the wealthiest groups will quickly be seen: 98 out of the total 193 votes were cast by the Equestrians and the first class. Not only that, but since the votes were taken in order of wealth beginning with the richest and working down to the poorest, often only an upper class vote needed to be taken to achieve a majority and decide an issue.

About the middle of the 3rd century B.C., however, a reorganization of centuries took place. Henceforth the first two wealth groups were assigned 88 out of a total of 373 votes, and the middle classes in the second, third, fourth, and fifth wealth groups held the balance of power. Nevertheless a system that gave 88 votes to 13,800 wealthy Romans and only 5 to 130,000 plebeians was hardly democratic. Certainly the mass of the people had virtually no influence in the Centuriate Assembly.

The Centuriate Assembly elected the consuls, praetors, and censors, voted on laws submitted to it, and declared war and ratified peace treaties. However since it could not *nominate* the magistrates it elected, nor initiate legislation, as the Republic developed it came to be mainly a body for electing magistrates.

The Tribal Assembly

Where the people *did* have some influence was in the Tribal Assembly. By the 3rd century B.C. measures passed by it were to have the force of law. The 35 "tribes" or electoral districts each had a single vote, but the number of voters in any one tribe varied greatly, and no quorum was necessary to register a decision for a whole tribe. Moreover, only 4 of the tribes

VOTING IN THE CENTURIATE ASSEMBLY

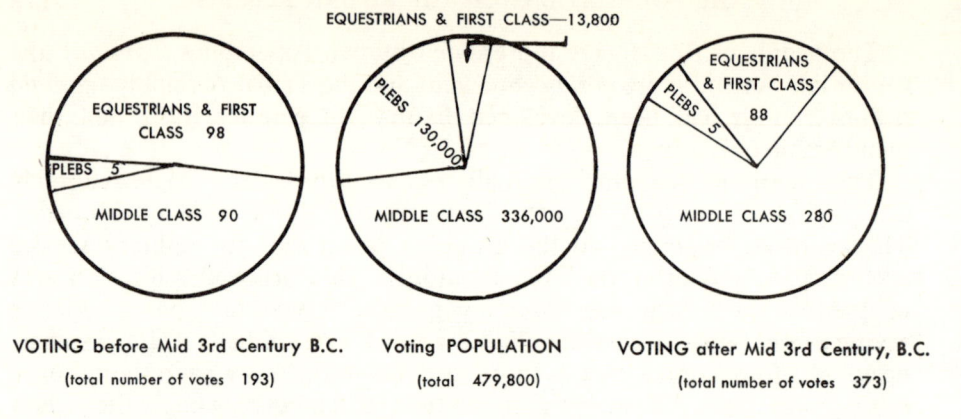

EQUESTRIANS & FIRST CLASS—13,800

- EQUESTRIANS & FIRST CLASS 98
- PLEBS 5
- MIDDLE CLASS 90

VOTING before Mid 3rd Century B.C.
(total number of votes 193)

- PLEBS 130,000
- MIDDLE CLASS 336,000

Voting POPULATION
(total 479,800)

- EQUESTRIANS & FIRST CLASS 88
- PLEBS 5
- MIDDLE CLASS 280

VOTING after Mid 3rd Century, B.C.
(total number of votes 373)

were made up of citizens living in Rome, whereas 31 came from outside the city. Ordinarily, then, citizens scattered over Italy had little voice in the deliberations of the Tribal Assembly. While it was more democratic than the Centuriate Assembly, the peculiar system of voting by tribes made it still a far cry from democracy.

The Tribal Assembly elected the aediles, quaestors, and tribunes, and was also a court for the trial of magistrates. Gradually it became the chief legislative body at Rome.

And so Rome's constitution evolved, shaped this way and that by political expediency, by the force of changing circumstances, just as Athenian democracy had been. As a matter of fact economic and political conditions at Rome in the 5th and 4th centuries B.C. bore a striking resemblance to those in Athens in the period between Solon and Cleisthenes. Rome was, in many ways, herself a city-state, and she too experienced a development from oligarchy to democracy, a conflict known as "The Struggle of the Orders."

Plebeians

The Struggle of the Orders At about the time of the abolition of the monarchy, the people of Rome (excluding slaves) were divided into two distinct social classes: the patricians and the plebeians. The patricians were the aristocrats, the great landowners who had the right to sit in the Senate, to hold the consulship and other magistracies, or to assume important religious offices. The plebeians or plebs, on the other hand, were made up of a number of smaller groups (city artisans, wealthy traders, and peasant farmers), and can be compared to the "Men of the Coast" and the "Men of the Hills" in Athens.

After 500 B.C. the power of the patricians increased at the expense of the plebs. The peasant farmers in particular suffered as a result of declining returns from agriculture, and when money replaced barter it was the aristocrats who were able to buy the new lands that Rome was acquiring, thereby increasing their estates. Moreover, the peasant was not permitted to mortgage his land. The only security he was allowed to put up was his person, whereas the owners of large estates could pay for improvements and afford to tide themselves over bad years. Many plebeians were simply sucked under by hard times. Some gave up their lands completely to become tenants on patrician estates. Some tried to work off their debts, and were punished for their failure by imprisonment, slavery, even execution. But a number managed to stick it out, determined at all odds to hold on to their independence. It was this determination that led them to adopt the tactics described on page 179-180.

The only advantage the plebs held was the fact that they were needed as soldiers to fight for the state. This meant that patricians were forced from time to time to surrender to plebeian demands for political equality if Rome were to be defended against foreign foes or preserved from civil war. The story of the concessions wrung from the patricians is a very complex one. The following table presents it in a simplified form, with the pleb victories listed opposite the traditional dates.

---

494 B.C. — Tribunes of the plebs first created.

457 B.C. — Number of tribunes increased to ten.

450 B.C. — Codification of the Law, the so-called Twelve Tables, in order that every citizen might know his rights and obligations. (See the Source Reading on pages 180-181).

445 B.C. — Intermarriage of plebs and patricians legalized, thereby removing the stigma of plebeian social inferiority. Probably only wealthy plebeians benefited by this change at first.

421 B.C. — Quaestorship opened to plebs.

367 B.C. — Formal right granted to plebs to be elected to one of the two consulships. Also an agrarian law limited the amount of arable land that could be held by any one individual.

366 B.C. — Curule aedileship (instituted by the patricians the year before) opened to plebs. It was a position of prestige for two of the aediles which included the right to occupy the curule chair, the seat reserved for consuls, dictators, and praetors.

356 B.C. — First plebeian dictator.

351 B.C. — Censorship opened to plebs; by 339 B.C. one censor had to be a plebeian.

337 B.C. — Praetorship opened to plebs.

326 B.C. — Enslavement or imprisonment for debt declared illegal and mortgaging of land permitted.

312 B.C. — Plebs eligible for the Senate.

300 B.C. — Plebs elected to higher priesthoods. Appeal allowed against sentence of capital punishment.

287 B.C. — Tribal Assembly's measures (*plebiscita*) to have the force of law even without Senate approval.

Thus in two centuries the plebs had won all their political objectives. The magistracies and the Senate had been opened to them, and measures passed by the Tribal Assembly had become law. In theory at least, Rome

THE MACHINERY OF GOVERNMENT IN CANADA TODAY

(see also p. 177)

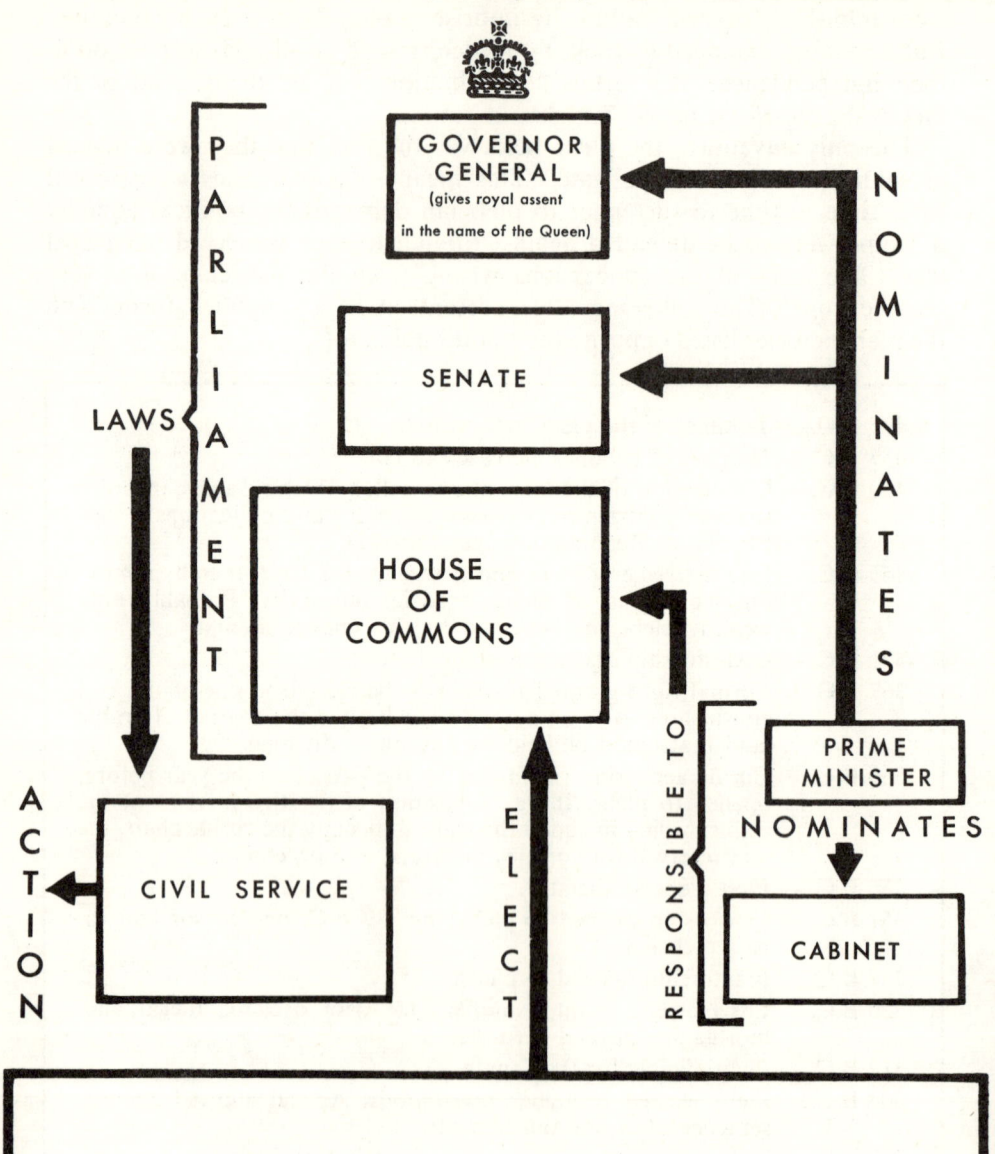

- GOVERNOR GENERAL (gives royal assent in the name of the Queen)
- SENATE
- HOUSE OF COMMONS
- PRIME MINISTER
- CABINET
- CIVIL SERVICE
- CITIZENS OVER 21

PARLIAMENT — LAWS → ACTION

NOMINATES

PRIME MINISTER NOMINATES → CABINET

RESPONSIBLE TO House of Commons

ELECT

CITIZENS OVER 21

THE MACHINERY OF REPUBLICAN GOVERNMENT IN ROME (121 B.C.)[2]

(see also pp. 170-174)

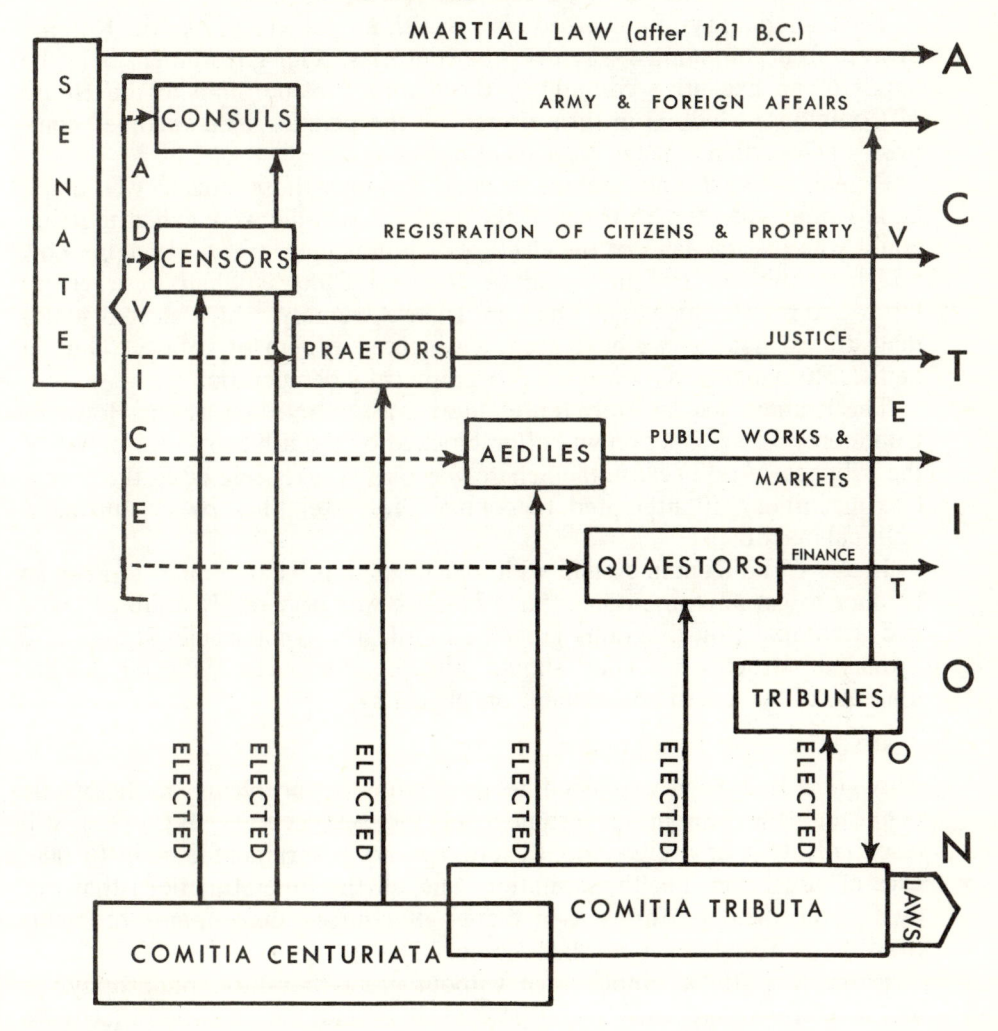

MARTIAL LAW (after 121 B.C.) → A

SENATE

ADVICE

CONSULS — ARMY & FOREIGN AFFAIRS → C

CENSORS — REGISTRATION OF CITIZENS & PROPERTY → T

PRAETORS — JUSTICE →

AEDILES — PUBLIC WORKS & MARKETS → I

QUAESTORS — FINANCE → O

TRIBUNES → N

ELECTED ELECTED ELECTED ELECTED ELECTED ELECTED

COMITIA CENTURIATA

COMITIA TRIBUTA

LAWS

[2] Adapted from F. R. Cowell, *Cicero and the Roman Republic* (Sir Isaac Pitman and Sons, London, 1948), Chart XIII, originally prepared by the Isotype Institute.

was a democracy. Nevertheless, like the Athenian democracy it did not give a vote to everyone, and there were certain contradictions in its working. Some of these practical difficulties will be more evident after a brief discussion of how the constitution actually worked.

A System of Checks and Balances The machinery of Roman government may become clearer if we compare it with our own.

The Canadian system is based on a British model which, like the Roman, evolved in an unplanned way over the centuries. Our government is in the hands of an executive committee, the Cabinet, responsible to the House of Commons, which is in turn elected by the people. The "people" comprise all Canadian citizens 21 years of age and over.

Roman voters did not include women, and only those citizens who could be in Rome when the 35 tribes of the Tribal Assembly were called together could vote on new laws or the election of junior magistrates. Imagine how undemocratic our system would be if every Canadian had to travel to Ottawa to register his vote! The Centuriate Assembly, which elected senior magistrates, had, as we have seen, a weighted system of voting. Thus in neither Roman assembly was the procedure truly democratic.

The Romans had no fully representative assembly such as our House of Commons. Indeed, action was often blocked by the arbitrary veto power of the tribunes. And because the Senate was only weakly related to the executive machinery, it attempted to control the Assemblies by organizing a political machine.

In short, the Roman system with its checks and balances had to *keep* in balance to get anything done. Such balance was very rarely attained. The two accompanying diagrams greatly simplify the comparable Roman and Canadian structures. They should also be compared with the earlier diagram of Athenian government on page 124.

4/SUMMARY AND CONCLUSION

Though it had been expanded to deal with the problems of the whole peninsula, the Roman government of the 3rd century B.C. was still essentially that of a city-state. There was no governmental body to take care of education, health, sanitation, and all the other functions that our own government performs as a matter of course. Only major problems from the peninsula ever reached Rome.

Rome was still a simple town without great literature, philosophy, or art, indeed without many social graces. It is said, for example, that there never was a barber in Italy before one arrived from Sicily in 300 B.C.

We have seen Rome develop from a primitive hill settlement harassed by her neighbours into the leading power in Italy. Rome's expansion was not what would today be called "imperialistic", since in many cases she was forced to protect herself and others from more warlike neighbours. When she conquered Italy, a federation with various gradations of citizens and allies was set up.

Rome tried to use the machinery of her city-state government, with some slight adaptations, to control the entire Italian peninsula. But this complicated Roman government was neither efficient nor democratic. What would happen to the system once the Romans set foot outside Italy? Could they adapt their government to the stresses and strains of empire?

(a)
SOURCE READINGS

APPIAN (A.D. 95-165) was a Romanized Greek who was active in the Roman civil service and wrote a *Roman History* in Greek. In the selection below he portrays the scene in the Senate in 280 B.C. when the Romans were considering the peace proposals made by Pyrrhus after he had defeated the consul Laevinus at Heraclea. Although Pyrrhus sent as his ambassador a famous Thessalian diplomat named Cineas, the Senate was persuaded to reject his overtures by Appius Claudius, "the first clear-cut personality in Roman history." His uncompromising speech on this occasion was memorized and quoted by patriotic Romans for centuries.

The Romans hesitated a long time, being much intimidated by the prestige of Pyrrhus and by the calamity that had befallen them. Finally Appius Claudius, surnamed the Blind (because he had lost his eyesight from old age), commanded his sons to lead him into the senate chamber, where he said: "I was grieved at the loss of my sight; now I regret that I did not lose my hearing also, for never did I expect to see or hear deliberations of this kind from you. Has a single misfortune made you all at once so forget yourselves as to take the man who brought it upon you, and those who called him hither, for friends instead of enemies, and to give the heritage of your fathers to the Lucanians and Bruttians? What is this but making the Romans servants of the Macedonians? And some of you dare to call this peace instead of servitude!" Many other things in the like sense did Appius urge to arouse their spirit. If Pyrrhus wanted the friendship and alliance of the Romans, let him withdraw from Italy and then send his embassy. As long as he remained, let him be considered neither friend nor ally, neither judge nor arbiter of the Romans.

The senate made answer to Cineas in the very words of Appius. They decreed the levying of two new legions for Laevinus and made proclamation that whoever would volunteer in place of those who had been lost should put his name on the army roll. Cineas, who was still present and saw the multitude jostling each other in their eagerness to be enrolled, is reported to have said to Pyrrhus on his return: "We are waging war against a hydra...."

N. Lewis and M. Reinhold, editors, *Roman Civilization: Selected Readings*, Vol. I, *The Republic* (Columbia University Press, 1951), p. 81.

(b)

LIVY (59 B.C.-A.D. 17) wrote a voluminous history of Rome in 142 "books", 107 of which have been lost. (The expression "book" as applied to ancient literature really means a roll of papyrus. Each papyrus roll contained a text that would amount to thirty to fifty pages in modern print.) The history is a patriotic account of Rome, written to glorify the Roman past. The following selection from the second book describes events traditionally attributed to the year 494 B.C. when the plebs, on returning from a campaign, refused to enter Rome. Five such secessions of the plebs are recorded between 494 B.C. and 287 B.C., though some, including this one, are probably fictitious.

Without orders from the consuls the plebeians withdrew to the Sacred Mountain beyond the river Anio, three miles from the city. ... In the city there was great panic; everything was at a standstill because of mutual apprehensions. The plebeians left behind feared violence from the senators, who in turn feared the plebeians remaining in the city, uncertain whether they should prefer them to stay or leave. "How long," they asked, "will the crowd of seceders remain quiet? What will happen if foreign war should break out in the meanwhile? Certainly the only remaining hope is harmony in the citizen body, and harmony must be achieved by fair means or foul." They decided to make their advocate Menenius Agrippa, an eloquent man, and a favourite of the plebeians because he was himself a plebeian born. When he was admitted into the camp he is said merely to have told this tale, in the unpolished old-fashioned style:

"Once when a man's parts did not, as now, agree together but each had its own program and style, the other parts were indignant that their worry and trouble and diligence procured everything for the belly, which remained idle in the middle of the body and only enjoyed what the others provided. Accordingly they conspired that the hands should not carry food to the mouth, nor the mouth accept it, nor the teeth chew it. But while they angrily tried to subdue the belly by starvation the members themselves and the whole body became dangerously emaciated. Hence it became evident that the belly's service was no sinecure, that it nourished the rest as well as itself, supplying the whole body with the source of life and energy by turning food into blood and distributing it through the veins." By thus showing that the plebeians' anger against the senators was like internal sedition in a body, he swayed the men's minds. Negotiations for concord were then undertaken. The terms included a provision that the plebeians should have their own magistrates, who should be sacrosanct and possess power to aid the common people against the consuls; it would not be lawful, moreover, for a patrician to hold this magistracy. In this way tribunes of the people were created.

A History of Rome from Its Origins to A.D. 529 as Told by the Roman Historians, prepared by Moses Hadas (Anchor Books, Doubleday, 1956), pp. 17-18.

(c)

THE TWELVE TABLES were a law code of Roman customs drawn up by a commission of ten patricians about 450 B.C. Until Rome was burned by the Gauls in 390 B.C., this code was inscribed on bronze tablets and exhibited in the Forum. The Twelve Tables were, in a sense, the Ten Commandments of the Romans; schoolboys were memorizing them while Greek children were learning their Homer. For a thousand years this code remained the foundation of Roman law.

TABLE III

When a debt has been acknowledged, or judgment about the matter has been pronounced in court, thirty days must be the legitimate time of grace. ...

Unless they make a settlement, debtors shall be held in bonds for sixty days. During that time they shall be brought before the praetor's court in the meeting place on three successive market days, and the third market day they shall suffer capital punishment or be delivered up for sale abroad, across the Tiber.

TABLE IV

Quickly kill ... a dreadfully deformed child.

If a father thrice surrender a son for sale, the son shall be free from the father.

THE FOUNDATIONS OF THE ROMAN REPUBLIC

TABLE VIII

If any person has sung or composed against another person a song such as was causing slander or insult to another, he shall be clubbed to death.

If a person has maimed another's limb, let there be retaliation in kind unless he makes agreement for settlement with him.

If he has broken or bruised a freeman's bone with his hand or a club, he shall undergo penalty of 300 *as* [bronze monetary measure] pieces; if a slave's, 150.

Any person who destroys by burning any building or heap of corn deposited alongside a house shall be bound, scourged, and put to death by burning at the stake, provided that he has committed the said misdeed with malice aforethought; but if he shall have committed it by accident, that is, by negligence, it is ordained that he repair the damage, or, if he be too poor to be competent for such punishment, he shall receive a lighter chastisement.

If theft has been done by night, if the owner kill the thief, the thief shall be held lawfully killed.

It is forbidden that a thief be killed by day ... unless he defend himself with a weapon; even though he has come with a weapon, unless he use his weapon and fight back, you shall not kill him. And even if he resists, first call out.

No person shall practice usury at a rate more than one twelfth [probably 8½%] ... A usurer is condemned for a quadruple amount.

... a person who has been found guilty of giving false witness shall be hurled down from the Tarpeian Rock. ...

TABLE X

A dead man shall not be buried or burned within the city.

Women must not tear cheeks or hold chorus of "Alas!" on account of funeral.

When a man is dead one must not gather his bones in order to make a second funeral. An exception [in the case of] death in war or in a foreign land. ...

Anointing by slaves is abolished, and every kind of drinking bout.

TABLE XI

Intermarriage shall not take place between plebeians and patricians.

Lewis and Reinhold, *Roman Civilization: Selected Readings,* Vol. I, *The Republic,* pp. 103, 104, 106, 107, 108, 109.

FURTHER READING

1. GENERAL

* BARROW, R. H., *The Romans* (Penguin, 1949)
BOAK, A. E. R., *A History of Rome to 565 A.D.* (Macmillan, 1955)
CARY, M., *A History of Rome down to the Reign of Constantine* (Macmillan, 1954)
† COWELL, F. R., *The Revolutions of Ancient Rome* (Thames and Hudson, 1962)
† * DUDLEY, D. R., *The Civilization of Rome* (Mentor, 1960)
† DURANT, W., *Caesar and Christ* (Simon and Schuster, 1944)
FOWLER, W. W. and CHARLESWORTH, M. P., *Rome* (Oxford, 1952)
FRANK, T., *An Economic History of Rome* (Johns Hopkins, 1927)
GRANT, M., *The World of Rome* (Weidenfeld and Nicolson, 1960)
† GRANT, M. and POTTINGER, D., *Romans* (Nelson, 1960)
GROSE-HODGE, H., *Roman Panorama* (Cambridge, 1944)
HEICHELHEIM, F. M. and YEO, C. A., *A History of the Roman People* (Prentice-Hall, 1962)
† MONTANELLI, I., *Rome: The First Thousand Years* (Collins, 1962)
ROBINSON, C. E., *A History of Rome from 753 B.C. to A.D. 410* (Methuen, 1956)
* ROSTOVTZEFF, M., *Rome* (Galaxy, 1960)
† * STARR, C. G., *The Emergence of Rome as Ruler of the Western World* (Cornell, 1953)
† STOBART, J. C., *The Grandeur that Was Rome* (Sidgwick and Jackson, 1961)

2. ROME'S EARLY NEIGHBOURS

BLOCH, R., *The Etruscans* (Thames and Hudson, 1958)
* PALLOTINO, M., *The Etruscans* (Penguin, 1955)
VON CLES-REDEN, S., *The Buried People: A Study of the Etruscans* (Hart-Davis, 1955)
WHATMOUGH, J., *The Foundations of Roman Italy* (Methuen, 1937)
WOODHEAD, A. G., *The Greeks in the West* (Thames and Hudson, 1962)

3. ROME AND HER INSTITUTIONS

ADCOCK, F. E., *Roman Political Ideas and Practice* (Michigan, 1959)
BLOCH, R., *The Origins of Rome* (Thames and Hudson, 1960)
† * COWELL, F. R., *Cicero and the Roman Republic* (Penguin, 1956)
FRANK, T., *Roman Imperialism* (Macmillan, 1914)
HOMO, L., *Primitive Italy and the Beginnings of Roman Imperialism* (Knopf, 1926)
──────, *Roman Political Institutions from City to State* (Knopf, 1929)
SCULLARD, H. H., *A History of the Roman World from 753 to 146 B.C.* (Methuen, 1961)
WOLFF, H. J., *Roman Law: An Historical Introduction* (Oklahoma, 1951)

7. The Fall of the Republic

What man is so indifferent or so idle that he would not wish to know how and under what form of government almost all the inhabited world came under the single rule of the Romans in less than fifty-three years?
POLYBIUS, *Histories*

It is due to our own moral failure and not to any accident or chance that, while retaining the name, we have lost the reality of a republic.
CICERO

How does the acquisition of world power change the pattern of a country's existence? It makes events *outside* its boundaries of vital importance. For Rome, this meant watching carefully the moves of the other four world powers of the time: the great empire of Carthage in the west, and the three Hellenistic monarchies of Egypt, Syria, and Macedon in the east.

Between 264 and 133 B.C. Rome conquered the Mediterranean world. There followed a triumphal procession of great generals—Marius, Sulla, Pompey, Caesar—whose legions won Rome eternal glory. But the cost of success came high. Rome paid with the fall of her Republic.

1/THE PUNIC WARS

Carthage proved to be one of Rome's most dangerous and stubborn foes. Situated in North Africa on what is now the Gulf of Tunis, the city of Carthage was ruled by an oligarchy of rich merchants whose ancestors had come, some time in the 9th century B.C., from the ancient Phoenician (= *Punic,* in Latin) city of Tyre.

All that is known of Carthaginian society does not add up to a very attractive picture, even when due allowance is made for the bias of Greek and Roman sources. Archaeological excavation has confirmed the gruesome tales of human sacrifice, particularly the death by burning of young children. Carthage's vast armies were made up of conscripted subjects from Libya, Sardinia, and Spain, as well as troops hired from Algeria and Morocco. For the general who succeeded in whipping this unruly mass into an efficient fighting machine there could be great glory; for the one who failed there might be crucifixion. Plutarch's estimate of the Carthaginians is probably not far wrong:

...a hard and gloomy people, submissive to their rulers and harsh to their subjects, running to extremes of cowardice in times of fear and of cruelty in times of anger; they keep obstinately to their decisions, are austere and care little for amusement or the graces of life.

In time Carthage came to dominate the western end of the Mediterranean and part of Sicily. As befitted a mighty commercial power, the Carthaginians mantained an imposing fleet which was in the habit of sinking at sight any foreign vessel that crossed their sea lanes.

Brigands on the northern tip of Sicily, finding themselves threatened by Syracuse, accepted aid from Carthage. Then an appeal was sent to Rome for deliverance from the Carthaginian garrison. Determined not to be shut out of the Straits of Messana, Rome answered the call. The result: war with Carthage.

The Raven The first war with Carthage lasted twenty-three years (264-241 B.C.). Rome realized from the beginning that if she were to push the Carthaginians out of Sicily she would sooner or later have to create a navy. Accordingly, in 260 B.C. the Senate called for the construction of 100 quinqueremes. The quinquereme was the pride of the Carthaginian navy, a one-deck vessel with from 20 to 60 oars, five men to an oar. Although heavier and slower than the trireme, it was better suited for ramming and sinking an enemy. There was another reason for Rome's decision to build quinqueremes. In a trireme every rower had to be a skilled oarsman; in a quinquereme one man directed the sweep and the other four had only to supply muscle power. Rome had no time now to waste in training multitudes of skilled oarsmen—and at that it was said that she trained her rowers sitting on benches on the dry land.

The First Punic War

The Roman fleet

The Romans now added an ingenious device to their quinqueremes. It was known as the *corvus* ("raven"), and was simply a gangplank thirty-six feet long and four feet wide with a heavy spike (the raven's "beak") on the underside at the outboard end. The corvus could be raised upright by means of ropes running through pulleys fastened to the mast. Equipped with this corvus, a quinquereme could close with an enemy vessel, drop the raven, which, with its spike, would hold fast to the enemy deck, and then send a boarding force rushing across the gangplank. The boarding force consisted of no fewer than 120 legionaries. In this way the Romans managed to convert sea battles into land battles and so make use of their prime resource, their magnificently disciplined army.

The untried Roman fleet met the mighty Carthaginian armada in 260 B.C. off Mylae, not far from Messana. As the Carthaginian admiral recklessly pressed a frontal ramming attack, he must have been stunned to hear the Roman gangplanks thud down. He was doubtless even more stunned at the final outcome of the encounter: Carthage lost 44 ships and 10,000 men in the engagement. The Senate, scenting victory and determined to win Mediterranean naval supremacy, pushed ahead with its ship-building programme. It was rewarded by three more naval victories. Finally a Roman expeditionary force was put ashore in North Africa.

But the Carthaginians were not going to let an invader molest their home land with impunity. The Romans suffered a crushing defeat at the hands of stubborn Carthaginian troops bolstered by a hundred elephants. Nor were the Carthaginians idle on the seas. After further naval defeats they crucified some of their admirals, and began to construct lighter ships, swifter and more seaworthy than the lumbering Roman quinqueremes with their clumsy gangplanks. By these means the Carthaginians retrieved their mastery of the seas and were at length able to establish their most brilliant general, Hamilcar Barca (Barca = "Blitz" or "Lightning") in Sicily. This young commander's lightning raids on the Italian coast convinced the Romans that their only hope of victory lay on the sea.

And so again Rome constructed a fleet, this time by calling on the wealthiest citizens to advance loans—repayable only in the event of victory! Meanwhile Carthage became so engrossed in the conquest of further lands in Africa that the Sicilian operation took second place. The final battle was fought off the western tip of Sicily in 241 B.C. The Romans sank 50 ships and captured another 70. Carthage was beaten.

Rome came out of the war having won her first province, Sicily, and with the western Mediterranean as her domain. The war had been costly enough: 500 ships and 200,000 men. But the gains were also great. Sicily brought to Rome an annual tribute of half a million bushels of wheat, and Romans came into even closer contact with the Greek culture of the colonies on Sicily. More important still, Rome had developed an excellent fighting machine, and her allies had stuck by her through all twenty-three years of the war. It was a good omen.

THE PUNIC WARS

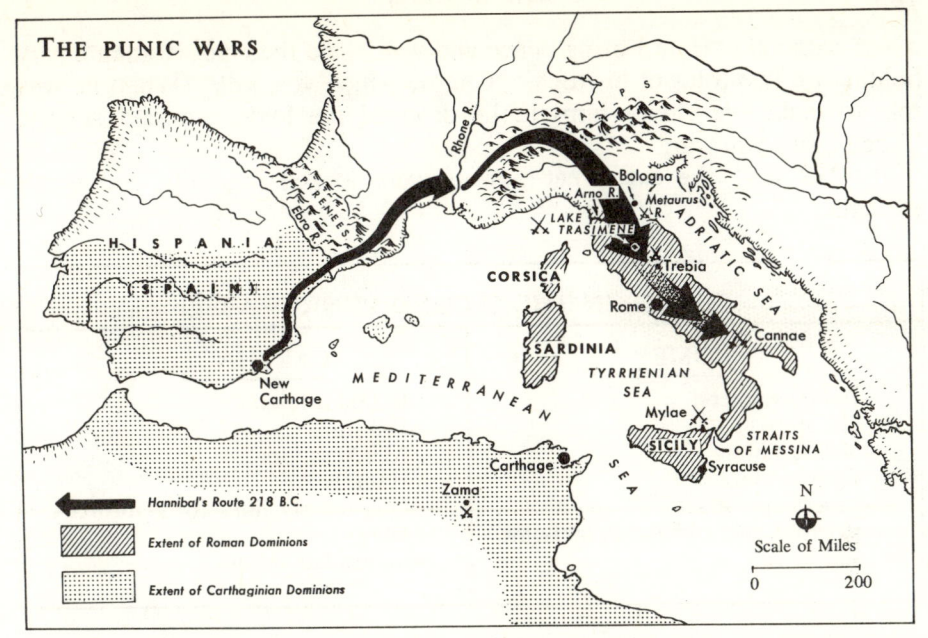

The Aggressor Unmasked In the years after the successful first round with Carthage, Rome conquered the Gauls in the Po valley and stole Corsica and Sardinia from Carthage while the Carthaginians were preoccupied with a revolt in Africa. But now there was a new worry. Carthage was beginning to expand in Spain, an expansion which was directed by that same Hamilcar who had been forced to capitulate to Rome in Sicily only a few years earlier. Hamilcar extended Carthaginian territory in Spain and trained the conquered Spaniards in warfare, a programme later carried on by his son-in-law, Hasdrubal.

Rome watched all these goings-on warily, hopeful that Carthage was not as aggressive as she appeared. Then in 226 B.C. both sides agreed to partition Spain along the line of the River Ebro. Still suspicious of Carthaginian intentions, Rome was deliberately provocative. She technically violated the treaty by allying herself with Saguntum, a city south of the Ebro in the Carthaginian section of Spain.

Now if Carthage had no aggressive intentions she could ignore the challenge, refuse to call Rome's bluff, and leave Saguntum alone. If, on the other hand, she was planning to use Spain as a springboard for an attack on Italy, she would not allow this alliance to continue. We can imagine that Rome watched her small Spanish ally with apprehension, and in 219 B.C. she got her answer. In that year Hamilcar's son, Hannibal, who at twenty-five had succeeded to the Carthaginian command, advanced against Saguntum and the city fell after a long siege. Rome, busy elsewhere, did not send aid, but after Saguntum was captured dispatched an embassy to Carthage demanding the evacuation of Saguntum and the surrender of Hannibal as a war criminal. "Here I bring peace and war," said the chief Roman envoy holding up two folds of his toga. "Choose which you will." When he was told to make the choice himself, he dropped the folds and said simply, "Then I give you war."

By 218 B.C. the two great powers were locked in a life and death struggle.

RIVALS IN THE SECOND PUNIC WAR	
ROME	CARTHAGE
—70,000 soldiers —220 ships —60,000 sailors —faithful allies —superiority at sea —patriotic spur of fighting on home soil	—80,000 soldiers —100 ships —25,000 sailors —Hannibal —long supply lines for reinforcements —unfamiliar terrain

Men Against Mountains Rome thought that hers would be the choice of battlegrounds; so she sent one army to southern Gaul and rushed another to Sicily, on its way to Africa. But she reckoned without the genius of Hannibal, a superb leader whose men would follow him anywhere. The Roman historian Livy draws a brilliantly-hued picture of the young commander.

Upon his first arrival in Spain Hannibal became the centre of attention in the whole army. The old soldiers thought that a younger Hamilcar had come back to them. They saw in him the same features, the same liveliness of expression, the same fire in the eyes. But shortly his resemblance to his father was only the least among their reasons for devotion to him. ... There was no leader under whom the army fought with greater confidence and daring. When danger was to be faced it was Hannibal whose spirit was the boldest, and in a crisis his strategy was the shrewdest. Under no hardship did his energy wane or his spirits flag. He could face heat or cold with equal endurance. His appetite for food and drink were controlled by hunger and not by pleasure. His waking and sleeping were not fixed by day and night. What time remained after the task in hand was done he gave to sleep, and this without any need of soft bed or quiet. Many a time he could be seen lying on the ground among the sentries and pickets off duty, covered only with a soldier's cloak. His dress was no different from that of his fellow-soldiers, but his weapons and his horses were of the finest. He was the best among cavalry and infantry alike, always the first to go into battle and the last to leave any clash of arms.

Never did Rome have a more dangerous foe.

Hannibal was convinced that if Rome were ever to be beaten it must be in Italy, where he could, by liberating her "unwilling allies", smash her power once and for all. His goal was clear: he must isolate Rome from the rest of Italy. To this end he moved swiftly. Commanding 40,000 infantry,

6000 cavalry, and 37 elephants, he left New Carthage (Cartagena) in southern Spain in April, 218 B.C., was north of the Ebro in June, and to the Rhone by mid-August. Just three days before the Roman army arrived to intercept him, he slipped across the Rhone and got clean away on his march to the Alps.

It was autumn when Hannibal reached the mountains, and the treacherous passes, inhabited by hostile tribes, were already deep with snow. Yet he grimly pushed on. Nothing else he ever did has so caught the imagination of the world.

Crossing the Alps

On the ninth day they arrived at the summit of the Alps.... Exhausted and discouraged as the soldiers were by many hardships, a snow-storm... threw them into great fear. The ground was everywhere covered deep with snow when at dawn they began to march, and as the column moved slowly on, dejection and despair were to be read in every countenance. Then Hannibal, who had gone on before the standards, made the army halt on a certain promontory which commanded an extensive prospect, and pointing out Italy to them... he told them that they were now scaling the ramparts not only of Italy, but of Rome itself; the rest of the way would be level or downhill; and after one or at the most two battles, they would have in their hands and in their power the citadel and capital of Italy....

But... above the old, untouched snow lay a fresh deposit of moderate depth, through which, as it was soft and not very deep, the men in front found it easy to advance; but when it had been trampled down by the feet of so many men and beasts, the rest had to make their way over the bare ice beneath and the slush of the melting snow. Then came a terrible struggle on the slippery surface, for it afforded them no foothold, while the downward slope made their feet the more quickly slide from under them.... But the baggage animals... would sometimes even cut into the lowest crust, and, pitching forward and striking out with their hoofs, as they struggled to rise, would break clean through it, so that numbers of them were caught fast, as if entrapped, in the hard, deep-frozen snow.

At last, when men and beasts had been worn out to no avail, they encamped upon the ridge, after having, with the utmost difficulty, cleared enough ground even for this purpose.... The soldiers were then set to work to construct a road across the cliff—their only possible way. Since they had to cut through the rock, they felled some huge trees that grew near at hand, and lopping off their branches, made an enormous pile of logs. This they set on fire, as soon as the wind blew fresh enough to make it burn, and pouring vinegar over the glowing rocks, caused them to crumble. After thus heating the crag with fire, they opened a way in it with iron tools, and relieved the steepness of the slope with zigzags of an easy gradient, so that not only the baggage animals but even the elephants could be led down. Four days were consumed at the cliff, and the animals nearly perished of starvation; for the mountain tops are all practically bare, and such grass as does grow is buried under snow.

At long last, after fifteen days, the tortuous crossing was accomplished. The great army had won a strange battle—but at what a cost. When Hannibal assembled his force on the sunny plains of northern Italy, there were left 26,000 infantry, 4000 cavalry, and 25 elephants.

Even if they had tried, it would have been impossible for the Romans to guard all the passes in the western Alps. Now that the crafty Carthaginian had slipped through they hoped to catch him at one of the river crossings in northern Italy. But they were stalking very wary game. And despite the

Perhaps no more striking procession ever wended its way than this—the thin line of Hannibal's troops silhouetted against the white-mantled Alps with the gigantic beasts lumbering in the lead, their fierce trumpeting echoing and re-echoing in the icy gorges as the sharp cries of commanders rang out above the muffled tramp of thousands of feet. The ancient historian Polybius records that in this agonizing crossing Hannibal's elephants "were of the greatest service to him; for the enemy never dared to approach that part of the column in which these animals were, being terrified by the strangeness of their appearance." And well they might be. Only the Carthaginians ever succeeded in breaking in African elephants, which are much more savage than the Indian species. How strange these 3rd century B.C. "tanks" must have looked in battle! Little wonder they were a terror to the Romans at the River Trebia. The Carthaginians' fearsome advantage was, however, short-lived. All but one of the elephants was dead by the next spring, Hannibal's mount alone surviving the war.

fact that they were fighting on their own home soil the Romans were destined to suffer only defeats. Three major ones followed in quick succession: Trebia (218 B.C.), Lake Trasimene (217 B.C.), and Cannae (216 B.C.).

The Master Strategist When it became clear to the Romans that Hannibal, not they, had chosen the battleground and that the battleground was Italy, the African expedition was cancelled and the combined Roman armies totalling 40,000 men assembled at the River Trebia one bitterly cold December morning. Before the Romans had even had their breakfast Hannibal sent a weak cavalry detachment across the river against them. It was easily defeated, and the Romans exultantly charged through the ice-cold river after it. They had fallen into Hannibal's trap. A strong Carthaginian force was lying in wait for them, and they were ambushed.

Trebia

Winter came and Hannibal rested at Bologna. Rome—who had lost 30,000 men at Trebia—licked her wounds and feverishly raised new armies. It was obvious that north Italy would have to be abandoned to Hannibal.

The next May Hannibal crossed the Apennines. Then misfortune struck him. In the marshes flooded by the River Arno he contracted opthalmia and lost the sight of one eye—yet still, riding high on his elephant, got his army through in four days. He proceeded to march south towards Rome, ravaging the land as he went, while an outnumbered Roman army of 25,000 trailed along in his wake declining battle. But once again Hannibal outwitted them. In the hills surrounding Lake Trasimene, the Carthagin-

Trasimene

ians waited in the early mists of morning until the unsuspecting Roman army marched into a narrow defile. Two hours later 15,000 Romans lay dead and most of the rest were captured.

Panic swept Rome. In their desperation the people elected a dictator, Quintus Fabius Maximus.

Fabius was soon nicknamed Cunctator ("Delayer"). His strategy was to avoid a pitched battle with the wily Hannibal until the Carthaginian could be forced to fight under conditions in which he could not use his cavalry. It was with this purpose that the patient Fabius dogged Hannibal's footsteps, always waiting for his chance, steadfastly refusing to aid Rome's allies whose lands the Carthaginians were laying waste. At last Fabius saw his opportunity. In the unfamiliar mountain country of Campania Hannibal found himself facing a pass commanded by Fabius—but what happened you will have to read in the Source Reading.

It was inevitable that these delaying tactics of Fabius would be criticized at Rome. Hannibal was wandering through Italy almost at will, and the wonder was that the allies still stood by Rome. Impatient Romans rejoiced to see Fabius out of office in 216 B.C., and readily granted permission to the eager consuls to force a battle.

Cannae

Late in the summer the consuls found Hannibal near Cannae. Here in the plains by the Adriatic, a perfect place for a cavalry engagement, the 50,000 Romans faced even odds: 40,000 Carthaginian infantry and 10,000 cavalry. Although the Romans were greatly inferior in their numbers of cavalry, they had their legions; and Hannibal well knew how strong these were. Accordingly he drew up his line in a crescent-shaped formation bulging out in the centre towards the Romans, while on the wings he stationed his cavalry. The heavy Roman infantry drove inexorably forward, forcing Hannibal's centre backwards until the crescent sagged into a hollow. On into this hollow they pushed, unaware that the Carthaginian cavalry, having routed the Roman wings, was about to strike again. With a wide sweep to the rear of the legionaries the Carthaginians completely surrounded the Romans, crushing them, vice-like, so closely together that they could scarcely move.

Cannae was a Roman slaughter-pen. 35,000 Romans perished; only 5700 Carthaginians were lost. With this shattering blow to Rome's prestige many towns in Samnium, Apulia, Lucania, and Bruttium went over to Hannibal. Nevertheless Latium, Umbria, and Etruria still remained loyal, so that Rome retained the loyalty of central and northern Italy. And Rome herself was safe. Hannibal did not have and could not get the siege engines necessary to storm her walls.

The Romans had been taught a lesson at a terrible price: they had learned that they must revert to Fabian tactics to wear down their enemy. Hannibal, on the other hand, now tried to surround Italy with a circle of foes. He encouraged the Gauls to the north and the Greeks to the south in

Sicily to be hostile; in the west he advised his home government to prosecute the war in Spain and Sardinia; and in the east he achieved an alliance with Philip V of Macedon. These various operations meant, of course, that he must continue on his own in Italy without reinforcements. He found himself free to roam in southern Italy, but whenever he ventured very far to the north he had difficulty feeding his troops because the Romans either burned the crops or carried them inside the walls of their fortified cities.

And so the see-saw contest continued, with neither side strong enough, as in the Peloponnesian War, to knock out the other. One factor could tip the balance—a Roman general the equal of Hannibal. He appeared in the person of Publius Cornelius Scipio, whose father had been killed fighting the Carthaginians in Spain.

The Balance Shifts   At his own request, young Scipio was appointed commander in Spain in 210 B.C. Technically he was not eligible for the imperium since he was only twenty-five and had held no rank higher than aedile, but the constitutional bars were waived. Assuming the rank of proconsul, this enterprising young commander proceeded to arm his troops more heavily, as well as to adopt a new troop formation similar to the one which Hannibal had used at Cannae. By 207 B.C. he had ended the Carthaginian empire in Spain. *Scipio*

But there was a grave omission. Scipio had not been able to prevent Hannibal's brother Hasdrubal from getting his troops out of Spain, and by the spring Hasdrubal had reached the Po valley with 30,000 men, bent on joining Hannibal in central Italy. The Romans now detailed one army to watch Hannibal and a second to try to intercept Hasdrubal. They had one crucial assignment: to keep the brothers apart. For once fate favoured the Romans. They captured the despatch riders Hasdrubal sent to Hannibal with a message arranging a meeting-place. Now they knew exactly where Hasdrubal planned to march, and they acted immediately. Leaving only a skeleton force in the south, the Roman commander there managed by forced marches to cover 240 miles in six days and link up with the northern army. Hasdrubal found himself faced with not one, but two Roman armies at the Metaurus River. At last, after eleven years, Rome savoured the taste of a major victory. *Metaurus*

A week later Hannibal received a grisly announcement of the disastrous defeat. Some Roman horsemen, riding hard past his camp, threw an object over the walls. It was his brother's head.

When Scipio returned from mopping up Spain in 206 B.C., he was elected consul for the next year. Fired by his conquests, he set about convincing the Senate that the war must be carried to Africa. Even though further desperate efforts to reinforce Hannibal were failing miserably, nothing, Scipio argued, short of the crushing of Carthage would leave

Although no one knows exactly what Hannibal looked like, he may have resembled this idealized portrait bust of him found at Capua. His eyes have the steely glint of the determined commander, but are rimmed with the weariness of a thousand marches; his mouth shows the sensitivity which made him beloved by his men, yet bears the marks of some deep disappointment. We can imagine that the profound sadness on this noble face has been etched there by the bitter knowledge of his failure as he looks back on the land of the enemy—the Italy he has come to love so well—after his recall to Africa.

Rome secure. In 204 B.C. a Roman expeditionary force of 30,000 sailed for North Africa, with Hannibal still in Italy.

Carthage was hard to break, and Scipio campaigned for over a year before the Carthaginians sued for peace. But while the terms were under discussion a hero returned: Hannibal landed in Africa. Livy says that the great commander left Italy with as much grief as most men feel on going into exile, and it was probably true. He had entered Italy in 218 B.C. when he was twenty-nine. Now it was 202 B.C. and he was forty-five. He had never been defeated in Italy; yet the war was over—or so it seemed.

Zama   Hannibal's arrival in North Africa gave the war party enough courage to make a last-ditch stand. It came at Zama. Hannibal faced Scipio, his 40,000 men equalling the Roman force although his cavalry was slightly weaker. Before the battle the two generals met and Hannibal proffered peace terms, which Scipio promptly rejected.

Scipio came near regretting his action. It was never safe to underestimate the skill of this master strategist, and the tactics Hannibal used at Zama were copied in tank battles in World War II. The great general ordered his weaker cavalry to pretend flight, in the hope that the Roman cavalry would be drawn away after it. Then he planned to throw all his infantry against Scipio, at the same time holding back a reserve force of veterans. For a time it seemed that he had manipulated the Roman cavalry into his scheme. They did rout the Carthaginian wings—but before Hannibal could get in a decisive blow with his reserves they returned to decide the battle. It was Cannae in reverse. Hannibal escaped, but his army was cut to pieces. The long war was over.

The peace terms were harsh. Carthage retained control over Tunisia, but in future had to have Rome's permission to make war; her elephants and her navy were surrendered; and an indemnity of 10,000 talents was to be paid in fifty annual instalments. She had been reduced to utter dependence on Rome.

Epilogue for Two Heroes The nobility of Hannibal has been allowed to overshadow the greatness of his opponent. "Conquered Hannibal has captivated the world's imagination, but the victorious Scipio stands aloof in his triumph," writes a modern biographer. After Zama, Scipio returned to Rome almost a living legend. "Africanus" was added to his name to perpetuate the memory of his greatest victory (compare Alexander of Tunis and Montgomery of Alamein)—an honour richly deserved, for in a very real sense Scipio was the founder of the Roman Empire. Back in Rome, however, he did not turn out to be a conspicuous success as a politician and statesman, and eventually the rivalries of the Senate forced him back into private life. The time was not yet ripe for an uncrowned king of Rome.

And what of Hannibal? In the years following he did his best to rebuild his shattered city's fortunes. But Rome could not forget or forgive, and ultimately he fled to Syria. Still the long arm of Rome reached out after him. Finally in 183 B.C. he committed suicide, "the noblest failure in antiquity."

Hannibal lost no major battle in Italy; he slaughtered at least 100,000 Roman soldiers in fifteen years; and he devastated the Italian countryside, destroying 400 towns in southern Italy alone. Yet he failed to win the war. Why?

For one thing, although Hannibal marched to within sight of Rome, he could not capture the city because of his lack of siege equipment. For another, Rome's Italian allies refused to join him in sufficient numbers to tip the balance against her. Moreover the home government was unable or unwilling to provide Hannibal with sufficient reinforcements to give him numerical superiority. And finally, the Romans never gave up, even when the outlook was blackest. When Hannibal reached the very gates of Rome in 211 B.C., the Romans were so confident of regaining the site of his camp that it was auctioned off—and sold at its normal price!

The Battle of Zama was a turning-point in the history of the ancient world. Henceforth Rome was all-powerful in the western Mediterranean. It was a significant beginning. If luck should be with her in the eastern Mediterranean, she might have a chance to impose on the world a unity unknown since the days of Alexander the Great.

2/THE NEW IMPERIALISM

Now that Rome was the strongest power in the Mediterranean, only the reckless would challenge her. In this case the reckless were to be found among the successors of Alexander the Great. Philip V of Macedon and

Antiochus III of Syria became engaged in a complicated series of wars with Rome, sorely trying her patience in the Middle East. Any brief account such as this has to oversimplify the many complex issues and diplomatic problems of this period. Every year envoys arrived from foreign kings, states, and cities, bearing tales and asking for aid. Somehow the Romans had to sort out the claims and counter-claims. It was all very complicated.

A sample of some of the many questions with which the Senate had to grapple in the first two or three decades of the second century B.C. would include such problems as the following: What was the relationship of King Attalus of Pergamum to the Celts of Galatia? What were the rights and the treaty obligations of each of the many rich and important free Greek cities of Asia Minor? Who had the rightful claim to the territories of Phrygia, Pisidia, Lycaonia? Was Antiochus of Syria entitled to lord it over Cilicia? Should the Ptolemies of Egypt possess South Syria? How could the Aetolian League be prevented from stirring up trouble for Rome in the Balkan peninsula and in the Near East? What lay behind the diplomatic manoeuvres of Philip of Macedon, how strong was he, was he ill-intentioned towards Rome and, if so, should he be checkmated and, if so, how? Was it true that Philopoemen, General of the Achaean League, had annexed the city of Sparta, destroyed its walls and sold thousands of its inhabitants into slavery and, if so, what was Rome's position under the Roman-Spartan treaty of 191 B.C.?[1]

The Third Punic War

Little wonder Rome finally lost both patience and restraint. Two events in 146 B.C. demonstrate how she had become imperialistic: in the east Corinth was sacked and burned, its treasures carried off and its inhabitants sold into slavery; in the west Carthage was destroyed in a third Punic War (149-146 B.C.) waged on a flimsy excuse, the inhabitants massacred or enslaved, and the very land on which the city had stood sowed with salt. There could be no doubt now about the mastery of the Mediterranean from one end to the other—especially when, in 133 B.C., the king of Pergamum died, leaving his kingdom to Rome. A few years later Pergamum was organized as the Roman province of Asia. It was the seventh province Rome had acquired within a century: Sicily and Sardinia-Corsica (227 B.C.), Hither and Farther Spain (197 B.C.), Macedonia (147 B.C.), Africa (146 B.C.), and Asia (129 B.C.).

The Greek historian Polybius (200-117 B.C.), whose *Histories* are quoted at the head of this chapter, saw Roman imperialism as an unexampled blessing, and wrote to show his fellow Greeks that further resistance to Roman strength was foolishness. But not all ancients approved of Rome's course. The new Roman diplomacy was strikingly criticized when two Roman legates who had been sent to the last king of Macedon returned to Rome in 172 B.C. The legates came under the fire of certain senators, and the criticism of their conduct may be read on pages 222-223 at the end of this chapter. Something had gone wrong with Rome. We must now consider the effects that her conquest of the Mediterranean world had on Rome herself.

[1] Cowell, *Cicero and the Roman Republic*, pp. 31-32.

Changes in the Wrong Direction Rome's victories from 264 to 133 B.C. affected her politically, economically, socially, and culturally. It is no exaggeration to say that Rome was never the same again once she had sallied forth on the conquest of the Mediterranean.

In the first place, the Roman system of government was not equal to the strains of empire; the changes were in the wrong direction. The expense involved in winning elections plus the outlay necessary for magistrates to conduct their unsalaried offices meant that the middle class was automatically ruled out from the magistracies. Consequently there arose a new aristocracy, an aristocracy of wealth and office. Moreover, the senatorial aristocracy was not willing to admit more families to its ranks. From 200 to 146 B.C. twenty-five senatorial families dominated the government to such an extent that only five men whose ancestors had not held office became consuls. The governor sent to the provinces each year was an unsalaried ex-consul or praetor assisted by a quaestor, and if the governor decided to enrich himself at the expense of the provincials not a great deal could be done about it. True, a Senate jury was set up to try cases of extortion, but senators were somewhat reluctant to condemn men of their own social standing.

Political consequences of empire

The fact was that the system of tax collection lent itself to great abuses. In many provinces the collection of taxes was undertaken not by a group of imperial civil servants but by a private company of professional tax-collectors (*publicani*) who bid for the privilege. The highest bidder won the tax contract, and any taxes collected over and above the bid were his source of income. Is it any wonder that the New Testament hardly ever mentions publicans without the accompanying phrase "and sinners"?

Along with the great wealth that poured into Rome came a change in Roman agriculture. During the second Punic War Hannibal had devastated southern Italy, and the small independent farmer, who was liable for military service, had been called away from his farm to serve for long periods in the army. When he returned, his land might have been seized for debt, or he might find it hopeless to compete with the slave labour and tribute grain which the conquests had brought to Italy from the provinces. Under such circumstances great Roman ranches (*latifundia*) mushroomed, financed by wealthy owners and manned by slaves, while earlier laws limiting the size of estates came to be openly disregarded.

Economic consequences

Where, then, could the landless peasants go? Even in the 2nd century B.C. all roads led to Rome. The city became a vast capital with a population of half a million, the landless who flocked there merging into a drifting unemployed mob which became the plaything of politicians. As the gap between rich and poor widened, certain Roman observers thought they detected a moral degeneration as well.

The collection of taxes, the mechanics of banking, the expanded commerce, all of these needed to be handled by a group devoted to business.

Social consequences

Originally the Equestrians had been the members of the cavalry, and had been enrolled in 18 centuries (see page 173). But in the 3rd century B.C. the censors had drawn up a new list of those who could afford cavalry service. These men came to be called the Equestrian Order, a group midway between the Senate and the people.

The Equestrians profited richly from Rome's conquest of the Mediterranean world. Singly or in companies they contracted to collect taxes, to supply the armies with equipment and stores, or to build roads, bridges, aqueducts, and temples. Many of them became well-to-do bankers and money-lenders. They were, however, blocked from membership in the Senate because a senator was forbidden to take public contracts or to engage in overseas trade. Hence for political influence they had to bide their time—and it was not long in coming. In the meantime, these

THE ROMAN REPUBLIC'S CONQUEST OF THE MEDITERRANEAN (265-44 B.C.)

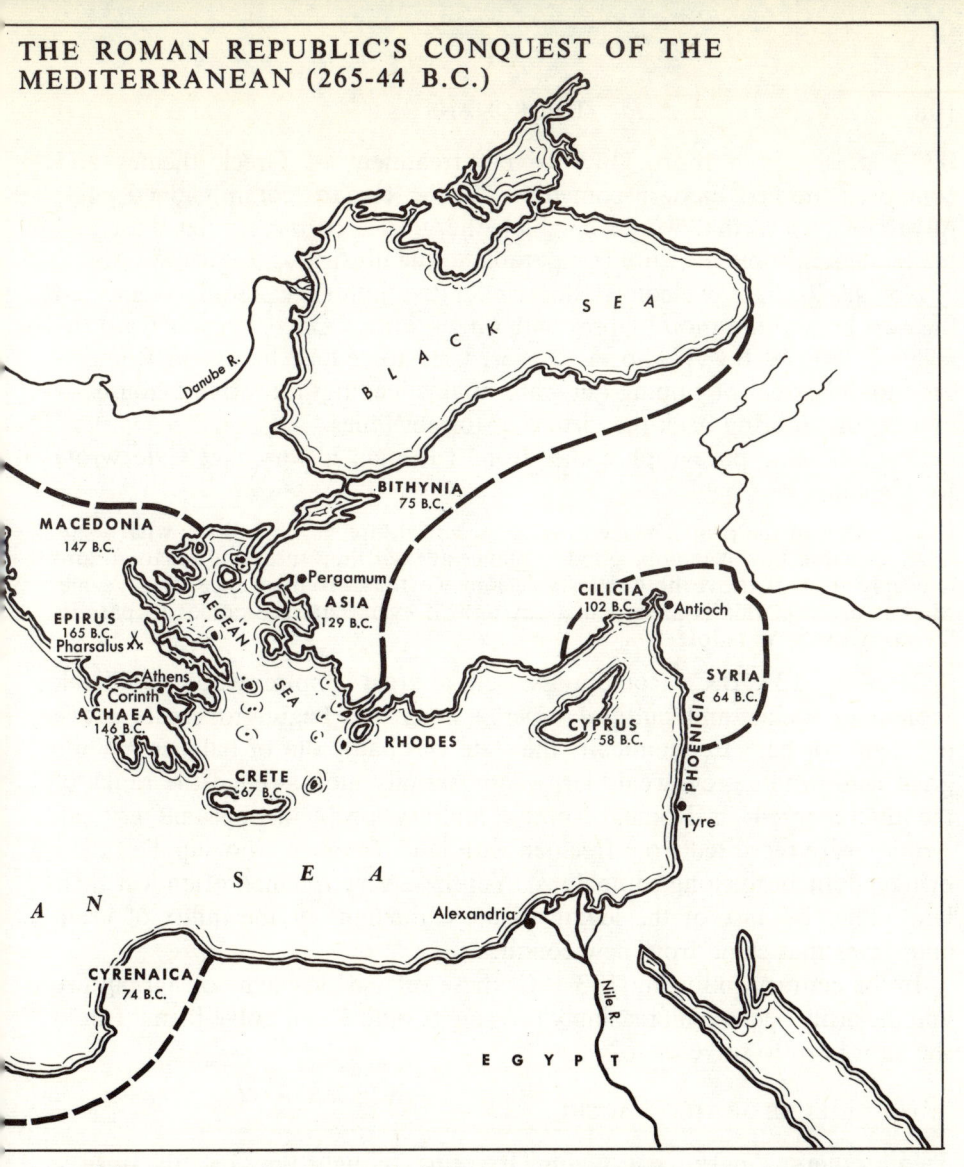

wealthy men could always build great houses filled with costly works of art and be waited on hand and foot by retinues of slaves who served elaborate banquets on silver plate.

Many of the trappings of these rich Equestrian homes were Hellenistic imports, a veneer of Greek culture on a Roman base. For the Romans admired the Greeks. As one of the Roman poets was to put it later, "Captive Greece took captive her rude conqueror." The upper classes admired Greek education, and, since schooling was a private affair, tried to give Hellenistic culture to their children through educated Greek slave-tutors in their households. Roman playwrights were strongly influenced by Greek plays. Livius Andronicus (284-204 B.C.), a Greek slave who was given his freedom in Rome, combined material from Homer and fragments of Greek drama in his Latin plays, whereas Plautus (254-184

Cultural consequences

B.C.) presented a thoroughly Roman treatment of Greek themes and composed the first modern comedies. Noble Romans commissioned poets to produce epics that would heighten the glories of the Roman past and make them in some measure comparable to the illustrious origins of Greece.

Not all Romans welcomed this Hellenistic influence. Some connected the new luxury and moral lapses with Greek ideas. One such was Cato the Elder (234-149 B.C.) who, as censor, tried to reduce luxury in Rome—even to the extent of ripping out water-mains leading into private houses—in an effort to bring back the virtues of former times.

The following paragraph comes from a Roman history that Cato wrote for his son.

I shall speak in the proper place of those accursed Greeks; I shall say what I saw in Athens, and how it may be good to glance at their literature, but not to go into it deeply. I shall prove how detestable and worthless is their race. Believe me, Marcus my son, this is an oracular saying: "if ever that race comes to pass its literature to us, all is lost."

Thus by 133 B.C. Rome possessed a great empire whose political, economic, social, and cultural repercussions had begun to create grave problems for her. Economically the state was badly out of balance. While great landowners grew wealthier, a landless city mob swelled the ranks of the discontented in Rome. Rome's military problems became critical. Armies were recruited from freemen with land, freemen who supplied their own equipment for long campaigns. Yet these very freemen often lost their land either because of the length of the campaigns or the influx of grain and slaves that came from new conquests.

In the century following 133 B.C. these related problems of agriculture and the army became increasingly pressing. Could Rome solve them? Could she afford *not* to solve them?

3/THE FAILURE OF THE GRACCHI

Two brothers, Tiberius and Gaius Gracchus, thought they had the answer to Rome's problems. But instead they ushered in a series of revolutions, which ended only with the fall of the Republic. What were the plans of the Gracchi that went so far wrong?

Unprecedented Actions Tiberius Sempronius Gracchus (162-133 B.C.) came from one of the noblest families of Rome. His grandfather, Scipio Africanus, had been the conqueror of Hannibal; his father had been a widely respected soldier in Spain and elsewhere; and his mother, Scipio's daughter Cornelia, was a woman of culture who employed Greek tutors for her children.

Tiberius Gracchus

Tiberius himself had served at the siege of Carthage in 146 B.C., and in 137 B.C. he was quaestor in Spain. There he was instrumental in achieving a settlement with some native Spanish tribes who had revolted against increasingly oppressive Roman rule. Because the Spaniards had come to respect his father greatly, they trusted Tiberius, and he was able to negotiate a treaty by which an army of 20,000 Romans was spared execution. Unfortunately, once the danger was over the Senate broke the

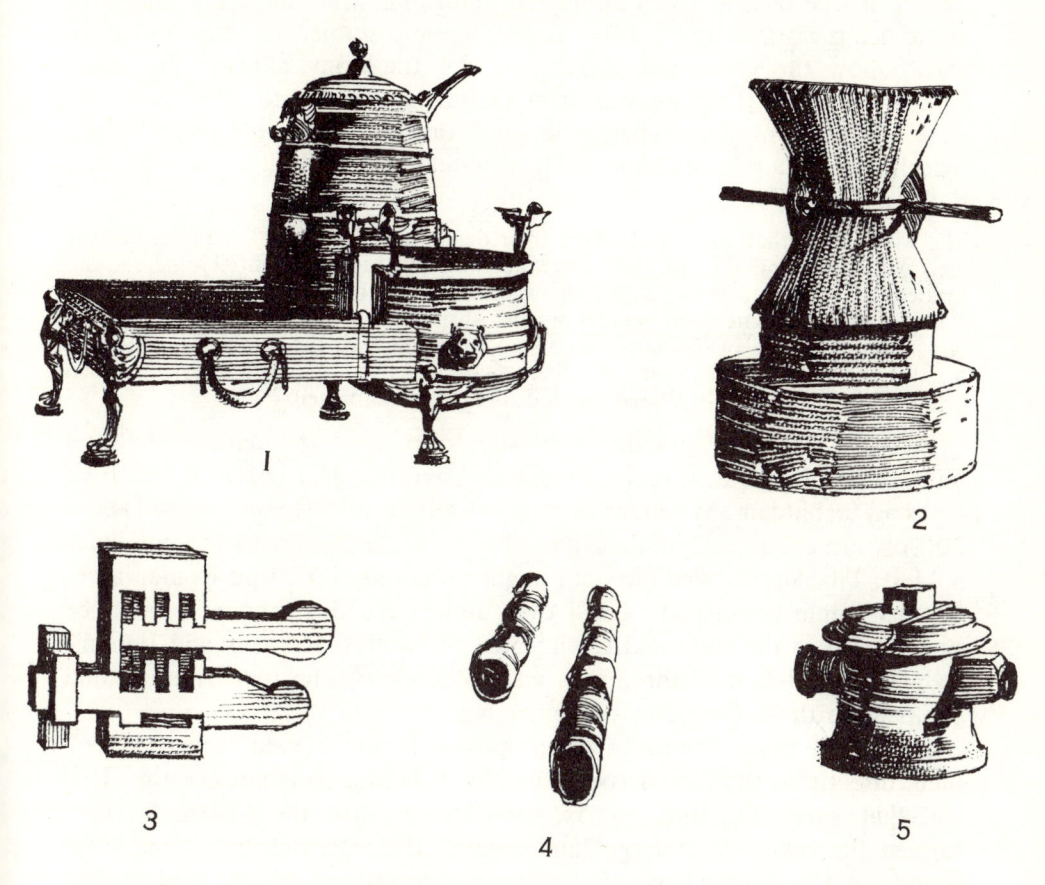

The great mechanical ingenuity of the Romans is evident in these five articles. The portable brazier (1) was a combination water heater and stove. The fire was built in the deep semi-circular pan hanging at the right, and a pot or kettle rested on the three winged figures around the rim. Since the walls of the brazier were hollow they could be filled with water. The hot water went into the large tank at the side, and could be drawn off by a tap. Grain poured into the hand-mill (2) was ground between an outer and an inner stone when the handles were turned, and the flour passed out at the bottom. Roman houses and apartments had fine strong doors which opened inwards. They were usually made of cypress, oak, or deal, with bronze or hardwood pivots at the top and bottom of the frame in place of hinges. Though the Egyptians had a primitive lock device and the Greeks a crude but operative key mechanism, it was the Romans who perfected the lock. The one shown here (3) employs the same principle used in many modern locks. The Romans brought water over long distances, and then distributed it within their cities by means of subterranean channels. Although the poor Italian, then as now, carried water from public fountains, it was supplied to the homes of the well-to-do through pipes of either clay or lead sheet bent around a core and soldered with lead at the seam. Because lengths were short, joints were numerous, as can be seen. The oval pipe at the right (4) probably shows the original shape, whereas water pressure has rounded the other. Even effective taps (5) were devised by the Romans to regulate the flow and prevent waste.

treaty, and Tiberius barely managed to escape from Spain. Back in Rome his success in extricating the doomed army made him so popular with the people that he was elected tribune in 133 B.C. at the age of thirty.

It was while Tiberius was on his way to Spain as quaestor that the dearth of free peasants was impressed upon him when he saw estate after estate being worked by gangs of slaves. Being a soldier, he must also have known how the system of recruitment for the army affected the small landowner. Now that he was tribune he determined to strengthen the state and the army by reviving the small farming class, and in forthright speeches he struck hard at the military system that exploited men who had nothing to fight for:

The wild beasts that roam Italy have their dens and lairs to shelter them, but the men who fight and die for Italy have nothing but air and light. Homeless and footless they wander about with their wives and children. In battle their generals exhort them to defend their sepulchres and shrines from the enemy: they lie. Not one among the host of Romans had his ancestral altar or the tomb of his fathers: it is for the wealth and luxury of others that they fight and die. They are called masters of the world, they have no clod of earth to call their own.

Perhaps, Tiberius thought, one way to remedy the agricultural and military problem was to revive the ancient agrarian legislation of 367 B.C. that had forbidden any tenant to rent more than 500 Roman acres (about 300 of our acres) of public land. But as a concession to present landholders, Tiberius decided the limit ought to be raised to 1000 Roman acres for any single household. Land over and above this amount was to be surrendered to the state and re-allocated to landless Romans and Italians. A land commission of three men was to supervise the redistribution and compensate those forced to surrender land.

Instead of first submitting his proposals to the Senate, Tiberius took them directly to the Tribal Assembly. Why he did so is not certain. It is true that those with most to lose were the senators themselves, for they formed the majority of large landowners. But Tiberius had some loyal friends in the Senate, and if, following normal procedure, he had first brought his bill to the Senate, it would probably have been given a fair hearing. Perhaps Tiberius feared the bill would get bogged down in long wrangles in the Senate, and he knew he had only one year in which to act. Whatever his reasons, his impatience was unwise. A constitutional crisis resulted, and the Senate countered by influencing one of the tribunes to veto the legislation.

Tiberius then took an unprecedented step: he persuaded the Assembly to depose the offending tribune on the grounds that he was thwarting the will of the people. This was done, and the new land law passed. The law was now binding on the Senate; but the senators tried to hold back the land commission by keeping operating expenses from them, whereupon Tiberius took yet another unprecedented action. When news arrived that

King Attalus of Pergamum had died, willing his wealth and kingdom to Rome, Tiberius blatantly interfered with the Senate's hitherto unchallenged control of finances and financial affairs by insisting that some of the wealth of Pergamum be allocated to help settle farmers on the redistributed lands.

At this point the tribunician elections for the next year took place. Tribunes, by established custom, held office only once; yet Tiberius stood for re-election. It was the last straw. The Senate lost all restraint and started a riot in which Tiberius and three hundred of his supporters were murdered.

One of the symbols of Roman citizenship was the toga, the most imposing and complicated garment ever created from a simple, uncut piece of material without the use of pins or buckles. Though generally made of fine white wool, the toga worn by certain magistrates, officials, and boys (*toga praetexta*) had a purple border. Underneath was worn a tunic similar to the usual costume of peasants and slaves. The toga underwent several changes throughout Roman history, the one pictured here dating from about the time of Julius Caesar. How was a toga draped?

Champion of the People The next year the Senate proceeded to condemn and execute more of Tiberius's supporters. But it did not try to halt the redistribution of land, which went on apace—perhaps proving that the Senate had objected more to Tiberius's methods than his aims. However, there was soon a new bone of contention. Many of the Latin and Federate allies did not want to surrender lands merely to provide farms for the unemployed of the city of Rome, and came to Rome to agitate. But instead of satisfaction, they got a law forbidding them to settle in Rome and expelling those that had already done so. A wise suggestion to grant all allies Roman citizenship was not acted upon, and the problem remained to fester.

Just as the Senate breathed easily again, who should turn up in Rome but Gaius Gracchus (153-121 B.C.). He was even more emotional and intense than his older brother had been, and was an electrifying and fearless orator. "Those worst of men have murdered the best of men, my brother!"

Gaius Gracchus

he thundered. Gaius had been quaestor in Sardinia in 126 B.C., but instead of staying there for his full term of office had returned suddenly to Rome in 124 B.C. to deliver a stinging indictment of those who milked the provinces. "Alone of all who went on the expedition," says Plutarch, "he had carried out a full and had brought home an empty purse, while others, after drinking up the wine they had carried out with them, brought back the wine jars filled again with gold and silver from the war."

Little wonder that when Gaius presented himself for the tribunate the country folk flooded in to Rome. "There came such infinite numbers of people from all parts of Italy to vote for Gaius, that lodgings for them could not be supplied in the city; and the Field not being large enough to contain the assembly, there were numbers who climbed upon the roofs and the tilings of the houses to use their voices in his favour."

As tribune, in 123 B.C., Gaius was determined first of all to avenge his brother's death. In a mounting crescendo he harangued the people.

Before your very eyes they beat Tiberius to death with cudgels; they dragged his dead body from the Capitol through the midst of the city to cast it into the river; those of his friends whom they seized, they put to death untried. And yet think how your constitution guards the citizen's life! If a man is accused on a capital charge and does not immediately obey the summons, it is ordained that a trumpeter come at dawn before his door and summon him by sound of trumpet; until this is done, no vote may be pronounced against him. So carefully and watchfully did our ancestors regulate the course of justice.

Gaius began by carrying through a measure declaring illegal any court with powers of capital punishment that was not set up by the people. When he went on to his economic reforms, it was soon evident that his programme was more ambitious again than that of Tiberius.

Without even being a candidate, Gaius was re-elected tribune for the next year (re-election had been legalized five years after the death of Tiberius) and proceeded to push hard for his reforms. Tiberius' land law was revived and amplified, but since much of the available land was by this time distributed Gaius also proposed to establish some new colonies in Italy and overseas in North Africa. Another law was enacted allowing the state to buy grain in bulk and sell it cheaply (at about 32 cents a bushel) to the poor of Rome. Roads were improved throughout the peninsula to facilitate the transport of agricultural produce from farm to market. A minimum age of seventeen was set for army service, and the state now furnished its soldiers with clothing and equipment free of charge. The collection of Asian taxes was put into the hands of wealthy Equestrians, and those who gained the tax contracts could not be prosecuted for extortion, a crime which, by definition, only magistrates could commit. The Equestrians were also given the conduct of the extortion courts, which had hitherto been the preserve of the Senate. But when Gaius made his most revolutionary proposal—to give the citizenship to the Latin allies—

both the Senate and the Roman mob opposed the measure for fear of having their influence swamped in Rome. Gaius was beginning to lose his support.

Determined to exploit the rift in Gaius's supporters, the Senate put up a rival tribune who promised the mob more of everything Gaius had proposed. The unscrupulous tactic worked. Gaius was not re-elected tribune in 122 B.C., and shortly thereafter he and 3000 of his followers were murdered by a senatorial gang in another outbreak of rioting.

The Real Tragedy It is hard to be moderate in judging the Gracchi. The tragic circumstances of their deaths and their essential nobility of character have remained a poignant chapter in Roman history. Why did they fail?

Perhaps history's final verdict on the two brothers must be that they failed because they were reformers in too much of a hurry. Losing patience with the senatorial oligarchy, they were prepared to try to make the Assembly the dominant force in politics. They tried to drive a wedge between the Equestrians and the Senate, and in so doing intensified class and political divisions. Henceforth two groups vied for power: the *Optimates* (the "Best") and the *Populares* (the "Popular Party"). The Optimates consisted of the members of a few aristocratic families. They backed a senatorial oligarchy and as a rule managed to control the Centuriate Assembly, thereby also controlling the election of consuls and praetors. The Populares were usually led by the Equestrians, who succeeded in dominating the Tribal Assembly on the pretext of defending the poor, and hence managed to elect the tribunes with the votes they gained from the urban masses. By thus splitting Rome up the centre, by creating conditions in which Romans slaughtered Romans, it may well be that the Gracchi did more than most to hasten the fall of the Republic.

This, then, was their real tragedy: they demolished once and for all the tradition of responsible senatorial government.

4/THE ARMY IN POLITICS

For the time being the triumphant Senate controlled events. But the bitterness ate deep into Roman life.

In 119 B.C. Gaius Marius (157-86 B.C.), a member of an equestrian family, was elected tribune. He had proved himself a courageous fighter in Spain and had become noteworthy as an officer fighting in North Africa against Jugurtha, king of Numidia. Marius had political ambitions as well, but since none of his ancestors had held high office he was looked upon by the senatorial class as an upstart. Rough soldier that he was, when he came to Rome in 107 B.C. to campaign for the consulship he bluntly poured scorn on the nobles.

Marius

Compare me, the "new man", my fellow citizens, with those proud nobles. . . . What they learn from handbooks I know from service. . . . My own belief is that men are born equal and alike: nobility is achieved by bravery. . . .

My expressions are not elegant; I don't care. Merit itself makes a sufficient show. It is they who need art to gloze baseness with rhetoric. I never learned Greek; I never wanted to, for Greek did little for the character of its professors. I did learn things far more useful to the state—to strike the enemy, to be vigilant on guard, to fear nothing except disgrace, to endure heat and cold alike, to sleep on the ground, to bear privation and fatigue at the same time.... They say I am vulgar and unmannerly because I cannot give a dainty dinner, that I have no entertainer or cook that costs more than a farm steward. I am happy to admit the charge, fellow citizens. From my father and other righteous men I learned that daintiness is appropriate to women, strenuousness to men, that good men ought to have more glory than riches, that weapons, not furniture, is the true ornament.

Not only did Marius manage to get himself elected consul by the Centuriate Assembly, but he persuaded the Tribal Assembly to go over the head of the Senate and give him the North African command as well. Here again Marius broke all precedent by recruiting a volunteer army in which there was no land-ownership qualification for service. With the successful conclusion of the war he returned to Rome in triumph, the champion of the Populares. There he was re-elected consul.

A Cure or A Disease? Although the Jugurthine War had no great strategic significance, it was important in another way. The new army of volunteers was a far different one from the traditional conscript citizen army with its property qualifications for service. The old distinction of front, middle, and rear line legionaries was abolished, and each man was now armed with a javelin and a short sword. The 30 maniples of the legion were regrouped into 10 *cohorts*, each 600 strong, in which men from twenty to forty-five served side by side for terms of as long as 16 years. In this way the legion itself was increased in strength to 6000. The entire army was revitalized by the reforms of Marius; its standards, the silver eagles of Rome, winged proudly everywhere.

The *legionary* (1) was the backbone of the Roman army, and was essentially a foot-soldier. His armour was simpler and less ornamented than that of the Greeks, the *cuirass*, for example, being made of loops of metal hinged in the back and clasped in the front so that it was much more flexible than its Greek prototype. The shield was usually shaped as in the illustration, its height purporting to be the length of a man's arm. The light-armed *velite* (2) wore a leather tunic in place of armour, and carried a sword, light spears, and an oval shield about 2 feet high. The sling (pictured here) was also used, with specially cast lead sling bolts (3) which were often inscribed with insulting messages to the enemy. The famous Roman short sword or *gladius* (4) was at its best about 22 inches long, double-edged and perfectly straight. Most of the Roman swords eventually came from Spain because of the excellence of Spanish manufacture. But the legionary's most important weapon was his heavy spear or *pilum* (5), whose iron head formed about one-third of its total length of 5½ feet.

The Romans possessed an extensive arsenal of siege engines. One of the oldest was the battering-ram, which was often employed under the protective cover of a movable roof (*testudo*). Fighting or scaling towers were used to bridge high walls, while sappers would undermine defences. The catapult (6) worked roughly along bow and arrow principles: the missile was propelled by two short arms working on the torsion principle supplied by two tightly twisted skeins of cords, and was guided on its way by a trough. The catapult could project a javelin about 500 yards. A massive, low-slung version of the catapult, the *ballista*, threw huge stones instead of javelins. The less powerful *onager* (7) was easier to build, and slipped large stones out through a sling when the arm reached the correct position. What is the significance of the name onager?

1

2

3

4

5

6

7

The new professional army was composed of soldiers whose trade was fighting and whose remuneration depended on the success and rewards of their commander. The Senate, in other words, was beginning to lose control of the military forces of the state, while at one stroke those twin problems of agriculture and the army found a measure of solution in the military reforms of Marius. But the solution was not as complete as it seemed. As it turned out, the cure was almost worse than the disease.

Unfortunately for Rome, Marius was not as good a politician as he was a soldier. He had saved Rome by preventing a new German invasion of Italy, and was so popular that he was repeatedly re-elected consul. But when the Populares persisted in catering to the rabble, the Senate called in Marius to suppress the regime. That done, the tough old man, six times consul, founder of the new Roman army, suffered a political eclipse and withdrew to Asia.

The alarming thing about these troubled years was the increasing influence of the army in politics. The new army found itself dependent for land and pensions on its general. And if the state would not recompense its soldiers, what might happen when a successful general came to Rome with an army at his back? This had not happened yet, but time was passing and the Senate's control was shaky.

A Horrible Revenge   If the early tendencies of Rome to grant eventual citizenship had only been followed, the war that finally erupted in Italy in 90 B.C. would have been unnecessary. But the Federate allies, encouraged by the holocaust of political murder and degenerate politics in Rome, finally made a last desperate bid for the citizenship so long denied them. Rome at last did grant citizenship to certain of the rebels, thus weakening the hard core of the rebellion, and the war ended in 88 B.C. with all Italy south of the Po—Latin and Federate allies alike—assured of a uniform Roman citizenship.

Sulla   A successful commander in the war against the Italian Federate allies was Lucius Cornelius Sulla (138-78 B.C.), an Optimate from an obscure family who had served as a cavalry officer under Marius. He was rewarded for his services in the war by election to the consulship, backed by the Optimates, and the Senate went on to give him the command against Mithridates VI, king of Pontus, who was bent on dominating all Asia Minor. Marius, jealous of Sulla's eastern command, intrigued to secure it for himself. Stung into retaliating, the enraged Sulla marched on Rome with his army and the Marians fled or were murdered.

Once again the army had been a deciding factor in politics. Now Sulla, secure in his command, left for the East; but in his absence from Rome the inevitable occurred. Back from Africa came a man bent on a horrible revenge—Marius, the great Marius, a pathetic sight in his degeneracy. "Filthy and long-haired he marched through the towns," says a Roman historian, "presenting a pitiable appearance, descanting on his battles . . . and his six consulships."

For the next four years there was a blood bath in Rome, although Marius himself died peacefully in 86 B.C. during his seventh consulship. When Sulla finally returned in 83 B.C. another civil war broke out. There could be little doubt about Sulla's course of action, backed as he was by 40,000 hardened veterans who looked for rewards, not to the state, but to him. In a savage reign of terror (see page 223) lists of "the proscribed" were posted in the Forum each day, and anyone who hunted them down was suitably rewarded. The victims—5000 in Rome alone, and uncounted thousands throughout Italy—were ruthlessly butchered, often after torture, and their estates confiscated in order to provide pensions and farms for Sulla's veterans. It is calculated that the civil wars, along with the strife with the Italian allies, cost about half a million Roman and Italian lives.

Turning Back the Clock   Although Sulla was appointed dictator for an unlimited period in 81 B.C., his main concern remained the restoration of the Senate to a position of pre-eminence in the state. War and massacres had cut its membership in half; Sulla now brought its numbers up to 600 by adding his supporters from senatorial families as well as 300 Equestrians. He also increased the number of quaestors to 20, and arranged for a steady supply of new senators by decreeing that all ex-quaestors were to become members of the Senate. In addition, senators were restored to their former jury eligibility and the Equestrians disqualified, while the tribunes, whom Sulla believed had achieved unmerited influence since the time of the Gracchi, had their power reduced and were made ineligible for any other office. Henceforth for ambitious young men the tribunate became a dead end, an office to be avoided.

Rome had already acquired an eighth and ninth province, Narbonese Gaul (121 B.C.) and Cilicia (102 B.C.). Now Sulla added a tenth: Cisalpine Gaul, in the fertile region between the Apennines and the Alps. He forthwith sent out a governor and garrison to guard his new province, for this troublesome region had long been a prey to raiders who periodically descended from the Alps. In order to provide enough governors for these provinces, the number of praetors was increased to eight; for praetors, along with the two consuls, automatically became provincial governors following their year of office in Rome.

Satisfied that he had curbed the power of the people and given the Senate a fresh chance to govern Rome, Sulla voluntarily retired in 79 B.C. Then, in this last remaining year of his life, he settled down on his country estate, married a young divorcee (his fifth wife), and wrote his autobiography. His epitaph, which he composed himself, was inscribed on his tomb: "No friend has ever done me a kindness and no enemy a wrong without being fully repaid." It is, as has been said, a boast that every gangster would like to make his own.

Part of the Etruscan legacy to Rome was the brutal practice of gladiatorial combat. Every sizeable city throughout the Empire (except in Greece) had its amphitheatre, and in this vast Colosseum in Rome as many as 50,000 spectators would crowd in, eager to watch men fight it out to the death. Gladiators were often condemned criminals, prisoners of war, or slaves, and for the successful contestant there was far greater fame and adulation than for a modern prize-fighter. Here a Samnite wearing a visored helmet, breastplate, and greaves advances with shield and sword on a Gaul, who is armed only with a net (in which he hopes to entangle his opponent) and trident. When one man was finally downed the crowd shouted "Habet!" ("That's got him!") and indicated their pleasure: waving handkerchiefs meant mercy; thumbs down, death. The Emperors Caligula and Claudius were devotees of gladiatorial combats, and Nero even allowed noblewomen to fight in the arena.

Very few Romans protested against these bloody spectacles, or the even bloodier beast shows in which hundreds of wild animals tore each other to pieces. A depth of depravity was finally reached in 107 when the emperor Trajan returned to Rome to celebrate his victories in Dacia. In an orgy of games that ran through 11,000 gladiators, Trajan in 123 days bested Augustus' record for his whole reign.

Sulla had made a concerted effort to restore the Senate to power. But his actions in turning his army loose on Rome and Romans had a more lasting effect than his constitutional changes. What he wanted was to prevent future generals from doing what he himself had done. In this he failed.

5/THE FIRST TRIUMVIRATE

With Sulla gone, it was an open question whether the Senate would be able to hold the reins of power firmly enough to govern Rome responsibly. In the course of the next generation four men were to show it could not.

Even as Sulla died, Cisalpine Gaul and Spain were already causing trouble. The man the Senate sent to Spain with sweeping powers to quell the disturbances was Gnaeus Pompeius (106-48 B.C.), a young Equestrian who, at twenty-three, had raised an army for Sulla and been jokingly given the title *Magnus* ("The Great"). Pompey was an indifferent public speaker, but a good athlete and swordsman and a fine soldier.

Pompey

The General and the Capitalist Pompey soon succeeded in pacifying the hot-headed Spaniards, and on his return from Spain he became associated with Marcus Licinius Crassus (112-53 B.C.). Crassus was a financier of great means shrewdly come by. He had been a veteran Sullan officer, and had been first in line to snap up the property of Sulla's victims. He also gained great fame for being on hand at fires that broke out in the crowded city of Rome, where he would stand idly by with his own private fire department until the owner had signed an agreement to sell the property for a song. He was soon a very wealthy man. In addition to his financial wizardry, Crassus had some military ability. Just at this time a violent slave revolt led by a Thracian named Spartacus was sweeping Italy, and one Roman legion after another fell before the slave army of 70,000. Finally the Senate appointed Crassus commander-in-chief, and with six legions he was able to kill Spartacus, crush the revolt, and crucify 6000 slave rebels along the Appian Way. Nevertheless it was Pompey who, having caught some of the remnants of the tattered slave army fleeing north through Etruria, boastfully claimed the credit for ending the war.

Crassus

Now Pompey and Crassus both approached Rome at the head of their armies. Each proclaimed his desire to be elected to a consulship, even though Pompey had not held the prerequisite offices and at thirty-seven was some six years under the legal age limit for the consulship. The Senate opposed both candidates, but armies spoke louder than laws. Pompey was granted a dispensation, and both men became consuls in 70 B.C. So ended the Sullan attempt to restore the prestige and influence of the Senate, erased by the first hand of force that was raised against it.

Pompey and Crassus quickly proceeded to restore all its powers to the tribunate and to revive the dormant censorship which Sulla had carefully contrived to control. The new censors promptly ejected sixty-four Sullans from the Senate in 70 B.C. A third measure broke the Senatorial monopoly of jury service by providing that two-thirds of each jury would henceforth be made up of Equestrians and other middle-class men.

The change in jury service had been given an impetus by the scandal surrounding a trial in Rome at the time. Gaius Verres, the ex-governor of Sicily, was brought to trial for extortion and all manner of crimes committed during his governorship. But Verres had powerful friends at Rome, and it appeared that he might be acquitted until Marcus Tullius Cicero (106-43 B.C.) was appointed prosecutor. Although Cicero came from an Equestrian family of limited means, he had received a good education and

Cicero

became a brilliant lawyer. His masterly indictment of Verres (selections from it may be read on page 224) was so damning that Verres went off into exile without waiting for the trial to end. The whole affair pointed up the need for a non-senatorial check on the administration of provinces, a check which was now provided by the presence of Equestrians on juries.

After their consulship in 70 B.C. both Pompey and Crassus retired to private life—though only briefly. In 67 B.C. the Senate gave Pompey an extraordinary command. For some years pirates had infested the Mediterranean, and had recently become so bold that the grain ships that plied between Sicily and Rome refused to sail. Pompey was granted enormous powers and told to clean out the pirates. He completed his assignment with such dispatch that the Senate sent him east on a second mission, to break once and for all the power of Mithridates of Pontus. By the time Pompey had completed his campaigns in the Middle East, he had extended the Roman Empire from the Mediterranean to the Euphrates, establishing a strip of provinces and allied states around the coast of Asia Minor from the Black Sea in the north to Syria in the south. The Middle East had not seen such peace and security since the days of Alexander the Great.

While Pompey had been absent in the East, however, his friend Crassus had not been idle. Once more eager for political recognition, he had formed a coalition with a young man who, some said, had escaped the Sullan proscriptions only because he was judged too wild and unstable to merit liquidation. The dissolute young man was Gaius Julius Caesar.

A Rising Star   Julius Caesar was born about 102 B.C., and grew into a magnetic youth whose charming eloquence was only matched by his genius for managing people. He seemed destined for politics: his uncle became consul, his father praetor, and most of his family had strong Optimate sympathies. Caesar, however, was even more closely linked to the Populares because his aunt had married Marius. It was natural that he should become a member of the Marian party, and though he escaped the Sullan dagger he did not escape Sulla's misgivings. The dictator saw in this disorderly young man "many a Marius"—and Caesar wisely fled Rome.

Caesar served in the army in Asia until Sulla's death. Then in 68 B.C. his star began to rise rapidly. He served as quaestor in Farther Spain, and after his return to Rome became closely associated with Crassus. Borrowing up to $1,000,000 from his patron, he staged lavish entertainments for the Roman people in order to win his election to the usual magistracies.

In 62 B.C. Pompey returned to Italy from the East, bent on obtaining three things: a *triumph* (that is, a victory parade granted by the Senate), senatorial ratification of his actions in the East, and rewards for his veterans. But instead of becoming a second Sulla he disbanded his army, and only then did he press the Senate to provide land for his 40,000 veterans. The Senate refused. In desperation Pompey turned to Crassus and Caesar, and the three of them, in 60 B.C., formed the so-called First Triumvirate.

The First Triumvirate

This head of a colossal statue now in the Palazzo dei Conservatori in Rome idealizes Julius Caesar as the serene ruler. We have, unfortunately, no contemporary portrait of him, and this particular representation, made over a century after his death, undoubtedly flatters the great man. It is, however, still possible to see in the sculpture the tall, well-formed physique, the dark, piercing eyes, and the fair complexion and somewhat full face that Suetonius, the historian (see page 216), described. Also evident, and perhaps more significant in a man born to lead, is a certain haughtiness, an aristocratic exclusiveness which won the respect —if not the love—of his fellow Romans.

Caesar wanted to be a consul with a chance at future military glory; Crassus wanted a share of the Asian taxes; Pompey wanted land for his veterans. These things, as consul in 59 B.C., Caesar was able to arrange.

The triumvirs, of course, had no intention of giving up power once Caesar's consulship was over. Accordingly they made sure that two of their supporters were elected consuls, banished some political opponents (including Cicero, though he was recalled the next year), and appointed Caesar governor of the Gauls for five years. They also provided free grain to all who requested it, with the result that before long the state found itself distributing over 123,000 tons of wheat a year. Thus was the cheap grain of the Gracchi converted into an unashamed dole, a powerful device for buying city votes.

In 56 B.C. a further agreement was made to extend Caesar's command for another five years, to make Pompey and Crassus consuls in 55 B.C., and then to give Pompey the western command and Crassus the eastern. Of the three generals, Caesar alone had great success. By conquering Gaul, he expanded Roman control to the Rhine and the North Sea.

He did more than that. During the ten years of Caesar's Gallic Wars it has been calculated that about 400,000 Gauls were slaughtered and as many more sold into slavery. The country was looted and the booty used either to pay off the army or to bribe politicians in Rome—wherever it went, it did not find its way into the Roman treasury. Julius Caesar was one of the first (and best) of the memoir-writing generals of history, but these unsavoury aspects of his wars are naturally not emphasized in the superb account of his campaigns that he wrote under the title of the *Gallic*

War. The fluency of his Latin and the spare clarity of his descriptions should not, however, blind later readers to his ruthless attitude. It is said that he made the equivalent of $40,000,000 out of the conquest of Gaul, and even so scrupulous a man as Cicero cleared about $110,000 in a single year as an honest provincial governor. Obviously the acquisition of empire was a very lucrative business, and the eventual extension of the benefits of Roman civilization was not—as the conquerors later chose to believe—the primary motivation for empire.

While Caesar was busy in Gaul, Crassus, who had always aspired to military glory, met an unhappy end. He had the misfortune to pick as his adversaries the Parthians, the most skilful archers in the world, one of whom treacherously terminated his turbulent career. Since, with the death of Crassus, Pompey became sole consul, the Senate worked to buy his support—probably hoping to eliminate Caesar and perhaps Pompey himself.

As Caesar's term drew to a close, moves were made in Rome to have him prosecuted before the courts for waging unauthorized war in Gaul. Pompey now threw in his lot with Caesar's enemies. Caesar said that if Pompey would disband his troops he would do likewise, a proposal acceptable to neither the Senate nor Pompey. Pompey was put in command of all forces in Italy, and on January 1, 49 B.C., the Senate decreed that unless Caesar surrendered his command he would be declared a public enemy.

Crossing the Rubicon   While Rome debated, Caesar had been waiting in his winter camp in Cisalpine Gaul, just across the Rubicon River which formed the northern border of Italy. He must have gone through some agonizing deliberation when he received the news that on January 7 the Senate had actually pronounced him a public enemy. But there was now no turning back: it was Caesar or the Republic.

On January 10 he led his army across the Rubicon and marched on Rome. Contrary to the expectations of the Senate, the citizens did not look upon Caesar's invasion of Italy as treason but hailed him as a great patriot. And behind him stood the formidable legions of the Gallic Wars.

Pompey and his army, accompanied by many senators, fled to Greece where they were defeated by Caesar at Pharsalus in 48 B.C. Pompey himself was subsequently killed in Egypt three days before Caesar reached Alexandria; but Caesar's fascination for the twenty-one-year-old Cleopatra kept him dallying in Egypt while the Pompeians grouped for further action in Asia Minor, Africa, and Spain. When the dust of battle cleared in 45 B.C., Caesar was the undisputed master of the Roman world.

6/ THE DICTATORSHIP OF CAESAR

Caesar was a different sort of dictator. There was no slaughter of his political enemies in Rome. "I am not moved," he wrote to Cicero, "by the fact that those whom I have let off free are said to have gone away to make war on me anew; I like nothing better than to be like myself and to let them be like themselves." To two of his political agents he wrote even more pointedly: "Let this be the new thing in our victory, that we justify ourselves by mercy and generosity." The transformation of Julius Caesar from a thoroughly unscrupulous politician into a great statesman is one of the most interesting studies in history.

To carry out his policies Caesar had to have power. This his armies gave him. But to cloak the power in legality Caesar accumulated a cluster of offices, titles, and honours—dictator, consul, tribune, Pontifex Maximus (high priest), censor, "father of his country," and so on. Only one title he dared not take: that of king.

Some of the awe inspired by the dictator Caesar comes out in a letter of Cicero's describing how he had entertained the great man (but was spared his entourage of 2000!) in 45 B.C. Cicero, who now lived in political retirement, had been a Pompeian whom Caesar forgave. He must have been more than a little uncomfortable as Caesar's host.

Everything was good and well served . . . and I think I made a good host. But the guest was not the sort to whom you would say, "Do stop in again on your way back"; once is plenty. Our talk was not serious but mainly literary. In a word, he was pleased and seemed to enjoy himself. . . . There you have the story of his visit or, as I may call it, his billeting.

P.S. When he was passing Dolabella's house, he mounted and paraded the whole of his armed guard to right and left, a thing he did nowhere else. I had this from Nicias.

Unfinished Business Caesar's programme for Rome was far-sighted and far-reaching. He did her the greatest service when he rid her of the cancer of civil war. He did not, however, restore the Republican constitution; it was paralysed in any case, crippled long before he came to power. The Assemblies and the Senate had little power, so little in fact that Caesar's future acts were ratified in advance. Nevertheless there were some notable reforms. The number of praetors, aediles, and quaestors were increased, partly to speed up municipal business and partly to provide rewards for Caesar's followers. The Senate's membership was increased from 600 to 900, a few of the new senators being drawn from outside Italy, but most of them being Equestrians from those parts of Italy that had won citizenship in 88 B.C. Their presence in Rome would serve two purposes: they would help to unite the country, and they would swamp the remaining Optimate opposition.

Caesar's reforms

In the city of Rome itself, Caesar disbanded certain plebeian guilds that had been misused as political clubs of ruffians who could be hired to terrorize the city. He began extensive building projects, which not only provided relief for unemployment but enhanced the beauty of Italy's capital city. And he cut the number of those receiving the grain dole by over one-half, bringing it down to 150,000. To take care of landless Romans, and his veterans as well, Caesar planned at least twenty colonies that were to provide homes for 100,000 citizens—a continuation of the work of

Many of the houses excavated at Pompeii were the *domus* type, a dwelling of great beauty in which Greek and Etruscan characteristics combined and developed into something new and distinctive. The main room, which we see here, was a sort of reception room called the *atrium*, where the inward slope of the tile roof channelled rain through the opening into the marble pool below. Beyond the atrium was a beautiful garden court surrounded by a colonnaded porch called a *peristyle*. Running off the peristyle and enclosing the whole section were the family rooms: sun room, dining-room, bedrooms, and lounge. Although a second storey might have some windows, the ground floor presented a windowless face to the outside world. Roman houses were sparsely furnished but elaborately decorated. Floors were resplendent with geometric mosaic tiles, while walls were splendidly adorned with thin slabs of marble contrasted with alabaster, porphyry, and mother of pearl, or with richly coloured geometric decorations and wall paintings. Stucco made of lime, sand, and fine marble dust was also widely used, and took a very high polish. In the city of Rome itself, however, multiple-family dwellings in tenement blocks called *insulae* were much more common than the single-family domus.

overseas settlement that Gaius Gracchus had begun. (Interestingly enough, Corinth and Carthage were among the new colonies.) Caesar well knew that the transplanting of these citizens was the surest means of spreading the Roman way of life throughout the provinces.

Other undertakings were envisaged for the whole peninsula: an artificial harbour for Ostia, a road across the Apennines, the draining of marshes. But these progressive projects were not accomplished in Caesar's lifetime. Neither did he see come into effect a municipal law that he ordered drafted to give all the communities of Italy uniform civic government. He did, however, witness the passage of two other vital laws. One required wealthy citizens to invest half their capital in land; the second provided that one-third of all Italian herdsmen should be made freemen, thereby alleviating the danger of another slave rebellion.

In imperial affairs, Caesar came to realize that generous grants of responsibility bred loyalty. Hence Roman citizenship was extended to Cisalpine Gaul and to certain provincial towns, while Latin rights were granted many others. In addition, the city of Rome attracted intellectuals to herself by granting full rights of citizenship to doctors, teachers, librarians, and scholars who came from the provinces to settle in Rome. Imperial financial matters were stabilized. Every debtor's obligations were reduced by one-quarter, and a new gold coinage was introduced. Not least important, the system of farming out provincial taxes was abolished and honest governors were appointed—a noteworthy step in establishing goodwill towards Rome.

Perhaps the most lasting and practical of all Caesar's reforms was his revision of the calendar. In place of the old lunar calendar of 355 days a new solar calendar of 365¼ days was introduced. This Julian calendar (as it came to be called) was introduced on January 1, 45 B.C., was slightly revised by Pope Gregory XIII in 1582, and is still the calendar we use today.

All these reforms were initiated in the brief sixteen months Caesar spent in Rome between 49 and 44 B.C. But his schemes were cut short. For despite all that he did for the common people of Rome there were those who took offence at his ways. Members of the old guard seethed with each new honour that was showered on him: the month Quinctilis was renamed Julius (July); his head appeared on coins; he was called "father of his country"; the dictatorship was granted him for life. And some of Caesar's utterances were not such as to reassure them.

Caesar's enemies

The Republic [he said] was nothing—a mere name without form or substance; ... Sulla had proved himself a dunce by resigning his dictatorship; and ... now his own word was law, people ought to be more careful how they approached him. Once when a soothsayer reported that a sacrificial beast had been found to have no heart—an unlucky omen indeed—Caesar told him arrogantly: "The omens will be as favourable as I wish them to be; meanwhile I am not at all surprised that a beast should lack the organ which inspired our finer feelings." ... During one of his Triumphs, he had ridden past the benches reserved for the tribunes of the people, and shouted in fury at a certain Pontius Aquila, who had kept his seat: "Hey there, Aquila the tribune. Do you want me to restore the Republic?" For several days after this incident he added to every undertaking he gave: "With the kind consent of Pontius Aquila."

THE ROMANS

The Ides of March

Soon the resentment boiled over. A conspiracy of 60 senators, jealous of Caesar's honours and particularly of his assumption of the dictatorship for life, plotted to assassinate him. Without any real plan of their own for the future, the conspirators stabbed the dictator to death on the Ides of March (the 15th), 44 B.C.

An untimely death often idealizes the victim. Nevertheless, five months after the assassination Cicero summed up what many others must have been too frightened to say:

> In that man were combined genius, a power of reasoning, memory, literary skill, accuracy, depth of thought and energy. He had performed exploits in war, which, though calamitous for the Republic, were nevertheless mighty deeds. Having for many years aimed at being a king, he had with great labour and much personal danger accomplished what he intended. He had conciliated the ignorant multitude by presents, by monumental buildings, by largesses and by banquets; he had bound his own party to him by rewards, his adversaries by a show of clemency. He had already brought a free city, partly by fear, partly by indulgence, to a habit of slavery.

We shall never know whether or not Caesar's plans for reforming the state would have been successful. We shall never even know the full extent of those plans. It is a reasonable guess, however, that the dictator realized that only through one-man rule could the complex problems facing Rome be dealt with efficiently. His death left the great question of who would succeed to his power. As one of Caesar's friends remarked to Cicero, "If Caesar, with all his genius, could not find a way out, who will now?"

7 / THE TRIUMPH OF OCTAVIAN

Antonius

Within hours of Caesar's death a bluff, unprincipled man moved forward, determined to seize the initiative. He was Marcus Antonius (82-30 B.C.), consul and lieutenant of the dead dictator. The Roman historian Suetonius (A.D. 69-150) describes Antony's part in Caesar's funeral:

> Instead of a eulogy the consul Antony caused a herald to recite the decree of the Senate in which it had voted Caesar all divine and human honours at once, and likewise the oath with which they had all pledged themselves to watch over his personal safety; to which he added a few words of his own.

Whatever the "few words" were, the populace of the city became so inflamed that the conspirators fled for their lives. For the moment, Antony held the power.

Octavian

Partners in Proscription But Antony had reckoned without a certain handsome eighteen-year-old, Caesar's grand-nephew Octavius (63 B.C.-A.D. 14). Octavius was in Illyricum at the time of Caesar's murder, and when the news reached him he decided to cross to Italy. Not until he landed at Brundisium did he learn that, according to Caesar's will, he had been adopted and designated as his heir.

The news galvanized Octavius into immediate action. As soon as he arrived in Rome he changed his name to Gaius Julius Caesar Octavianus and demanded his share of Caesar's estate from Antony. Antony was evasive—and perhaps with good reason, for some say he had already embezzled the inheritance. When Octavian asked why the conspirators had not been punished he again hedged. Antony underestimated the lad. He possessed wisdom beyond his years, and set to work like a seasoned politician to win the loyalty of his great-uncle's troops. Cicero and the Senate also underestimated him. They planned to use him to get rid of Antony, who had meanwhile gone to Cisalpine Gaul and been declared a public enemy. But Octavian was playing for the highest stakes. He demanded the consulship. When the Senate refused it, he marched on Rome at the head of eight legions of Caesar's veterans.

The Senate's plans for Antony also went awry. Marcus Aemilius Lepidus, the governor of Nearer Spain and a former consul with Caesar in 46 B.C., now chose to back Antony. The combined forces of Antony and Lepidus could easily overwhelm Octavian—but would their men fight against Caesar's heir? It was safer to parley. So it was that in 43 B.C. these three—Octavian, Antony, and Lepidus—formed the Second Triumvirate. In complete cold blood the triumvirs sat down in Rome to chart their course. Caesar's ghost must have wept to see his adopted son reject his idealism for the brutal pragmatism of Marius and Sulla.

The Second Triumvirate

As soon as the triumvirs were together alone they wrote the names of those to be killed.... They traded their own relatives and friends for liquidation, both then and later when they made new lists, one after another, proscribing some for enmity or mere friction, some because they were friends of enemies or enemies of friends, or very wealthy.... The triumvirs in their need levied very heavy contributions on the common people and women, and contemplated imposts on sales and rents.... The number of senators condemned to death and confiscation was about 300, and of the equestrians about 2000. The lists included brothers and uncles of the proscribers, and also some of the officers, who had been at odds with their superiors or fellow officers.... twelve (some say seventeen) of the most powerful they decided to send men to kill at once. Among these was Cicero.

Octavian had courted Cicero's favour when he came to Italy to claim his inheritance. ("The young man must be flattered, used, and pushed aside," was Cicero's monumental misjudgment.) But Cicero had come out of retirement to attack Antony in a slashing series of speeches called the *Philippics*. His death, then, was assured, and Octavian cynically assented to it. "Caesar's heir," says the best modern historian of these years, "was no longer a rash youth but a chill and mature terrorist."

The first task was accomplished; resistance was wiped out in Italy. Now the triumvirs turned to Caesar's murderers, the conspirators, who had fled to Greece. They were defeated by Antony and Octavian at Philippi in 42

The magnificent granite aqueduct almost 100 feet high at Segovia, Spain, demonstrates the skill perfected by Roman engineers. The aqueduct is still in use, and, as can be seen, has been converted to a kind of Christian shrine by the addition of a cross and a statue of the Virgin.

B.C. Then the triumphant three carved up the empire: Octavian took the West, Antony the East, and Lepidus Africa.

Partnerships sealed with blood are not often long-lived, and the triumvirate was no exception. Octavian had to find land in Italy for 100,000 veterans. But he also had to defeat Pompey the Great's son, who had seized Sicily and Sardinia. This he did. Lepidus, however, now stepped up to claim Sicily and ordered Octavian to leave, whereupon his own troops began to desert to Octavian. And so this hapless triumvir, a high born but ineffective man, was shunted into retirement in 36 B.C. Octavian spared his life, but kept a close eye on him until his death twenty-four years later.

Cleopatra

A Fatal Fascination Octavian had no other rivals in the West—but there was still Mark Antony. After Philippi Antony had gone to the East to raise money, and amongst other treasures wanted that of the Ptolemies of Egypt. Undoubtedly Antony had already met the famous Cleopatra (69-31 B.C.) when he visited Egypt with Caesar. Besides, Caesar had brought her to Rome, though she returned to Alexandria after the assassination. It was in 41 B.C. that she came at the triumvir's summons to Tarsus, and there occurred one of the most famous confrontations in all history.

[She sailed] up the river Cydnus in a barge with gilded poop, its sails spread purple, its rowers urging it on with silver oars to the sound of the flute blended with pipes and lutes. She herself reclined beneath a canopy spangled with gold, adorned like Venus in a painting, while boys like Loves in paintings stood on either side and fanned her. Likewise also the fairest of her serving-maidens, attired like Nereïds and Graces, were stationed, some at the rudder-sweeps, and others at the reefing-ropes. Wondrous odours from countless incense-offerings diffused themselves along the river-banks. Of the inhabitants, some accompanied her on either bank of the river from its very mouth, while others went down from the city to behold the sight. The throng in the market-place gradually streamed away, until at last Antony himself, seated on his tribunal, was left alone. And a rumour spread on every hand that Venus was come to revel with Bacchus for the good of Asia.

Antony was to fall completely under the spell of this Venus, and little wonder. Not that she was incomparably beautiful, for there is evidence that she was not. But, says Plutarch,

> converse with her had an irresistible charm, and her presence, combined with the persuasiveness of her discourse and the character which was somehow diffused about her behaviour towards others, had something stimulating about it. There was sweetness also in the tones of her voice; and her tongue, like an instrument of many strings, she could readily turn to whatever language she pleased, so that in her interviews with Barbarians she very seldom had need of an interpreter, but made her replies to most of them herself and unassisted, whether they were Ethiopians, Troglodytes, Hebrews, Arabians, Syrians, Medes or Parthians. Nay, it is said that she knew the speech of many other peoples also, although the kings of Egypt before her had not even made an effort to learn the native language, and some actually gave up their Macedonian dialect.

Antony was not at first so infatuated with Cleopatra that he could not leave her for an unsuccessful four-year campaign to subdue the Parthians. But about 36 B.C. he married her, though he had not yet divorced his fourth wife Octavia, Octavian's sister.

In 34 B.C. Antony celebrated a triumph over the Armenians at Alexandria—not at Rome where all patriotic Romans did such things. Then, seated high on a great throne like a Hellenistic king, with Cleopatra at his side, he announced that Cleopatra's son Caesarion was Julius Caesar's rightful heir, and that henceforth Caesarion would rule Egypt and Cyprus along with his mother. Antony's children by Cleopatra were given other kingdoms. The implication was, of course, that Octavian was a usurper, though it is doubtful that Antony and Cleopatra actually planned to claim the West.

In 32 B.C. Octavian carried his differences with Antony to the Senate. But the two consuls, along with 300 Senators, fled Rome to join Antony. Antony reacted by sending letters of divorce to Octavia, whereupon Octavian declared war on Cleopatra (but not Antony). Now the propaganda machine in Rome worked at fever pitch to blacken the reputations of Antony and Cleopatra for all time. The struggle for power that had long been building up between Antony and Octavian must be masked. If Romans were to be persuaded to fight against the popular Mark Antony, it would be necessary to portray him as the dupe of an Oriental sorceress bent on world conquest. Octavian's propaganda harped on one theme: this was no private Roman contest; this was a war of ideas, East versus West.

Antony dared not invade Italy, and both armies sailed for Greece. Antony had some 500 ships to Octavian's 400; each had about 90,000 men. But though the armies took up positions near Actium they never

TIME CHART FOR ROME AND GREECE
(1400 - 30 B.C.)

DATES B.C.	ROME	GREECE
1400	Terramare folk in Po valley	Knossos destroyed; Mycenae supreme on mainland
1300		
1200		Siege of Troy
		Dorian invasions
1100		Fall of Mycenae
		Greek migrations to Asia Minor
1000		
900	Villanovans at Bologna	City-states
	Etruscans in Etruria	
800		Homer in Chios
	Greeks in Sicily and southern Italy	Colonization around Mediterranean and Black Seas
700	Traditional founding of Rome	
		Age of Tyranny
600	Etruscans at Rome	Draco codifies Athenian law
		Solon's reforms at Athens
		Pisistratus becomes tyrant
500	Expulsion of Kings	Cleisthenes and the triumph of democracy
	"Struggle of the Orders" begins	Persian Wars
		Themistocles and Confederacy of Delos
400	Rome entered by Gauls	Pericles and Peloponnesian War
	Twelve Tables	
		Philip II of Macedon conquers Greece
300	Final victory of the Plebs	Alexander the Great conquest the East
		The Hellenistic Age
	Rome supreme in Italy	
200	Rome conquers Carthage	Philip V of Macedon
	Rome turns to eastern Mediterranean	
	Destruction of Corinth and Carthage	Macedonia becomes a Roman province
	A Century of Revolution: Gracchi, Marius	
100	Revolt of Italian allies	
	Sulla	

THE FALL OF THE REPUBLIC 221

fought. When the Roman fleet blockaded Antony, his troops, unhappy at the prospect of fighting fellow Romans for the queen of Egypt, began to desert in droves.

Finally early in September (31 B.C.) a naval battle took place. It was not much of a battle. Only a few ships were engaged, and Cleopatra's squadron fled to Egypt, followed by Antony with 40 ships. The rest of the fleet surrendered, as did the land army. The next year Antony and Cleopatra committed suicide. Octavian ruled the world.

In the early summer of 29 B.C. Octavian returned to Rome and celebrated a magnificent three-day triumph. The wars were over at last.

8/SUMMARY AND CONCLUSION

When the Romans started on a career of world conquest, they were militarily successful from one end of the Mediterranean to the other. But as the appetite for conquest increased there came also a brutalization of Roman character. The flood of slaves and tribute into Italy ruined most of the small independent farmers. The government of the city was slow to adjust to the responsibilities of empire, and by 133 B.C. Rome's related military and agricultural problems cried out for solution.

The Gracchi made the fatal mistake of attempting to justify the means by the end. Too impetuous, they divided Rome against herself and ushered in a century of civil war. This, combined with the new army of Marius, resulted in the competition of successful generals—Marius and Sulla, Pompey and Caesar—for political power. Caesar managed to give Rome efficient one-man rule, but was removed before he could achieve his reorganization. By 29 B.C. his heir, Gaius Julius Caesar Octavianus, had used the magic name of Caesar to enhance his own political brilliance and had emerged the unquestioned leader of Rome and her empire.

The Republic had not measured up to the strains of empire. Would one man be able to do any better?

SOURCE READINGS

(a)

PLUTARCH (see page 100), in his life of the Roman dictator Fabius Maximus, describes the stratagem Hannibal used in 217 B.C. to slip out of a mountain pass where Fabius thought he had him trapped. Amongst his plunder Hannibal had thousands of head of cattle. Plutarch tells us how he used these to escape Fabius's trap.

He gave orders to take about two thousand of the cattle which they had captured, fasten to each of their horns a torch consisting of a bundle of withes or faggots, and then, in the night, at a given signal, to light the torches and drive the cattle towards the passes, along the defiles guarded by the enemy. As soon as his orders had been obeyed, he decamped with the rest of his army, in the darkness which had now come, and led it slowly along. The cattle, as long as the fire was slight, and consumed only the wood, went on quietly, as they were driven, towards the slopes of the mountains, and the shepherds and herdsmen who looked down from the heights were amazed at the flames gleaming on the tips of their horns. They thought an army was marching in close array by the light of many torches. But when the horns had been burned down to the roots, and the live flesh felt the flames, and the cattle, at the pain, shook and tossed their heads, and so covered one another with quantities of fire, then they kept no order in their going, but in terror and anguish, went dashing down the mountains, their foreheads and tails ablaze, and setting fire also to much of the forest through which they fled. It was, of course, a fearful spectacle to the Romans guarding the passes. For the flames seemed to come from torches in the hands of men who were running hither and thither with them. They were therefore in great commotion and fear, believing that the enemy were advancing upon them from all quarters and surrounding them on every side. Therefore they had not the courage to hold their posts, but withdrew to the main body of their army on the heights, and abandoned the defiles. Instantly the light-armed troops of Hannibal came up and took possession of the passes, and the rest of his forces presently joined them without any fear, although heavily encumbered with much spoil.

Plutarch's Lives, translated by B. Perrin, Loeb Classical Library (William Heinemann, 1915), III, 137-138.

(b)

LIVY (see page 179) in Book 42 of his *History of Rome* tells this story from the year 172 B.C. Marcius and Atilius, Roman legates to Greece, are reporting on their mission to King Perseus of Macedon. In 171 B.C. Rome declared war on Perseus and by 168 B.C. he was soundly beaten.

On their return to Rome, Marcius and Atilius reported the results of their mission to the senate in the Capitol. They prided themselves on nothing so much as the way in which they had hoodwinked the king by agreeing on a truce and holding out hopes of peace. He was so fully provided with all the means of war, while they themselves had nothing ready, that all the strategic positions could have been occupied by him before an army was sent over into Greece. The interval of the armistice, however, would make the war an equal combat; he would no longer have the advantage of preparation, and the Romans would begin the war better equipped in every way. In addition, they had succeeded by a clever stroke in disrupting the Boeotian League, so that it could never again be united in support of the Macedonians.

A good many of the senators approved of these actions as showing very skilful management. The elder senators, however, and others who had not forgotten the moral standards of earlier days, said that they failed to recognize the traditional Roman character in these negotiations. "Our ancestors," they said, "did not conduct their wars by lurking in ambush and attacking by night, nor by feigning flight and then turning back unexpectedly upon the enemy when he was off guard, and they did not pride themselves on cunning more than on true valour. It was their custom to declare war before waging it, sometimes even giving the enemy notice of the time and place where they would fight.... This is the Roman scrupulousness, unlike the cunning of the Carthaginians or the cleverness of the Greeks, who pride themselves more on deceiving an enemy than on overcoming him by force. Occasionally more can be gained for the time being by craft than by courage, but an enemy's spirit is broken for good only when you have compelled him to confess that he has been overcome not by tricks nor by accident, but after a hand-to-hand trial of strength in a just and righteous war."

Such were the views of the older senators, who were not well-pleased with the new overcunning diplomacy; but that portion of the senate which preferred expediency to honour won out, with the result that the previous mission of Marcius was approved.

Lewis and Reinhold, editors, *Roman Civilization: Selected Readings*, Vol. I, *The Republic*, pp. 191-192.

(c)

PLUTARCH, in the following selection from his life of Sulla, describes the dictator's methods on his return to Italy in 83 B.C.

When Sulla learned that the greater part of the enemy had been destroyed and the remainder had fled to Antemnae, he came to Antemnae at dawn. When 300 of the inhabitants sent a herald, he promised them safety if they would do some mischief to his other enemies before coming over. They took him at his word, attacked their fellows, and many on both sides were cut down. Nevertheless Sulla collected these and other survivors to the number of 6000 in the hippodrome, and convoked the senate in the nearby temple of Bellona. As he began to speak, those assigned to the task began to butcher the 6000. Naturally the shrieks of such a multitude being slaughtered in so small a space carried, and the senators were startled. Sulla continued his speech with a calm and unconcerned expression, and bade the senators pay attention to his speech and not busy themselves with what was going on outside: some naughty people were being admonished at his orders. Even the stupidest Roman could now realize that they had changed tyrants, not escaped tyranny. . . .

Slaughter now became Sulla's business, and murders without number or limit filled the city. Private animosities doomed many who had no relations with Sulla; he consented to gratify his associates. Young Gaius Metellus made bold to ask Sulla in the senate what end there would be to these evils and at what point he might be expected to stop. "We do not ask you," he said, "to free from punishment those you are resolved to kill, but to free from suspense those you are resolved to save." Sulla said he did not yet know whom he would spare, whereupon Metellus said, "Then tell us whom you are going to kill." . . . At once, without communicating with any official, Sulla proscribed 80 persons. Despite general indignation he proscribed 220 more on the following day, and as many again on the third day. In a public address on the subject he said he was proscribing as many as he could remember; those who escaped his memory for the present he would proscribe another time. Further, he penalized humanity with death, proscribing any who harboured or protected a proscribed person, making no exception of brother, son, or parents: the prize for killing a proscribed person was two talents, if a slave murdered his master or a son his father. What seemed the greatest injustice of all, he cancelled the civil rights of sons and grandsons of proscribed persons and confiscated their property. Proscriptions were the rule not only in Rome but in every city of Italy; neither temple of god nor hearth of hospitality nor ancestral hall was unstained by bloodshed; husbands were slaughtered in the arms of their wives, sons in the arms of their mothers. The victims of political passion or private animosity were nothing compared to those slaughtered for their property. Even the executioners were moved to say that his fine house killed this man, his garden that, his warm baths the other. Quintus Aurelius, who had no political connections but thought that his only concern with the misfortune was to condole with those affected by it, walked into the Forum and read the list of the proscribed. He saw his own name, and said, "Too bad. My Alban farm has condemned me." He had not gone far before he was overtaken and massacred. . . .

Hadas, *History of Rome . . . as Told by the Roman Historians*, pp. 58-60.

(d)

CICERO (106-43 B.C.) was appointed prosecutor in 70 B.C. of Gaius Verres who from 73 to 71 B.C. had been the infamously cruel and dishonest governor of Sicily. Verres openly boasted that he had amassed three fortunes: one for himself, a second to repay his friends and patrons, and a third (the largest) to bribe the jury that tried him for extortion. The following selection from Cicero's *Second Speech against Verres* makes clear the collusion between Verres and Carpinatius, the manager of a tax-collecting company in Sicily, as well as Verres' attempt to make certain that the company's records contained nothing incriminating. Although Verres escaped into exile, Nemesis finally caught up with him in 43 B.C. when he was proscribed by Mark Antony, who coveted the works of art this rogue had stolen from Sicily.

I was at Syracuse looking through the company's accounts kept by Carpinatius.... With the accounts open in my hands, I suddenly caught sight of some erasures that suggested recent injuries to the tablets. As soon as this suspicion struck me, I transferred my eyes and attention to these special items. There were sums entered as received from *Gaius Verrucius son of Gaius*; but whereas up to the second *r* the letters were plainly untouched, all after that were written over an erasure; and there was a second, a third, a fourth, a large number of items of the same character.

Since these erasures on the tablets manifestly indicated some conspicuously villainous and dirty proceeding, I proceeded to ask Carpinatius who this Verrucius was with whom he had such extensive money transactions. The man hesitated, shuffled, went red in the face.... I stated my charge before Metellus [Verres' successor as governor of Sicily], saying that I had inspected the company's accounts; that they included a large one, with a great many entries, under the name of Gaius Verrucius; and that by comparing the months and years I had discovered that this Verrucius had had no account with Carpinatius either before the arrival of Gaius Verres or after his departure. I demanded therefore that Carpinatius should tell me who this Verrucius was ... where he was, where he came from, and why the company's slave who wrote up the accounts always went wrong at one particular point when he wrote the name Verrucius....

If Carpinatius would not answer me then, will you answer me now, Verres, and say who you suppose this Verrucius is who is almost one of your own clan? I see the man was in Sicily during your praetorship, and the account is enough to show me that he was rich, so it is out of the question that you in your own province were not acquainted with him. Or rather, for the sake of brevity and clearness, step forward, gentlemen, and unroll this facsimile transcript of the accounts, so that instead of following the tracks of his veracity the world may now see it at home in its lair. Do you see the word *Verrucius*? Do you see how the first letters are all right? Do you see the last part of the name, the tail-bit of *Verres* there sunk in the erasure like a pig's tail in the mud? Well, gentlemen, the accounts are what you see they are; what are you waiting for, what more would you have? You yourself, Verres, why are you sitting there doing nothing? Either you must show us Verrucius, you know, or you must confess that Verrucius is you.

Lewis and Reinhold, *Roman Civilization: Selected Readings,* Vol. I, *The Republic,* pp. 363-364.

FURTHER READING

1. THE PUNIC WARS
    ADCOCK, F. E., *The Roman Art of War under the Republic* (Harvard, 1940)
    † CHARLES-PICARD, G. and C., *Daily Life in Carthage at the Time of Hannibal* (Allen and Unwin, 1960)
    † * COTTRELL, L., *Enemy of Rome* (Pan, 1962)
    DE BEER, G., *Alps and Elephants* (Bles, 1955)
    SCULLARD, H. H., *Scipio Africanus in the Second Punic War* (Cambridge, 1930)
    * WARMINGTON, B. H., *Carthage* (Praeger, 1960)

2. A CENTURY OF REVOLUTION: THE GRACCHI TO OCTAVIAN
    † BUCHAN, J., *Julius Caesar* (Davies, 1932)
    † ──────, *Augustus* (Hodder and Stoughton, 1937)
    † HASKELL, H. J., *This Was Cicero* (Knopf, 1942)
    † ──────, *The New Deal in Old Rome* (Knopf, 1947)
    HILL, H., *The Roman Middle Class in the Republican Period* (Blackwell, 1952)
    MARSH, F. B., *A History of the Roman World from 146 to 30 B.C.* (Methuen, 1963)
    * SCULLARD, H. H., *From the Gracchi to Nero* (Praeger, 1961)
    SMITH, R. E., *The Failure of the Roman Republic* (Cambridge, 1955)
    * SYME, R., *The Roman Revolution* (Oxford, 1960)
    * TAYLOR, L. R., *Party Politics in the Age of Caesar* (California, 1961)
    WALTER, G., *Caesar: A Biography* (Scribners, 1952)
    VOLKMANN, H., *Cleopatra: A Study in Politics and Propaganda* (Elek, 1958)

8. The Roman Empire and Roman Civilization

Others, doubtless, will mould lifelike bronze with greater delicacy, will win from marble the look of life, will plead cases better, chart the motions of the sky with the rod and foretell the risings of the stars. You, O Roman, remember to rule the nations with might. This will be your genius —to impose the way of peace, to spare the conquered and crush the proud.
VIRGIL, *Aeneid*

The power of Rome was now in the hands of one man: what would he do with it? Would he try to set back the constitution a hundred years? Would Roman culture match her political developments? Would the Empire expand or contract? Such are the questions that must be asked in a brief survey of Roman imperial history from 29 B.C. to A.D. 476.

1/THE PRINCIPATE

Octavian faced an awful dilemma. He could retire, as Sulla had, in which case there would likely be a flare-up of civil war. Or he could wield complete power openly, in which case he might suffer his great-uncle's fate. Or he might do neither. He could, by a slow process of trial and error, feel out the degree to which public opinion would tolerate his exercise of power.

In 27 B.C. Octavian was thirty-five years of age and consul for the seventh time. That same year he appeared before the Senate, surrendered his powers, and asked that the Republic be restored. In an autobiographical inscription forty-one years later (some selections from which appear on page 266) Octavian wrote, "I handed the commonwealth over from my own control to the free disposal of the Senate and the people of Rome." The Senate voted him proconsular power with control of the army and the title "Augustus", meaning the revered or consecrated one. The month Sextilis was also renamed Augustus in his honour.

Yet despite appearances, Augustus did not really restore the Republic. The *Princeps* (First Citizen), as he was now called, was anxious to have Romans *believe* he was restoring republican government. In other words he was using propaganda. Outwardly there was a certain balance between the Princeps and the Senate, and magistrates were once again elected yearly; but by his control of the armed forces, by his management of public finance, and by his absolute powers of veto, the Princeps was in reality supreme. Furthermore, his unique position as the unifying symbol of the Empire greatly strengthened his hand.

To give this arrangement time to settle, Augustus kept himself away from Rome in the Western provinces for the next few years. After he arrived back in Rome he was taken seriously ill in 23 B.C., and upon his recovery decided he ought once more to clarify his position in the state in order to reconcile the senatorial die-hards to his rule. Consequently he resigned as consul, but received in compensation the tribunician power for life. Henceforth the tribunician power was really the basis of the Princeps' power. By it he could bring measures before the people, and summon and consult the Senate. Thus a legal veneer gave a pleasing face to Augustus' military power.

Something for Everybody Augustus had four main groups to please. The Roman people wanted peace and prosperity; the aristocracy wanted a share in government; the Equestrians wanted to become more prosperous; and the army wanted to have some guarantee of grants of land and money when its veterans were discharged. The first two groups were satisfied by the constitutional settlement, and to win over the people even more Augustus set up a government department in Rome to supervise the collection, storage, and distribution of provincial grain. Employment was provided by a programme of public works which included new docks, granaries, baths, theatres, libraries, and aqueducts. And the city of Rome itself was divided into fourteen districts and given a long-overdue fire department and police force.

To allow the senators to share in the work of the government, commissions were instituted to supervise the care and maintenance of temples and

This handsome marble statue of Augustus, found in the villa of his wife Livia at Prima Porta, well sums up what the Princeps meant to the Roman world. One set of symbols on the richly decorated breastplate (originally gilded and enamelled) portrays Earth rejoicing in the blessings of peace and prosperity, while the emperor's stance, with outstretched arm and sceptre, indicates that he was addressing the troops who assured this peace. The face, although somewhat idealized by the Greek sculptor of 20 B.C. who carved it, has a certain majestic individuality complemented by the quietly self-confident angle of the head. The child at Augustus' feet probably represents Cupid, an allusion to the emperor's alleged descent from the goddess Venus.

aqueducts and to keep roads throughout Italy in repair. In addition, while the Princeps himself assumed direct control of the frontier provinces, he turned over to the Senate the administration of the more highly civilized provinces that were farther removed from invasion.

Since, however, the government of the Empire required the services of more men than the Senate could supply, Augustus employed Equestrians in the civil service, from which they had been mostly excluded under the Republic. In this way a salaried professional civil service developed. The Equestrians were also used as governors in less important provinces (such as Judaea), and as financial agents in others. Henceforth governors were paid a salary—the most practical incentive to conscientious conduct—and all provincial officials were expected to govern efficiently.

Finally, having learned well the lessons of the civil wars, the Princeps now made up his army by voluntary enlistments. The first step he took was to cut the total armed strength from over 60 legions to 28 (168,000 men), the discharged veterans being settled in Italy or the provinces. Some twenty-eight new colonies were founded for them in Italy alone. At the same time conditions of service and pay were improved for those still in the army. The recruits were Roman citizens, and non-citizens were granted the citizenship on enlisting in the legions. The term of service was set at sixteen (later twenty) years, with a regular pension on discharge to ensure that generals would no longer have to enter politics to secure rewards for their men at the end of a campaign. Of particular importance in the

THE ROMAN EMPIRE'S EXPANSION (44 B.C.-180 A.D.)

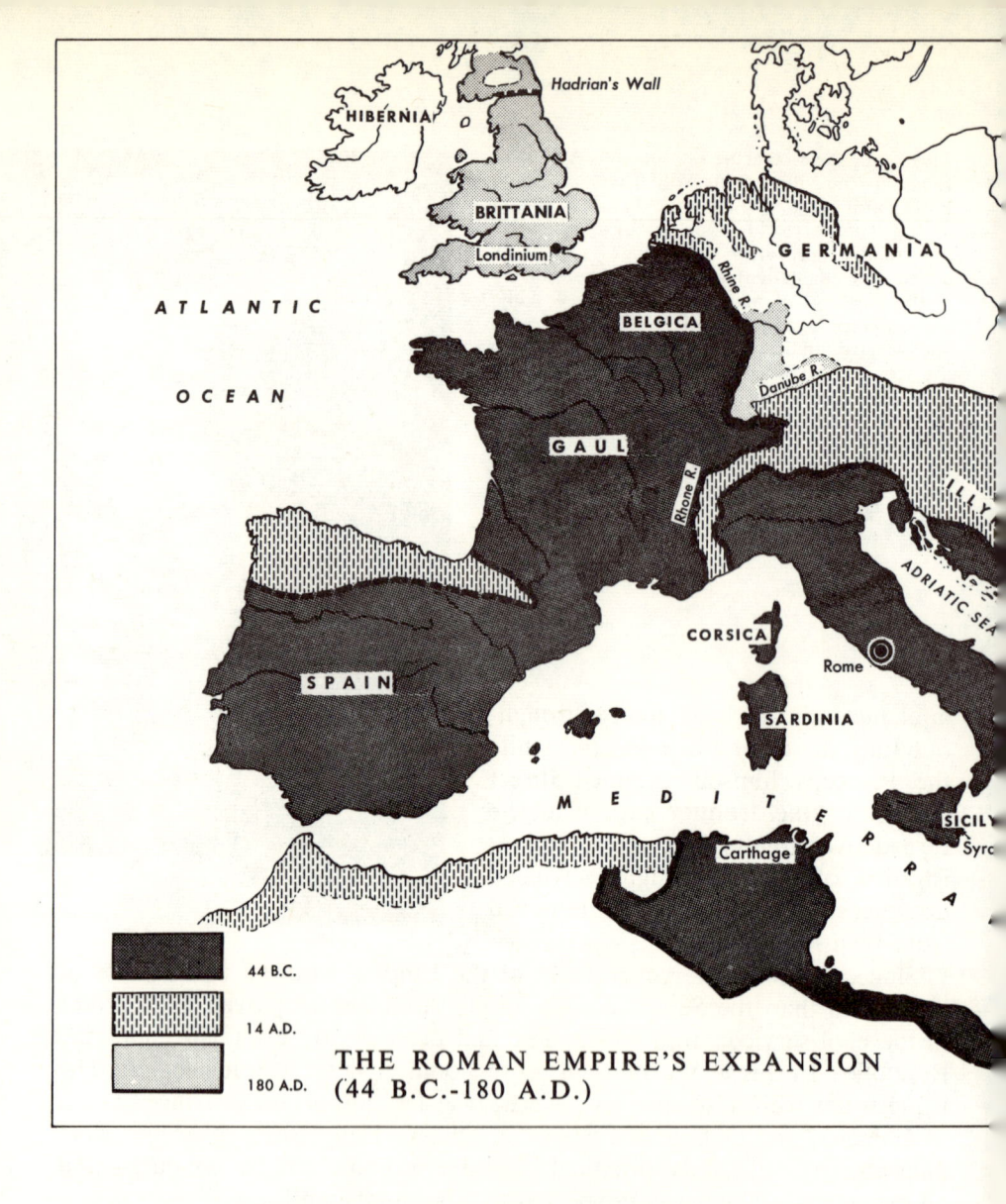

- 44 B.C.
- 14 A.D.
- 180 A.D.

Augustan military establishment was the imperial bodyguard, the so-called Praetorian Guard of 9000 men recruited exclusively from Italy.

But more than governmental reorganization was needed. Augustus felt that Rome should also undergo a moral regeneration. To this end the ancient temples were repaired and cults were revived, while in the Orient there was even worship, along Hellenistic lines, of the Princeps himself. In Rome, the "genius" of Augustus (the divine spirit that watched over him), was worshipped, and although Augustus was not in favour of direct emperor-worship, doubtless many of the ordinary people lost sight of any distinction between worshipping the Emperor and worshipping his "genius". In an attempt to restore a sound family life, Augustus had laws passed to encourage marriage and discourage childlessness. Nor was he oblivious to the plight of slaves. The story goes that one day a slave of Vedius Pollio dropped and broke a costly cup of myrrh at a banquet being given by his master for Augustus. As punishment for his clumsiness, Vedius ordered the slave thrown to man-eating carp; whereupon Augustus, countermanding the order, called for all of Vedius's precious cups and calmly broke them one after another before his eyes.

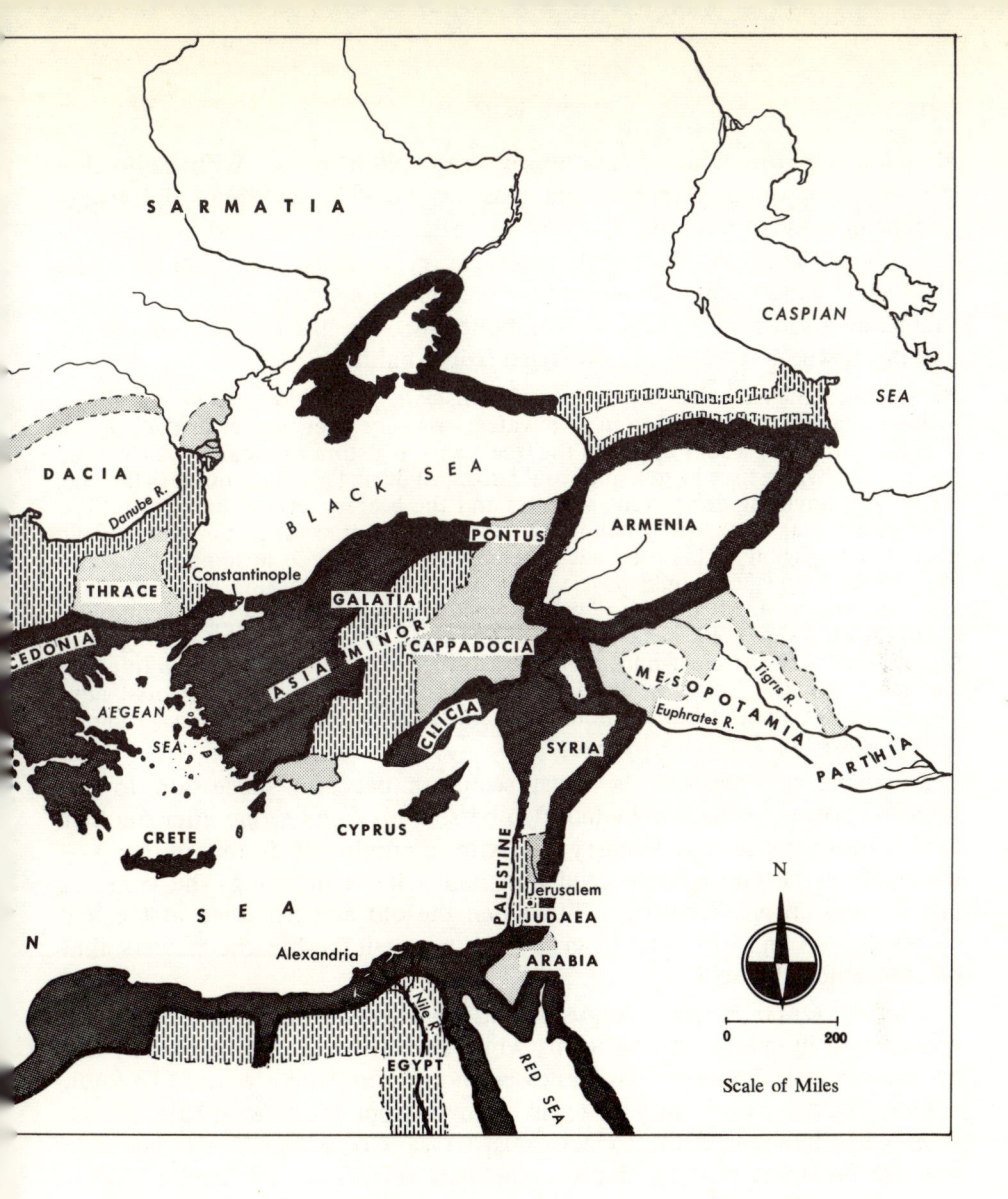

In these and various other ways Augustus sought to create efficiency and political stability, and to restore the old-fashioned Roman virtues of self-discipline, a sense of responsibility, and a respect for tradition and authority. For Augustus, like his great-uncle before him, seemed to

Character of Augustus

abandon the role of a ruthless opportunist and to accept with dedication that of philosopher-emperor. He liked to be with children, and had a delightful sense of humour. Once at a public ceremony he said to a nervous petitioner, "You look as though you were trying to offer a small tip to an elephant." And though he lacked the personal magnetism of Julius Caesar there was a touch of homely virtue about him. He had come from a small country town, and all his life practised frugal habits.

He seldom touched wine between meals; instead, he would moisten his throat with a morsel of bread dunked in cold water; or a slice of cucumber or the heart of a young lettuce; or a sour apple off the tree, or from a store cupboard. . . . He often needed more sleep than he got, and would doze off during his litter journeys through the City, if anything delayed his progress and the bearers set the litter down.

Augustus was remarkably handsome, and of very graceful gait even as an old man; but negligent of his personal appearance. He cared so little about his hair that, to save time, he would have two or three barbers working hurriedly on it together, and meanwhile read or write something, whether they were giving him a haircut or a shave. He always wore so serene an expression, both talking or in repose, that a Gallic chief once confessed to his compatriots: "When granted an audience with the Emperor during his passage across the Alps, I would have carried out my plan of hurling him over a cliff, had not the sight of that tranquil face softened my heart; so I desisted."

Augustus realized that he must secure a peaceful succession to the imperial power; so he designated members of his family, one after another as they predeceased him, to carry on. After a number of disappointments—the relatives he chose to succeed him had a habit of dying—he came at last to his stepson, Tiberius. Thus when the old emperor died at the age of seventy-six in A.D. 14, Tiberius was granted for life the powers that his stepfather had held.

The Old Order Passes   Augustus was a statesman of the first rank. He was able to maintain the old forms while exercising the new powers. But his great system did have its weaknesses and, in time, these were to be fatal.

One weakness was that which makes any system of one-man rule vulnerable: what happens if the successor is a tyrant or a fool? Another flaw was the apparent sharing of power between Princeps and Senate. What would happen if the Senate proved incapable of exercising power, and preferred rather to fawn upon the Princeps? A third weakness is best expressed in the words of the Roman historian Tacitus (55-120), who lamented the way in which Augustus had wrought the ruin of the old aristocracy.

But when he had won the soldiery by bounties, the populace by cheap corn, and all classes alike by the sweets of peace, he rose higher and higher by degrees, and drew into his own hands all the functions of the Senate, the magistrates and the laws. And there was no one to oppose; for the most ardent patriots had fallen on the field, or in the proscriptions; and the rest of the nobles, advanced in wealth and place in proportion to their servility, and drawing profit out of the new order of affairs, preferred the security of the present to the hazards of the past. . . . Thus a revolution had been accomplished. The old order had passed away; everything had suffered change. The days of equality were gone: men looked to the Prince for his commands.

Even so the Augustan system was probably the best that could be devised under these difficult circumstances. After all, the Empire stretched from the North Sea to the Red Sea and had a population of 70 million. Over it all for forty-one years ruled this one man, this Caesar who had grown from a cold-blooded politician into the architect of the Roman Empire. Is it any wonder that one of the first official acts of his successor, Tiberius (14-37), was to arrange for the deification of Augustus?

2/ THE ROMAN PEACE

The Roman emperors who followed Augustus are a fascinating lot, touched, as some of them were, by scandal if not by madness. Now that the Republic was gone, free discussion and political parties were no more. Hence ancient Roman historians tended to concentrate on a few themes: the emperor's character, his relations with the Senate, and especially gossip and intrigue connected with the court. Unfortunately they were not greatly concerned with the other side of the coin, and left it to modern archaeologists to tell us much about life in the provinces. But however interesting these emperors may be, it should be remembered that individually they were less important than the system over which they presided.

ROMAN EMPERORS (27 B.C.-A.D. 337)

	27 B.C.-A.D. 14	—Augustus	
Julio-Claudians	14-37	—Tiberius	
	37-41	—Caligula (Gaius)	
	41-54	—Claudius	
	54-68	—Nero	
	68-69	—Galba	
	69	—Otho	
	69	—Vitellius	
Flavians	69-79	—Vespasian	
	79-81	—Titus	
	81-96	—Domitian	
	96-98	—Nerva	
	98-117	—Trajan	
	117-138	—Hadrian	
Antonines	138-161	—Antoninus Pius	
	161-180	—Marcus Aurelius	
	180-192	—Commodus	
	193	—Pertinax	
	193	—Didius	
	193-211	—Septimius Severus	
	211-217	—Caracalla	
	217-218	—Macrinus	
	218-222	—Elagabalus	
	222-235	—Severus Alexander	
	235-284	—"Barracks Emperors"	
	284-305	—Diocletian	
	306-337	—Constantine	

Sudden Deaths When Tiberius, the first of the Julio-Claudians, became emperor at fifty-two, he was a completely embittered man. He had had to wait a long time for power; yet he lacked Augustus' gifts for winning support, and probably felt inadequate to exercise it once it was finally his. He tried to do what he thought Augustus would have done as new circumstances arose, but he was never very comfortable, as is shown by the fact that he reserved the right to abdicate.

Tiberius had no desire to extend the empire. He merely wished to consolidate what Augustus had annexed. For this reason he lengthened

Tiberius

the terms of governors (one term of twenty-five years is recorded), hoping to promote efficient government in the provinces. But as the reign went on he became increasingly moody, and when his adopted nephew Germanicus died during an eastern campaign it was rumoured that he had been poisoned at the emperor's command. Tiberius, so it was said, wanted his son Drusus to ascend the throne without having to vie with any competitors. Germanicus' widow and her children were, of course, shunted aside. Drusus, however, also died, whereupon the emperor made himself even more unpopular by absenting himself from Rome to live at his villa on Capreae (Capri). When he died of a stroke at the age of seventy-eight there was much rejoicing in Rome, and the Senate absolutely refused to vote him divine honours.

Caligula   Tiberius was succeeded by his great-nephew, Gaius. Nicknamed Caligula ("Little Boots"), this son of Germanicus had been brought up as a child in army camps, suffered from epilepsy, and probably had spells of insanity. He became pathologically suspicious of anyone who seemed to be at all eminent, and made it plain to the Senate that he could do nicely without them. Finally the Praetorian Guard assassinated him early in 41. Thus died an emperor who had been heard to remark morosely that if only humanity had one neck it could be severed in a single blow!

As the Praetorian Guard ransacked the royal palace after Caligula's murder, one of them saw a pair of feet sticking out from beneath a curtain.
Claudius   The gentleman-in-hiding turned out to be no less than the dead emperor's uncle, Claudius. This unattractive man, who had suffered poliomyelitis in childhood, was forthwith proclaimed emperor by the Guard.

During his reign which began so inauspiciously the Empire was extended in the East and in Africa. Most famous of all Claudius's campaigns—which were doubtless undertaken as one means of popularizing his reign—was the invasion of Britain in A.D. 43. In the government of the Empire, Claudius sincerely tried to co-operate with the Senate after the fashion of Augustus and to treat it with deference. But he also added a number of men from the provinces to the Senate, and went so far as to denounce the senators as "yes men".

If you disapprove [of this measure] find another solution, but here and now . . . it is unbecoming to the majesty of this assembly that one man alone, the consul-designate, should repeat the remarks of the consuls word for word as his opinion and that the rest of you utter only one word, "I agree," and then when you leave say, "We debated."

The fact was that the Senate had become increasingly inadequate, with the result that the emperor overshadowed and overruled it. He further alienated it by organizing a centralized civil service, the members of which owed their positions to him and hence were independent of the Senate. Other administrative reforms were undertaken by Claudius, who also stepped up the efficiency of the courts.

Claudius carried out two projects originally envisaged by Julius Caesar —the construction of a harbour at Ostia and the draining of some of Italy's marshy areas. Moreover, he copied the illustrious Julius in making generous grants of citizenship to certain parts of the Empire. For Claudius knew his history (he had in fact written some history before becoming emperor), and was convinced that in the past Rome had flourished because she had constantly readjusted her constitution and had not been afraid to admit foreign elements into the citizen body. Nevertheless, by trying to raise the status of the provinces Claudius undermined the position and influence of the old governing class at Rome still further.

Claudius may have known his history, but he did not know much about women. His third and fourth wives proved such unfortunate choices that he finally executed one after she insisted on going through a form of public marriage with her lover. The other, the notorius Agrippina, was his own niece (he was her third husband). She persuaded Claudius to adopt her son by a previous marriage, Nero, and began to intrigue to get him the throne. Then in 54 Claudius died very suddenly, after—so it was said— eating a dish of mushrooms prepared by his loving wife.

One Man's Savagery At last Agrippina had what she wanted: her son emperor at an early enough age (he was sixteen) for her to influence him. Now she could become mistress of the world. But matters did not turn out quite as she had planned. Nero's advisers encouraged him to be independent of her, and for a few years the emperor gave over the management of affairs to the Senate, led by his tutor Seneca the Younger (5 B.C.-A.D. 65).

Nero

As time went on Nero began to hate his dominating mother, and eventually to plot her death. At this he was nothing if not persistent. He began by depriving her of her bodyguard and so harassing her with lawsuits that she left Rome. Three times he tried poison, and three times she escaped its effects. Then the floor above her bedchamber was made to collapse while she was asleep; yet still she survived. Finally an ingenious ship was built, a ship that would disintegrate and drown her. But Agrippina swam ashore. When news of her miraculous escape was brought to Nero he thought of the perfect solution. He pretended that the messenger had come to assassinate him and ordered his mother put to death. Even so, the records tell us, he was not rid of his ambitious mother. It is said that thereafter her ghost haunted him.

When news of the murder reached Rome, it was accepted as having been forced on the emperor. Now Nero felt free to indulge his every whim. He made public appearances as a poet, musician, actor, and charioteer, and posted men to flog the audiences if they did not applaud loudly enough. Hollywood notwithstanding, Nero did not fiddle while Rome burned. It is true that Rome burned in 64; but the emperor was away when the fire started, and he hurried back to take energetic steps to

The excellent pavement (*pavimentum*) of the highway here is proof of the advanced Roman road-building techniques. The earliest Roman roads were made of logs and planks, but later most road-beds had large stones in the bottom and worked up to smaller stones on the top, sometimes (as here) with a middle layer of hard gravel covered by a layer of fine limed gravel. On the surface, closely fitted limestone slabs sloped slightly to the edges, with drains at both sides. Roman road-beds were at least 40 inches deep—somewhat deeper than modern ones—and highways varied in width from 5 to 23 feet. In all, over 47,000 miles of roads were constructed. In the background a man ploughs with oxen on a Roman *villa*. His plough breaks and turns the soil easily because of the coulter blade ahead of the share-beam. Various Roman methods of transport can also be seen: a donkey, a two-wheeled country cart, and behind it a heavy wagon and a Roman chariot.

bring it under control and to provide relief for those who had been burned out of house and home. Even so, public suspicion had been aroused by Nero's conduct, and when the rumour flew that the emperor himself was responsible for the fire he thought it prudent to provide a scapegoat.

To allay the rumour Nero fastened the guilt and inflicted exquisite tortures upon a people hated for their wickedness, vulgarly called Christians. The name was derived from Christ, who was executed by Pontius Pilate in the reign of Tiberius. Checked for the moment, the mischievous superstition broke out again, not only in Judaea, the source of the evil, but even in Rome, into which everything infamous and abominable from all quarters flows and flourishes. First some were seized and made confession, and then upon their information a huge multitude were convicted, not so much for the crime of arson as for their hatred of the human race. Mockery was added to their deaths. They were covered with animal skins and torn to pieces by dogs, many were crucified or burned, and some were set afire at nightfall to serve for illumination. For the spectacle Nero offered his gardens, and he presented horse races in addition. In the dress of a charioteer he himself mingled with the crowd or stood up in his sulky. Hence, though the victims were guilty and deserved extreme punishment, nevertheless they aroused compassion, for it was not for the public good but for one man's savagery that they were being destroyed.

So unpopular did Nero become as a result of his growing despotism, and so obsessed was he with the ideas of emperor-worship that had come out of the East, that he decided to get away from it all for a while and to visit the birthplace of Hellenic culture. Accordingly, in 66, he undertook a tour of Greece. It was a triumphal tour too: the Greeks awarded him no fewer than 1808 first prizes in artistic and athletic competitions! By the time Nero arrived back in Rome, however, rebellion was brewing, and in 68 the Praetorian Guard and the armies revolted. Wailing "What an artist is lost!", the emperor committed suicide—or rather had one of his own freedmen finish him off when he lost his nerve. The last of the Julio-Claudians had died at the age of thirty.

The Secret of Empire   Now ensued a frantic grab for power. As Tacitus put it, "The secret of empire was revealed: an Emperor could be made elsewhere than at Rome." The year 69, the "year of the four emperors" as it became known, saw four men bid for the titles as Galba, Otho, Vitellius, and Vespasian occupied the imperial throne in quick succession. The Praetorian Guard made and unmade the first three. The fourth, Titus Flavius Vespasianus, had served in Britain, had been governor in Africa, and was campaigning in Judaea in 69 when the legions in the East and along the Danube frontier proclaimed him emperor. Once more civil war erupted in Italy, until in June 70 Vespasian arrived in Rome, bringing peace with the sword. The fact that Vespasian had two sons, Titus and Domitian, meant that the Flavian dynasty, as it is called, solved for the time being the troublesome problem of the succession.

Vespasian

Vespasian was from a country family, and throughout his life exhibited the traits of solid plebeian stock. For one thing he was somewhat tight-fisted—a timely blessing for a Rome whose coffers had been rifled by the wanton prodigalities of Nero. Vespasian overhauled the tax system, increased provincial tributes, and insisted on honesty in financial affairs. He also made himself censor, and as holder of that office saw to it that men from the Italian towns and western provinces were appointed to the Senate.

The frontiers were stabilized when rebellions were suppressed in Germany and Gaul. In Judaea Titus captured Jerusalem in 70, destroying the

Temple and carrying its treasures off to Rome. The peace and prosperity of Vespasian's later years were well symbolized by the Colosseum, that vast Flavian amphitheatre built on the site of Nero's Golden House and capable of accommodating 50,000 spectators. But what a sad contrast there was between the tastes of Greek and Roman audiences. The Greeks had sat for hours to watch the plays of their leading dramatists. The Romans went quite literally for blood, as this account by a spectator clearly shows.

Chance had led me to the Amphitheatre at midday; I was expecting games, jokes, those interludes in which the spectator's eye is rested from seeing human blood flow. But the opposite happened: the morning's contests had by comparison been humane. Now, no more trifling: these were sheer slaughters. The gladiator had nothing with which to cover himself; all parts of his body were exposed to blows, and no encounter failed to leave its wounds. Did the majority of spectators prefer this kind of combat to that which the usual or extra pairs of gladiators undertook? How could they but prefer it? No helmet, no shield against the sword. What need of protective armament or of a struggle according to rules? These are only ways of delaying the kill. In the morning, men had been delivered to lions and bears; at midday it was to the spectators that they were thrown. After having killed, the combatant must fight again to be killed in his turn; even the victor was earmarked for death. For the contestants there was one outcome only, death, carried out by fire and sword. And all this occurred during the intermission.

Titus   Vespasian's son Titus was a popular emperor, though his short reign was marked by several tragedies, including the famous eruption of Mount Vesuvius that buried the cities of Pompeii and Herculaneum in August, 79. A graphic eye-witness account of this catastrophe may be read on pages 268-269.

This moving plaster cast from Pompeii is the mould of a victim in the very throes of being smothered during the eruption of Vesuvius in A.D. 79. The process by which such moulds have come down to us is an interesting one. Actually most of the volcano's victims were killed, not by the initial hail of pumice, but by deadly carbonic acid fumes which accompanied the dense black cloud of ash and condensed steam. As the victims fell the ashes continued to sift down on them, eventually mixing with steam or rain to harden into a mud shell around the corpse. Thus when the bodies decomposed more than 1800 years ago they left empty moulds that have graphically preserved their death struggles, and which, when filled with plaster, form casts. By this method excavators have preserved countless mute testimonies to the agonies suffered by men, women, children—even dogs—on that August day. When the sky cleared three days later, a city had disappeared, buried in a volcanic mantle 20 feet deep.

Domitian, who became emperor at thirty, was a man full of cruelty and, *Domitian* like Tiberius, of resentment at being held back. He increasingly dominated the state, ignoring the Senate completely in the latter part of his reign. He put Equestrians into positions hitherto reserved for senators, and tried to uplift the moral tone of Roman society by restoring the old gods—though this last involved persecuting minority religions. His final years were nothing short of a reign of terror, as nobles were accused, executed, and dispossessed on the flimsiest charges of treason. Again violent means bred a violent end. In 96 palace conspirators murdered the last of the Flavians.

Tacitus, looking back on the reign of this sadistic man, saw these years as a disgraceful blot on Rome's history.

Domitian fancied that the voice of the Roman people, the liberty of the Senate, the conscience of the human race were obliterated; he banished teachers of philosophy and exiled every noble pursuit, so that nothing honourable might anywhere be encountered. We did provide a wholesale example of submission; as a former age saw the extreme of liberty, so we witnessed the extreme of servitude, when investigations abolished free communications. We should have lost our memories as well as our voices if it were as easy to forget as to keep silent.

"Five Good Emperors" The assassins of Domitian selected a sixty-six-year-old senator, Nerva, as the new emperor. But even though he tried to please the senators and keep the goodwill of the army there were still plots against him. And so, with no son of his own, Nerva realized that he must secure a successor. He chose a wise expedient: he adopted one—a *Trajan* distinguished soldier, Trajan, whom he had just recently appointed governor of Upper Germany. This adoption was one of the most significant events in the history of the Empire, for it inaugurated a system that provided an orderly and able succession for almost a century.

Trajan was a Spaniard, the first emperor of provincial origin, and one of the very best to occupy the imperial throne. By waging wars in Dacia (101-106) and Parthia (114-116), Trajan extended Roman territory north across the Danube and east to Mesopotamia. The great warrior emperor was also a fine administrator. Under him the system of state maintenance for poor children in Italy, initiated by Nerva, was extended. This scheme provided for loans to farmers of up to one-twelfth of the value of their land, for which they paid interest rates of 5% to their local communities. The local communities then used this income to furnish grants to the parents of poor children—a sort of family allowance arrangement.

Provincial misgovernment was simply not tolerated. Trajan personally scrutinized public expenditures in the provinces, and an example of the careful attention he devoted to provincial affairs can be read on pages 267-268. Fortunately for Trajan, the booty from Dacia paid for the vast building projects he undertook, as well as for the state loans to farmers and all the other imperial expenditures such as the lavish shows and spectacles he loved.

Trajan was admired by every class. He won the plebs by fine shows. He won the legionaries by his endurance. (Who could help worshipping an emperor who tore up his own clothes to bind your wounds?) And he won the Senate by treating them with respect, always sending back to them reports on his campaigns and insisting that any peace terms with an enemy must be ratified by them. When he died of a stroke at the age of sixty-four while returning from his Parthian campaigns he had not formally adopted a successor, but his wife said that on his deathbed he had indicated his choice: his second cousin, Hadrian, the governor of Syria.

Hadrian   Hadrian was also a remarkably good emperor. He went on the army's route marches, disdaining the use of a vehicle, and was an accomplished architect, builder, and surveyor. Under his competent direction, taxes were revised and the system of tax-farming was practically ended; a new code of Roman law was drawn up and circuit judges were appointed for Italy; and Equestrians were made the holders of many administrative posts previously held by freedmen. For most of his reign Hadrian managed to maintain good relations with the Senate, so good in fact that for twenty years no conspiracies were formed against him.

Hadrian did not wish to expand the Empire; indeed he allowed it to contract in the Middle East. Yet he believed he must keep the frontier armies at a keen fighting pitch so that outsiders would not be tempted to attack, and to this end he spent at least half of his reign touring the Empire. It was to prevent raids along Britain's northern frontier that what has been called "the greatest piece of military engineering in antiquity" was built, a wall 73 miles long running from the Tyne to the Solway.

Antoninus Pius   Antoninus Pius, Hadrian's choice as his successor, may have been the most universally liked and respected of Roman rulers. Far from being a citizen of the Empire like Hadrian, Antoninus Pius spent nearly all his time in Italy. This handsome and cheerful man won the praise of Roman historians for his wise, if unspectacular, rule. But modern historians see the reign as a time of slackness during which the army was allowed to grow lax. For while there was calm within the Empire, without seethed masses of barbarian tribes awaiting the day of Roman weakness.

Marcus Aurelius   The second of the Antonines, and the last of the so-called Five Good Emperors, was Marcus Aurelius. He was a philosopher who believed war was "robbery". Yet, ironically, war dominated much of his reign as Parthians in the east, barbarians in the north, and a plague all strained the resources of the Empire. Even so, a more serious strain still was to follow.

Chaos Descends   Marcus Aurelius was the first emperor of the 2nd century to have sons of his own; hence the adoptive principle was abandoned and his son Commodus became emperor in 180. Commodus was a thoroughgoing degenerate. He abandoned his father's conquests, gave himself up to riotous living in Rome, and was assassinated on the last day of the year 192.

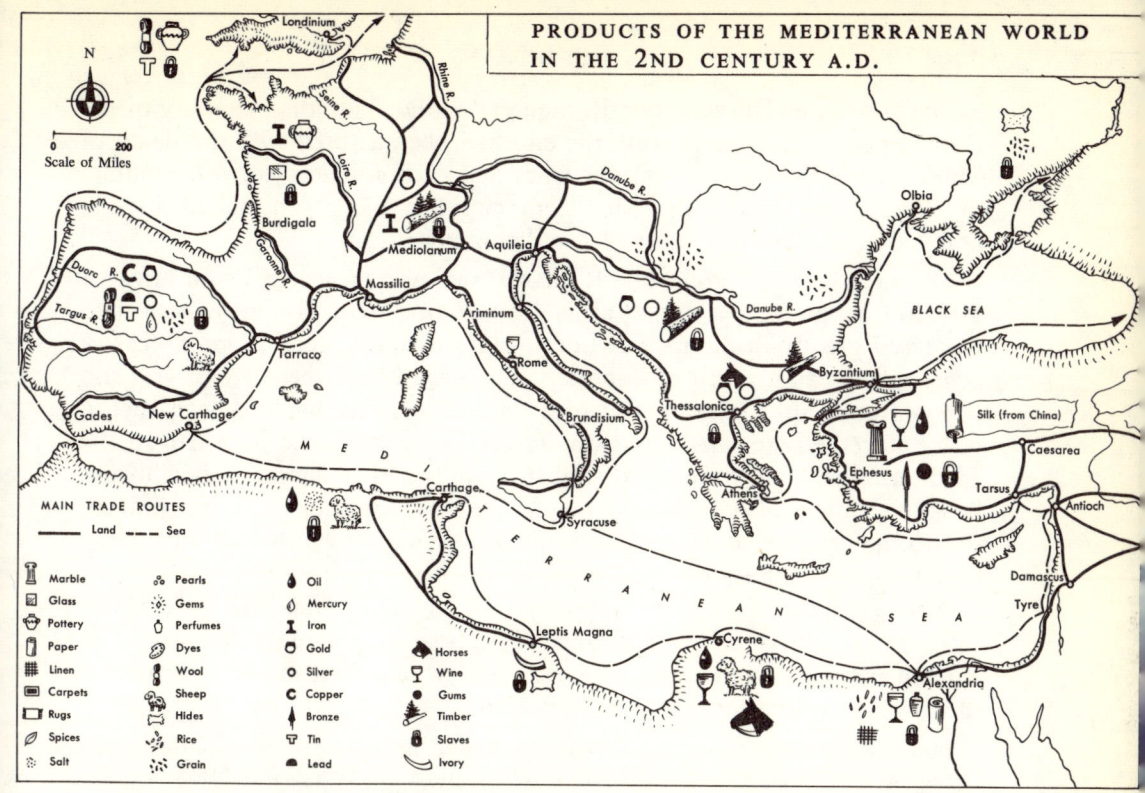

PRODUCTS OF THE MEDITERRANEAN WORLD IN THE 2ND CENTURY A.D.

MAIN TRADE ROUTES — Land ---- Sea

Marble, Glass, Pottery, Paper, Linen, Carpets, Rugs, Spices, Salt, Pearls, Gems, Perfumes, Dyes, Wool, Sheep, Hides, Rice, Grain, Oil, Mercury, Iron, Gold, Silver, Copper, Bronze, Tin, Lead, Horses, Wine, Gums, Timber, Slaves, Ivory

   The nominee of the Praetorian Guard, Pertinax, lasted exactly 87 days before he too was struck down. The Guard then auctioned off the Imperial Crown, and it was bought by Didius. But once more, as in 69, the armies put forward their respective commanders, and the one with the largest army of all, Septimius Severus with the twelve Danube legions, got to Rome first. He proved himself to be a competent administrator. Secure in power, he discharged the troublesome, predominantly Italian Praetorian Guard and reconstituted it with a handpicked force of 15,000 men from the frontier provinces, taking the further precaution of stationing a legion near Rome to keep an eye on both Italy and the new guard. The emperor himself was of Libyan-Phoenician descent and spoke Latin with a Punic accent. It was now obvious to all that Italy took second place in the Empire.
   Septimius Severus' son and successor, Caracalla, continued the trend to military autocracy and the further reduction of Italy's privileged position. In 212 Caracalla completed the long process of extending the citizenship. Henceforth all freedmen within the Empire were given the citizenship; only slaves were now non-citizens. But Caracalla, in his turn, was murdered, to be succeeded by a praetorian, Macrinus, who ruled for a year until he was supplanted by the fourteen-year-old Elagabalus. The praetorians killed him in 222, whereupon his cousin, the thirteen-year-old Severus Alexander, provided such weak rule that he was unable to hold back Persian and Germanic invaders. In 235 his troops murdered him also.

So formally ended what one Roman called the "boundless majesty of the Roman peace", though in truth the end had been assured with the death of Marcus Aurelius in 180. Border wars and civil conflicts notwithstanding, the lands surrounding the Mediterranean Sea had been united in one political system for a quarter of a millennium.

Diversity under Unity The Roman Peace was protected by an army of 400,000 men—not a large army considering that the provinces had increased greatly in number. Under Augustus there were 28 provinces; a century later as a result of conquest and subdivision there were 45 (and two centuries after that there were over 100). There has not been such a peace since. Rome tolerated diversity under the unity she imposed on the Mediterranean, and trade and industry flourished under the efficient and beneficial imperial government bequeathed by Augustus. So far, indeed, did Romanization go under the great peace that Roman coins have been found in modern Sweden.

About the year 150, a Greek rhetorician named Publius Aelius Aristides (117-185) delivered an enthusiastic oration in praise of Rome. The Roman peace was not as ideal as Aristides would have us believe; nevertheless, the world has not seen its like again.

The whole world, as on a holiday, has doffed its old costume—of iron—and turned to finery and all festivities without restraint. All other competition between cities has ceased, but a single rivalry obsesses every one of them—to appear as beautiful and attractive as possible. Every place is full of gymnasia, fountains, gateways, temples, shops, and schools. ... Greek and barbarian can now readily go wherever they please with their property or without it. It is just like going from their own to their own country. Neither the Cilician Gates nor the narrow, sandy approaches through Arabia to Egypt present any danger. Nor do impassable mountains, vast stretches of river, or inhospitable barbarian tribes. For safety, it is enough to be a Roman, or rather, one of your subjects. ...

You have surveyed the whole world, built bridges of all sorts across rivers, cut down mountains to make paths for chariots, filled the deserts with hostels, and civilized it all with system and order. ...

One might thus appraise and evaluate the state of things before your rule and under you: before it, they were all mixed up topsy-turvy, drifting at random. But with you in charge, turmoil and strife ceased, universal order and the bright light of life and government came in, laws were proclaimed, and the gods' altars acquired sanctity. ...

Yet there is another side to all this. It is true that the Roman solution to the political problems of the ancient world was an advance on the Greek—but at a cost. The tragedy was Rome's loss of personal liberties. There is something terrifyingly reminiscent of 20th century dictators in a statement by the Stoic philosopher Epictetus (55-135), whom the emperor expelled from Rome.

A soldier, dressed like a civilian, sits down by your side and begins to speak ill of Caesar, and then you too, just as though you had received from him some guarantee of good faith in the fact he began the abuse, tell likewise everything you think, and the next thing is—you are led off to prison in chains.

3 / THE LONG DECLINE

Despite its outward appearance of prosperity, the Empire was beginning to show signs of strain by the 2nd century. Cities were beginning to decline economically, taxes mounted, epidemics raged, and famines were widespread. And still the expenses of the state mounted. In the late 2nd century Marcus Aurelius was even forced to auction the crown jewels to find war funds. Imperial control over cities and provinces, which had begun as an emergency measure to end financial mismanagement, gradually hardened into permanent imperial direction. A long decline had set in that was to last from 180 to 476.

Rust in the Empire "Our history now descends from a kingdom of gold to one of iron and rust." Thus a contemporary Roman historian describes the transition from the emperor Marcus Aurelius to his brutal successors. With the 3rd century military absolutism triumphed as the army made and unmade a bewildering succession of "Barracks Emperors." In the middle years of this century, of some twenty-six emperors only one died a natural death. In the year 260 the ruling emperor faced no fewer than eighteen rivals. Blood drenched the Roman Peace.

In these dark days the emperors had even begun to hire Germans and Persians to do their fighting for them, so depleted had the legions become. The decline of cities continued. Between 235 and 253 some 83 cities of Asia Minor issued their last coins, and between 253 and 268 some 107 more followed suit. To meet the crisis, the imperial government imposed crushing burdens on an already impoverished middle class. Town councils were ordered to pay up back taxes on deserted farms, and municipal offices had to be forced on unwilling citizens. "Shall I have to be a member of the local council?" was a stock question put to an Egyptian oracle in these years.

Diocletian

Temporary relief came with Diocletian and Constantine in the late 3rd and early 4th centuries. The imperial organization was tightened up, particularly the economy, which now became thoroughly state-controlled. Professions became hereditary: farmers could not leave the land nor city officials their posts. The Empire was divided by Diocletian into two parts, the emperor of each part being called an *Augustus* and his administrative assistant and eventual successor a *Caesar*.

Constantine

This scheme did not work well, and on Diocletian's abdication after he had suffered a stroke in 305 civil war raged amongst the Augusti and the Caesars. In 306 Constantine, the son of a general then defending Britain, was proclaimed emperor by his troops; but it was to take him another sixteen years to become sole ruler. In the course of his campaigns, he won his most memorable victory near Rome at the Milvian Bridge in 312. According to a contemporary historian, on a day sometime before the battle, "about the hour of noon ... the Emperor had seen with his own eyes, so he said, the victorious emblem of the Cross formed out of light, up in the sky above the sun, and near it the words: 'Through this sign you will conquer'." Though he was not baptized a Christian until he lay on his deathbed a quarter of a century later, there can be little doubt that Constantine was a believer from that epoch-making day when his troops went forth to battle with the name of Christ emblazoned on their shields.

Once he had made himself supreme in the Empire, this outstanding general, administrator, and legislator became more and more like an Oriental monarch and less and less like a Roman emperor. Henceforth he was not *Princeps*, First Citizen, but *Dominus*, the absolute lord and master, and those granted audiences had to kiss the hem of his robe. The army was increased to 650,000, many warriors from barbarian tribes being hired along the frontier to replenish ebbing voluntary manpower. Wages were controlled and price ceilings set, with death the penalty for evasion, although this edict could not be enforced.

Such recovery as there was under Diocletian and Constantine was bought at too high a price. The state controlled everything; the state was absolute. When Constantine transferred his capital from Rome to Byzantium, to that New Rome called "Constantinople," he in effect abandoned the West.

Storm in the West After Constantine's death civil war festered in the Empire for sixteen years. Meanwhile, although the Empire remained united in theory, the final split between the West and East had begun. Moreover, in the 4th century the westward migrations of the Mongolian Huns were pushing other barbarians before them, the fierce Orientals inspiring sheer terror in all who lay in their path. Here is what they were like according to Gothic and Roman historians of the time.

They have . . . a shapeless lump rather than a head, and pinholes rather than eyes. They are short in stature, broad-shouldered . . . and have large, thick necks, always stretched in pride. Their limbs are compact and strong; they are quick in bodily movement, and bow-legged so that you might imagine them to be beasts with two legs. At birth the faces of male children are disfigured with a sword. They grow up, therefore, with no comeliness, a stunted and foul race, scarcely human, their very language having little resemblance to human speech. . . . They never shelter themselves under roofed houses and have no settled abode, but are homeless and lawless, perpetually wandering with their wagons, which they make their homes. . . . They wear . . . garments of the skins of field-mice . . . and a sort of helmet made of the skins of wild-rats patched together. . . . After a tunic is once put around their necks, however worn it becomes, it is never taken off or changed until, from long decay, it becomes so ragged as to fall to pieces. . . . There is not a person who cannot remain on his horse day and night. . . . These men, in short, live in the form of humans, but with the savagery of beasts.

The Visigoths The first to feel the Hunnish onslaught were the Ostrogoths (East Goths), who lived north of the Black Sea. They were incorporated into the Hunnish state, whereupon their neighbours the Visigoths (West Goths), north of the Danube, desperately petitioned the Roman emperor to be allowed to cross the Danube and settle within the Empire. Their very desperation made it clear that they would force the frontier if their petition

The Huns, whose strange Oriental features struck such terror into their Roman foes, probably looked much like the modern nomadic Mongols pictured here. The early Mongoloids have been described by a contemporary Roman historian as a race who "avoid [houses] as people ordinarily avoid sepulchres. . . . and accustom themselves to bear frost and hunger and thirst from their very cradles. . . . None of them plough, or even touch a plough-handle . . .; in fact they seem to be a people always in flight."

were refused. And so it was granted. For days they passed before the wondering gaze of the Roman outposts, men, women, children, cattle, on rafts, in canoes, some hanging on to planks or barrels—an entire people on the move.

The Romans now had a new nation on their hands, a people that could not be dispersed but would have to be allowed to settle down under their own king and their own laws. To give the Visigothic king a place in the imperial administration the Romans proclaimed him one of their own generals. But the Visigoths were treacherously mistreated by the Romans, and eventually turned on their hosts. In a surprise attack the barbarian cavalry conquered the Roman infantry at Adrianople, and both the emperor Valens (364-378) and his army of the East were destroyed. The gate had been forced open. The barbarians whom the Romans had been

Adrianople

holding back for centuries on the Rhine-Danube frontier finally burst in upon the Empire.

Before these mass invasions occurred, however, an insidious infiltration had been taking place. For some time barbarians had been hired as mercenaries by Rome to bolster up the armies for which her own citizens would no longer volunteer. These Germanic troops often served under their own kings or generals, so that they formed virtually private armies within the Roman ranks. And should their leader be killed, they were as often tempted to fight against Rome as for her. Thus though Rome's armed forces may have totalled some 400,000 troops as contrasted with the largest opposing German tribal army of 40,000, the Roman armies were spread thin along the frontiers and their loyalty was, in many cases, highly dubious.

Valens' successor, the emperor Theodosius the Great (379-395), did manage for a while to buy peace at a heavy price. The Goths were settled within the Empire, but as self-governing allies who, for a subsidy, would furnish troops for Rome's army. And although Theodosius finally succeeded in controlling the whole Empire, when he died the next year (395) the Empire split apart for good.

Alaric Upon Theodosius' death the Visigothic prince, Alaric, felt that he was cheated out of what he had assumed would be his place in the Empire. Enraged and humiliated, he ravaged Thrace and marched to the very walls of Constantinople. There he was bought off by the title of commander of the army and told to move westward into the rugged Balkan country of Illyria. But Italy was too tantalizingly near. In 410, for the first time since the Gallic invasion of the 4th century B.C., barbarians entered Rome as Alaric and his hordes pillaged that great city. In far away Bethlehem a Roman-educated monk mourned the shattered capital: "The lamp of the world is extinguished, and it is the whole world which has perished in the ruins of this one city."

Alaric marched south, but died shortly afterwards. His people then returned north, moved into southern Gaul, and by 415 had pressed on into Spain, where they established a kingdom that was to last for three centuries.

The Vandals When the Visigoths came to Spain they found another people, the Vandals, already there. While the Visigoths had been wandering around in the Balkans, the Vandals had crossed the Rhine and plundered their way across Gaul to finally settle in Spain. With the arrival of the Visigoths they were again uprooted. On they went, crossing the Strait of Gibraltar to Africa, where, under their king, Gaiseric, they set up a North African kingdom from which they launched piratical expeditions along parts of the Mediterranean coast. In 455 they made their famous raid on Rome when for two weeks they sacked the city, an act of "vandalism" for which they became immortal. After Gaiseric's death in 477 Vandal power declined rapidly, and their kingdom was crushed by a Roman general from Constantinople some two generations later.

The Burgundians—a people "mild in character, harsh in voice . . . [who] greased their hair with rancid butter"—had crossed the Rhine about the same time as the Vandals. But they did not wander as far afield and eventually settled down near Lake Geneva and in the Rhone valley, an area that came to be known as Burgundy.

Meanwhile the cause of all the trouble, the Huns, had been sweeping steadily westward under their mighty leader Attila, "the Scourge of God" (434-453). He was, says a historian of the time,

> . . . a man born to shake the races of the world, a terror to all lands, who in some way or other frightened everyone by the dread report noised abroad about him, for he was haughty in his carriage, casting his eyes about him on all sides so that the proud man's power was to be seen in the very movements of his body. A lover of war, he was personally restrained in action, most impressive in counsel, gracious to suppliants, and generous to those to whom he had once given his trust. He was short of stature with a broad chest, massive head, and small eyes. His beard was thin and sprinkled with grey, his nose flat, and his complexion swarthy, showing thus the signs of his origins.

Attila forced the East Roman emperor to pay a large sum of gold and promise an annual tribute before he would move on. Attila was, however, checked in Gaul in the summer of 451 at the battle of Chalôns near Troyes, turned back by an army that was largely German. Visigothic recruits fighting for Rome managed to halt the motley army of Attila (which also included German auxiliaries) in a rather inconclusive day-long struggle. This battle did not destroy Attila's army. It merely deflected the Huns southward into Italy, where Pope Leo I managed to persuade Attila to withdraw without attacking Rome—doubtless partly because the Huns were suffering from disease and famine at the time. Actually, despite the terror it instilled, the Hunnish menace had already passed its peak by Attila's day, so much so that when he died in 453 his "empire" fell to bits. Fierce though their sudden ravages had been, the Huns had, in point of fact, never been a very serious threat to European civilization.

Two other groups of Germanic barbarians also established kingdoms during these years: the Franks and the Anglo-Saxons. Since, however, these kingdoms alone of the barbarian states were destined to become long-lived medieval monarchies, they may best be discussed when we come to deal with the Frankish and Anglo-Saxon states on pages 277-335.

In Italy, it was now the various barbarian generals who were the real rulers since they commanded the barbarian troops that Rome hired to fight for her. Little was left of Rome's greatness.

> Time has dried our tears, and save for a few old men, the rest, born in captivity and siege, no longer regret the liberty of which the very memory is lost. But who could believe that Rome on her own soil fights no longer for her glory, but for her existence, and no longer even fights but purchases her life with gold and precious things.

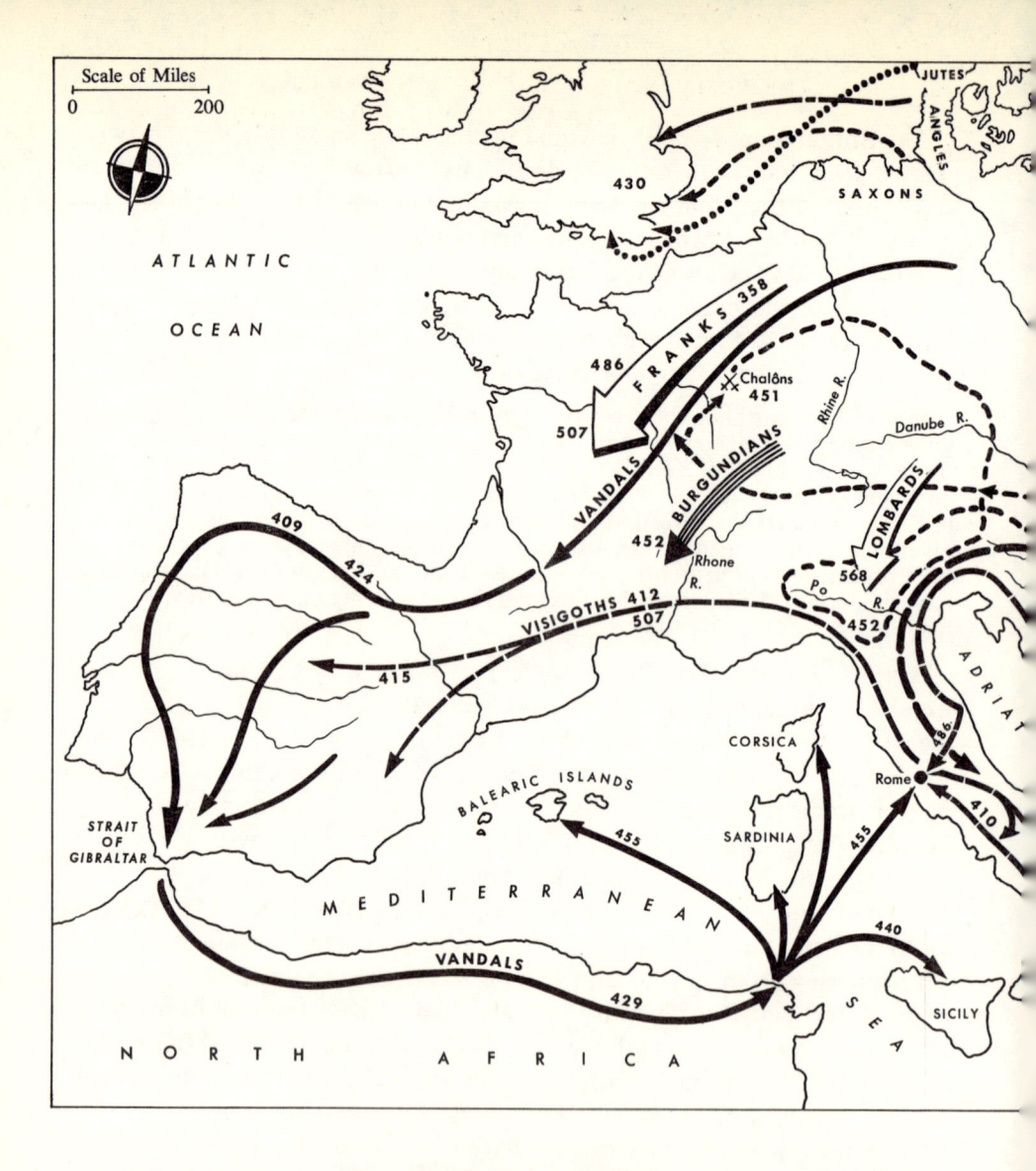

Finally in 475 one of the mercenary commanders had his young son Romulus proclaimed emperor under the name of Augustulus, "the baby emperor". The next year the barbarian troops revolted against the father, but this time the leader of the uprising, Odoacer, did not think it necessary to assume the title relinquished by the puppet emperor. Instead he had the Roman Senate write to Constantinople that there would be no further need for a western emperor if he, Odoacer, were given authority to rule the West on behalf of the East Roman emperor. This authority was duly granted.

A German king now ruled Italy. The Western Roman Empire was ended in all but name.

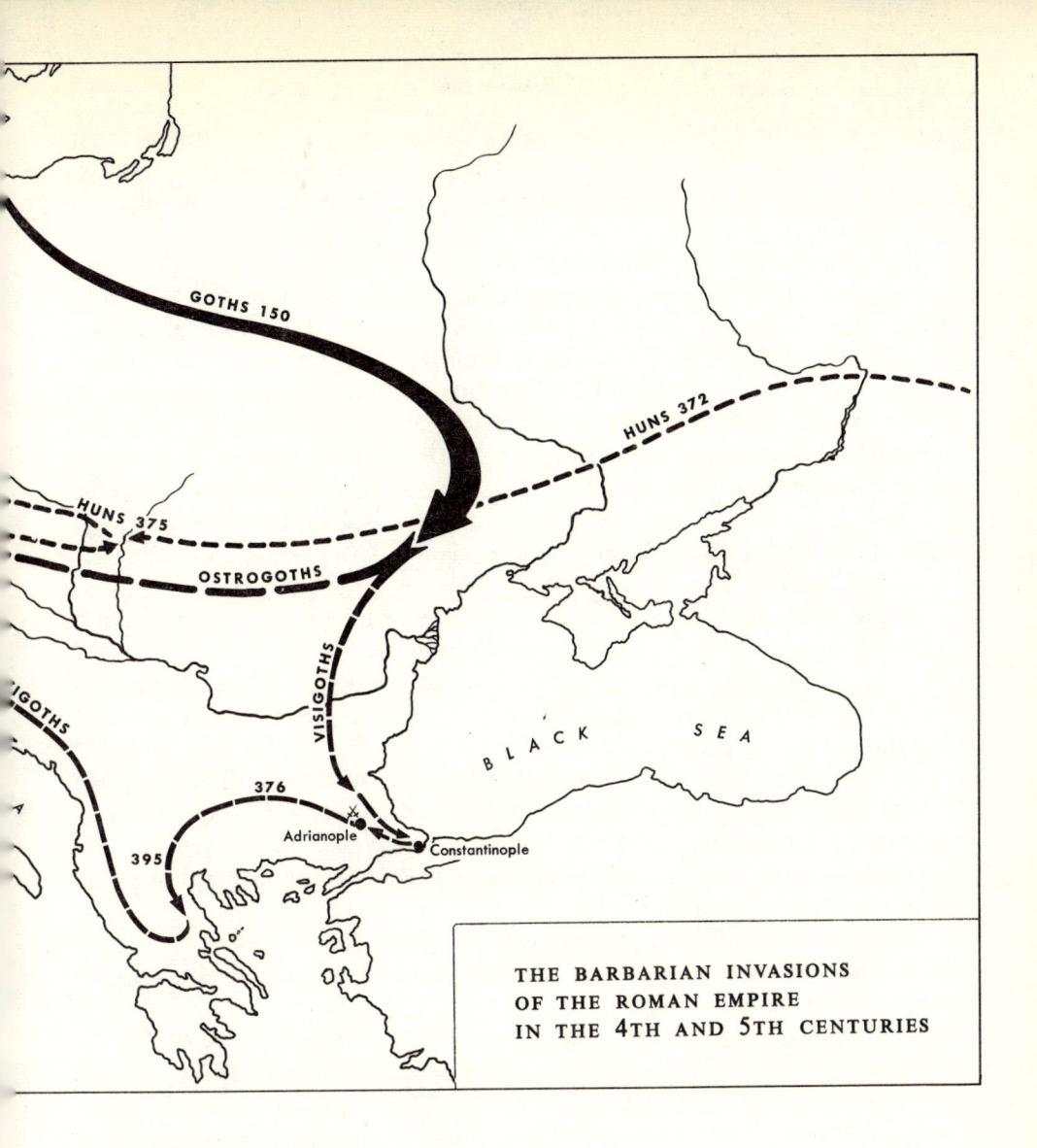

THE BARBARIAN INVASIONS
OF THE ROMAN EMPIRE
IN THE 4TH AND 5TH CENTURIES

4/THE FALL OF ROME IN THE WEST

In the last quarter of the 18th century, one of the most famous of English historians, Edward Gibbon, wrote his *History of the Decline and Fall of the Roman Empire* in six fat volumes. He was only one, though the best known, of many who have tried to puzzle out the reasons for Rome's fall. It may therefore be worthwhile—if only to spur you on to read further on the subject—to mention a few modern historical explanations.

For such a complex and controversial subject as Rome's decline it is natural that many explanations should be suggested. Among the more famous are soil exhaustion, the racial watering down of a superior Roman stock, the decay of morals, malaria and other plagues, the problem of the succession to the imperial throne, and a "failure of nerve" manifesting itself in a collapse of will-power—all of which are now looked upon as unsatisfactory explanations. All told, historians have offered at least fifty different theories.

As often as not the particular theory presented tells more about the person presenting it than about the decline of Rome. A geographer from California detects a major climatic change that produced prolonged droughts in sections of the Western Empire. A German scholar blames Rome's fall on the extermination of the élite. A White Russian says it came about as the result of a class-conscious alliance of ignorant soldiers and peasants who overthrew the aristocracy. A Frenchman writing after World War II deplores Rome's renunciation of compulsory military service. A premier of a Canadian province criticizes Rome's centralization of power. And in 1962 a high school coach said Rome fell partly because of too many spectator sports.

While all these theories of decline are interesting (and some do contain a kernel of truth), they really describe symptoms of a deeper set of ills. Just as a headache and rising temperature signal the onset of infection, so a variety of things went wrong with Rome that were merely outward manifestations of her decline.

In our attempt to get at the fundamental causes of Rome's difficulties two points must be kept in mind: first, no one final answer is possible for such a complex phenomenon as the decline of a mighty civilization; second, any valid answers must explain why the Eastern Roman Empire outlasted the Western one by a thousand years.

There seem to be three interrelated groups of causes which were basic to Rome's decline: political and military, technological and educational, and economic and social.

The Seeds of Destruction One of the persistent themes in Rome's long story is the failure of the civil power to control the army. The rivalry of the great soldiers of the late Republic led inexorably to the fateful days when the troops themselves made and unmade emperors. Was this trend towards military dictatorship caused by Roman imperialism? Did her very rise cause her to fall?

In a sense it did. Rome eventually held a world empire, yet Rome herself had only a city-state government. When this comparatively elementary type of administration broke down under the weight of empire, the Republic gave way to the military dictatorship of one man. Imperialism cost Romans their republican institutions. And it cost them more. As Roman citizenship was gradually extended, the new citizens were less than eager to serve in the army. They preferred to enjoy the fruits of their new status in peaceful pursuits.

Citizenship and recruitment

Once all Italians had acquired citizenship the army had to be recruited largely from the provinces, and the provincials *did* enlist, since army service was one way to earn citizenship. But, ironically enough, as the areas of citizenship spread wider the recruiting areas shrank alarmingly. The towns and cities who were given rights of citizenship ceased to supply soldiers. And when ultimately in 212 Caracalla granted citizenship to all freeborn men within the Empire, the chief inducement for enlistment, at least for the better class of men, was removed. At the same time a marked decline in the population of the Empire (from 70 to 50 million) aggravated the problem of obtaining recruits.

It is not surprising, then, to find a Roman historian of the 3rd century writing that the army was coming to consist of those "most vigorous and violent elements who are usually obliged to make a living by banditry." The once proud Roman army began to look like a collection of thugs, making and unmaking "Barracks Emperors". Thus, paradoxically, Rome's policy of extending the citizenship to those she had conquered brought her military disorder and ruin. The inferior Roman armies, riddled with foreign mercenaries, must have been regarded with increasing contempt by the hardy barbarians watching and waiting across the frontiers.

Education and technology

A second underlying cause of Rome's decline was her technological and educational failure. Technology may be defined as the application of science to industry, and in their development of inventions the Romans were markedly weak. For a long time this weakness could be partly ascribed to Rome's steady supply of slaves. Slavery discourages invention: why bother to replace manpower by machines as long as there is a cheap and plentiful supply of human muscles? But when the supply of slaves declined after the wars of the Republic and the use of freed labour increased, there was still no large-scale technological advance. It cannot, then, have been slavery alone that blighted its development.

This delicate Roman pitcher (only 5½ inches high) was manufactured of semi-transparent greenish glass about A.D. 200. Glass was probably invented by the Egyptians: there are examples of their work at least 3500 years old. The Romans so perfected the Egyptian art of glass-making that glassware came to replace pottery in the home. Glass-blowers took a special pride in their art, as can be seen in this charming creation where tiny blue and white threads of semi-liquid glass have been applied to form a dainty pattern against the blown form.

The failure in technology seems to have been due, at least in part, to the type of education Romans received. Undue emphasis was placed on rhetoric, that is, training in speaking and writing, in order to produce lawyers, administrators, and more teachers of rhetoric. Any interest in science was severely practical; the aim was to construct something substantial such as a bridge or an aqueduct. Romans were not attracted by the kind of abstract scientific speculation that leads to inventions. For some reason no Roman ever, for instance, invented an efficient system of harnessing horses, nor did any Roman ever make the logical transition from stamping coins in a mint to inventing a printing press. In any case, technology was simply not considered respectable work for a gentleman. "All mechanics," sneered Ciçero, echoing Aristotle, "are engaged in vulgar trades; for no workshop can have anything liberal about it." Instead of setting up scientific laboratories the Romans built baths.

Economics

But even had the educational system been such as to promote technology, it is still by no means certain that there could have been a Roman Industrial Revolution. This brings us to the third underlying cause of Rome's decline, her failure to keep her economy expanding along with her Empire. A decentralization of industry occurred when, because of the high costs of transportation and a shortage of labour, industry exported itself rather than its products. That is, industries moved from the cities out to the villages or large country estates where workers and raw materials were more plentiful, with the result that trade became local rather than international. Then too, capitalists now invested heavily in land and parcelled out their vast country estates to tenant farmers (*coloni*) who paid rent in produce and managed a bare existence on these self-sufficient "manors."

Even if the Romans had been technically capable of an industrial revolution, the masses within the Empire were too poor to purchase enough goods to stimulate a machine industry. The coinage was increasingly debased, a process that was hastened by the heavy drain on the Empire's precious metals through the cash purchase from Egypt, India, and China of such luxuries as ebony, ivory, pearls, spices, perfumes, and silks. Heavier taxation was a hindrance rather than a help to recovery. Large western landowners, who were the high imperial officials, made matters worse by granting their own class certain tax exemptions. It was a vicious circle. The more expensive the Empire became, the more repressive the machinery of government had to be to grind out the needed taxes.

One final question remains. Why did the Western Empire collapse while the Eastern one survived? Or, to put the question another way, which causes of decline apply more to the West than to the East?

The Survival of the East Roman Empire

First, the East was able to recruit citizen armies from the stalwart hillsmen of Asia Minor, whereas unstable military dictatorships had greatly diminished loyalty and patriotism in the West. This factor, plus the widening of the citizenship, resulted in a western army made up largely of toughs and barbarians. Second, while some of the richest areas of the East—Asia

Minor, Syria, and Egypt—were virtually sealed off from invasion, the lush farmlands of Gaul and the Po valley were easily accessible to marauding tribes. Third, although lack of technological development was equally marked in both parts of the Empire, such a lack was far more serious for the West because of its unhealthy economic state combined with a labour shortage. As has already been noted, top Western officials aggravated these bad conditions by exempting many of their own class from taxes, whereas in the East the senior officials did not play favourites. And finally, when the West had lost Africa, Spain, and Gaul to the barbarians, only impoverished Italy remained, woefully lacking in both manpower and natural resources.

The disintegration of the Western Empire is, therefore, hardly to be wondered at. With the East using its reservoirs of men and money to shove and bribe the barbarians westward, the West could not escape the brunt of the invasions. Eastern Roman civilization thereby preserved itself for another millennium.

5/CIVILIZATION UNDER THE EMPIRE

The Roman conquest of the Mediterranean world meant that the culture of many peoples, particularly that of the Greeks, was to receive widespread expression. The resulting Roman mixture, though inferior to the Greek, was in itself no mean culture. Greek philosophy, art, science, and literature came west, and, given Latin renderings, became the basis of western Europe's later intellectual development. We will restrict our consideration of Roman culture almost entirely to the imperial period.

Roman culture evolved against a background of Hellenistic philosophy. Epicureanism and Stoicism both had adherents in Rome, with Stoicism becoming the more important of the two until by the end of the 1st century after Christ it was justly famous as the "religion" of educated Romans. Its chief prophets were Seneca the Younger, a Spaniard, and Epictetus, a Greek. Seneca felt that, as he said, "the soul's entire struggle is with the flesh that oppresses it." His application of such a principle may seem amusing to us, for he denounced heating houses with furnaces, said that to take a bath was to violate nature, and thought no one should use feather pillows or bother to shade his head from the sun. But Seneca was mostly concerned with the moral problems of the noble class. He did not hesitate to condemn the Roman "games" (see page 236) with their bloodshed and appeal to the baser instincts, where, as he said, even in the intermissions "men were strangled lest people be bored."

Stoicism

Stoicism made a particular appeal to the emperor Marcus Aurelius, whose philosophy was of a gloomier, more pessimistic kind. His pessimism mirrored his fate. By nature quiet, shy, and a gentle scholar, Marcus Aurelius found himself spending most of his life in hard frontier fighting. His *Meditations*, often written in his tent after a long day of marching or battle, were an expression of his Stoic creed and a reminder to himself of his exacting duty as emperor.

Stoicism with its emphasis on self-discipline was well suited to be the philosophy of the Roman ruling class. However, it developed religious ideas and overtones that had been no part of Greek Stoicism, and that were not as profound as those of Christianity.

Practical Feats of Art   Roman art was a blend of Etruscan, Greek, and Oriental traditions. We may characterize it as city art, that is, art intended for urban convenience or decoration.

Architecture   Augustus boasted that he found Rome brick and left it marble, and it is true that before his time brick, cement, and wood had been the chief building materials. Of these, cement ought to be specially noted, because it was a characteristic Roman material made from volcanic residue found near Rome and mixed with lime. Whereas Greek architecture had striven for harmony and balance, the Roman aim was grandness. The use of the arch, learned from the Etruscans, permitted the Romans to construct huge buildings which became more immense as the Empire continued. And although these colossal structures have often been interpreted as mere monuments to the vanity of the emperors, some were highly practical.

At least two types of building originated with the Romans: those great palaces called the "baths", and the aqueducts. The baths were magnificent edifices, on whose evidence alone we can fairly say that the Romans were the cleanest people in antiquity. Although every conceivable type of bath was available—hot, cold, swimming, tub, and so on—these baths were much more than bathing places. Enclosed gardens, promenades, gymnasia, lounges, libraries, and museums enhanced the interior of the greatest baths, such as those of the emperor Caracalla. They were an attempt to blend physical culture and intellectual curiosity under one roof. And what a roof! The central hall very closely resembled the great concourses of our North American railway stations, such as Toronto's or Ottawa's Union Stations. The great engineering problem involved in enclosing and roofing over such a grandiose structure without using a forest of columns was magnificently solved by means of the arch and the vault.

Rome's aqueducts, too, were built solidly and well, and by the end of the 1st century A.D. were supplying Rome, then a city of over 1,000,000 people, with 100 gallons of water per person per day. Perhaps the most famous aqueduct, because it is the best preserved, is the one called the Pont du Gard near Nîmes in France. It still stands, a monument to both Roman engineering skill and provincial administration.

Sculpture   Roman sculpture attempted to portray real persons rather than the idealized types of the Greeks. For example, there is nothing in Greek sculpture to compare with the depiction of individual personality that we find in the statue of Augustus from Prima Porta. The Romans also excelled in a new kind of sculpture, the equestrian statue. But as the Empire grew

The original Pantheon was built in Rome during the reign of Tiberius as a temple for the worship of all the gods of the Empire, and was remodelled under Hadrian. The result is the marvellous rotunda pictured here in this striking painting by Giovanni Paola Panini (1691-1765), a rotunda 142 feet wide and 142 feet high topped by a deeply coffered dome with a 30-foot aperture at the crown. Placing a hemisphere on a cylinder was a neat feat of Roman engineering, and was made possible by the use of horizontal layers of brick laid in thick cement to form the dome. This dome is the largest that has ever been built, and has served as the model for such famous domes as Hagia Sophia's in Constantinople, St. Peter's in Rome, and the Capitol's in Washington.

older, the trend in sculpture, as in architecture, was to the immense. A good example of this is the huge head of Constantine (now set up in the Palazzo dei Conservatori in Rome), which towers some eight feet high yet is only a fragment of what must have been a colossal statue. It does show a certain force, combined with the belief that a symbolic human form could be more impressive because of its sheer mass. Most of us, however, would prefer the graceful beauty of the *Discus Thrower* to the ungainly bulk of the *Constantine*.

Murals painted on stucco were a favourite wall decoration. They were often intended to create the illusion of more space, and those recovered seem to be mostly copies of Greek or Hellenistic pictures. The art of portraiture was widely practised, whether the subject was an emperor, a wanted criminal, or a lover who hoped his picture would be carried next someone's heart.

Proficient artists were these Romans, whose highest artistic development lay in the practical feats of engineering that decorated the capital and provincial cities of the Empire.

As also befitted a practical people, the Romans applied their science but did not make it a subject of academic study. They were accurate surveyors. The great engineering accomplishments we have just described bear testimony to this skill, as does the remarkable Roman road system. How many of us who use "highways" realize that they are named for those majestic causeways which the Romans raised up above the level of the surrounding land? In tunnelling, mining, and drainage no ancient people can bear comparison with the Romans.

Science

The two greatest scientific treatises of the Roman era were written by Ptolemy of Alexandria (90-168) and Galen of Pergamum (130-200). Ptolemy's works summarized all ancient scientific knowledge of astronomy and geography. Contrary to many other scientists of the time, Ptolemy held that the earth was stationary at the centre of the universe (a theory that was to hold up further astronomical progress for some time). But Ptolemy was also, and more importantly, the first to locate places on the earth's surface by latitude and longitude. Galen, on the other hand, was a doctor in Rome who treated up to 3000 patients a year. His chief contributions were to the sciences of anatomy and physiology. His dissections of apes led him into some errors when he made inferences as to human anatomy, but even so he formulated a theory as to what causes disease.

Both Ptolemy and Galen have been mentioned because they lived and wrote in the 2nd century Empire. But they were actually Greeks, representing more the Greek speculative genius than the practical Roman mind.

Literature

Classics Fit for a Queen Those of you who are studying Latin need little reminder that the Romans produced an extensive literature. We have already come across several famous literary figures—Julius Caesar writing of his campaigns in Gaul, Cicero defending the Republic in his political treatises and letters, Tacitus writing his history of the first imperial century, to mention but three. Only a few of us will ever read Latin authors in their native tongue; but all of us should remember that the modern literatures of England, France, Spain, and Italy would be incomprehensible without the Latin literary tradition.

A E K T

ㅅ E ㅅ ㅜ

The Romans borrowed their letter forms from the Greeks, probably through the Etruscans. Letters from the inscription on Trajan's Column in Rome of about A.D. 114 (upper line) are typically Roman in design, combining beauty with power and function in a form that must have evolved slowly over many centuries. The Roman letter is the foundation for most letter forms — in some cases virtually unchanged. Roman cursive writing of about 50 B.C. (lower line) was designed for speed in transcription.

Literary production was at its height in the Ciceronian (mid 1st century B.C.) and Augustan Ages. To the Romans, rhetoric was the most important part of formal education, and when Julius Caesar said of Cicero that he had advanced the boundaries of the Latin genius he meant that Cicero had attained perfection in the art of public speaking. Throughout Roman schools emphasis was laid on oratory, the pupils being taught both to write and speak discourses based on Cicero's models.

Cicero also wrote essays on literary criticism and philosophy, essays that drew heavily on Greek sources. Perhaps most interesting of all are his more than 800 extant private letters, which, because they were not intended for publication, throw a candid light on his times. Cicero's literary creations became classics for generations to come. Centuries later a young English princess, who was to become Queen Elizabeth I, had read nearly all his works before she was sixteen.

With the Augustan Age three poets came to the fore—Virgil (70-19 B.C.), Horace (65-8 B.C.), and Ovid (43 B.C.-A.D.18). Virgil was a gentleman farmer until in 42 B.C. he lost his estate to Octovian, who was confiscating land to provide for the veterans of Philippi. Virgil was to spend the rest of his life in Rome or Naples in the literary circle of the wealthy Maecenas, a man who may be described as Augustus' minister of propaganda, and the patron of many a struggling artist. Although he wrote pastoral poems as well as an agricultural handbook in verse, Virgil's greatest work was the *Aeneid*, the story of the founding of Rome by Aeneas. This magnificent epic in twelve books weaves together fiction and fact in Homeric fashion in order to demonstrate that all events of the Roman past found their fulfilment in the peaceful and prosperous age of Augustus. It established Virgil as one of the world's great poets. Posterity has always been grateful to Augustus for disregarding Virgil's deathbed request that the unpolished *Aeneid* be burned.

Virgil

Horace, another poet in Maecenas' circle, was the son of a freedman. He received an excellent education, and although he fought at Philippi against Octavian and Mark Antony he was pardoned, and after Virgil's death became Augustus' poet laureate. His most famous work is the four books of *Odes*, 103 lyric poems using Greek metres and dealing with an astonishing variety of subjects, from a banquet to the defeat of Cleopatra.

Horace

Ovid, the least important of the three, is chiefly remembered for his *Metamorphoses*, a Roman *Arabian Nights* consisting of 250 myths and legends about miraculous transformations. But he specialized in gay elegies, which finally got him into trouble with Augustus. He died in exile on the Black Sea, where he was banished for "a poem and a mistake".

Ovid

One Law for All One non-literary branch of writing must also be singled out for some emphasis. This is the body of law that the Romans originated and wrote down. Under the Empire one universal system of law, built up through centuries, gave equal justice to all Roman citizens and yet made allowance for local custom. As one emperor wrote to his ambassador, "When in doubt follow the law of the local city."

The law also tended increasingly to protect the poorer and weaker against the stronger and richer. Poor men in a province who could not obtain anyone to advocate their cause were provided for: "No man must be overborne by the influence of his opponents, for that would merely bring discredit on the governor." What was said on behalf of St. Paul at Jerusalem is eloquent testimony to the fairness of Roman law: "It is not the custom of the Romans to hand over any man before the accused prisoner has met his accusers face to face, and has had an opportunity of defending himself upon the charge laid against him."

Roman law is still a vital force. While 70 million people lived under it in the days of the Empire, some 870 million today have legal systems directly traceable to it. The following main principles of law we owe to the Romans:

> 1. General principles of justice should be applied without prejudice to all cases.
> 2. Human rights ought to be based on a universal moral law.
> 3. The spirit is more important than the letter of the law.

This noble system of law is generally thought to have been ancient man's most successful instrument for securing justice. Perhaps, then, Rome's law was her most enduring achievement.

But the world owes Rome a greater debt. She provided the environment in which Christianity could be born and could grow. In this, as in many other ways, she proved to be the civilizer of Europe.

6/THE TRIUMPH OF CHRISTIANITY

Much superstition was embedded in Roman, as in Greek, religion. The Greek gods and goddesses were imported and given Roman names: Zeus became Jupiter; Aphrodite, Venus; Poseidon, Neptune; and so on. Trees, brooks, even sewers had their deities. As the Empire expanded there were foreign gods to be placated and oracles to be consulted. Emperors came to be deified on their death—although it should be emphasized that "emperor-worship" did not mean praying to some dead ruler. Rather it involved a commemorative ceremony carried out in a patriotic spirit.

Many cults found their way into Rome from the Orient, from Egypt, Asia Minor, Syria, or Persia. Of all these foreign imports, Mithras, the sun-god, was the most popular, since Mithraism adopted Zoroaster's eternal struggle between good and evil and made a powerful appeal to the soldiers of the Empire.

No objection was raised by the government to this multiplicity of cults. A citizen might worship as he pleased—provided, of course, that there was nothing subversive in his religion, in other words, provided his religion did not affect his loyalty to the state. Men and women whose religion was practised in secret gatherings, usually at night, were naturally subject to suspicion. A religion whose rites were not public or a religion that seemed to be narrowly nationalistic Rome felt—perhaps understandably—that she could not permit, simply because such exclusiveness and political affiliation seemed to challenge the imperial power. For this reason Judaism was suspect, as was the religion that appeared to the Romans to be an offshoot of Judaism, Christianity.

A Mightier than Caesar   On a hill outside Jerusalem, about the year 30, there occurred a crucifixion. The Roman legionaries whose job it was to nail up the condemned criminals probably soon forgot the grim incident. But the world never did. For one of the three men crucified that day was the founder of Christianity, a carpenter named Jesus of Nazareth.

Aside from a few references by ancient Roman historians writing long after the events, our evidence for this man's life and deeds comes from the New Testament, and almost entirely from four of its books written in the second half of the 1st century: the Gospels according to Matthew, Mark, Luke, and John. These Gospels tell us that Jesus of Nazareth went up and down his native Galilean countryside preaching to crowds and performing miracles of healing. His was not a strictly intellectual appeal, such as that of Socrates. Rather he taught the common people by means of parables, simple stories that enshrined great moral or spiritual truths such as the equality of *all* men before God:

Behold a sower went forth to sow	*Matthew* 13:3
If a man have an hundred sheep, and one of them be gone astray	*Matthew* 18:12
A certain man went down from Jerusalem to Jericho, and fell among thieves	*Luke* 10:30
Consider the lilies how they grow	*Luke* 12:27
A certain man had two sons	*Luke* 15:11
Two men went up into the temple to pray	*Luke* 18:10
I am the true vine, and my Father is the husbandman	*John* 15:1

In time—and despite the fact that he made a clear distinction between "the things that are Caesar's" and "the things that are God's"—some of his followers believed that he was the long-awaited Messiah, come to unshackle them from Rome. Apparently Jesus thought of himself as standing squarely in the Jewish prophetic tradition, teaching that his blood would seal a New Covenant (see page 74) and saying that he had come not to destroy the Law but to fulfil it. But he went further in his insistence on a "new commandment . . . that you love one another."

You have heard that it was said, "An eye for an eye and a tooth for a tooth." But I say to you, Do not resist one who is evil. But if any one strikes you on the right cheek, turn to him the other also; and if any one would sue you and take your coat, let him have your cloak as well; and if any one forces you to go one mile, go with him two miles. Give to him who begs from you, and do not refuse him who would borrow from you.

You have heard that it was said, "You shall love your neighbour and hate your enemy." But I say to you, Love your enemies and pray for those who persecute you, so that you may be sons of your Father who is in heaven; for he makes his sun rise on the evil and on the good, and sends rain on the just and on the unjust. For if you love those who love you, what reward have you? Do not even the tax collectors do the same? And if you salute only your brethren, what more are you doing than others? Do not even the Gentiles do the same? You, therefore, must be perfect, as your heavenly Father is perfect.

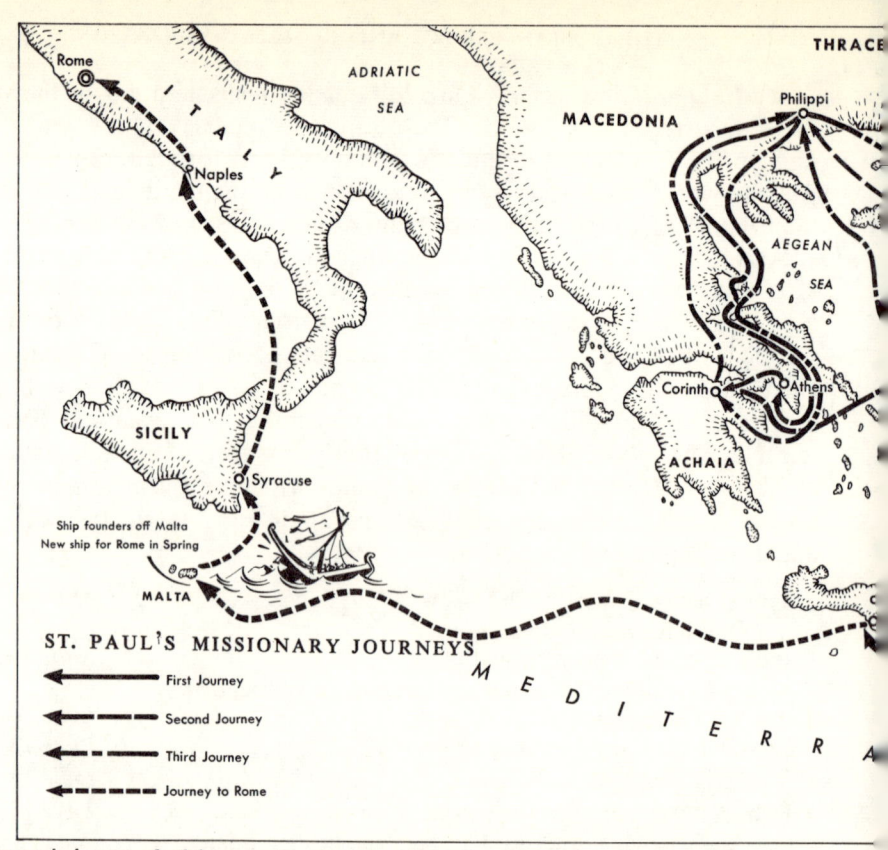

ST. PAUL'S MISSIONARY JOURNEYS
- ← First Journey
- ←--- Second Journey
- ←-·- Third Journey
- ←--·-- Journey to Rome

The ministry of this master teacher appeared to be short-lived. When certain religious authorities in Jerusalem accused him of sedition against Rome and had him brought to trial before Pontius Pilate, the Roman procurator of Judaea, he was condemned to die after the manner of a common criminal, by the slow torture of crucifixion. He died on the gallows of that day, the cross, forsaken by all his followers except for some sorrowing women. Despite all the hope of a Messiah, the great mission was apparently over, ended in the ultimate defeat of the tomb.

Three days later the body was missing, and shortly thereafter the disciples became convinced that they had witnessed their risen master. Cowards were transformed into dauntless missionaries as these rough men proclaimed that it had been a divine Messiah, no less than the Son of God Himself, who had walked with them during his earthly life.

From such an unlikely beginning, in a backward part of the Roman Empire, sprang the Christian church. Its earliest meeting is described in the first chapter of the book of *Acts*.

Then they returned to Jerusalem from the mount called Olivet, which is near Jerusalem, a sabbath day's journey away; and when they had entered, they went up to the upper room, where they were staying. . . . In those days Peter stood up among the brethren (the company of persons was in all about a hundred and twenty) . . . .

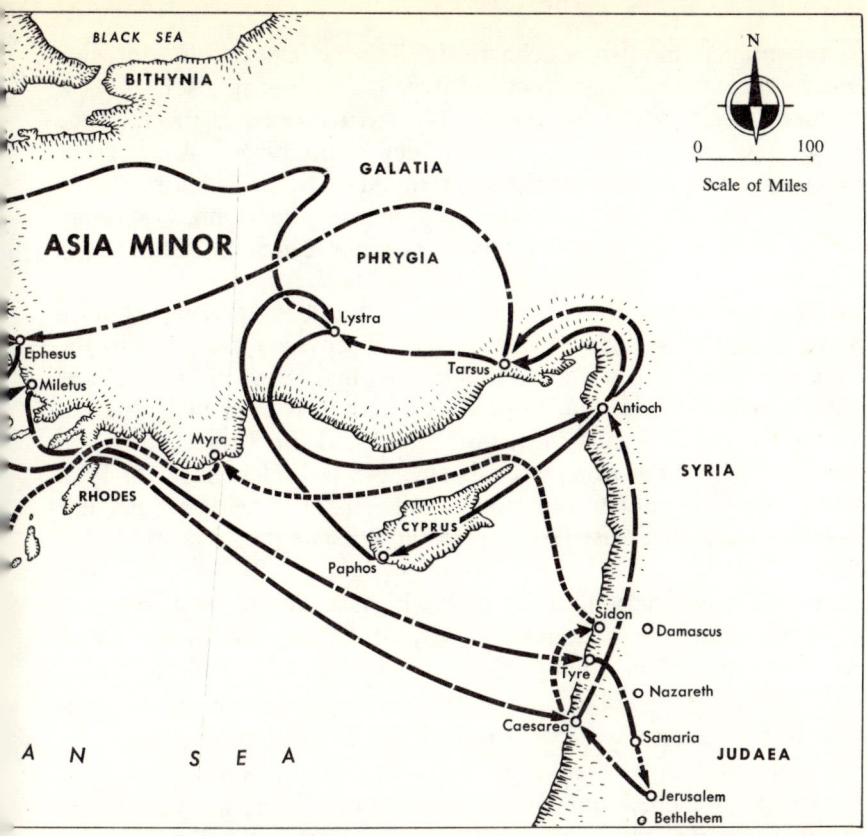

An Apostle to the Gentiles   Within a few years of the crucifixion, a man named Paul, a Roman citizen born in Tarsus, joined the church. This Jew, whose Aramaic name was Saul, was certainly not an imposing looking figure, a "man of small stature, thin-haired upon the head, crooked in the legs, of good condition of body with eyebrows joining. . . ." At first Paul had been a violent opponent of Christianity, and had led persecutions of Christians in Jerusalem. But one day while on his way from Jerusalem to Damascus to institute fresh persecutions he saw a vision of Jesus and was converted. Soon he himself began to preach the gospel of Christ to fellow Jews.

Paul

Some three years later (about the year 36) Paul met Peter, the leading figure in the church in Jerusalem. At this time the members of the church were mainly converted Jews, but at Antioch there had grown up a Christian group which included both Jews and Gentiles. In fact it was at Antioch that the word *Christian* was first coined to describe a follower of Jesus.

The Antioch church commissioned Paul and Barnabas to preach the gospel, and in the ten years between 48 and 58 three great missionary journeys were undertaken. Paul and his fellow preachers travelled some 8000 miles, founding churches and preaching and writing in Greek. In doing so, they rendered into Greek certain Old Testament concepts such as sin, grace, repentance, baptism, eucharist, faith, hope, and charity, infusing these Hebrew tenets with a fresh depth of meaning so that in a sense a new pattern of ideas was created. St. Paul's task was to translate the message of Jesus, who had spoken Aramaic, into Greek terminology. As a result, Jesus, who was called Master or Teacher in Aramaic, was addressed in Greek as *Kyrios* (Lord); and whereas he had been the Hebrew Messiah (the anointed Servant of God), he now became *Christos* (Christ, the Anointed One).

Paul's missionary journeys

As Paul and his companions passed through the leading cities of Asia Minor on their first missionary journey, they discovered that it was the Gentiles, not the Jews, who welcomed their teachings. Jew though he was himself, Paul came to believe that to be a Christian it was not first essential to observe all the requirements of Jewish ritual and the Law. A council of the church at Jerusalem accepted this suggestion, and henceforth Paul preached that "there is neither Jew nor Gentile, there is neither bond nor free . . . for all are one in Christ Jesus." He had become the "Apostle to the Gentiles."

On his second missionary journey through Asia Minor Paul came at length to Troas. Here he had another vision: "There stood a man of Macedonia, and prayed him, saying, 'Come over into Macedonia, and help us'." Few decisions can have been as fateful as Paul's. Now the gospel was going overseas; Christianity was passing from the Orient into Europe.

In Macedonia and in Greece Paul had notable successes, though at Athens, where he preached one of his most quoted sermons (*Acts* 17: 22-31), he was not well received. A few years later a third missionary journey was undertaken.

About the year 58, Paul was back in Jerusalem and was arrested by the Roman authorities because of a disturbance which broke out when he was in the Temple. He was imprisoned, first at Caesarea in Judaea, and later in Rome, where he was under a kind of house arrest which allowed him to write letters to Christian communities outside Italy. Then, possibly because no one had come from Judaea to Rome in order to accuse him, he was mysteriously released—but not for long. He was soon rearrested and executed, probably during the Neronian persecutions following the fire of 64. So perished Paul, the great apostle who had felt himself called to be the light of the Gentiles and to spread Christianity in Europe as well as in Asia. For "although without Paul Christianity had reached Gentiles and would have reached more Gentiles, it is Paul more than any other man who was responsible for the fact that Christianity was not a Jewish sect but an independent body with an independent life."[1]

A few years later the Jerusalem church itself disappeared when the Romans suppressed the Jewish revolt of 67-70. From this point onwards the church became predominantly Gentile. Christians continued to obey

---

[1] A. D. Nock, *St. Paul* (Thornton Butterworth, 1938), pp. 246-247.

the Ten Commandments, but they abandoned obedience to the Jewish ritual code and they slowly replaced observance of the Sabbath, the last day of the week, by commemoration of the Resurrection on the first day of the week, Sunday.

The Blood of the Martyrs By the 60's the Christians were regarded by the Roman authorities as adherents of a new religion rather than as just another Jewish group. One of the earliest non-Christian historians to mention the existence of the new faith is Tacitus, who has already been quoted on page 234. About the same time the governor of Bithynia, the younger Pliny, expressed his alarm when he learned that this "crazy and unrestrained superstition" had spread to the villages of the Empire. In fact even the country folk were beginning to forsake the old Roman gods! Pliny's letter asking for the advice of the emperor Trajan, together with the emperor's reply, is printed at the end of this chapter, and shows that there was no organized persecution of Christians at this time (though if you admitted you were one you might be executed). Such persecutions as there were usually resulted from discontent caused by bad harvests, a flood, or an unpopular governor. The Christians were convenient scapegoats.

The earliest conversions had been made amongst the humble and downtrodden of the Empire—slaves, poor city workers, and women. By the 2nd century, however, Christianity had penetrated the middle and upper classes as well. Christians had even served in the army under Marcus Aurelius, though they were never very numerous in the legions and must often have had qualms about rendering military service unto Caesar. There was, for instance, the case in 298 of a certain Marcellus.

On October 30, at Tingis [Tangier], Marcellus . . . was brought into court, and it was officially stated: "Marcellus, a centurion, has been referred to your authority by the governor Fortunatus. There is at hand a letter dealing with his case, which at your command I shall read."

When this was read, Agricolanus [the deputy prefect] said, "Did you say what appears in the official records of the governor?" Marcellus answered, "I did." Agricolanus said, "Were you in service with the rank of centurion first class?" Marcellus answered, "I was." Agricolanus said, "What madness possessed you to renounce your oath of allegiance and to speak as you did?" Marcellus answered, "There is no madness in those who fear God." Agricolanus said, "Did you say each of the things contained in the official records of the governor?" Marcellus answered, "I did." Agricolanus said, "Did you throw away your arms?" Marcellus answered, "I did. For a Christian, who is in the service of the Lord Christ, ought not to serve the cares of this world."

Agricolanus said, "The acts of Marcellus are such that they must receive disciplinary punishment. Accordingly, Marcellus, who was in service with the rank of centurion first class, having declared that he has degraded himself by publicly renouncing his oath of allegiance, and having, moreover, put on record insane statements, it is my pleasure that he be put to death by the sword."

When he was being led to execution, Marcellus said, "Agricolanus, may God be kind to you!" And after he said these words he was killed by the sword, and obtained the glory of martyrdom he desired.

Strangely enough, the greater the persecution the more converts to Christianity. "The blood of the martyrs," it has been written, "is the seed of the church." The 3rd century, the Empire's period of stress and strain, saw savage persecutions—and still the number of Christians increased, increased so greatly, in fact, that at a meeting in Milan in 313 the emperor Constantine amplified an Edict of Toleration of 311 by granting Christianity toleration and an equal status with other religions. Christianity was still a minority religion, but its triumph was assured. Constantine went on in 321 to declare Sunday, the day originally dedicated to the sun-god, a legal holiday and a day of prayer. Christians in the army were permitted to attend religious services, and the army was even to recite a prescribed prayer suitable for both Christians and pagans. Constantine's law stated:

Edict of Toleration

All judges and city people and all craftsmen shall rest on the venerated day of the sun. Country people, however, may freely attend to the cultivation of their land, since it frequently happens that no other day is so opportune for sowing grain . . . or setting out vines. . . .

This is not to say that Christianity became the state religion, nor that competing religions were banned. It was not until the year 380 that the emperor Theodosius the Great declared pagan worship illegal, a menace to the official religion of the Empire, Christianity, and enjoined all subjects to embrace "the religion delivered by the Apostle Peter to the Romans." Christianity may not, as Gibbon believed, have caused Rome's decline; but it was a symbol of the passing of ancient civilization.

A Union of Strengths What factors were responsible for this wide and rapid spread of Christianity? This is a far-reaching question, and whole books have been written in an attempt to answer it. All that we can do here is suggest a few of the main factors favouring the growth of Christianity.

To begin with, the new religion combined the special strengths and aptitudes of Jews, Greeks, and Romans. From the Jews came monotheism and the overruling idea of religion not just as the observance of ritual but as a whole way of life. There was little in the Roman world that provided the spiritual satisfaction Christianity offered. Oriental cults, or Caesar-worship, or philosophical speculation were no match for the impelling message which instilled such a strong social conscience that Christians actually tried to love their neighbours as themselves. Indeed the very joy with which Christians faced death in the arena must have given many a Roman pause to think.

From the Greeks came the system for presenting the new religion in the form of theology (Greek *theos* = God; *logos* = a treatise), that is, a coherent series of answers to certain problems arising from the attempt to formulate a precise statement of the faith. The Hellenistic Jew, Paul, began

Built on the Aventine Hill in Rome in the early 5th century A.D., the Church of Santa Sabina is a typical Christian basilica and one of the oldest to survive. It was modelled after Roman basilicas, which were usually used for markets or law courts. The walls are supported by Corinthian columns taken from a pagan building, and at the end of the nave stands the altar, bathed in light from the windows in the semicircular apse behind it. The roof is wooden, and the whole interior of the basilica is colourfully decorated with frescoes and mosaics which have been restored, along with other parts of the church, to their original appearance.

this work when he presented the faith in Greek to the Gentiles. Then as time went on and the first hopes that Christ would soon reappear on earth to set up a kingdom proved illusory, Christians became less antagonistic toward their pagan neighbours and more willing to try to work out a reasoned statement of beliefs that would appeal to intellectuals. Justin Martyr (see page 269), for example, set about proving that Christianity and Greek philosophy were not incompatible, and that the new religion was not a danger to the Roman Empire.

Finally, from the Romans came a magnificent organization: the Empire. The church, working within the political subdivisions of the Empire, modelled its own administrative units on them. In time the successors of St. Peter, the bishops of Rome (see page 274), came to exercise a certain authority, so that the church, like the Empire, had its chain of command and carefully linked administration. No other religion could boast such a complete and efficient organization. Moreover, in the beginning Rome had allowed a fair amount of religious toleration, and—more important—had provided a singe state that facilitated ease of communication. It has been said that travel in the Age of the Antonines was safer, more rapid, and more comfortable than it was to be again until the age of steamships and railways.

Early Christian writers clearly saw what the Roman Empire had done for the faith.

God willed peoples of discordant tongues, kingdoms of conflicting laws, to be brought together under one empire, because concord alone knows God. Hence He taught all nations to bow their necks under the same laws and to become Romans. Common rights made all men equal and bound the vanquished with the bonds of fraternity. The City is the fatherland of all humanity, our very blood is mingled, and one stock is woven out of many races. This is the fruit of the triumphs of Rome; they opened the doors for Christ to enter.

Thus in four short centuries a remarkable thing had happened. Who could have predicted that when "a decree went out from Caesar Augustus that all the world should be taxed," ensuing events, like the everyday occurrence of the birth of a baby, would raise up a force mightier than all of Caesar's legions?

7/SUMMARY AND CONCLUSION

The wonder is not that Rome fell, but rather that she lasted so long.

By the 2nd century B.C. a rather ugly kind of imperialism had begun to contaminate the Roman state, leading, nonetheless, to the rule of the Mediterranean by a just and efficient imperial government. By the 2nd century after Christ there were signs that the western half of the Empire was beginning to crack up—though it took centuries more for this to finally happen, whereas the Empire persisted in the East for another 1000 years.

We shall never know whether Roman culture could have matched the Greek, for it died on the vine. Perhaps state control stifled genius. Nevertheless Rome's civilization followed her legions, and lived on when the political empire had gone. A common language, Latin, affected for all time the modern languages of Italy, Spain, Portugal, France, Britain, and the pockets of Roman civilization in Europe—Switzerland and Rumania. And the memory of Rome's majesty lived on. "At first," said a Gothic chieftain, "I longed to destroy and beat down the Roman Empire, but after experience had shown me that the Goths because of their unbridled savagery could not obey laws . . . I chose instead to seek the glory of restoring completely the Roman name and of buttressing it by using the strength of the Goths, in the hope that later ages might know of me as the restorer of Rome."

Later ages did look back to Rome, mostly because of her success in doing what the Greeks could not—in managing an empire and inventing a world citizenship. But when Rome accomplished this, her people paid with their individual political rights. By the 3rd century A.D. armies made and unmade emperors at will—the logical conclusion to a long process

TIME CHART FOR THE ROMAN EMPIRE
(30 B.C. - A.D. 476)

	WEST	EAST
B.C.	Formal end of Republic; Octavian becomes Augustus	Octavian annexes Egypt Birth of Jesus of Nazareth[1]
A.D.	Tiberius Caligula Claudius and invasion of Britain Nero and persecution of Christians in Rome Year of the Four Emperors Vespasian to Trajan	Crucifixion of Jesus Missionary journeys of Paul and others Jewish revolt culminating in Roman destruction of Jerusalem Jerusalem ceases to be centre of Christian Church
100	Trajan; Empire at its greatest extent Hadrian Antoninus Pius Marcus Aurelius Commodus Pertinax and Didius Septimius Severus	Dacian and Parthian Wars Parthian War; plague
200	Macrinus to Severus Alexander "Barracks Emperors"	Caracalla grants citizenship to all freeborn of the Empire Persecutions of Christians throughout the Empire Diocletian formally divides the Empire into Western and Eastern sections Persecutions of Christians throughout the Empire
300	Constantine	Victory of the Milvian Bridge followed by toleration of Christianity throughout the Empire Constantine reunites Empire and founds new capital at Constantinople Romans defeated at Adrianople by Visigoths Empire permanently divided after the death of Theodosius the Great
400	Romans evacuate Britain Visigoths sack Rome and establish kingdom in Spain Huns under Attila invade Italy Vandals establish kingdom in N. Africa and sack Rome Romulus Augustulus Western Emperor	East Roman Empire manages to shove and bribe the barbarians to the West and so survives
476	Odoacer rules Italy on behalf of Emperor at Constantinople	

[1] Scholars now set the date of the birth of Jesus between 10 and 4 B.C. The dating of the year A.D. 1 was first arrived at by an abbot, Dionysius Exiguus, who in the 6th century calculated that Christ had been born during the 754th year. We now know that these calculations were in error, hence the strange situation of Christ having actually been born B.C.

stretching back to the time when Marius created a professional army independent of the state. Small wonder the Senate was reduced to a rubber stamp. Yet even so, and even despite political terrorism, the imperial system of Rome was the most ambitious and successful up to that time.

It has been said that the Romans, building on the Greek political experience, "showed how man could combine a citizenship both local and imperial." That story ought to interest us, because in the 20th century the problem remains basically the same. How can we combine a citizenship both national and international? How can we do this without making the mistake of the Greeks and Romans—without destroying our own liberties in the process?

SOURCE READINGS

(a)

AUGUSTUS (63 B.C.-A.D. 14), a few months before his death, had an autobiographical inscription set up in Rome and throughout the Empire. The copy, in Greek and Latin, which has survived was inscribed on the walls of a temple in Ancyra (modern Ankara, Turkey), whence it gets its name of the *Monumentum Ancyranum*. From the 35 paragraphs of this most famous ancient inscription here are a few samples.

At the age of nineteen on my personal initiative and at my personal expense, I raised the army by means of which I restored the liberty of the commonwealth, suppressed by the tyranny of a faction. . . .

. . . Three times I celebrated triumphs . . . twenty-one times I was acclaimed *imperator* for victories in battle. . . . In my triumphs there were led before my chariot nine kings or sons of kings. . . . When I made this record I had been thirteen times consul, and I had held tribunician power for thirty-seven years. . . .

Three times I gave a show of gladiators in my own name, and five times in the names of my sons or grandsons; in these shows there fought about 10,000 men. Twice I gave the people, in my own name, an exhibition of athletes brought together from all quarters, and I gave a third in the name of my grandson. . . . Twenty-six times, in my own name, or in the names of my sons and grandsons, I arranged hunts of African wild beasts for the people, in the circus, or the forum, or the amphitheatres. . . .

In my sixth and seventh consulships, when I had extinguished the fires of civil war after receiving by common consent absolute control of affairs, I handed the commonwealth over from my own control to the free disposal of the Senate and the people of Rome. For this service done by me I received the title of Augustus by decree of the Senate, and the door-posts of my house were officially covered with laurels; a civic crown was put up over my door, and a golden shield was placed in the new Senate-house, with an inscription recording that it was a gift to me from the Senate and people of Rome in recognition of my valour, my clemency, my justice, and my fulfilment of duty. After that time I took precedence of others in dignity, but I enjoyed no greater power than those who were my colleagues in any magistracy. . . .

When I wrote this record, I was in my seventy-sixth year.

From *Alexander to Constantine: Passages and Documents Illustrating the History of Social and Political Ideas, 336 B.C.-A.D. 337*, translated by E. Barker (Clarendon Press, 1956), pp. 225-229.

(b)
PLINY THE YOUNGER (A.D. 61-113) was sent as governor to Bithynia by the emperor **TRAJAN** (98-117). From Pliny's letters we have an excellent picture of provincial administration under a good emperor. The letter that follows, together with Trajan's answer, is self-explanatory. It was written about the year 112.

Pliny to Trajan

It is my habit to refer to you, Sire, all matters on which I am in doubt. Who can better guide my hesitation or instruct my ignorance? I have never been present at the hearing of legal cases concerned with Christians; and I am therefore ignorant of what it is that is usually punished or investigated, and to what extent this is done. I have hesitated greatly about several problems. Is there any distinction to be made on the ground of age, or are even the youngest to be dealt with in just the same way as the older? Is pardon to be granted to repentance, or does a man who has once been a Christian derive no benefit from having ceased to be one? Is the name of Christian itself to be punished, even if it is not attended by any crime, or is it the crimes that go with the name that are to be punished?

Meanwhile, and for the time being, this is the line which I have followed in dealing with persons who were brought before me as Christians. I have asked them, "Are you Christians?" If they confessed that they were, I have asked them the question a second and a third time, threatening them as I did so with punishment; and then, if they persisted in their confession, I have ordered them to be executed. I had no doubt in my mind that—apart from their belief, and whatever it might be—such inflexible pertinacity and obstinacy ought in any case to be punished. There were also others who showed a similar folly, but whom, as they were Roman citizens, I remitted for trial in Rome.

Eventually, as the proceedings continued, and the range of offences, as usually happens, grew wider, a number of different problems arose. An anonymous list was put before me which contained the names of many persons who denied that they were or had been Christians, and who, repeating the words I dictated to them, invoked the gods, made their supplication . . . to your image (which I had ordered to be brought into court for the purpose along with the statues of the deities), and cursed the name of Christ; none of which things, I am told, any real Christian can be made to do. I therefore thought that these persons ought to be acquitted. Others, who had been named by an informer, first said that they were Christians and then denied that they were; they *had* been, they said, but they had ceased to be —some of them three years back, some of them many years ago, and some even as far back as twenty years. All of them venerated your image and the statues of the gods; and they, too, cursed the name of Christ. They stated that the whole of their fault, or error, had consisted in their habit of meeting before dawn on an appointed day, and singing in turn among themselves a hymn to Christ as their God; they had also bound themselves by oath not to the commission of any crime, but to refrain from theft or larceny or adultery, from any breach of faith, and from refusing to acknowledge a debt to a creditor; after which it was their habit to depart and then to meet again for the purpose of taking food—but food of an ordinary sort and an innocent character; but they had ceased to do even this after the issue of the edict in which, acting under your instructions, I had forbidden the meetings of clubs, and brotherhoods. This made me think it all the more necessary to discover what truth there was in their statement, and I even used torture for the purpose on two of their serving maids, called deaconesses, but I found no evidence of anything except a crazy and unrestrained superstition, and so I postponed the hearing of the case and proceeded at once to consult you.

The matter seemed to me to be important enough for such consultation, especially in view of the number of the persons who were involved. There are many of all ages and every rank, and also of both sexes, who already are, or will be, implicated. The contagion of the superstition has spread not only in cities, but also through villages and the countryside; and yet it still seems possible to arrest and correct it. Certainly it is a fact that the temples, which had been almost deserted, have begun to be attended again, and that the regular services, which had not been held for a long time, are being revived; animals for sacrifice, which had previously found very few purchasers, are now on sale in abundance. This makes it easy to guess what a number of persons can be brought back to the right path if the way is eased for repentance.

Trajan to Pliny

You have followed, my dear Pliny, the right line of action in trying the cases of the Christians who had been brought before you. No general rule can be laid down in any definite terms. Christians are not to be sought out; if they are brought into court and found guilty, they are to be punished—but with this reservation, that any person who has denied that he is a Christian, and has given actual proof by making supplication to our gods, should be pardoned as a penitent even if he is under suspicion in regard to the past. Anonymous lists ought not to be regarded in dealing with any offence: they are the worst of precedents, and they do not agree with the spirit of our age.

Barker, *From Alexander to Constantine*, pp. 249-252, slightly adapted.

(c)

PLINY THE YOUNGER, in this letter to the historian Tacitus, describes the escape of his mother and himself from Misenum the day after the celebrated eruption of Vesuvius on August 24, A.D. 79. Misenum was a naval base on the bay of Naples, and was situated about twenty miles from the volcano.

It was now the first hour of day, but the light was still faint and doubtful. The adjacent buildings now began to collapse. . . . Then at last we decided to leave the town. The dismayed crowd came after us; it preferred following some one else's decision rather than its own; in panic that is practically the same as wisdom. So as we went off we were crowded and shoved along by a huge mob of followers. When we got out beyond the buildings we halted. We saw many strange and fearful sights there. For the carriages we had ordered brought for us, though on perfectly level ground, kept rolling back and forth; even when the wheels were chocked with stones they would not stand still. Moreover the sea appeared to be sucked back and to be repelled by the vibration of the earth; the shoreline was much farther out than usual, and many specimens of marine life were caught on the dry sands. On the other side a black and frightful cloud, rent by twisting and quivering paths of fire, gaped open in huge patterns of flames; it was like sheet lightning, but far worse. . . .

Soon thereafter the cloud I have described began to descend to the earth and to cover the sea; it had encircled Capri and hidden it from view, and had blotted out the promontory of Misenum. Then my mother began to plead, urge, and order me to make my escape as best I could, for I could, being young; she, weighed down with years and weakness, would die happy if she had not been the cause of death to me. I replied that I would not find safety except in her company; then I took her hand and made her walk faster. She obeyed with difficulty and scolded herself for slowing me. Now ashes, though thin as yet, began to fall. I looked back; a dense fog was looming up behind us; it poured over the ground like a river as it followed. "Let us turn aside," said I, "lest, if we should fall on the road, we should be trampled in the darkness by the throng of those going our way." We barely had time to consider the thought, when night was upon us, not such a night as when there is no moon or there are clouds, but such as in a closed place with the lights put out. One could hear the wailing of women, the crying of children, the shouting of men; they called each other, some their parents, others their children, still others their mates, and sought to recognize each other by their voices. Some lamented their own fate, others the fate of their loved ones. There were even those who in fear of death prayed for death. Many raised their hands to the gods; more held that there were nowhere gods any more and that this was that eternal and final night of the universe....

It lightened a little; this seemed to us not daylight but a sign of approaching fire. But the fire stopped some distance away; darkness came on again, again ashes, thick and heavy. We got up repeatedly to shake these off; otherwise we would have been buried and crushed by the weight....

MacKendrick and Howe, *Classics in Translation,* Vol. II, *Roman Literature* (University of Wisconsin Press, 1952), pp. 365-366.

(d)

JUSTIN MARTYR (100-165) began his career as a wandering professor of Platonic philosophy. After his conversion in 130 he became one of the earliest, and most famous, apologists for Christianity, and when he went to Rome some twenty years later he set up a school and wrote several defences ("Apologies") of Christianity addressed to Marcus Aurelius. He was eventually condemned to death and beheaded along with six other Christians. The extract below is from the *First Apology,* written about 150, and describes early Christian worship and the sacraments.

All who are convinced and believe that what is taught and said by us is true, and promise that they are able to live accordingly, are taught to pray and with fasting to ask forgiveness of God for their former sins; and we pray and fast with them. Then they are brought by us to where there is water, and they are reborn in the same manner as we ourselves were reborn. For in the name of God, the Father and Lord of the universe, and of our Saviour Jesus Christ, and of the Holy Ghost, they then are washed in the water....

After thus washing the one who has been convinced and has given his assent, we conduct him to the place where those who are called the brethren are assembled, to offer earnest prayers in common for ourselves, for him who has been enlightened, and for all others everywhere.... When we end our prayers we greet each other with a kiss. Then bread and a chalice of wine mixed with water are brought to the one who presides over the brethren; and he takes it, and offers up praise and glory to the Father of the universe, through the name of the Son and the Holy Ghost.... When he has finished the prayers and thanksgiving, all the people present express their assent, saying "Amen."... And when he who presides has celebrated the eucharist [thanksgiving], and all the people have expressed their assent, those called among us deacons allow each one of those present to partake of the bread and wine and water for which thanks have been given, and they bring it also to those not present.

And on the day called Sunday there is a gathering in one place of all who dwell in the cities or in the country places, and the memoirs of the Apostles or the writings of the prophets are read as long as time allows. Then when the reader has finished, he who presides gives oral admonition and exhortation to imitate these excellent examples. Then we all rise together, and offer prayers; and, as stated before, when we have ended our praying, bread and wine and water are brought. ... And each receives a share and partakes of the food for which thanks have been given, and through the deacons some is sent to those not present. The prosperous, if they so desire, each contribute what they wish, according to their own judgment, and the collection is entrusted to the one who presides. And he assists orphans and widows, and those who are in need because of illness or any other reason, and those who are in prison, and strangers sojourning with us; in short, all those in need are his care.

Lewis and Reinhold, *Roman Civilization: Selected Readings,* Vol. II, *The Empire* (Columbia University Press, 1955), pp. 589-590.

FURTHER READING

1. POLITICS AND THE DECLINE OF THE EMPIRE

BOAK, A. E. R., *Manpower Shortage and the Fall of the Roman Empire in the West* (Michigan, 1955)

* CHAMBERS, M., *The Fall of Rome: Can It Be Explained?* (Holt, Rinehart and Winston, 1963)

* HAYWOOD, R. M., *The Myth of Rome's Fall* (Apollo, 1962)

* KAGAN, D., *Decline and Fall of the Roman Empire: Why Did It Collapse?* (Heath, 1962)

† * KATZ, S., *The Decline of Rome and the Rise of Mediaeval Europe* (Cornell, 1955)

† LISSNER, I., *Power and Folly: The Story of the Caesars* (Cape, 1958)

* LOT, F., *The End of the Ancient World and the Beginnings of the Middle Ages* (Torchbook, 1961)

* MOSS, H. ST.L. B., *The Birth of the Middle Ages, 395-814* (Oxford, 1963)

PARKER, H. M. D., *A History of the Roman World from A.D. 138 to 337* (Methuen, 1958)

SALMON, E. T., *A History of the Roman World from 30 B.C. to A.D. 138* (Methuen, 1959)

THOMPSON, E. A., *A History of Attila and the Huns* (Oxford, 1948)

† WADDY, L., *Pax Romana and World Peace* (Chapman and Hall, 1950)

WALBANK, F. W., *The Decline of the Roman Empire in the West* (Cobbett, 1946)

2. ROMAN CIVILIZATION AND SOCIETY

BAILEY, C., editor, *The Legacy of Rome* (Oxford, 1923)

† * CARCOPINO, J., *Daily Life in Ancient Rome* (Penguin, 1956)

† CHARLESWORTH, M. P., *The Roman Empire* (Oxford, 1951)

† COWELL, F. R., *Everyday Life in Ancient Rome* (Batsford, 1961)

* DILL, S., *Roman Society from Nero to Marcus Aurelius* (Meridian, 1956)

* ———, *Roman Society in the Last Century of the Western Empire* (Meridian, 1958)

* GRANT, M., *Roman Literature* (Penguin, 1958)

† * HAMILTON, E., *The Roman Way to Western Civilization* (Mentor, 1957)

† JOHNSTON, M., *Roman Life* (Scott, Foresman, 1957)

† * MACKENDRICK, P., *The Roman Mind at Work* (Anvil, 1958)

* MATTINGLY, H., *Roman Imperial Civilisation* (Anchor, 1959)

ROWELL, H. T., *Rome in the Augustan Age* (Oklahoma, 1962)

STARR, C. G., *Civilization and the Caesars: The Intellectual Revolution in the Roman Empire* (Cornell, 1954)

* WHEELER, M., *Rome Beyond the Imperial Frontiers* (Penguin, 1955)

3. CHRISTIANITY

* BAINTON, R. H., *Early Christianity* (Anvil, 1960)
  CARRINGTON, P., *The Early Christian Church*, 2 vols. (Cambridge, 1957)
  CROWNFIELD, F. R., *A Historical Approach to the New Testament* (Harper, 1960)
  GILMOUR, G. P., *The Memoirs Called Gospels* (Clarke, Irwin, 1959)
† * GOODSPEED, E. J., *A Life of Jesus* (Torchbook, 1956)
† * ——————, *Paul* (Apex, 1947)
  GRANT, R. M., *The Sword and the Cross* (Macmillan, 1955)
  KEE, H. C. and YOUNG, F. W., *Understanding the New Testament* (Prentice-Hall, 1957)
  LATOURETTE, K. S., *A History of Christianity* (Harper, 1953)
* NOCK, A. D., *St. Paul* (Torchbook, 1963)
  PEROWNE, S., *Caesars and Saints* (Hodder and Stoughton, 1962)

UNIT IV

MAKING THE MIDDLE AGES

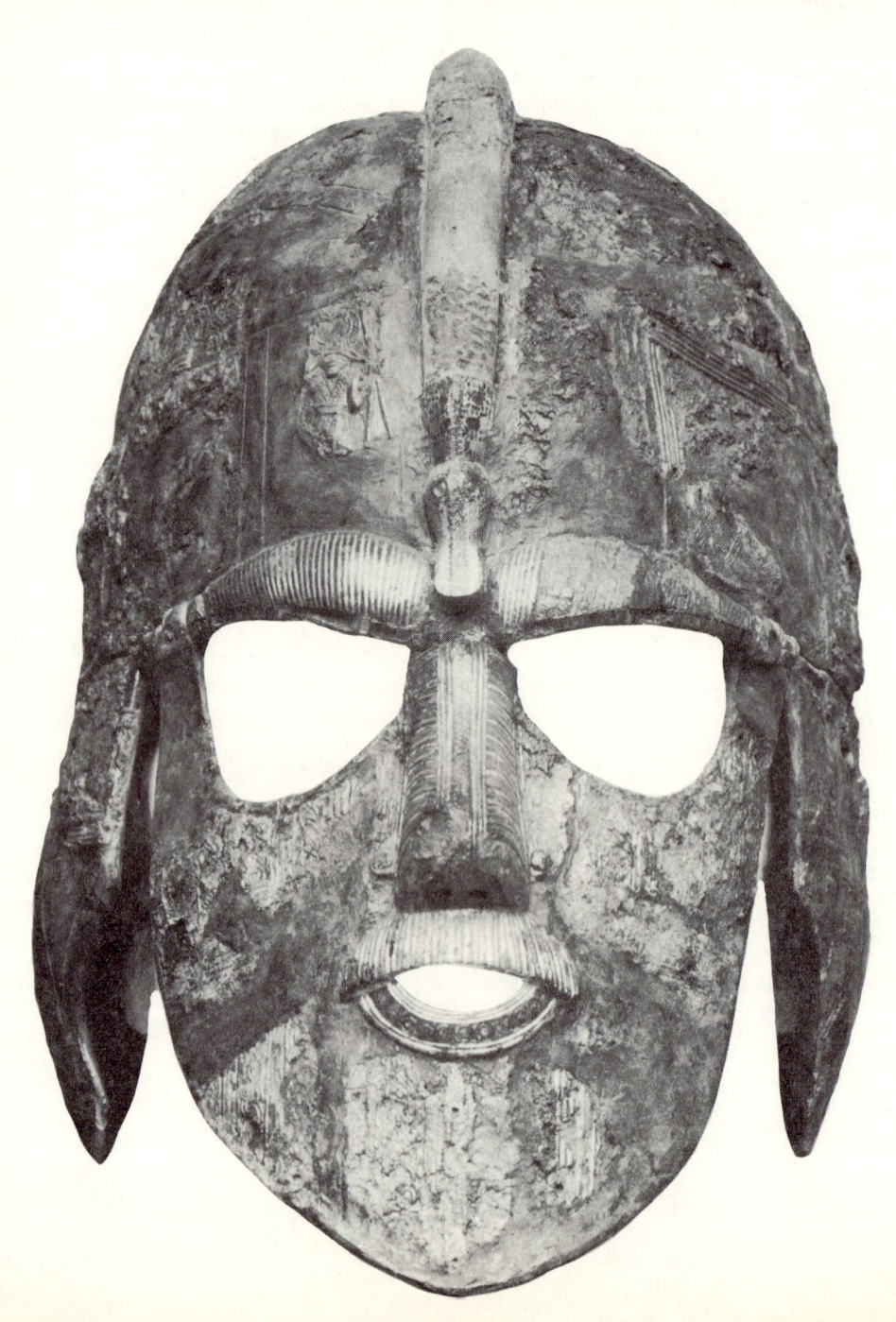

9. The Birth of European Civilization

Then all the faithful Romans, seeing what a pillar of defence he was, and what love he had for the holy Roman Church and its vicar, unanimously, at the will of God, and of St. Peter, the doorkeeper of the Kingdom of Heaven, cried out with a great shout: "To Charles, most pious Augustus, crowned by God, great and pacific emperor, Life and Victory!"

THE CORONATION OF CHARLEMAGNE
AS REPORTED BY THE *Liber Pontificalis*,
WRITTEN ABOUT 816

In the year of our Lord 476, the emperor of the Eastern Roman Empire officially recognized the German Odoacer as ruler of the Western Roman Empire. Thus though the Western Empire was gone in everything but name, the Eastern emperor was able to claim that Roman rule continued, while the German kings were pleased to associate with their rule not only the lustre of ancient Rome but also the approval of the Christian Church.

The map on page 280 shows the Europe that we are about to study. It is a Europe different from that which existed in the glorious days of the Roman Peace. In fact we are about to witness the rise of an entirely new type of civilization, with traditions composed of three different elements: Christian, Germanic, and Roman. The resulting mixture is called European.

1/MAPPING OUT THE MIDDLE AGES

When a 17th century Dutch teacher of history came to deal with the period we are now studying, he saw that it fitted in between ancient times and his own "modern" age. He therefore published a textbook with the long title *History of the Middle Age from the Times of Constantine the Great to the Capture of Constantinople by the Turks,* and very soon the term "Middle Age" or Middle Ages was commonly accepted for the long period from 330 to 1453. Modern historians do not necessarily follow these chronological limits, but they still use the term Middle Ages.

How may we divide up this span of roughly a thousand years? Though it must be realized that all chronological divisions are artificial, a few guideposts can still be erected to point the way through the Middle Ages. Thus in this and the following Unit the dates 800, 1204, 1302, and 1400 have been selected as significant:

This fearsome looking helmet was once worn by a 7th century Anglian king. It was discovered, disintegrated, in an untouched funeral mound where a vast hoard of treasure was unearthed at Sutton Hoo in Suffolk, England, in 1939. Once the helmet had been patiently reconstructed from hundreds of tiny fragments of decayed metal, it was found to have been made of iron, with a silver-plated crest and a covering over the main part of thin embossed sheets of silvered bronze.

> 800—Charlemagne crowned Roman Emperor
> 1204—Capture of Constantinople by the Fourth Crusade
> 1302—First meeting of the French Estates-General
> 1400—Death of Geoffrey Chaucer, English poet

In this Unit and the next the themes of the creation of European civilization during the Middle Ages and of the transition to modern times will be developed. Here again special attention should be paid to the problem of government. Would the men of the Middle Ages be able to succeed where the ancient legislators had failed? Or would their society, in looking back to ancient times, stagnate? Would the Middle Ages, or a part of them, be "Dark Ages" during which the light of learning guttered out and a long night descended over Europe?

The answers to these questions will not be found easily. We can make a beginning by seeing how and why a genuine European civilization was created.

2/THE CHRISTIAN ELEMENT

It soon became evident that if the Christian Church were to preserve its unity—indeed, to survive—amidst the secular pressures of a mighty Empire, it would have to organize on a world-wide scale. Accordingly in 325 Constantine convoked the first universal Church council at Nicaea when he summoned over 300 bishops to work out a statement of the Church's beliefs.

The bishops were the heads of the Church in every important city of the Empire, and their coming together at the Council of Nicaea marked them out as the collective authority of the Church. Naturally, certain bishops, those of the oldest and largest Christian communities—Alexandria, Rome, Antioch, and Constantinople—came to exercise special authority; and from the 3rd century on the bishops of Rome insisted on formal recognition of their absolute supremacy, basing their claim on Christ's words to St. Peter as recorded in *Matthew* 16: 18-19. Their contention was simply that Christ had designated Peter as His successor on earth with unlimited power in matters of faith and morals, and hence each bishop of Rome must succeed to this same power.

By the 5th century these bishops of Rome came to call themselves *papa*, which we translate "pope". Despite the fact that their supremacy was denied by certain Eastern bishops who later broke with Rome to form the Greek Orthodox Church, the Church in the West became powerful, monopolistic, and wealthy. In a word, Christianity became fashionable. It began to attract the wealthy and the ambitious as converts, with the result that some voices were raised against the Church's increasing worldliness. St. Jerome (340-419), for instance, wrote bitterly of a certain type of priest:

All their care is for their clothing, their scents and odours, the close and even fitting of their shoes. The curling iron has left its traces in their crisped locks, their fingers flash with rings, and they scarce venture to go a-tiptoe lest the puddles in the street should soil their feet.

Such criticisms as this came from Christians who often wanted to shut out all the comfortable temptations of the world. The early heroes of the faith had been martyrs. Now they became hermits and monks.

Rules for Salvation   Hermits—men who believed that the wickedness of the world could be escaped by retiring to a cave or the desert—had been generally known in the East by the 4th century. One of the more picturesque of these ascetics was St. Simeon Stylites (395-461), who spent thirty years near Antioch living on top of a pillar some sixty feet high. Sometimes he fasted for forty days at a time, even tying himself to his pillar in order not to sit or lie down. Less continuous was the bodily punishment practised by a certain Northumbrian hermit.

Since his cell stood upon a river's side, he was wont to dip and plunge himself in the flowing water oftentimes, and continue there singing of psalms as long as he could abide for cold, the water now and then coming up to his hips, and now and then to his chin. In the winter season, when pieces of ice half-broken dropped down on every side of him, which he had broken to plunge into the river, diverse men seeing him said; "It is a marvellous matter and a strange case, brother Birthelm (for so he was called), that you can possibly suffer such bitter and sharp cold." Whereupon he answered simply (for he was but a simple and sober-spirited man), "I have known places colder than this."

Such extreme behaviour was not, of course, general, nor was it the sort of physical and mental tension that many could stand.

A more reasonable type of meditative life had been instituted in Egypt in the 4th century. There an establishment was built with bare rooms in which a number of men might live together under strict regulations of obedience, silence, manual labour, and religious exercises. Such an institution came to be called a monastery, and the whole system was called monasticism. It was introduced into Italy, and during the 4th and 5th centuries it spread throughout western Europe until the monasteries, like the Church they had protested against, became successful, wealthy, and worldly. Soon the reformers themselves were in need of reform.

The greatest figure in Western monasticism was St. Benedict of Nursia (480-543). The son of rich parents, he fled from the wickedness of Rome in disgust and took to living in a cave in the hills. Many times he was tempted to abandon his holy life, but finally, "seeing hard by a dense thicket of nettles and briars, [he] threw off his garment and cast himself naked amid the sharp thorns and stinging nettles, and rolling in them for a space came out with his body torn and wounded." From then on his temptations left him.

St. Benedict

The cloister was often the place where the younger members of the monastic community attended lectures and the elder monks studied. Here, too, the brethren might be allowed, at certain set hours, to converse together. This particular cloister, which is now part of the Lutheran Church of the Redeemer in Jerusalem, may be as old as the 8th century—in fact it is probably part of one of the monasteries whose construction Charlemagne aided by sending large sums of money to Palestine. Once frequented by Benedictine monks, it has been reconstructed on several occasions, the last time being during the period of the Latin Kingdom of Jerusalem.

As Benedict's fame spread he gained many followers, and to regulate their life he built a monastery at Monte Cassino, the site of an ancient pagan temple on a hill between Rome and Naples. In the year 529 he drew up a set of regulations called a *Rule*, some of whose provisions can be read at the end of this chapter. The Benedictine Rule was not fanatical. It bound the monk to obey the will of an abbot, who was elected by all the brothers to administer the monastery. The monk must observe poverty; he could own no property—"neither a book, nor tablets, nor a pen—nothing at all." He must observe chastity, that is, moral purity. And he must vow not to leave the monastery except at the abbot's command.

Monastic life

The monk's day began early. In the winter, prayers were said at about half past three in the morning, and throughout the day there were other religious services. Bedtime came shortly after sunset. The time between services was spent in study and such manual labour as working in the fields, the latter being Benedict's answer to the problems of health created by excessively long periods of meditation and contemplation among the early hermits. The monks lived in dormitories and carried out their daily tasks in silence, since idle chatter was strictly forbidden. One meal a day, simple but adequate, sufficed during winter. In summer there were two meals, but the schedule was stricter and there was more work and less sleep.

The monastery of Monte Cassino became the model for many others. From this time on, monks studied, prayed, and led economically productive lives. The purpose of the Rule, however, was not to serve society with the labours of the brothers. Its purpose was simply to give the monks a safe and practicable way to save their souls.

Those living under the rule (*sub regula*) were called *regular* clergy; those living in the world (*in saecula*) were called *secular*. Rivalry and jealousy grew up between these groups, so much so that on occasion the monasteries appealed to the pope at Rome to grant them exemption from the control of the bishop. As often as not the pope, anxious to curtail the power of some bishop, granted such exemptions. Under these circumstances the monasteries frequently pursued their way independently, and eventually turned into something St. Benedict had never intended—islands of culture and wealth.

"Servant of the Servants of God" Gregory I, the Great, came to the papal throne in 590, having renounced a brilliant career in the Roman civil service to become a simple monk. Because he checked the papacy's decline in political power, Gregory has often been called the "Father of the Medieval Papacy." He tightened up the administration of the Church and established the form and order of church worship. "Gregorian chant" is the name used for that type of church music which, though not invented by Gregory, was established on a firm basis by his reorganization of choir schools.

Gregory the Great

When the patriarch of Constantinople took the high-sounding title of "Bishop of Bishops," Gregory countered with the title "Servant of the Servants of God." But Gregory also wrote, "I know of no bishop who is not subject to the Apostolic See [i.e. the papacy] when a fault has been committed." Though such a grandiose claim was not accepted in the East, Gregory summed up in his person and programme the high claim of papal supremacy, and he is chiefly remembered for his insistence on that supremacy.

By the 6th century, then, the Church in the West was a highly organized, efficiently managed institution centred in Rome and thoroughly Roman. The local units were cities, and Latin was its official tongue. In both the Church and the monasteries there was a respect for authority as strong as that accorded any emperor. But unlike the Empire, the Church was not decaying. In fact it was confidently asserting high political claims.

Let us now examine the Germanic element of European civilization up to the 6th century. Was it, too, forged on the Roman anvil?

3/THE GERMANIC ELEMENT

The barbarian invasions of the western Roman Empire were not, as we know, sudden events. For centuries barbarian peoples had been wandering into the Empire and settling down. They knew that across the long and weakly-held Roman frontiers lay the tempting prosperity of rich farm lands and flourishing towns, which contrasted so strongly with their own deep forests. Naturally, then, these Germans—Ostrogoths, Visigoths, Vandals, Burgundians, and Franks—wanted to share this wealth. They did not understand Roman civilization, but neither did they want to destroy it.

Nevertheless as they made their homes inside the frontiers they brought a new sort of society to Western Europe.

Government to the Clash of Spears Because the sources of information are extremely limited, it is not easy to describe the society of the invaders. About all we have to go on are some passages in Julius Caesar's *Gallic War* and the *Germania* of Tacitus, the latter written about the end of the 1st century A.D.

We know that the Germans were not mere nomads. They were divided into tribes, but pursued settled agriculture. There was a noble class, a class of freemen, and a class of slaves. Public discussions on important matters affecting the entire tribe were decided by a general assembly of men of military age. "If a proposal displeases them," Tacitus records, "the people roar out their dissent; if they approve, they clash their spears." Usually the public business so decided involved a declaration of war. Other public business was considered unimportant.

Two features of Germanic society are worth emphasizing. These are the element of personal attachment and the administration of justice.

Each Germanic chieftain was the head of a war band known as a *comitatus*. Lesser warriors served under him as their chosen war chief. The chief swore to lead his "companions" well, to share booty with them, and never to desert them, while his warriors swore undying allegiance to him. In Tacitus' words:

> On the field of battle it is a disgrace to the chief to be surpassed in valour by his companions, to the companions not to come up to the valour of their chief. As for leaving a battle alive after your chief has fallen, that means lifelong infamy and shame. To defend and protect him, to put down one's own acts of heroism to his credit—that is what they really mean by "allegiance". The chiefs fight for victory, the companions for their chief.

This bond of personal attachment is the most important Germanic element of the entire Middle Ages.

In administering justice, the Germans made extensive use of the oath. If the crime was not too serious, the accused could clear himself by having his oath of innocence supported by "compurgators" (those who swore with him that he was honest). If, on the other hand, guilt was decided, damages could be inflicted through a fine, the cost depending on the value placed by Germanic society on the injured individual. For example, a Roman was fined more for robbing a German than a German for robbing a Roman.

In major crimes where compurgation was not deemed a suitable method of arriving at the truth, the accused had to prove his innocence by undergoing the Judgment of God through the *ordeal*. The ordeal might consist of such trials as walking through fire, carrying a red-hot iron in the bare hand, or plunging an arm into a kettle of boiling water, the assumption being that the innocent would be protected by God and the guilty revealed. After the ordeal the wounded limb was bandaged. If the wounds were healing when the bandages were removed a few days later (and remember all this was before antiseptics), the man was innocent. If they were festering he was guilty, and might be beheaded, hanged, or exiled.

There were several refinements of these ordeals. The suspect could, for instance, be bound hand and foot and thrown into cold water. Since water was supposed to cast out impure substances, he was guilty ("impure") if he floated. If he drowned, he at least died innocent! Later in the Middle Ages trial by battle became a popular form of ordeal. A duel was an exciting way to settle guilt, and provided a good show to boot.

Despotism Tempered with Assassination None of the Germanic kingdoms were destined to be long lived. In Italy, where the Ostrogoths reappeared after the collapse of the Hunnish empire, they were able under a brilliant ruler, Theodoric (493-526), to conquer Odoacer and establish a kingdom. But although Theodoric's administration was a good balance between Germanic and Roman traditions, it broke down soon after his death. In Spain the Visigothic state perished after some three centuries, and in North Africa the Vandal kingdom lasted only a century.

Why did these barbarian states pass away so quickly? Partly because they were a minority—thousands of Germans among the Roman millions—and hence a kind of parasitic military aristocracy living off the conquered population. And partly because one other factor prevented them from forming a single society with the Romans. They were opposed to the Church of the Empire.

In the 4th century many Christians had accepted the teachings of an Alexandrian priest, Arius, who declared that Christ, the son of God, was lesser than God and of a different nature from Him. This doctrine was opposed by Athanasius, the bishop of Alexandria, who taught that God and the Son were of the very same substance and nature, and protested that Arianism relegated Christ to a secondary place. Constantine sought to settle this dispute at Nicaea, and the outcome was that the Athanasian views were proclaimed to be the orthodox ones and were embodied in the Nicene Creed. *Arianism*

As long as the emperor himself was orthodox, Arianism was suppressed. But some of Constantine's successors were Arians (one was even pagan), and it so happened that most of the Germanic invaders of the empire were converted to Christianity by Arians. Accordingly, with the Goths, Vandals, and Burgundians all Arian Christians, the orthodox Christians who looked to Rome clung to their religion as a distinctive trait marking them off from their overlords. "The barbarian flood," as one of them said, "would break itself against the rock of Christ." So it was that the bishops remained in a position of strength as the representatives of orthodoxy and the leaders of the conquered population. All they needed was a barbarian champion; and he was soon to appear.

Of far more importance than the other Germanic kingdoms was the Frankish state. The Franks, a Germanic people, settled along the North Sea and the Rhine River. Unlike the other Germans they did not abandon *The Franks* their homeland, but merely expanded it to take in Gaul, which stretched from the Pyrenees Mountains in the south to the Rhine and included modern Belgium, part of Holland, and western Germany. The Frankish kingdom was, therefore, much larger than modern France: both French and Germans can and do claim the Franks as their ancestors. The fact that Gaul was the most completely Romanized of the western provinces of the Empire had, as we shall see, its effect on the Franks.

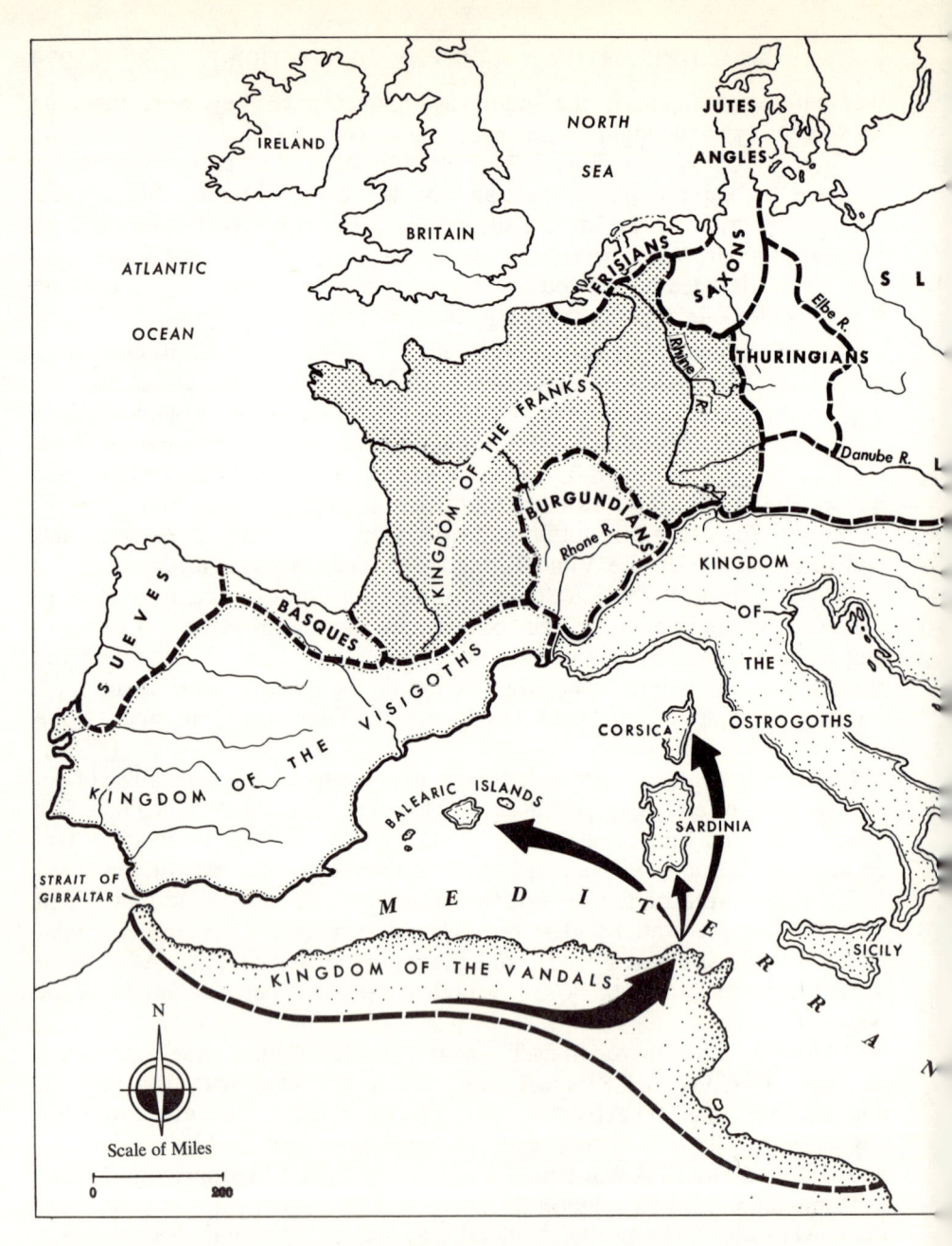

THE GERMANIC KINGDOMS AND THE BYZANTINE EMPIRE 527 A.D.

Clovis

    In 481 a youth of fifteen, Clovis by name, became king of one Frankish tribe. A few years later this shrewd ruler began a career of conquest to round out the Frankish kingdom in Gaul. In the beginning he was not a Christian, but he became a convert to the orthodox brand of Christianity espoused by Rome and the bishops. Now his wars became crusades, aimed not only at conquering Gaul but also at blotting out heresy. But Clovis was no saintly Christian conquerer. He became king of the Franks through murder and treachery—though, since it was murder and treachery practised against heretics, the Church of that day approved. It has been said of Clovis's descendants, as it might be said of Clovis himself, that their rule was "despotism tempered with assassination."

The Merovingians

The dynasty founded by Clovis was called the Merovingian, supposedly after Meroveg, a half-mythical royal founder. Clovis died in 511, and was succeeded by his four sons. They took Roman titles as their father had done before them, stamped their coins with the Roman emperor's profile, and pretended to be regents for an emperor in faraway Constantinople. But this testimony to the veneration of a continuing Roman tradition was all sham.

Clovis's sons quarrelled over the four parts into which their father had divided Greater Gaul. The civil wars and brutality of Merovingian Gaul are well portrayed by a contemporary, Bishop Gregory of Tours, a passage from whose *History of the Kings of the Franks* may be found on pages 299-300. These civil wars weakened the authority of the crown so drastically that the Merovingian rulers are known as the "do-nothing kings." To discover how much society at large was disrupted by all these dynastic quarrels we must briefly examine the way of life of the masses.

A Roman Veneer The interesting thing about Merovingian Gaul is that the ancient Roman traditions were not destroyed. Rather they became intertwined with Germanic ones. The long-haired Frankish king was elected by his warriors, elevated on their shields, and called *Princeps*. Roman titles were assumed by court officials, such as the *comes stabuli*, the count of the stables, who took care of the horses. The unwritten German customary laws were set down in codes to stand alongside the written Roman law. The 112 old Roman cities (a "city" consisted of a town plus its surrounding area) continued to be the units of local administration under a local count or duke who collected taxes, dispensed justice, and led the troops of his city to war. As each city had a count, so it also had a bishop, usually the descendant of a Roman senatorial family. These bishops, through the Church, linked the old society of Rome with the new Frankish society.

The majority of the population probably was little affected by this alleged continuity with Rome. Even though the old Roman taxes were still collected by the new Frankish masters, society remained mainly agricultural. Trade within Gaul petered out—though some continued along the Mediterranean coast—and ordinary men lived and died without having travelled more than 20 miles from home. The people led such circumscribed existences that they were far more concerned with the count's local government than with which Frankish murderer was at the moment occupying the throne. The brutal world of high politics scarcely affected the ordinary merchant, craftsman, or townsman.

In our study of both the Christian and Germanic aspects of Europe's development up to the 6th century, we have noted the persistence of Roman traditions. Now let us complete the circle by a closer examination of this indispensable link, the Roman element of European civilization.

4/THE ROMAN ELEMENT

After 476 the two Roman Empires dwindled into one. That one empire now revolved around a glamorous capital—Byzantium (Constantinople), the "New Rome" founded by Constantine in 330. This remnant of Rome still pretended to exercise authority over both East and West. Germanic kings often claimed to be the deputies of the Byzantine emperor, and Roman emperors continued to reign at Constantinople from the time of Constantine until 1453.

The Byzantine Empire

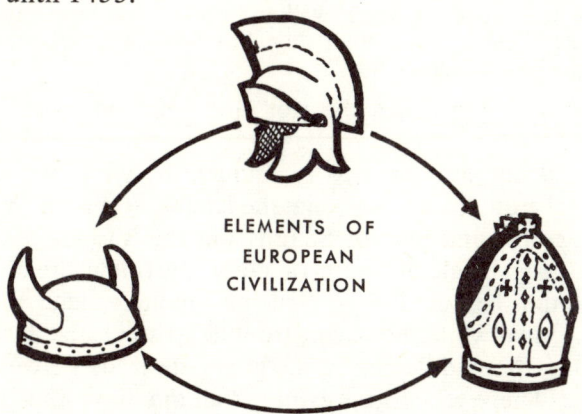

ELEMENTS OF EUROPEAN CIVILIZATION

Despite the unbroken succession from Old Rome to the New, a characteristic Byzantine civilization developed. Hellenistic and Oriental influences were still strong in the East, and these blended with western Roman traditions. Whereas the culture of the barbarians tended to overlay that of the Romans in the West, Roman laws and institutions were preserved in the East. And while a long economic decline destroyed western wealth, the richness and opulence of Constantinople became synonymous with the adjective "Byzantine." The port of Constantinople was the New York and London of its day rolled into one.

Finally, while the orthodox Church of Rome triumphed in the West, the sects and churches of the East became embroiled in endless theological disputes. It is hard for us today to appreciate the extent to which such quarrels over belief divided men. The furore aroused by the controversy between Arians and orthodox Catholics, for example, has been reported by a 4th century patriarch in Constantinople:

Ask a tradesman how many obols he wants for some article in his shop and he replies with a disquisition on generated and ungenerated being. Ask the price of bread today, and the baker tells you "the Son is subordinate to the Father". Ask your servant if the bath is ready and he makes answer "the Son arose out of nothing". "Great is the only begotten" declare the Catholics, and the Arians reply "But greater is He that begot."

A Greater than Solomon By the 6th century the Byzantine Empire was in its golden age, and the emperor Justinian the Great (527-565) was prepared to use all its rich resources to establish himself as a Christian Caesar. Arianism was to be suppressed; all of the old Empire was to be ruled in the interests of a uniform Christianity. Assisting him was his wife Theodora, an ex show-girl of great strength of character. At a crucial moment early in his reign, Theodora saved Justinian's throne for him by this ringing declaration that she, at least, would stay and face the rioting citizens.

If there were left me no safety but in flight, I would not fly. Those who have worn the crown should never survive its loss. Never will I see the day when I am not hailed Empress. If you wish to fly, Caesar, well and good, you have money, the ships are ready, the sea is clear; but I shall stay. For I love the old proverb that says "The purple is the best winding-sheet."

The emperor stayed, put down the riot, and went on to rule for another 33 years.

Justinian's first ambition was to reconquer the western provinces from German rule. North Africa was snatched back from the Vandals, Italy from the Ostrogoths, and part of Spain from the Visigoths. Although he was not able to regain all of Spain or Gaul, Justinian created himself an empire so huge that he could boast that the Mediterranean was once more a Roman lake. Meanwhile, however, trouble arose on the eastern frontiers with the Persians, the middlemen in the great silk trade between China and Constantinople. There was also pressure from the Slavs and Bulgars across the Danube.

Under these circumstances it probably would have been better had Justinian concentrated on stemming the tide of invaders from the East instead of wasting his resources in a vain attempt to turn back the clock in the West. Actually far more important than his conquests are Justinian's legal reforms and legislative work.

Justinian's legal reforms Justinian decided to compile and edit all the ancient Roman laws and legal opinions, and to accomplish this task a great commission was appointed. The following table shows the result. Although Justinian's *Corpus*

CORPUS JURIS CIVILIS.
(The Body of Civil Law)

1. The Code —statutory laws of the Roman emperors; published in 529, revised in 534 in 12 "books" containing 4652 laws.

2. The Digest —3 million lines of legal literature condensed into 150,000; published in 533 in 50 "books" with 432 subject divisions.

3. The Institutes—a textbook for the study of law.

4. The Novels —Justinian's new laws issued in a body at the end of his reign.

preserved the legal genius of Rome in the East, it had little effect on the West at the time. Only with the 12th century did western Europe come to adapt Germanic law to it.

In government the emperor, of course, was supreme—the elect of God, head of the Church and the state. Christianity (of the Eastern variety) was thus made to sanction the autocratic monarchy of the Byzantine Empire. Elaborate costumes, rich court ritual, and a huge imperial civil service all contributed to the "cloth-of-gold and red tape" administration whose staggering expenses crippled the Empire's feeble economy in later days.

The fame of Justinian the builder rivals that of Justinian the lawmaker. Incredible though it may seem, the most famous of his beautiful architectural works still stands in Constantinople—the great 6th century basilica of Hagia Sophia (Holy Wisdom), built in the short space of five years at enormous cost. From the outside it appears plain, almost shabby; but inside, its walls glitter with mosaics—the Byzantine equivalent of painting —small pieces of coloured glass or stone fitted together to make designs and pictures. To give the huge dome height and allow it to enclose a square space, it is raised up on yet another dome from which the top has been sliced off to leave only corner supports called "pendentives". Forty arched windows around the lower lip of the great dome give, from inside, the impression that it is floating on air.

This brilliant solution to the problem of erecting a dome over a square has inspired architects ever since. On the day of the mighty temple's dedication Justinian is said to have exclaimed, "Glory be to God who deemed me worthy of this deed! I have outdone thee, Solomon."

A Bulwark against Barbarians From the 6th century onwards the Byzantine Empire was mainly important as Europe's bulwark against invaders from farther east. Justinian's successes, however, were almost cancelled out by the high price at which they were bought. In Italy the Germanic Lombards seized parts of the peninsula, and continued to nibble at the remaining Byzantine territory there for another two centuries. In the east the Persians, Slavs, and Bulgars mounted a series of campaigns, managing before they were eventually checked to defeat the Empire time and again and even to besiege Constantinople itself. By the 8th century the Byzantine Empire had become a Greek empire, stripped of all its Western possessions.

Though it failed to uphold the political heritage of Rome, the Byzantine Empire did preserve the cultural traditions of both Rome and Greece against the day when they would be returned to a West that had forgotten them. There they would be fused with the western Christian tradition and the Germanic traditions of the barbarian kingdoms. These three, in ferment, produced a new European civilization, but not until after this evolving civilization had faced its own trial by ordeal in the 7th and 8th centuries.

5 / THE CRISIS OF THE SEVENTH CENTURY

Mohammed

About the year 570, some twenty years before Gregory the Great came to the papal throne, there was born in the Arabian city of Mecca the prophet Mohammed. Orphaned at an early age, he later became a camel driver in the caravans of a wealthy widow whom he subsequently married. Once he was a man of means he devoted his leisure to religious speculation, and began to have visions in which he believed both the angel Gabriel and God spoke to him. Mohammed converted a small group in Mecca to his beliefs. Persecutions followed, and in the year 622 he led his disciples north to the city of Medina. This migration is known as the *Hegira,* and is taken as marking the beginning of the first year of a new era. Ten years later Mohammed returned in triumph to Mecca, dying the same year (632).

The Koran

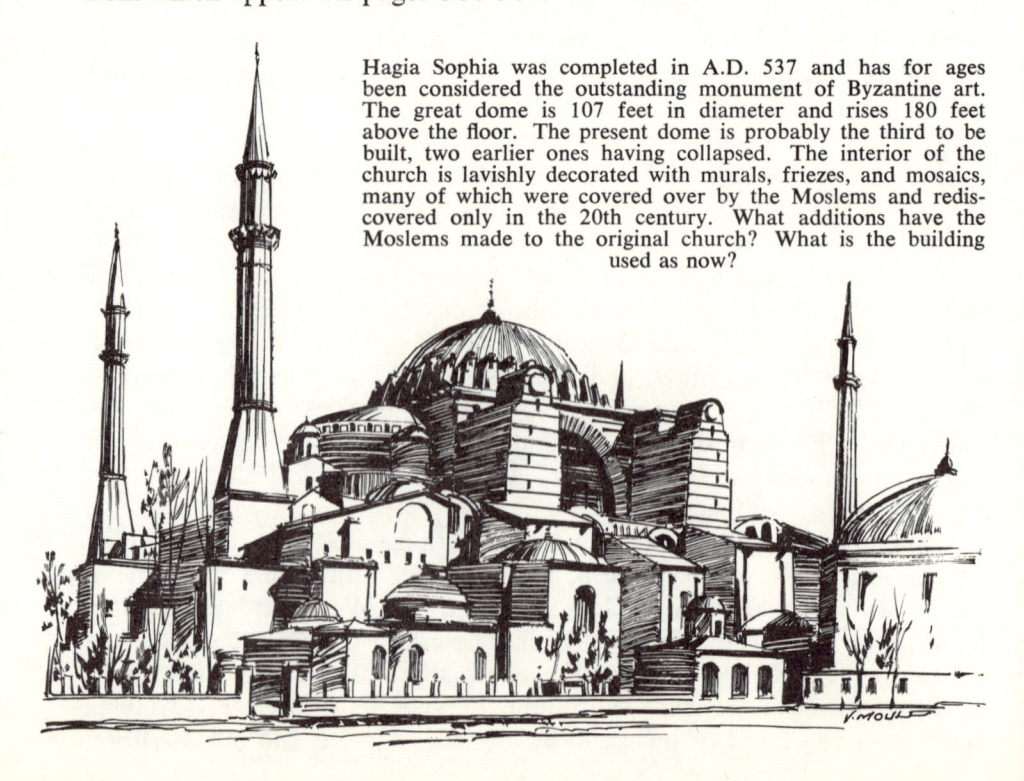

The Whirlwind of Islam   Mohammed's inspirations came to him in bits and pieces from time to time. In his years of study and contemplation he had learned a good deal about both Judaism and Christianity, and strands of these religions appear in his teaching. Mohammed called the new religion *Islam* (submission to God), while the follower of Islam is a *Moslem* (he who professes God). After the Prophet's death his revelations were collected into a book somewhat shorter than the New Testament, the 114 chapters of the *Koran* (Arabian *Quaran*= recitation), a few selections from which appear on pages 300-301.

Hagia Sophia was completed in A.D. 537 and has for ages been considered the outstanding monument of Byzantine art. The great dome is 107 feet in diameter and rises 180 feet above the floor. The present dome is probably the third to be built, two earlier ones having collapsed. The interior of the church is lavishly decorated with murals, friezes, and mosaics, many of which were covered over by the Moslems and rediscovered only in the 20th century. What additions have the Moslems made to the original church? What is the building used as now?

The Pendentive method of supporting a dome is illustrated here in a simple diagram. The rim of the dome rests upon what is in effect a second and larger dome, from which segments have been sliced to form four arches bounding a square. This Eastern system of dome and vault building was in sharp contrast with post-and-lintel construction. How did a dome on pendentives enable the architect to secure a lofty and unobstructed interior space?

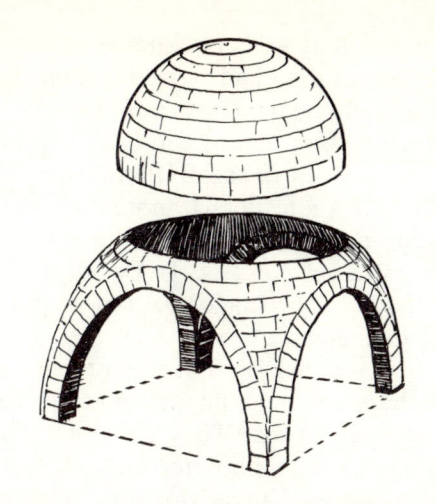

Much of Islam's success has been due to its simple beliefs. Here is no elaborate theology. At the centre is an uncompromising monotheism: "There is only one God and Mohammed is His Prophet." The Moslem must pray five times daily, kneeling and facing Mecca; he must give alms to the poor; he must fast daily during the holy month of Ramadan; and, if at all possible, he must make at least one pilgrimage to Mecca. Mohammed did *not* teach that Moslems ought to wage a Holy War against unbelievers, but this fanatical belief was developed after the Prophet's death.

The new Moslem faith enjoyed an amazingly rapid adoption, and was followed by a whirlwind drive for territory. By 634 Arabia was conquered and Palestine and Syria had been invaded. The Persian Empire fell shortly afterwards, and to the west Moslems galloped out across the North African desert to take Carthage by 695. In 711 the Berber leader of western North Africa, Tarik, led the Moslems past the great rock of *Gebel-el-Tarik* (the mountain of Tarik, or Gibraltar) on their way into Spain. And still the Moslems swept on, not to be checked in Europe until 733, when they were turned back by the heavy hand of Charles Martel at the Battle of Tours near Poitiers in modern France.

Meanwhile Moslem forces had pushed eastward from Persia to penetrate India, and even, by the early 8th century, to probe into the western areas of China. Actually the most decisive check to Moslem expansion came probably not at Tours, but some 2000 miles to the east. Leo III (717-741), known as the Isaurian, seized power in the Byzantine Empire at the very moment when the capital was under Moslem siege, and became the saviour of Constantinople by turning back the Moslems and eventually driving them out of Asia Minor.

Leo the Isaurian

Nevertheless, even though checked at Poitiers and at Constantinople, the Moslems had conquered from one end of the Mediterranean to the other. "In a few decisive battles they had won, with insignificant armies, an empire that was comparable to that of Alexander the Great; and this, even though they had started out on their conquests with no premeditated plan of campaign." How and why?

For one thing Arabs had been moving out to the frontiers for centuries, and many lived in the Byzantine and Persian empires or in border vassal states. Syria and Persia, moreover, were worn out by long wars and were ripe for the picking, while the Arabian desert's periodic droughts were forcing out her nomads. In Egypt, which was a Byzantine province, certain unorthodox Christians welcomed the conquerors as saviours from the Orthodox persecution of Constantinople. Internal divisions within Spain also bettered Moslem chances of success there.

Perhaps the most important factor in Moslem success was the tolerance of the conquerors. They did not surge into a country wielding a sword with one hand and holding out the Koran with the other. In fact they came more as Arabs than as Moslems. The conquered were not always converted or killed. Christianity and Judaism were recognized as genuine (if misguided) religions, and their adherents could, by paying a tax, retain their own faith and customs—although if one wished to rise in the service of the state or engage in business and commerce conversion was recommended.

Nevertheless, the Moslem raiding parties were halted in 733. To discover what stopped them in the West, we must return to the Franks and the Merovingians.

6/THE EIGHTH-CENTURY SYNTHESIS

Mayors of the Palace

The "do-nothing kings" of 7th century Gaul were, as their name suggests, an uninspiring lot. One lived to the ripe old age of thirty-eight, but others barely survived their teens. The real power behind the throne in the state was an official known as the "mayor of the palace." He was a household official of the king, and his control of administration, taxes, and lands made him the one indispensable man. This development is clearly expounded by the secretary and friend of a later Frankish king:

> The real power and authority in the kingdom lay in the hands of the chief officer of the court, the so-called Mayor of the Palace, and he was at the head of affairs. There was nothing left the King to do but to be content with his name of King, his flowing hair and long beard; to sit on his throne and play the ruler; to give ear to the ambassadors that came from all quarters, and to dismiss them, as if on his own responsibility, in words that were, in fact, suggested to him, or even imposed upon him. He had nothing that he could call his own beyond this vain title of King, and the precarious support allowed by the Mayor of the Palace in his discretion, except a single country-seat, that brought him but a very small income. There was a dwelling house upon this, and a small number of servants attached to it, sufficient to perform the necessary offices. When he had to go abroad, he used to ride in a cart, drawn by a yoke of oxen, driven, peasant-fashion, by a ploughman; he rode in this way to the palace and to the general assembly of the people, that met once a year for the welfare of the kingdom, and he returned home in like manner. The Mayor of the Palace took charge of the government, and of everything that had to be planned or executed at home or abroad.

What increased the power of the mayor of the palace was the rise of a new aristocracy of landowners—an aristocracy hardly new in one sense, since it fitted in with the older Roman and Germanic social patterns. In a situation where the only wealth was landed wealth, the mayor of the palace might well tend to usurp royal prerogatives by virtue of his control of the king's estates. Naturally, then, political cliques desirous of controlling the state would try to enlist the aid of the mayor of the palace.

This was roughly the situation when Charles Martel, a mayor of the palace, defeated various rivals and became leader (*dux*) of all the Franks in 714.

The Hammer and His Cavalry Charles Martel (714-741) faced grave threats: in the south the Frankish dominions were being invaded by the Moslems; in the north loomed the Saxons. What sort of Merovingian army could be called on to stem the tide? Charles could summon the mounted nobles, supplemented by a general levy of all able-bodied men; but the former were few, and the latter, though many, were unequipped and undisciplined. To cope with the Moslems and Saxons Charles needed a new and effective army. How should it be made up, and where could he get it?

In the first place Charles modelled his new army on the traditional cavalry. Stirrups, which had just recently been introduced into western Europe, allowed a mounted man to stand, so that he could wield lance and sword while riding. Such mounted cavalry, well armoured, could make short work of infantry. But a horse, armour, and equipment were expensive, and their proper use required considerable practice. Charles needed men with wealth and free time.

Having decided the kind of army he needed, Charles devised the means of getting it. He enlisted warriors who swore fidelity to him. Then he gave estates to these *vassals* of his, estates that would be worked by *serfs* in order that the vassals might be free to fight. Each estate was called a *beneficium* or a benefice. Charles even forced the Church to grant benefices, so that there would be a large enough fighting force.

Tours

Recent research has cast some doubt on the supposition that a heavy-armed Frankish cavalry was the decisive factor when Charles turned back the Moslems at the Battle of Tours—a feat for which he earned the name of Martel ("Hammer"). In fact it is possible that neither side fought on horseback. But it is probably true that cavalry played a vital role in Martel's later campaigns when he conquered part of Saxony. It is also very likely that he was the first military genius in the West to realize the possibilities inherent in stirrups, and to create thereby a new type of warfare based on a system of landholding. At any rate, Charles Martel had under his command a loyal, well-trained fighting force of cavalry with which to deter the attacks of rival nobles.

COMPARATIVE EMPIRES:—

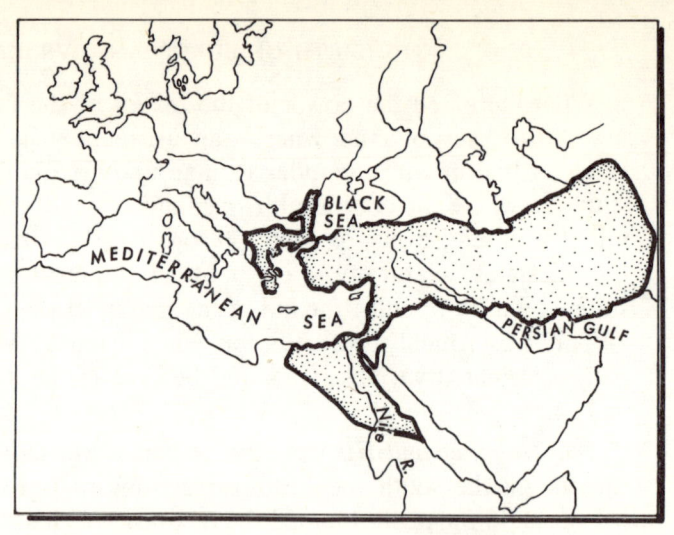

ALEXANDRIAN (323 B.C.)

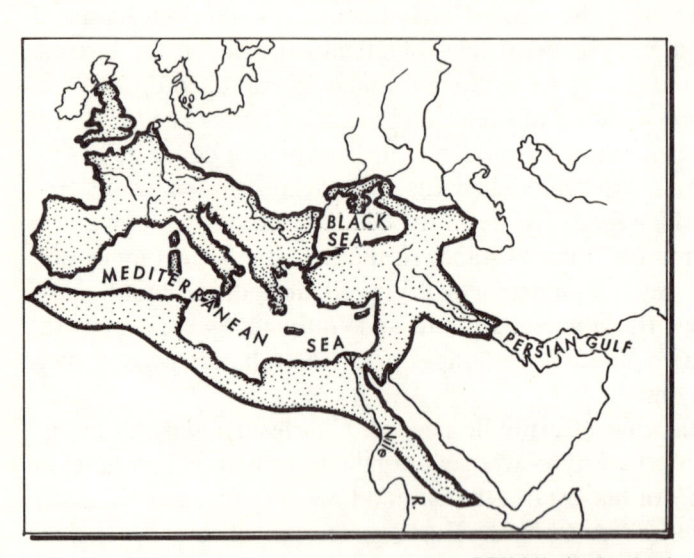

ROMAN (A.D. 180)

The Joining of Two Traditions Charles Martel was succeeded as mayor of the palace by his son Pepin the Short (741-768). Pepin's reign is notable because it marked an alliance with the papacy. But why should the papacy ally itself with this northern kingdom instead of with the rich domain of the Byzantine emperor in the East?

The answer to this question takes us back once more to Leo the Isaurian. Besides being an accomplished general, Emperor Leo III considered himself a religious reformer. In 726 he decreed that his subjects must no longer use *icons*, that is, statues, mosaics, and painted images, in religious worship. Leo hoped thereby to end his people's belief in the miracle-working power of icons, as well as to reduce the wealth of the Byzantine monasteries—whose riches rivalled the emperor's—by removing their many icons.

Iconoclasm

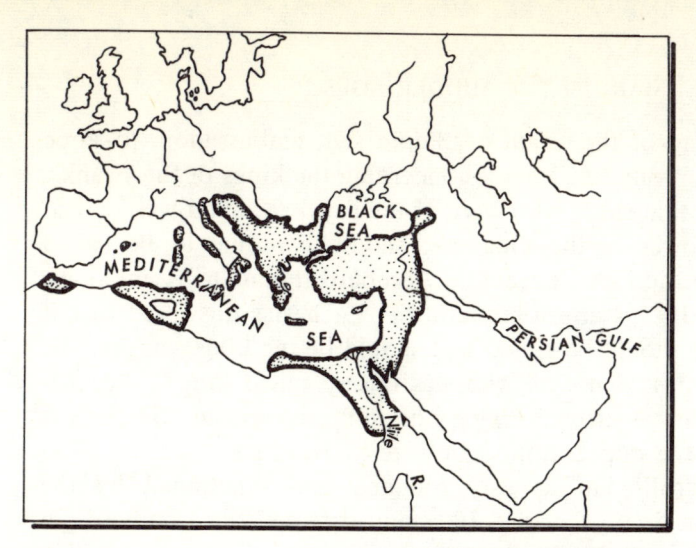

BYZANTINE (565)

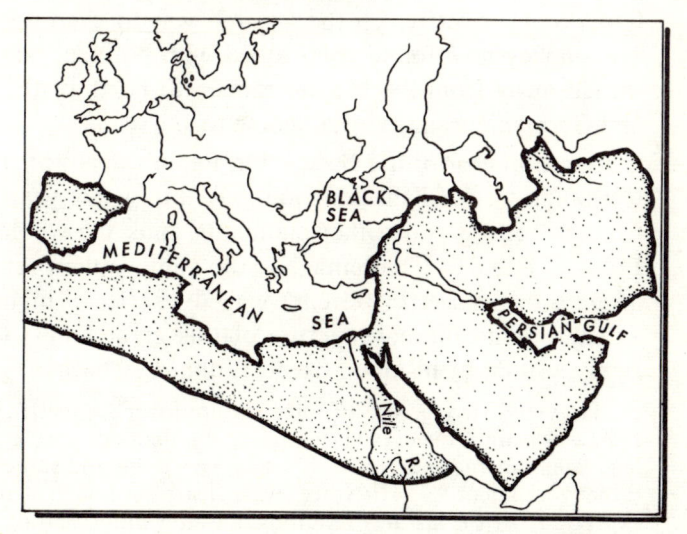

MOSLEM (733)

In the West, however, image worship was not a serious problem, and any interference in this realm was strongly opposed by the popes both on theological grounds and as an unwarranted invasion by a secular ruler into the sphere of religion. Having denounced iconoclasm (icon-breaking) and thus alienated the Byzantines, Pope Gregory III (731-741) was forced to seek help elsewhere against the Lombards who now threatened to take over the remaining Byzantine territory in Italy and were even besieging the city of Rome. He appealed to Charles Martel; but Charles, who had already received Lombard help against the Moslems, sent no aid. The danger subsided when the Lombards withdrew.

Thus inconclusively ended these early overtures between the papacy and the Franks. Then in 751, ten years after he became mayor of the palace, Pepin decided to appeal to the pope to end the Merovingian dynasty and

Pepin the Short

make him in fact king of the Franks. "Pepin sent ambassadors to Pope Zacharias," writes a chronicler, "to ask concerning the kings of the Franks, whether it were good that those who were of the royal race and were called kings but had no power in the kingdom, should continue to be called kings." The pope, in the face of renewed Lombard threats to Rome, gave the expected reply: "By his apostolic authority Zacharias answered that it seemed better that he should be called and should be the king who has the power in the kingdom, and not he who was falsely called king." Accordingly, Pepin the Short was elected *king* by his subjects, and in 754 he was duly consecrated by the pope. Both acts were innovations.

Thus was the Carolingian dynasty founded and sanctioned by the Church. When Pepin finally invaded Italy and defeated the Lombards in 756 he gave new territory to the Papacy, thereby founding the so-called Papal States over which the Church was to exercise authority until 1870. The papacy now found itself allied with political power, a power that had shifted away from the Mediterranean to north of the Alps. The Christian and Germanic traditions had come together.

New Emperor of the West   On Pepin's death his two sons, Charles and Carloman, divided the kingdom. Within a few years Carloman died, and Charles, setting aside other claimants, took the whole domain for his own.

Charlemagne   From 771 to 814 Charlemagne (the name is the French version of *Carolus Magnus* or Charles the Great) waged a series of highly successful wars.

There is an intimate picture of this great king (768-814) in the biography written by his close friend Einhard. Charlemagne was

> ... large and strong, and of lofty stature [over six feet]; ... the upper part of his head was round, his eyes very large and animated, nose a little long, hair fair, and face laughing and merry; ... his neck was thick and somewhat short and his belly rather prominent.... His voice [was] clear, but not so strong as his size led one to expect.... At the last he even limped a little with one foot.... Physicians ... were almost hateful to him, because they wanted him to give up roasts, to which he was accustomed, and to eat boiled meat instead. In accordance with the national custom, he took frequent exercise on horseback and in the chase ... and often practised swimming, in which he was such an adept that none could surpass him. ... He used to wear the national, that is to say, the Frank, dress—next his skin a linen shirt and linen breeches, and above these a tunic fringed with silk; while hose fastened by bands covered his lower limbs, and shoes his feet, and he protected his shoulders and chest in winter by a close fitting coat of otter or marten skins. Over all he flung a blue cloak.... He was temperate in eating and particularly so in drinking ... but he could not easily abstain from food, and often complained that fasts injured his health.... He had the gift of ready and fluent speech and ... was such a master of Latin that he could speak it as well as his native tongue; but he could understand Greek better than he could speak it.... He took lessons in grammar ... and other branches of learning. He also tried to write, and used to keep tablets and blanks in bed under his pillow, that at leisure hours he might accustom his hand to form the letters; however, as he did not begin his efforts in due season, but later in life, they met with ill success.

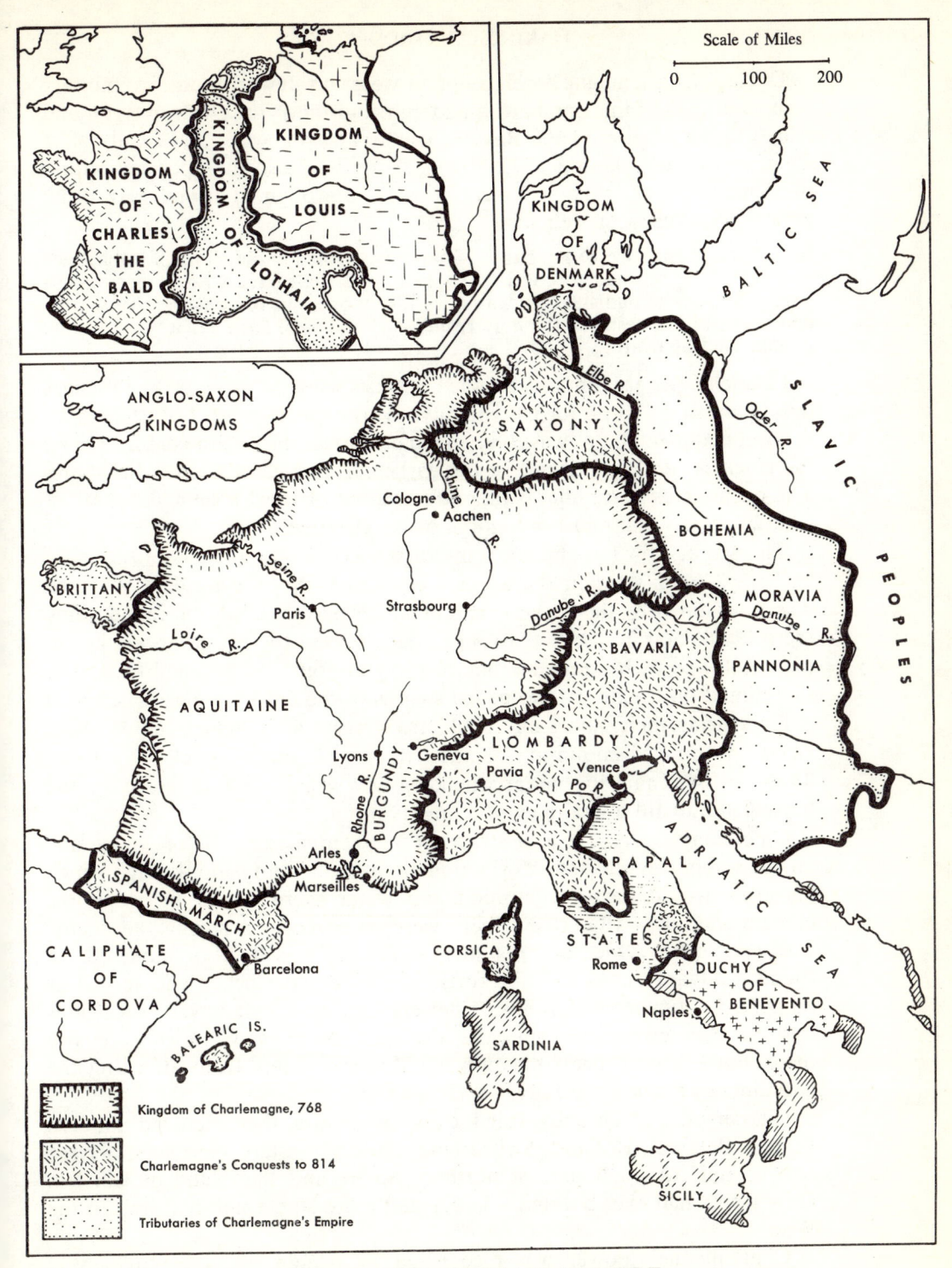

THE CAROLINGIAN EMPIRE

During Charlemagne's reign there were no fewer than 54 military campaigns. First the Lombard kingdom of northern Italy was overrun and added to his empire. Then Bavaria was treated likewise, and finally after 32 years of campaigning (772-804) Saxony was also incorporated within the Empire. Charlemagne's methods were thorough, if brutal. His edict gave the Saxons a clear choice:

If any one of the race of Saxons ... shall have wished to hide himself unbaptized and shall have scorned to come to baptism, and shall have wished to remain a pagan, let him be punished by death. ... If any one, out of contempt for Christianity, shall have despised the holy Lenten fast and shall have eaten flesh, let him be punished by death.

Charlemagne's army

. Charlemagne also anticipated the Crusades when he crossed the Pyrenees into Spain to do battle for Christianity. This expedition of 778, though an unsuccessful campaign, was to be much celebrated in a 12th century poem, the *Chanson de Roland*, one of the earliest examples of French literature. The poem gives a graphic, if fanciful, account of the Moslem slaughter of Charlemagne's rearguard in a gorge of the Pyrenees.

The secret of Charlemagne's military success was the organization of his army. Each summer there was a campaign, the starting date of which was determined by the earliest time at which pasture could be found along the route of march. Over the rough roads moved the proud armed horsemen, followed by baggage wagons carrying armour, clothes, and food. The sun glinted on the heavy two-edged swords of the cavalrymen as they rode along bearing their lances and bucklers. These hardy fighters could shoot arrows without dismounting as they thundered down upon the enemy. Then on foot came the infantrymen, by now greatly outnumbered by the horsemen but still a vital part of a mighty army.

Charlemagne was a master strategist and campaigner. Before beginning an operation, he collected information about the countryside through which his army would pass—what food and water there would be, what the climate was like, what rivers there were to cross, how strong the enemy was, and so on. Once an area such as Saxony was finally conquered, he found it advantageous to build forts, which kept the peace and served as outposts for further expansion. Charlemagne even sent naval expeditions into the Mediterranean against the Moslems. No great victories were won there, but Frankish fleets did prevent Moslem raiders from attacking the coastlines of France and Italy.

Charlemagne eventually ruled over an empire that included modern France, Belgium, Holland, Switzerland, most of western Germany, a large part of Italy, a small part of northern Spain, and the island of Corsica. Only Rome had ever boasted a larger and more stable political unit in the West.

Charlemagne's power was recognized by Alcuin, the most influential scholar in the West, who in 799 wrote a letter to him summing up the

On Christmas Day, A.D. 800, in St. Peter's basilica in Rome, Pope Leo III crowned Charlemagne as "Emperor". Leo III had become pope five years earlier and was having trouble holding his position. In 799 he was attacked by his enemies who attempted to tear out his tongue and eyes; but he escaped from Rome to Paderborn and the protection of Charlemagne, who sent him back to Rome with a powerful escort. Though crowned as emperor, Charlemagne did not treat Rome as his capital and never returned to it again.

world situation. Alcuin said that up to that time the three great leaders had been the pope, the Byzantine emperor, and the Frankish king. But now the situation had altered: the papacy had lost prestige (the pope had been beaten up by a street mob in Rome), and the Byzantine crown had been taken over by a female usurper. Then he went on:

Now in the third place is the royal dignity which our Lord Jesus Christ has reserved to you so that you might govern the Christian people. This dignity surpasses the other two in power; it excels them in wisdom, and exceeds them in regnal dignity. It is now on you alone that rests the support of the churches of Christ, on you alone that depends their safety; on you, avenger of crimes, guide to those who err, consoler of the afflicted, exalter of the good.

The climax of Charlemagne's career came on Christmas Day in the year 800. Pope Leo III had been having trouble hanging on to his throne, and when Charlemagne marched to Rome to aid him he was crowned by the Pope in St. Peter's basilica on December 25. The papal version of this momentous event is quoted at the head of this chapter—but that is only one version. Einhard, Charlemagne's biographer, wrote of this coronation that his master "received the title of Emperor and Augustus, to which he was so averse that he remarked that had he known the intention of the Pope, he would not have entered the Church on that day, great festival though it was."

What the two versions boil down to is probably this: Charlemagne did indeed want the imperial crown, but not from the hands of the pope. It is significant that in 813 Charlemagne crowned his son Louis *himself.*

In the year 812 the Byzantine emperor finally recognized Charlemagne as emperor of the West. Thus symbolically was completed the 8th century synthesis wherein a *German* king was crowned by a *Christian* pope as *Roman* emperor. A new European civilization compounded of these three elements was beginning.

7/SUMMARY AND CONCLUSION

Between A.D. 476 and 800 the Christian, Germanic, and Roman elements merged into one stream of civilization. The Christian Church perfected its organization and defined its beliefs, and the bishops of Rome came to exercise authority over the whole Western Church. The monastic system was organized as a protest against the Church's increasing worldliness. Yet in the monastic system as in the Church, the organization demonstrated a respect for authority as great as that rendered any emperor.

The various German tribes who infiltrated the Roman Empire from the 3rd century on brought with them their own traditions of government and law. Of all the Germanic kingdoms set up in the West the longest lived was the kingdom of the Franks, whose ruler, Clovis, founded the Merovingian line. Clovis and his descendants attempted to maintain the useful fiction of continuity with Rome by pretending to exercise the ancient Roman political authority and customs as latter-day Roman magistrates.

The Roman element persisted in Constantine's "New Rome" at Constantinople, on the site of the ancient Greek colony of Byzantium. A characteristic Byzantine civilization with both Greek and Roman elements flourished there, including unorthodox versions of Christianity. The Byzantine emperor, Justinian, tried unsuccessfully to restore unity to the 6th century Empire, but his most important work was his transmission of the old Roman law in a systematically edited body of civil law. Justinian, however, began to exhaust his Empire by the extravagance of his campaigns, his courts, and his great building programme.

In the next century all three of these elements faced their greatest challenge in the eruption from Arabia of the followers of Mohammed. Within a century they swept around the Mediterranean basin, and were checked only in the 8th century by Charles Martel and the Franks in southern France. The heavy cavalry which became the strong striking arm of the Franks was the result of a new system probably invented by Charles

Martel, a system of military service in return for the use of land granted by the all-powerful mayor of the palace. It was only with Charles's son Pepin that the formal designation of "king" was transferred to the mayor of the palace to express his pre-eminence, a designation given official consecration by the pope. Thus the new dynasty, the Carolingian, was closely allied with the papacy, and Christian and Germanic traditions were joined.

Pepin's son, Charlemagne, created in western Europe an extensive and ably administered empire. Finally on Christmas Day 800 the pope crowned Charlemagne as Roman emperor in the West, an act that completed a synthesis of Christian, Germanic, and Roman elements. A new European civilization was arising from the ashes of the old Roman Empire.

SOURCE READINGS

(a)

ST. BENEDICT (480-543) fled from Rome and from his noble family at Nursia to become a hermit. His example attracted so many disciples that in 529 at Monte Cassino he drew up his *Rule,* a set of regulations for a community devoted to worshipping God. The Benedictine Rule has been called "a model of practical and spiritual wisdom combined, . . . the greatest document of the whole Middle Ages." In all, the Rule has 73 chapters.

2. *What kind of man the Abbot ought to be* An Abbot who is worthy to rule over the monastery ought always to remember what he is called, and correspond to his name of superior by his deeds. For he is believed to hold the place of Christ in the monastery, since he is called by His name, as the Apostle saith: "Ye have received the spirit of the adoption of children, in which we cry Abba, Father." And, therefore, the Abbot ought not (God forbid) to teach, or ordain, or command anything contrary to the law of the Lord. . . . .

3. *Of calling the Brethren to Council* As often as any important matters have to be transacted in the monastery, let the Abbot call together the whole community, and himself declare what is the question to be settled. And, having heard the counsel of the brethren, let him consider within himself, and then do what he shall judge most expedient. We have said that all should be called to council, because it is often to the younger that the Lord revealeth what is best. But let the brethren give their advice with all subjection and humility, and not presume stubbornly to defend their own opinion; but rather let the matter rest with the Abbot's discretion . . . .

6. *Of the Practice of Silence* Let us do as saith the prophet: "I said, I will take heed to my ways, that I sin not with my tongue, I have placed a watch over my mouth; I became dumb and was silent, and held my peace even from good things." Here the prophet sheweth that if we ought at times to refrain even from good words for the sake of silence, how much more ought we to abstain from evil words, on account of the punishment due to sin. Therefore, on account of the importance of silence, let leave to speak be seldom granted even to perfect disciples, although their conversation be good and holy and tending to edification; because it is written: "In much speaking thou shalt not avoid sin"; and elsewhere: "Death and life are in the power of the tongue." . . . But as for buffoonery or idle words, such as move to laughter, we utterly condemn them in every place, nor do we allow the disciple to open his mouth in such discourse.

34. *Whether all ought alike to receive what is needful* As it is written: "Distribution was made to every man, according as he had need." Herein we do not say that there should be respecting of persons—God forbid—but consideration for infirmities. Let him, therefore, that hath need of less give thanks to God, and not be grieved; and let him who requireth more be humbled for his infirmity, and not made proud by the kindness shewn to him: and so all the members of the family shall be at peace. Above all, let not the evil of murmuring shew itself by the slightest word or sign on any account whatsoever. If anyone be found guilty herein, let him be subjected to severe punishment.

36. *Of the Sick Brethren* Before all things and above all things care is to be had of the sick, that they be served in very deed as Christ Himself, for He hath said: "I was sick, and ye visited Me." And, "What ye have done unto one of these little ones, ye have done unto Me." And let the sick themselves remember that they are served for the honour of God, and not grieve the brethren who serve them by unnecessary demands. ...

48. *Of the daily manual labour* Idleness is an enemy of the soul; and hence at certain seasons the brethren ought to occupy themselves in the labour of their hands, and at others in holy reading. We think, therefore, that the times for each may be disposed as follows: from Easter to the first of October, let them, in going from Prime in the morning, labour at whatever is required of them until about the fourth hour. From the fourth hour until near the sixth let them apply themselves to reading. And when they rise from table, after the sixth hour, let them rest on their beds in perfect silence; or if any one perchance desire to read, let him do so in such a way as not to disturb any one else. Let None be said in good time, at about the Middle of the eighth hour: and then let them again work at whatever has to be done until Vespers. And if the needs of the place, or their poverty, oblige them to labour themselves at gathering in the crops, let them not be saddened thereat; because then are they truly monks, when they live by the labour of their hands, as did our fathers and the Apostles. ...

54. *Whether a Monk ought to receive letters, or tokens* By no means let a monk be allowed to receive, either from his parents or any one else, or from his brethren, letters, tokens, or any gifts whatsoever, or to give them to others, without permission of the Abbot. And if anything be sent to him, even by his parents, let him not presume to receive it until it hath been made known to the Abbot. But even if the Abbot order it to be received, it shall be in his power to bid it be given to whom he pleaseth; and let not the brother to whom it may have been sent be grieved, lest occasion be given to the devil. Should any one, however, presume to act otherwise, let him be subjected to the discipline of the Rule.

58. *Of the Discipline of receiving Brethren into Religion* To him that newly cometh to change his life, let not an easy entrance be granted, but, as the Apostle saith, "Try the spirits if they be of God." If, therefore, he that cometh persevere in knocking, and after four or five days seem patiently to endure the wrongs done to him and the difficulty made about his coming in, and to persist in his petition, let entrance be granted him, and let him be in the guest-house for a few days. Afterwards let him go into the Novitiate, where he is to meditate and study, to take his meals and to sleep. Let a senior, one who is skilled in gaining souls, be appointed over him to watch him with the utmost care, and to see whether he is truly seeking God, and is fervent in the Work of God, in obedience and in humiliations. Let all the hard and rugged paths by which we walk towards God be set before him. And if he promise steadfastly to persevere, after the lapse of two months let this Rule be read through to him, with these words: "Behold the law, under which thou desirest to fight. If thou canst observe it, enter in; if thou canst not, freely depart." If he still stand firm, let him be taken back to the aforesaid cell of the Novices, and again tried with all patience. And, after a space of six months, let the Rule be again read to him, that he may know unto what he cometh. Should he still persevere, after four months let the same Rule be read to him once more. And if, having well considered within himself, he promise to keep it in all things, and to observe everything that is commanded him, then let him be received into the community, knowing that he is now bound by the law of the Rule, so that from that day forward he cannot depart from the Monastery, nor shake from off his neck the yoke of the Rule, which after such prolonged deliberation he was free either to refuse or to accept. . . .

66. *Of the Porter of the Monastery* At the gate of the Monastery let there be placed a wise old man, who knoweth how to give and receive an answer, and whose ripeness of years suffereth him not to wander. . . .

The Monastery, however, ought if possible to be so constituted that all things necessary, such as water, a mill, and a garden, and the various crafts may be contained within it; so that there may be no need for the monks to wander abroad, for this is by no means expedient for their souls. And we wish this rule to be frequently read in the community, that none of the brethren may excuse himself on the plea of ignorance.

The Rule of St. Benedict, translated by D. O. H. Blair (Abbey Press, Fort Augustus, Scotland, 5th edition, revised and corrected, 1948), pp. 15, 25, 39, 97, 101, 123, 125, 137, 139, 145, 147, 171.

(b)

GREGORY OF TOURS (538-594) was the son of a wealthy Gallo-Roman family, and was made bishop of Tours in 573. He is most famous for his *History of the Kings of the Franks*, written, he said, "to hand down the memory of the past to future generations." Here he describes the methods used by Clovis to become king of all the Franks.

While Clovis was sojourning at Paris, he sent secretly to the son of Sigibert, saying: "Thy father is grown old, and is lame of one foot. If he were to die, his kingdom would fall to thee of right, together with our friendship." The prince, seduced through his ambition, plotted his father's death. One day Sigibert left Cologne and crossed the Rhine, to walk in the forest of Buchau. He was enjoying a midday repose in his tent when his son compassed his death by sending assassins against him, intending so to get possession of the kingdom. But by the judgment of God he fell himself into the pit which he had treacherously digged for his father. He sent messengers to King Clovis announcing his father's death in these terms: "My father hath perished, and his kingdom and treasures are in my power. Come to me, and right gladly will I hand over to thee whatever things may please thee from his treasure." Clovis answered: "I thank thee for thy goodwill, and request of thee that thou show all to my envoys; but thou shalt keep the whole." On the arrival of the envoys, the prince displayed his father's treasure, and while they were inspecting its various contents, said to them: "In this coffer my father used to amass pieces of gold." They answered: "Plunge thy hand to the bottom, to make sure of all." He did so; but as he was stooping, one of them raised his two-edged axe and buried it in his brain; so was his guilt towards his father requited on himself. When Clovis heard that Sigibert was slain, and his son also, he came to Cologne and called all the people together, addressing them in these words: "Here ye what hath befallen. While I was sailing the Scheldt, Chloderic, son of my cousin, was harassing his father, and telling him that I desired his death. When his father fled through the forest of Buchau, he set bandits upon him, delivering him over to death. But he in his turn hath perished, stricken I know not by whom, while he was showing his father's treasure. To all these deeds I was in no wise privy; for I could not bear to shed the blood of my kindred, holding it an impious deed. But since things have so fallen out, I offer you this counsel, which take, if it seemeth good to you: turn ye to me, and live under my protection." At these words the clash of shields vied with their applause; they raised Clovis upon a shield, and recognized him as their king. Thus he became possessed of the kingdom of Sigibert and of his treasures, and submitted the people also to his dominion. For daily the Lord laid his enemies low under his hand, and increased his kindgom, because he walked before Him with an upright heart, and did that which was pleasing in His sight.

The History of the Franks by Gregory of Tours, translated by O. M. Dalton (Clarendon Press, 1927), II, 78-79.

(c)

The KORAN is the Bible of Islam, and consists of the revelations of Allah to MOHAMMED (570-632), although its final version was not established until after Mohammed's death. The Koran is greatly revered: a true believer would not put it at the bottom of a pile of books but always on top, and he would not drink or smoke while reading it aloud. It is not supposed to be translated. Moslems must read the Koran in Arabic, and Arabic scholars tell us that it is impossible to translate it without a loss in "both artistic merit and shades of meaning." The Koran is divided into chapters (*suras*), of which a very few are given below. The opening chapter (1: 1-7), repeated in Arabic twenty times a day by millions of Moslems all over the world, is the "Lord's Prayer" of Islam.

> In the Name of God, the Merciful, the Compassionate.
>
> Praise be to God, the Lord of the Worlds,
> The Merciful One, the Compassionate One,
> Master of the Day of Doom.
>
> Thee alone we serve, to Thee alone we cry for help.
> Guide us in the straight path
> The path of them Thou hast blessed.
> Not of those with whom Thou art angry
> Nor of those who go astray. (1: 1-7)
>
> Muhammed is the Messenger of God,
> and those who are with him are hard
> against the unbelievers, merciful
> one to another. Thou seest them
> bowing, prostrating, seeking bounty
> from God and good pleasure. Their
> mark is on their faces, the trace of
> prostration. That is their likeness
> in the Torah, and their likeness
> in the Gospel .... God has promised
> those of them who believe and do deeds
> of righteousness forgiveness and
>   a mighty wage. (48: 29)

THE BIRTH OF EUROPEAN CIVILIZATION

It is not piety, that you turn your faces
   to the East and to the West.
   True piety is this:
to believe in God, and the Last Day,
the angels, the Book, and the Prophets,
to give of one's substance, however cherished,
   to kinsmen, and orphans,
the needy, the traveller, the beggars,
   and to ransom the slave,
to perform the prayer, to pay the alms. . . . (2: 177)

And fight in the way of God with those
who fight with you, but aggress not: God loves
   not the aggressors.
And slay them wherever you come upon them,
and expel them from where they expelled you;
persecution is more grievous than slaying. . . .
Fight them, till there is no persecution
and the religion is God's; then if they
give over, there shall be no enmity
   save for evildoers. . . .
Whoso commits aggression against you,
do you commit aggression against him
like as he has committed against you;
and fear you God, and know that God is
   with the godfearing. (2: 190-191, 193, 194)

J. A. Williams, editor, *Islam* (George Braziller, 1961), pp. 13, 39, 44, 46.

FURTHER READING

1. GENERAL: THE EARLY MIDDLE AGES

* BARK, W. C., *Origins of the Medieval World* (Anchor, 1960)
  BROOKE, Z. N., *A History of Europe from 911 to 1198* (Methuen, 1951)
  BURNS, C. D., *The First Europe* (Allen and Unwin, 1947)
* DAWSON, C., *The Making of Europe* (Meridian, 1956)
  DEANESLY, M., *A History of Early Medieval Europe from 476 to 911* (Methuen, 1960)
* SOUTHERN, R. W., *The Making of the Middle Ages* (Grey Arrow, 1959)
† * SULLIVAN, R. E., *Heirs of the Roman Empire* (Cornell, 1960)
† * WALLACE-HADRILL, J. M., *The Barbarian West: The Early Middle Ages, A.D. 400-1000* (Torchbook, 1962)

2. THE CHURCH AND MONASTICISM

* BAINTON, R. H., *The Medieval Church* (Anvil, 1962)
† * BALDWIN, M. W., *The Mediaeval Church* (Cornell, 1953)
* DAWSON, C., *Religion and the Rise of Western Culture* (Image, 1958)
† * KNOWLES, D., *Saints and Scholars: Twenty-Five Medieval Portraits* (Cambridge, 1962)
* MCCANN, J., *Saint Benedict* (Image, 1958)
* RAND, E. K., *Founders of the Middle Ages* (Dover, 1957)
* TAYLOR, H. O., *The Emergence of Christian Culture in the West: The Classical Heritage of the Middle Ages* (Torchbook, 1958)

3. THE BYZANTINE EMPIRE

* BAYNES, N., and MOSS, H. ST.L. B., *Byzantium: An Introduction to East Roman Civilization* (Oxford, 1961)
† DOWNEY, G., *Constantinople in the Age of Justinian* (Oklahoma, 1960)
† * HUSSEY, J. M., *The Byzantine World* (Torchbook, 1961)
  OSTROGORSKY, G., *History of the Byzantine State* (Blackwell, 1956)
  RICE, D. T., *The Byzantines* (Thames and Hudson, 1962)
* RUNCIMAN, S., *Byzantine Civilization* (Meridian, 1956)
* URE, P. N., *Justinian and His Age* (Penguin, 1951)
* VASILIEV, A. A., *History of the Byzantine Empire, 324-1453*, 2 vols. (Wisconsin, 1958)

4. ISLAM

† * ANDRAE, T., *Mohammed, The Man and His Faith* (Torchbook, 1960)
  CRAGG, K., *The Call of the Minaret* (Oxford, 1956)
* GIBB, H. A. R., *Mohammedanism: An Historical Survey* (Galaxy, 1962)
* GUILLAUME, A., *Islam* (Penguin, 1956)
  HITTI, P. K., *History of the Arabs* (Macmillan, 1958)
* ———, *Islam and the West* (Anvil, 1962)
† * LEWIS, B., *The Arabs in History* (Torchbook, 1960)

5. MEROVINGIANS AND CAROLINGIANS

  DILL, S., *Roman Society in Gaul in the Merovingian Age* (Macmillan, 1926)
  WALLACE-HADRILL, J. W., *The Long-Haired Kings and Other Studies in Frankish History* (Methuen, 1962)
† * WINSTON, R., *Charlemagne: From the Hammer to the Cross* (Vintage, 1960)

10. The Feudal Organization of Society

The court of France met and judged that the king of England should be deprived of all the lands which he had held of the king of France up to that time, because they (the kings of England) had, for a long time past, neglected to do services for those lands, and had on practically no occasion been willing to comply with their lord's summons.

RALPH OF COGGESHALL WRITING OF THE SENTENCING OF KING JOHN TO THE LOSS OF HIS FRENCH LANDS BY THE COURT OF PHILIP AUGUSTUS IN 1202

It has been said that if Al Capone, the notorious Chicago gangster of the 1920's, had lived in the 10th instead of the 20th century, he would probably have ended up as a count—if not a duke. If we can imagine the armoured car as a mailed horse and the tommy-gun as a lance, then we are well on the way to understanding the outward trappings of society in the Middle Ages.

1 / THE CAROLINGIAN EMPIRE

Charlemagne continued Merovingian government, modifying it as circumstances required. His claim to be God's anointed representative made him feel quite capable of exercising authority over both the state and the Church. He thought nothing of telling parish priests what to preach and congregations how to sing.

> He was most particular in the observance of the Christian religion ... in which he had been brought up since childhood. ... When he was well enough, he always attended services morning and evening and watched carefully to see that all was done properly. Quite often he ordered the sacristans to see that the place was decent. He provided many sacred vessels of gold and silver and enough priestly vestments to make sure that no cleric, however humble, had to appear unrobed. Finally, he gave much attention to correct reading and psalmody; for he was an expert, although he never read in public, and sang only in unison or to himself.

The secular government was that of an agricultural society. The king owned the land and divided it up like a private estate, each division being managed by one of his personal servants known as a "count". There were about 300 of these divisions or "counties".

Charlemagne feared—and with good reason—that the counts, who administered the outlying districts and who came from the strongest local families, might come to regard their positions as hereditary and set up private states in opposition to his central government. He therefore created a new official, the *missus dominicus*, or royal messenger, to supervise the counts. Travelling in pairs, the *missi dominici* made a yearly circuit of their particular district of the empire and submitted a report to the emperor; and each year the various royal messengers had their circuit changed to prevent collusion between them and the counts. Charlemagne was well aware of the danger of dissension in his empire.

Under the Carolingian administration, the officers at the centre of the government also came into greater prominence. What function these men had originally fulfilled, and the later development of their office, may be listed as follows:

OFFICE	ORIGINAL FUNCTION	LATER DEVELOPMENT
Seneschal	steward of the king's estates	viceroy of the king, commanding the army; manager of the king's courts and administration
Constable	chief groom of the king's household	chief commander of the army
Marshal	assistant groom of the king's household	high military figure
Chancellor	priest able to write, who was therefore the king's secretary	head of the royal secretariat or chancery

In addition to developing cavalry, the Franks had learned how to use iron, although they did not discover how to harden it until the 9th century. They had iron-tipped spears and iron swords but of poor quality, and their wooden shields were often iron-rimmed. The best known weapon was the *francisc,* the curiously shaped throwing axe shown here. This typical Frankish warrior wears a surcoat or jerkin, and long pants or *trews* with strips of leather wound about the legs.

These officials, who along with the king comprised the central government, travelled with the ruler. Wherever the royal entourage was at a particular moment, there was the "capital" of the empire.

Charlemagne did not develop a permanent army or navy, but relied instead on the general levy of freemen. Although the cavalry was increasing in importance, not every freeman could afford the required horse and equipment. Those who were able to buy them were provided with a benefice. Those who could not qualify for such a benefice might prefer to become the serfs (from the Latin *servus,* a slave) of those who could.

Charlemagne attempted to avoid the inherent dangers of this situation by trying to preserve free status. Small farmers were allowed to band together and send one mounted representative to the wars. The trouble was that the one man on horseback became more important than the four or more who sent him off, and the ones who stayed at home found themselves continuing, even after the war, to work the land for their mounted lord.

Thus the Carolingians tended to develop on the one hand an aristocracy of mounted fighters, and on the other a large unfree peasant class.

The Steel of the Heathen Glistens   Even before the Moslem invasions of Europe, the Graeco-Roman cities had fallen on evil days; and along with the decay of the Western cities went a decline in their sea trade. The Vandal conquests of North Africa in the 5th century, which had given them control of the narrows between Tunis and Sicily, had cut trade between the eastern and western halves of the Mediterranean. The West had few commodities that the East wanted anyway, and when the Moslems surrounded the Mediterranean they avoided Christian ports. Exports dwindled, industries became mere local businesses in the shells of the old cities, and money no longer circulated freely. Henceforth the crumbling towns sheltered only a bishop or a count with his household, while the majority of the people lived on the land in the country. In this agricultural society there grew up a small noble class, and a very large peasant class which ranged from free to slave status. Charlemagne could hold the downgrading of the peasants in check, but under later weaker rulers they were to be increasingly exploited.

Scholars have recently devoted a great deal of attention to economic policy under Charlemagne, and now give the emperor high praise for his attempts to revive the sagging Carolingian economy. He issued a standard silver coinage and recalled and forbade the circulation of money coined before his time. (During the 7th century, 800 mints are known to have operated in Gaul alone.) Charlemagne also encouraged trade with the Frisians, Saxons, and Slavs, although this trade was on a rather limited scale.

Charlemagne's economic policy

Nevertheless, the complex economic changes in Europe from the fall of the Western Roman Empire to the Carolingian Empire may be summed up as a "recession towards a natural economy," that is, a falling off of industry and commerce until the economy became predominantly agrarian. The emperor himself gave the closest scrutiny to the management of his royal estates as may be seen from the text of his Decree concerning Villas, part of which is translated on page 332. If there had not been fresh barbarian invasions of Europe, and if Charlemagne's policies had been pursued by his successors, an economic revival might possibly have been effected in Europe. But such was not to be.

Here are eight sentences taken from contemporary chronicles which describe Europe after Charlemagne.

The steel of the heathen glistened. . . . A hundred and twenty ships of the Northmen ravaged all the country on both sides of the Seine and advanced to Paris without meeting any opposition. . . . The Saracens . . . slaughtered all the Christians whom they found outside the walls of Rome. . . . Michael, bishop of Regensburg . . . gathered his troops and joined the other Bavarian nobles in resisting an invasion of the Hungarians. . . . The cities are depopulated, the monasteries ruined and burned, the country reduced to solitude. . . . Every man does what seems good in his own eyes. . . . The strong oppress the weak; the world is full of violence against the poor and of the plunder of ecclesiastical goods. . . . Men devour one another like the fishes in the sea.

In the evolution of ships, the Roman merchantman (*corbita*) appeared as a dignified round ship (1) which, although slow, sailed excellently before the wind because of the high stern. Unlike the sleek Greek triremes, the corbita carried no oars, depending on simple sail arrangements with usually only one mast. The sail was square and was composed of many strips of canvas sewn together. The Roman ship was deep, fully decked for the protection of cargoes, and had a bowsprit at the front, a feature not employed again for approximately 1000 years. The Viking long-ship (2) was probably the only vessel of significance to appear throughout the Middle Ages. Strong and graceful, it was superbly designed and eminently seaworthy, with oars supplying additional power.

The Vikings

To anyone with a knowledge of the past, the 9th and 10th centuries must have looked like the 4th and 5th centuries all over again. Charlemagne's death was followed by two centuries of attack from without by Norsemen, Magyars, and Moslems, and civil war between the rival claimants for the Carolingian Empire gnawed at the realm from within. But there was this difference between the situation in the 4th and 5th centuries and that in the 9th and 10th. Whereas the earlier barbarians had been tolerated, the later invaders were resisted desperately.

The Danes, Swedes, and Norwegians were known as Norsemen (Northmen) or Vikings. About the 9th century a combination of circumstances, including population pressure and the desire of turbulent elements to be free of their own kings, combined to drive the Vikings out of Scandinavia. Sailing their long, graceful ships across seas and up rivers, they attacked Europe, looting and burning towns and monasteries. The Vikings also swept the northern seas, going eventually as far as Iceland, Greenland, and even the coast of North America. (Interestingly enough, archaeological confirmation of Viking settlement in Newfoundland was found in 1962 when eight house sites and a smithy were discovered on the island's northern tip.) It was against such intrepid raiders that the Anglo-Saxon king, Alfred the Great of Wessex (871-899), won everlasting fame.

Our main interest in the Vikings, however, centres around their campaigns in France. They continually attacked Paris, and sailed all around the Frankish kingdom, even passing through the Strait of Gibraltar into the Mediterranean. In 885 some 40,000 Vikings with 700 vessels besieged

Paris, but were bought off with 700 pounds of silver and a chance to ravage the land of the Burgundians, with whom the emperor was unfriendly at the time. Although the Vikings did not actually take Paris they did settle down along the channel coast in an area eventually known as Normandy.

The Vikings, of course, left economic devastation in their wake. But they had another far-reaching effect on the Carolingian state. The Carolingians were, in the beginning, quite unable to cope with Viking hit-and-run raids. Gradually, however, they learned how to meet the problem: by building fortified strongholds to delay the raiders, and, should they bypass these, by providing local forces of well trained cavalrymen who could sally forth to meet the Vikings as they were returning to their ships laden with booty. Thus what Charles Martel had begun the successors of Charlemagne continued. They relied more and more on a professional fighting class of mounted men.

At the same time as northern Europe was being harassed by the Vikings, it was also being attacked in the east and south. In 895 the Magyars, a Turkish people, arrived in what is now Hungary, and shortly after swept westward, eventually getting as far as western Germany and the Rhine valley. Meanwhile from Crete, North Africa, Spain, and the Balearic Islands came Moslems who conquered Sicily and southern Italy, and forced their way into the Rhone valley of southern France.

A Sorry Lot Why were these latest barbarians able to raid almost at will? Where was the strength of Charlemagne's Carolingian Empire? Charlemagne's immediate successor, Louis the Pious (814-840), found his later years saddened by quarrels amongst his four sons. Finally in 842 an event occurred that destroyed any pretence by the eldest son, Lothair, to be sole emperor: Lothair's brother, Louis the German, and his half-brother, Charles the Bald, met to seal an alliance. Each leader swore a solemn oath in the tongue of the other's army—Louis in Old French, Charles in Old High German—so that all might understand. Thus, in their use of two distinct languages, these Strasbourg Oaths provide the first symbolic appearance of the French and German nations. A year later (843), Lothair agreed with Louis and Charles to partition the empire by the Treaty of Verdun. The political unity of Western Christendom had finally been destroyed.

The Strasbourg Oath

This tripartite division, however, did not prevent further wars, and the Carolingian line proved to be a sorry lot. In place of "the Hammer" and "the Great" we now hear of "the Pious," "the Bald," "the Fat," "the Stammerer," "the Simple." In the face of weak rulers and division from within, attack from without could scarcely be resisted. Now the old idea of a great new empire on the foundations of Rome was a mockery; henceforth each man gave his loyalty to the man who could protect him. In such circumstances the strong held the whip hand, and kings and emperors might count for very little.

VIKINGS, MAGYARS, AND MOSLEM[S]

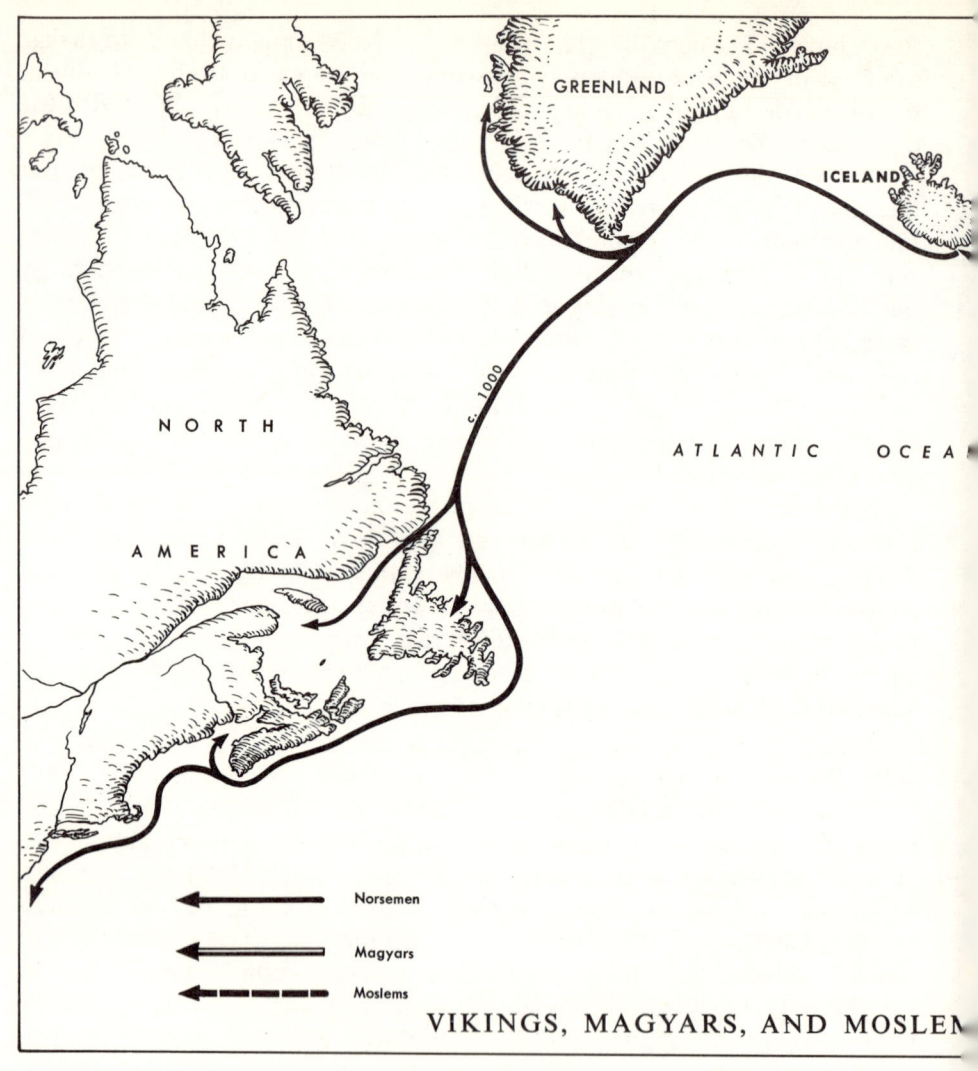

2/FEUDALISM

The word "feudalism" was first used by scholars after the French Revolution of 1789 to describe the ancient institutions that had been overthrown. Since then the adjective "feudal" has often been used as a term of abuse to describe a political or economic system regarded as antiquated or backward. Actually feudalism proper flourished only in medieval Europe, and it was a political, not an economic, organization.

Like much else in the Middle Ages, feudalism drew on various traditions. In the late Roman Empire it had been customary for a humble man to secure the protection of a powerful lord by *commending* himself to him. The poorer man, the client, placed his person and goods at the disposal of the richer man, his patron, who would grant him land and protection in return for his services. These services might include menial work as a household servant or some more exalted administrative function such as a stewardship. Some specially picked clients, however, were expected to perform military service in their patron's retinue. Most historians agree that it is at this point that the Germanic and Roman traditions inherent in feudalism come together.

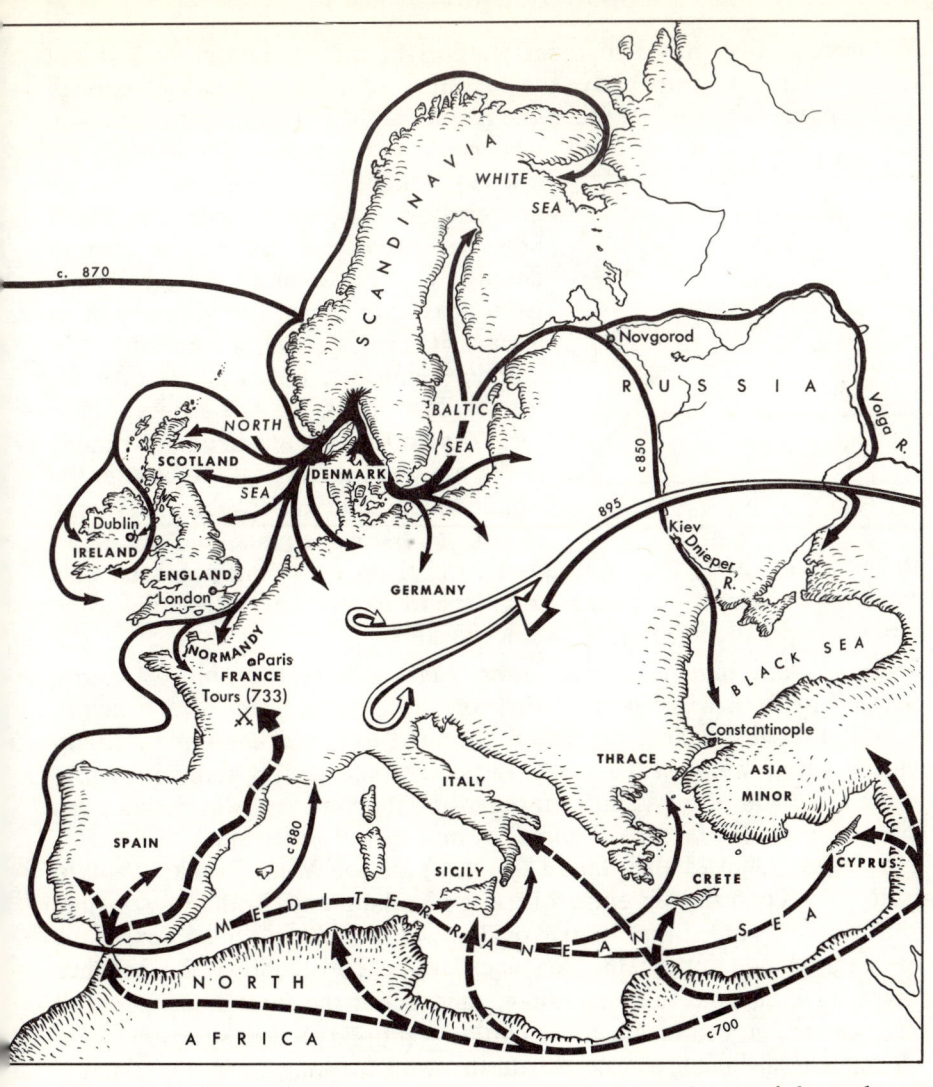

As we have already seen, the Germanic barbarians who invaded the Roman Empire were strongly attached to their war chieftain through a *comitatus* or war band. We have seen the manner in which Charles Martel raised a strong cavalry force by granting benefices to certain cavalrymen who became his vassals, and how under Charlemagne certain peasants were downgraded into serfdom while a few became cavalrymen.

We need notice only two main features in this evolution. In the first place, to belong to a rank above the peasantry a man had to be rich enough to own a horse—hence in France the name *chevalier* (knight) for the man who could afford a *cheval*. Second, the lowliest knight who had just enough property to support *himself* could not hope to protect it unless he surrendered it to another more powerful man, from whom, in return for his services as a vassal (French *vassal*, a free servant or military retainer), he would receive his former land back as a benefice. Thus there developed a regular feudal pyramid: at the top was the king, at the bottom the simple knight, and in between a hierarchy of knights, barons, counts, and dukes.

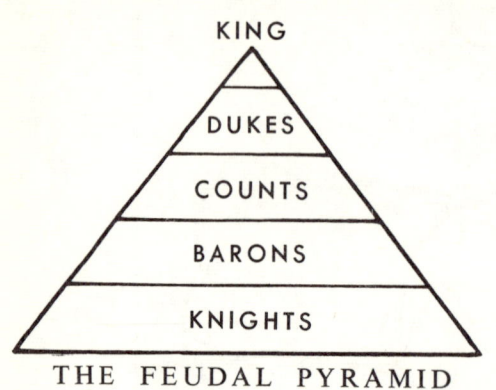

THE FEUDAL PYRAMID

(pyramid: KING / DUKES / COUNTS / BARONS / KNIGHTS)

Below all of these and quite apart from them came the peasants, in a bewildering variety of gradations from free to slave. They will be discussed separately under the heading of "Manorialism."

From Lifetime Benefice to Hereditary Fief The relationship between a vassal and his lord was based, at first, on the lifetime grant of a benefice. The lord granted the land; the vassal promised to render mounted military service. As time went on it must have proved increasingly difficult to prevent the adult son of a vassal from regarding the benefice, which had been granted only for his father's lifetime, as his rightful inheritance.

Thus when Charles the Bald (840-877) of the West Franks went to Italy to be recognized as emperor (after Lothair and Lothair's son had both died), he decreed that if any royal vassal died during his absence, the vassal's son should inherit the benefice. By the 10th century the benefice seems to have become fully hereditary, and about the same time the word *fief* (of Germanic origin, meaning cattle or property) was substituted for the Roman term benefice. The two terms were for long used interchangeably, and the landholding relationship in which they figured is called the "feudal system" (Latin *feodum*=a fief)—although countries and even regions differed so much in their feudal practices that it ought not to be called a "system". We must understand, then, that the term "feudal system" denotes only a rough generalization of Europe's landholding practices reduced to their lowest common denominator.

Under feudalism the lord and his vassal were each parties to a contract, that is each owed something to the other. This contract was duly solemnized by the ceremony of homage and the oath of fealty. The ceremony that took place in 1127 at the court of William, count of Flanders, was typical, and may be read in the translated words of a Flemish chronicler given in one of the Source Readings for this chapter.

What did each party to the feudal contract owe the other? The chief obligations were as follows:

LORD OWED VASSAL	VASSAL OWED LORD

1. protection from foes
 	
2. justice in the lord's court, and consultation with all vassals for advice
 	
3. respect for family and personal interests; see "incidents" under *Vassal Owed Lord*
	1. military service, usually for 40 days
	
	2. castle-guard — garrison duty in lord's castle
	
	3. suit to court — attendance at lord's court to obtain justice or give advice
	
	4. customary aids
	   a) ransom for a captured lord
	   b) payment on knighting of lord's eldest son
	   c) payment on marriage of lord's eldest daughter
	
	5. extraordinary aids — payment when lord went on crusade or some other extraordinary undertaking
	
	6. purveyance — hospitality to lord and his entourage
	
	7. incidents
	   a) relief — inheritance tax for a fief on death of lord or vassal
	   b) wardship — lord's guardianship of a minor heir of a fief
	   c) marriage — lord's right to name a husband for the heiress of a fief
	   d) escheat — lord's right to recover a fief if there was no heir
	   e) forfeiture — lord's right to recover a fief in event of vassal's disloyalty or default of feudal obligations

This neat table of obligations should not make you forget that great variation was possible from fief to fief and from state to state. Moreover, a man might well be the vassal of two or more lords, or the lord of certain vassals and a vassal of other lords at the same time. The king of England, though supreme lord in his own land, was a vassal of the king of France for the duchy of Normandy; while the French count of Champagne was famous because as lord he was owed the service of 2036 knights, yet was himself the vassal of nine lords.

Liege Homage   What would happen if a vassal holding land of two lords found them at war with one another? To whom would he owe allegiance? In such circumstances a vassal chose between *liege* homage and ordinary homage, the liege lord being the one to whom his obligations were more binding. The complications that could still arise even after such distinctions were drawn are well illustrated by the perplexed John of Toul, who, in the early 13th century, took his oath of homage to Beatrice, countess of Troyes, in the following terms:

I, John of Toul, make it known that I am the liege man of the lady Beatrice, Countess of Troyes, and of my most dear lord, Theobald, count of Champagne, her son, against all persons living or dead, except for my allegiance to lord Enjorand of Coucy, lord John of Arcis and the count of Grandpré. If it should happen that the count of Grandpré should be at war with the countess and count of Champagne on his own quarrel, I will aid the count of Grandpré in my own person, and will send to the count and the countess of Champagne the knights whose service I owe them for the fief which I hold of them. But if the count of Grandpré shall make war on the countess and the count of Champagne on behalf of his friends and not in his own quarrel, I will aid in my own person the countess and count of Champagne, and will send one knight to the count of Grandpré for the service which I owe him for the fief which I hold of him, but I will not go myself into the territory of the count of Grandpré to make war on him.

With such conflicting claims as these, and we have given only a single example, you can readily see that many lords could assert a large degree of independence. A lord with a strong, well garrisoned castle could probably hold out against even his king for five or six weeks, by which time the attacking army's forty days' military service would have run out and the besiegers would melt away. In this way the noble with a strong castle became practically independent.

When Knights Were Bold   Fighting was a knight's sole occupation. But war was a sport in which dead knights were dead losses: captive ones could be ransomed. Hence even serious wars were not very bloody during a great part of the Middle Ages. When 600 knights fought against 800 others at the Battle of Lincoln (1217), one knight was unfortunate enough to be killed!

When he was not on campaign, the knight lived with his lady in his uncomfortable castle. It usually consisted of only two rooms, the hall in which the knight administered his fief and ate his meals, and the chamber,

Even the noble had little privacy in the Middle Ages. His great hall, the main room of his castle or manor house, was the centre of activity, and here he received his vassals, presided over his court, and ate with his family and retainers. The hall itself was, by modern standards, more picturesque than luxurious. A large open fire supplied heat, the grimy smoke staining the rafters as it slowly seeped out through a hole in the roof. The walls were usually hung with pennants and shields, along with trophies won in battle, tournaments, or the hunt. In Spain and southern France contact with the Moslems suggested the idea of laying carpets on the floor, but in the north rushes were more common. Besides, they provided a convenient place to dump the left-overs after a meal—scraps soon ferreted out by the ever-present dogs. Dinner entertainment is being presented in the hall pictured here, as musicians on an open gallery accompany the juggler who is balancing a pair of banners.

which was the family's private room and bedchamber. Most castles before the 12th century were wooden and windowless. If the castle was stone there could be a roaring fire, but as chimneys were often non-existent keeping warm must have been a choking experience.

On an ordinary day the knight rose about dawn or even earlier to get the day's business (such as judging cases or consulting his manorial officials) out of the way so as to be able to go hunting. About two or three o'clock in the afternoon came the main meal of the day—not a balanced menu of proteins and carbohydrates, but a combination of meat, poultry, and pastry, the whole washed down with gallons of beer, ale, and wine. The extensive use of spices and pepper covered up the rank flavour of tainted meat before the days of refrigeration. Table manners were hardly genteel. One was not supposed to pick one's teeth with a knife, to spit, or to put one's feet on the table—until the meal was over.

After-dinner recreation was simple, consisting of songs by wandering minstrels or the antics of tumblers or dancing bears. The knight usually went to bed soon after darkness, his eyes smarting from the smoky torches which provided the only lamps of that day. "It seems likely," writes a modern medievalist, "that if one of us were offered the choice between spending a winter night with the lord or his serf, we would choose the comparatively tight mud hut with the nice warm pigs on the floor."

You may well ask how the noble class found time for such a comparatively useless existence. Did no one work for a living? The answer is, of course, that the noble class was a very small minority supported, quite literally, on the bent backs of the peasant majority who formed about 90% of the population living on manors.

3/MANORIALISM

The manor was the economic basis for feudalism. Manorialism in its origins went back as far as feudalism, and, like it, drew on Roman and Germanic traditions.

The late Roman Empire, because of its declining money economy, was responsible for manorialism's first taking root in the 3rd century. The Roman landowner had been able to personally supervise work on his *villa*. He decided what should be planted, where it should be planted, and what work his peasant labourers (*coloni*) ought to perform. The Germanic tribes also had a system of distributing land by rank, land which was worked by the lower classes and any persons incapable of fighting. Soon these two agrarian systems—the Germanic and the Roman—fused.

As time went on the great landowners became so involved in war and politics that they could no longer supervise their estates. Meanwhile the peasants fell into fixed habits passed on from one generation to another. Changes in political circumstances, too, brought changes to the economy. Because the cities provided a ready market for food, the Roman *latifundia* had been farmed for profit. But in the early Middle Ages trade and industry declined, with the result that estates became more and more self-sufficient until they turned into the economic units known as *manors* (from the Latin *manerium,* an agrarian estate with its fields, buildings, and inhabitants). Manorialism was as diverse as feudalism itself, and in describing it only a rough generalization can be made. For example, on any manor there might be half a dozen different grades of peasants, according to their type of land tenure. Let us concentrate on the *serfs*.

The serf was almost a slave. He could not leave his lord's manor for another, nor could he move to a town. Under a ruthless lord he might be treated with less consideration than the cattle of the fields. Under a kindly lord he might be protected, in some measure, by local custom, which decreed that he could not be dispossessed of his land. Since the lord was himself a vassal, he probably spent considerable time away from the manor, which consequently was administered by his *bailiff* or *steward*, assisted by a *provost* or *reeve* chosen from among the peasants. The serfs laboured so that their lord could fight on behalf of his own particular lord, from whom he held the manor as a fief.

Almost the only safe generalization that can be made about manorialism is to say that the three-field system prevailed in northern Europe and the two-field system in southern Europe. Within these two general regions diversity abounded.

About the year 1000 the most usual type of manor in northern Europe probably looked much like the one in the diagram on page 316. Note that the peasants lived in the village and went out to the fields each day. Unlike modern North American farmers, they did not live on the fields they farmed. The fields of the Middle Ages were often "open", that is, unfenced, but were subdivided into smaller fields called *shots*. The shots were divided again into long strips which the peasants farmed, each strip averaging 40 rods x 4 rods (160 square rods = 1 acre). Forty rods (220 yards) is a furlong (a furrow long), that is, the average distance oxen could pull a plough without resting. The peasant's strips were scattered, but made up a total of about 30 acres, while the lord of the manor usually reserved, also in scattered strips, about one-sixth to one-third of the best land of the manor. This was known as his *demesne*. The scattering of the strips seems to have arisen from two considerations. It meant, first, that good and bad land on the manor would be equally shared; and second, that as more land was cleared every peasant would get his fair portion.

Actually, when we speak of the three-field system, the word "field" is

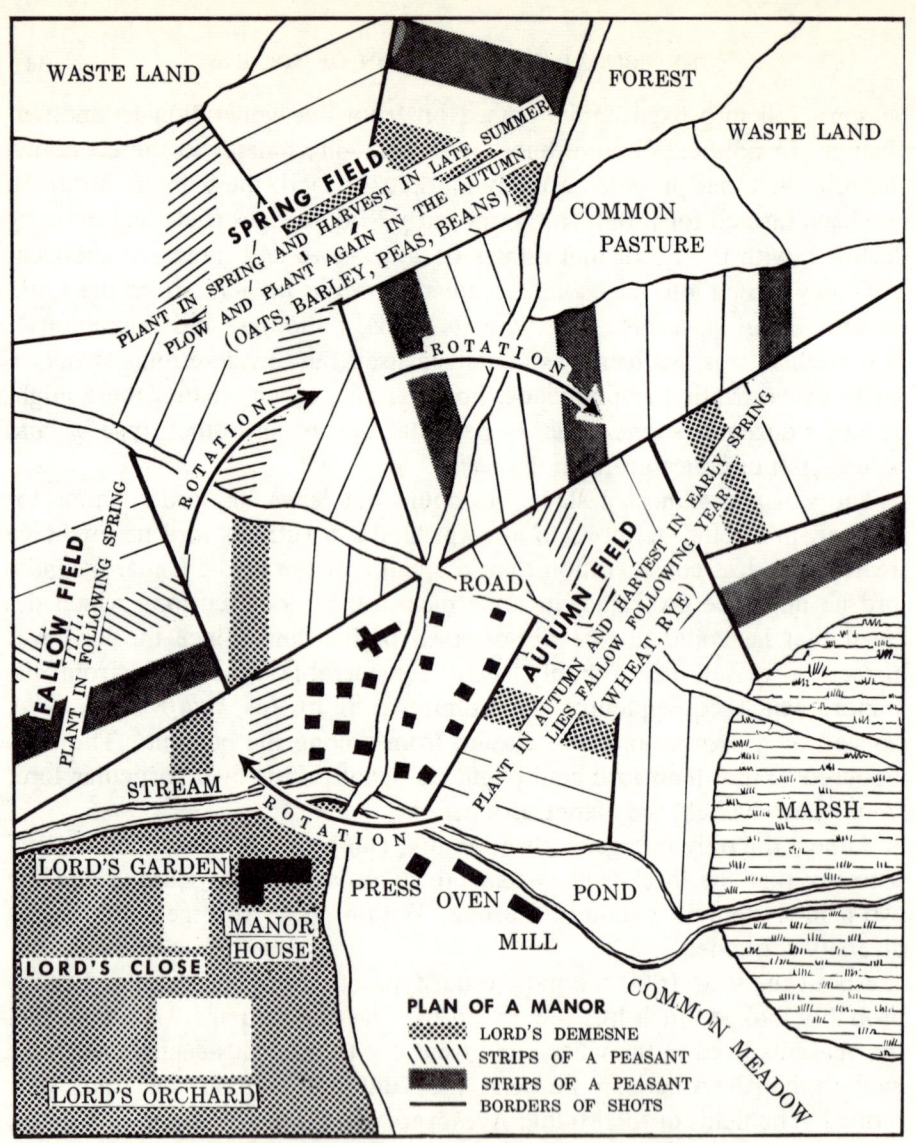

C. G. STARR AND OTHERS, A HISTORY OF THE WORLD (RAND MCNALLY, CHICAGO, 1960), VOL. 1

misleading, since each field was a combination of many smaller divisions (the shots) which we today would call fields. The medieval fields were not neatly arranged, but depended on the kind of soil, contour of land, crops, and type of plough used on any particular manor. Under the three-field system the total area of the manor was divided up into three main groups of shots. One of these was the spring "field", planted in the spring; one the fall "field" planted in the autumn; one the fallow "field" in which nothing was planted that year.

The advantages of the three-field system over the two-field system were considerable, for medieval agriculture was not at all scientific and the commonest device to restore fertility to a field was simply to let it lie fallow for a year. Since the fallow field was ploughed twice in June, under the two-field system for a total of 300 acres, 450 acres would be ploughed for 150 acres of crop thus:

TOTAL CROP—150 ACRES

PLOUGHED ONCE AND PLANTED —150 ACRES CROP	PLOUGHED TWICE & LET LIE FALLOW —300 ACRES PLOUGHING

TOTAL PLOUGHING—450 ACRES

TOTAL CROP—200 ACRES

PLOUGHED ONCE AND PLANTED IN SPRING —100 ACRES IN CROPS	PLOUGHED ONCE AND PLANTED IN FALL —100 ACRES IN CROPS	PLOUGHED TWICE AND LET LIE FALLOW —200 ACRES PLOUGHING

TOTAL PLOUGHING—400 ACRES

If the same area were divided into three strips of 100 acres, the result, for the same total area, would be 50 acres less ploughing and 50 acres more crop than with the two-field system.

The Discovery of Horsepower The three-field system was flourishing in northern Europe by the year 900. It is important to note, however, that this was an agricultural innovation which depended on technological progress, and that there had been continued progress in technology even after the decline of the Roman Empire in the West. Thanks to the Germanic invaders of Europe, a number of epoch-making technological innovations came into western Europe during the so-called "Dark Ages".

One very important new device was the heavy wheeled plough. The Roman plough had been suited to the lighter, drier Mediterranean soils and could not cultivate the heavier soils found in northern Europe. But the new plough, mounted on wheels, was sturdier and well suited to the rich northern river valleys. In southern Europe the Roman farmer had worked the fields in one direction and then cross-ploughed them at right angles to the first furrows in order to scratch deep enough into the soil. But with the heavy wheeled northern plough only one direction of cultivation was required. In this way there developed those long strips of land on the northern manors, where in the fertile, moist soil agricultural production could be greatly increased.

Along with the heavy plough went another set of technological innovations which were, perhaps, the most revolutionary of all. The substitution of horses for oxen in plough teams resulted in an increase in the speed of cultivation comparable only to that brought about by the 20th century replacement of horses by machines. Since horses were faster animals, fewer teams were needed to plough the same area; and because less time was required in the operation, changing weather conditions were not such a hazard. Before the full potential of horses could be utilized, however, certain inventions were necessary.

The Romans had not used horseshoes, with the result that many a broken hoof had rendered an animal useless. Previously, too, there had been only a yoke system of harness, and teams had been hitched side by side. Then about the year 900 the horseshoe, the modern tandem harness, and the horse collar were in use in northern Europe. Horses could now get better traction on stony soil (which would hurt the oxen's feet) and could combine their power by being harnessed one behind the other, while the horse collar allowed them to pull without being strangled by a yoke strap across their windpipe. A team that could pull only 1000 pounds with the antique yoke could pull three or four times that weight when harnessed with the horse collar.

Horses never entirely replaced oxen. They were not as strong and were more susceptible to disease. Moreover they were more expensive to keep because they were grain-fed animals as opposed to the hay-fed oxen. If horses were to be used, then, some surplus grain would have to be grown. This could only be done in northern Europe where the summers were wet enough to allow a spring planting (harvested in the late summer) as well as a fall planting (harvested early the next summer). In southern Europe no more than one crop a year could be produced. Hence horsepower was suited to northern Europe and the three-field system, while oxen survived mainly in the two-field areas.

So it was that a combination of climatic features and technological innovations produced an expanding agricultural economy in northern Europe. As far as power was concerned, the horseshoe, tandem harness, and horse collar "did for the 11th and 12th centuries what the steam-engine did for the 19th." But they did more than that. They made a greater quantity of food available in the north, and that food was supplemented by the legumes planted in the spring in addition to oats and barley. These legumes—peas and beans—provided proteins to supplement the carbohydrates already supplied by grain. Northern Europe's population soared, partly as a result of the improved nutrition traceable to the two-crop three-field system which was made possible by the heavy plough pulled by horsepower. All these elements had interlocked by 900. Truly a medieval agricultural revolution had occurred.

With the Carolingians, the political centre of gravity in the West had shifted away from the Mediterranean and to the north. By the 10th century the economic centre of gravity had followed it.

Much Toil and Fatigue The manor's staple crops were wheat, rye and oats. The peasant could usually supplement his own meagre harvests by the use of the common pasture land, his access to vineyards and orchards, and produce from his small vegetable plot beside his cottage. He might own an animal, a few fowls, and a beehive or two. In fact in all things except such commodities as salt (which came from mines, or from marshes along the sea coast) or luxury goods from the East, the manor formed its own self-sufficient community.

The serf, then, sometimes had a few homely privileges. But he bore a heavy burden of obligations to his lord. Did the lord owe the serf anything at all?

The chief obligations of manor lord and serf may be listed as follows:

LORD OF MANOR OWED SERF	SERF OWED LORD OF MANOR
1. good government according to local custom 2. justice in the manorial court 3. fair treatment according to the Church's teachings: "The great must make themselves loved by the small. They must be careful not to inspire hate. The humble must not be scorned; if they can aid us, they can also do us harm. You know that many serfs have killed their masters or have burnt their houses." (See also the Source Reading on pages 333-334.)	1. week-work — 3 days' work each week, with 2 oxen, on lord's demesne 2. boon-work — extra work at harvest 3. cartage — bringing firewood to manor house, hay to barns, harvest to granaries 4. corvée — making roads, repairing bridges, digging ditches, cleaning out moats 5. merchet — marriage tax for permission to marry his own lord's serf 6. formariage — tax for permission to marry another lord's serf 7. head tax — general annual tax 8. tallage — arbitrary tax on each serf 9. customary taxes — payments of farm produce at Easter and Christmas 10. heriot — death duty when a serf died: the best piece of furniture or best beast 11. banalities — fees for use of lord's mill, bake-oven, winepress, brewhouse, etc. 12. tolls for use of lord's bridges, roads, market, etc. 13. suit to court at the manor court

It will readily be seen that the serf, unlike the vassal, was in no position to oppose his lord. Even though it is unlikely that any individual serf bore *all* these obligations, we must remember that he could escape neither of his taxes to the Church—the *tithe* (one-tenth of his produce) and the *mortuary tax* (death duty of second-best piece of furniture or beast). All told, a serf's life must have been dreary in the extreme. Perhaps the most poignant portrayal of its drudgery is given to us by the poet in the selection at the end of this chapter.

One of the difficulties in describing medieval peasant life is caused by the lack of sources, or rather the lack of objective sources. Most of the medieval literature that mentions the peasant does so in such scornful remarks as these:

— Peasants are those who can be called cattle.
— They have such hard heads and stupid brains that nothing can penetrate them.
— He was large and marvellously ugly and hideous. He had a huge head blacker than coal, the space of a palm between his eyes, large cheeks, a great flat nose, large lips redder than live coals, long, hideous and yellow teeth. His clothing and shoes were of cowhide, and a large cape enveloped him. He leaned on a great club.
— The devil did not want the peasants in hell because they smelled too badly.
— I do not know of a meaner people than the peasant.
— The rustic is best when he weeps, worst when he is merry.

The serf's lot was only bearable because it was the only life he knew. True, there were holidays and fairs to go to, as well as the occasional feast at the manor house. And there was a final consolation: the Church told the serf that God had ordained his lowly estate. Such an attitude is obvious in these words written by a medieval saint to a monk about to set out on a journey to save his sister from an unjust serfdom.

What concern is it of monks—men who have resolved to flee the world—what does it matter to them, who serves whom in the world, or under what name? Is not every man born to labour as a bird to flight? Does not almost every man serve either under the name of lord or serf? And is not he who is called a serf in the Lord, the Lord's freeman; and he who is called free, is he not Christ's serf? So if all men labour and serve, and the serf is a freeman of the Lord, and the freeman is a serf of Christ, what does it matter apart from pride—either to the world or to God—who is called a serf and who is called free?

Although this exalted attitude may have provided some consolation, one can hardly escape the feeling that it was the religious justification of the upper class view that, as Aristotle had long ago believed, some men were marked out for slavery. The medieval version of this theory was put succinctly by the 13th century Spanish mystic Raymond Lûll in a brutally frank sentence from his *Book of the Order of Chivalry*:

It is seemly that the men should plough and dig and work hard in order that the earth may yield the fruits from which the knight and his horse will live; and that the knight, who rides and does a lord's work, should get his wealth from the things on which his men are to spend much toil and fatigue.

That is the serf's real epitaph.

There are medieval castles and manor houses still in existence. But because the houses of the peasantry were rarely built of stone or any other substantial material, hardly any have survived. They were poor little huts like the one pictured here, usually built of sticks packed with mud or clay mixed with bits of straw or hair. The roof was thatched with straw or reeds, and the eaves projected well beyond the walls to keep the rain from softening them. Smoke escaped through a hole in the roof, and often through doors and windows as well. Judging from the oxen, which field system, two- or three-, do you think these serfs farmed?

By the 10th century, then, a feudal society had been built up. It was mainly an agricultural society consisting of a handful of nobles supported by a multitude of serfs, a mere fleck of foam on a vast ocean. How long could this situation continue? What forces would be strong enough to burst asunder the bonds of feudalism, at last lifting the serf from the depths of manorialism? We shall again look back to the 10th century when we examine the medieval transition to modern times.

Now that the establishment and the institutions of feudal society have been described, let us study those political units which comprised the three great feudal monarchies: France, England, and the German Empire. The following table lists only the most important monarchs of these states and the dates of their reigns.

FRANCE	ENGLAND	GERMAN EMPIRE
Hugh Capet (987-96)		Otto I (936-73) Henry III (1039-56)
	William the Conqueror (1066-87)	Henry IV (1056-1106)
Louis VI (1108-37)		
Louis VII (1137-80)	Henry II (1154-89)	Frederick I, Barbarossa (1152-90)
	Richard I (1189-99)	
Philip II, Augustus (1180-1223)	John (1199-1216)	Frederick II (1215-50)

4/THE FEUDAL MONARCHIES

The steady conflict between the mighty feudal lords tended to reduce the power of the kings, until often a king was weaker than his great vassals. The Carolingians, for instance, had preserved an elective kingship. Much as a particular king might want to pass on the throne to his son, the nobles kept insisting on election to the throne.

When the nobles elected Hugh Capet king of France in 987, the "France" over which he ruled was composed only of Paris and the country around it, some 4250 square miles today called the Ile de France. This was Hugh Capet's fief, the duchy of France. Other duchies such as Normandy, Aquitaine, Burgundy, and Lorraine were ruled by Hugh's vassals, although they often refused to recognize the overlordship of Hugh, or later of Hugh's descendants. The nobles had no intention of creating a new dynasty; but by demanding, soon after his accession, that the nobles elect his eldest son as his associate, Hugh founded the Capetian dynasty. The result was that up until 1227 no monarch of France died before he saw his successor safely crowned as king designate.

The Capetians

Yet Hugh Capet and his successors were not powerful figures. Their great vassals stood up to them, and lorded it over the countryside from their lowering castles. The king of France could not even travel from Paris to Orléans without the permission of the particular lord whose fortress controlled that road. The cavalier disregard of the king's wishes was well demonstrated by a certain French count:

... Aldebert, count of Périgord ... carried on war against the town of Poitiers and, through a great slaughter, came out the winner, especially because its citizens rashly attacked him before they should have. He also took by siege the town of Tours, accepted its surrender, and gave it to Fulk, count of Anjou. The latter, however, lost it shortly after through a crooked trick of the viscount and of the citizens, so that Count Eudes of Champagne recovered it once more.

And while Aldebert was besieging the town, King Hugh and his son Robert did not in any wise dare to provoke him to war, but they sent him this message: "Who appointed you count?" they said. And Aldebert replied to them: "Who appointed you kings?"

The Monarchy Survives If the kings were so weak, why did the nobles tolerate them? For one thing the monarchy was not strong enough to threaten the vassals' interests. More important, the Church supported the Capetians, and feudal theory dictated that there must be some overlord. Why not, then, accept the most harmless overlordship then available, that of the Capetians?

So the monarchy survived, and by the reign of Louis VI (called "the Fat" because in later years he even had to be hoisted onto his horse) the royal authority was being asserted over the great fiefs. When the Duke of Aquitaine felt death approaching, he entrusted his daughter and heiress, Eleanor, to Louis. In this way the great duchy of Aquitaine, comprising a

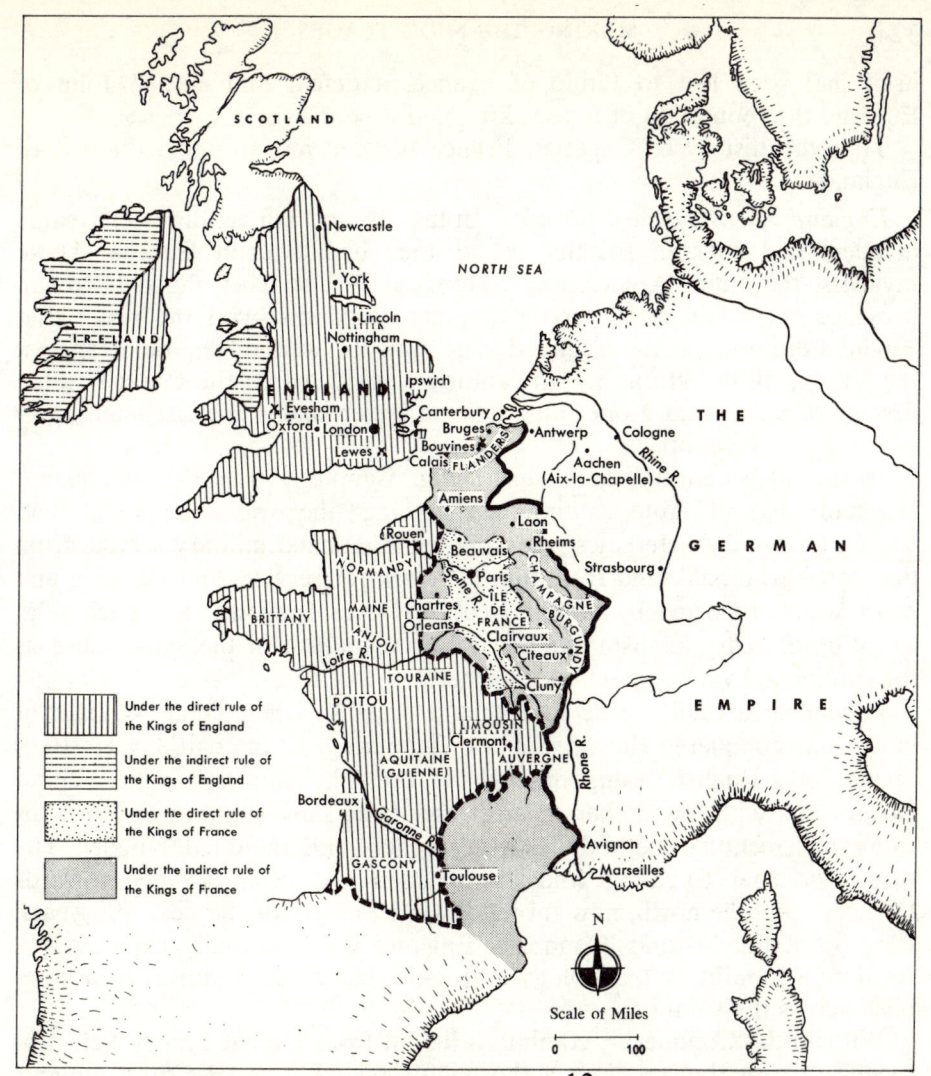

ENGLAND AND FRANCE IN THE 12TH CENTURY

third of the territory of medieval France, was added to the Capetian domain, and the Capetians became the greatest of French feudal lords.

However, although Louis VI married off Eleanor of Aquitaine to his son Louis VII, she subsequently divorced him, and two months later, at twenty-six, married the sixteen-year-old Henry of Anjou (the future Henry II of England). With her went her lands. It was inevitable that the weakened Capetians should regard the English with jealous eyes.

Philip II, known in history as Philip Augustus, realized that England was now the chief threat to his power. Accordingly he spent much of his life stirring up trouble for the kings of England, finally succeeding in engaging both Richard the Lionheart and the notorious King John (sons of the high-spirited Eleanor of Aquitaine) in warfare. By 1204 Normandy itself had been lost to Philip of France, a defeat that earned John of England the reputation of a poor king and a poor fighter.

Thus the history of Capetian France became intertwined with that of England.

England Turns Toward France   In the 5th and 6th centuries Germanic invaders had overrun Britain just as they had overrun Europe. These invaders, the Jutes, Angles, and Saxons, at first split up the old Roman province of Britain into a score of petty kingdoms, with the result that England did not become a unified state until the wars of resistance against the Vikings in the 9th and 10th centuries. King Alfred the Great was the first to rule England alone, although he never really exercised authority over the entire country.

In the 11th century the Danish king, Canute (1016-35), conquered England. Under Canute, as under earlier kings, the Anglo-Saxon state had certain feudal characteristics: the king could demand military service from the thanes who held land from him in return for service, and the king and court were supported by manors scattered over England. But such relatively unformed feudalism was not really feudalism in the same sense as the continental variety mentioned earlier.

Hastings   Feudalism actually came to England in 1066 when Duke William of Normandy conquered Harold, the last Saxon king. At the battle of Hastings about 7000 English, using methods of the 7th century, fought without cavalry and with few archers against about the same number of Normans using 11th century methods, that is, archers and mounted knights. The result was that England, who up till then had been oriented towards Denmark and the north, now turned toward France for the next 500 years. Because of the Norman Conquest, England was to rapidly experience a revolution in military technology that had taken three centuries to accomplish across the Channel.

William the Conqueror, coming as he did from a great French fief, proceeded to turn England into a thoroughly feudal state. His 5000 knights insisted on their share of the conquest, and, as each county was conquered, William was forced to allot fiefs in each district. Hence, unlike France, England was not divided up among the great nobles into compact blocks. Instead, the fiefs were scattered about, a factor which turned out to be greatly to the crown's advantage. Moreover, both the king and his vassals built many great castles, so that by 1150 there were about 1200 castles in England. Yet William had no intention of letting his vassals get the upper hand, and in 1086 he had all the landholders of England take a great oath to him as liege lord. Thus in various ways William the Conqueror built up a strongly feudal English state.

Nevertheless, not even William the Conqueror was able to prevent a baronial revolt. Such revolts continued after his death (1087), and finally resulted in the anarchy of the weak Stephen (1135-54).

The Greatest English King In 1154, however, there came to the throne perhaps the greatest of all English kings, Henry II, the great-grandson of the Conqueror. Henry spent much of his reign fighting against his four sons, particularly Richard and John, whom Philip Augustus was not at all averse to turning against their father. On the other hand Henry, who ruled England, a section of Ireland, and about two-thirds of France, refused to trust his sons with a share in his so-called "Angevin Empire."

Henry II

The dust of Henry's wars of empire has long since settled down, until the very names of his battles no longer evoke recognition. What is far more important is the way in which Henry II increased the royal power in England. Under Stephen it had reached its lowest point since the Conquest. Now Henry set about revitalizing it—and he was just the man for the job. The royal energy and restlessness knew no bounds, as a distracted member of Henry's entourage well knew:

If the king has promised to remain in a place for a day—and particularly if he has announced his intention publicly by the mouth of a herald—he is sure to upset all the arrangements by departing early in the morning. As a result you see men dashing around as if they were mad, beating their packhorses, running their carts into one another—in short giving a lively imitation of Hell. If, on the other hand, the king orders an early start, he is certain to change his mind, and you can take it for granted that he will sleep until midday. Then you will see the packhorses loaded and waiting, the carts prepared, the courtiers dozing, traders fretting, and everyone grumbling. People go to ask the maids and the doorkeepers what the king's plans are, for they are the only people likely to know the secrets of the court. Many a time when the king was sleeping a message would be passed from his chamber about the city or town he planned to go to, and though there was nothing certain about it, it would rouse us all up. After hanging about aimlessly for so long we would be comforted by the prospect of good lodgings. . . . But when our courtiers had gone ahead almost the whole day's ride, the king would turn aside to some other place where he had, it might be, just a single house with accommodation for himself and no one else. I hardly dare say it, but I believe that in truth he took a delight in seeing what a fix he put us in. After wandering some three or four miles in an unknown wood, and often in the dark, we thought ourselves lucky if we stumbled upon some filthy little hovel. There was often a sharp and bitter argument about a mere hut, and swords were drawn for possession of lodgings that pigs would have shunned.

Henry was a man of no mean education. He could speak French and Latin and could understand most of the other west European languages. Foreign scholars were encouraged to come to his court and, as his secretary wrote, "with the king of England there is school every day, constant conversation of the best scholars and discussion of questions."

Henry undertook a far-reaching reorganization of his realm, and in this project his legal reforms were most important. The royal courts were given wide jurisdiction, and the jury system was instituted. Previously the only way to get a criminal case into court had been through a formal accusation by the injured party or by one of his friends or relatives. Naturally people were slow to accuse a powerful man, and if a murdered man had no friends or relatives an accusation could not be brought. Henry changed this system by ordering twelve men in each district to appear before the royal justices to state on oath whether anyone in their district was suspected of committing a crime. What happened next is best told in the words of Henry's law.

And whoever is found by the oath of the aforesaid men to have been accused or publicly known as a robber or murderer or thief, or as a receiver of them since the lord king has been king, shall be seized; and he shall go to the ordeal of water.... The lord king also wills that those who make their law [by ordeal] and are cleared by the law, if they are of very bad reputation, being publicly and shamefully denounced by the testimony of many lawful men, shall abjure the lands of the king ... and thenceforth not return to England, except at the mercy of the lord king; they shall there be outlaws and shall be seized as outlaws if they return.

It should be noted that all the property of a convicted criminal was confiscated; hence Henry's legal innovations brought welcome revenue to the Crown. Nevertheless, Henry had shown his suspicion of trial by ordeal a half century before the Church moved to abolish it in 1215.

By this so-called "jury of presentment", the ancestor of our grand jury, and by other reforms, Henry II gave the government the initiative in both criminal and civil prosecutions. Here we have the beginning of what is called *common law*, that is, the law common to all men and applied throughout the whole kingdom. Now, instead of the great variety of feudal custom being applied helter skelter, the judges in the expanded royal courts began to apply the king's law, and this common law began to supplant all other varieties of law. The birth and development of the common law is one of the proudest accomplishments traceable to Henry II and England in the Middle Ages.

It was in these fateful and formative years that the English-speaking peoples began to devise methods of determining legal disputes which survive in substance to this day. A man can only be accused of a civil or criminal offence which is clearly defined and known to the law. The judge is an umpire. He adjudicates on such evidence as the parties choose to produce. Witnesses must testify in public and on oath. They are examined and cross-examined, not by the judge, but by the litigants themselves or their legally qualified and privately hired representatives. The truth of their testimony is weighed not by the judge but by twelve good men and true, and it is only when this jury has determined the facts that the judge is empowered to impose sentence, punishment, or penalty according to law. All [this] might seem very obvious, even a platitude, until one contemplates the alternative system which still dominates a large portion of the world. Under Roman law, and systems derived from it, a trial in those turbulent centuries, and in some countries even today, is often an inquisition. The judge makes his own investigation into the civil wrong or the public crime, and such investigation is largely uncontrolled. The suspect can be interrogated in private. He must answer all questions put to him. His right to be represented by a legal adviser is restricted. The witnesses against him can testify in secret and in his absence. And only when these processes have been accomplished is the accusation or charge against him formulated and published. Thus often arises secret intimidation, enforced confessions, torture, and blackmailed pleas of guilty. These sinister dangers were extinguished from the Common Law of England more than six centuries ago. By the time Henry II's great-grandson, Edward I, had died English criminal and civil procedure had settled into a mould and tradition which in the mass govern the English-speaking peoples today. In all claims and disputes, whether they concerned the grazing lands of the Middle West, the oilfields of California, the sheep-runs and gold-mines of Australia, or the territorial rights of the Maoris, these rules have obtained, at any rate in theory, according to the procedure and mode of trial evolved by the English Common Law.[1]

Great administrator that he was, Henry II was anything but a likeable person. He and his two Angevin successors, Richard and John, bore down hard on their English nobles, collecting feudal dues up to the limit and doing everything in their power to increase the royal might and to clip the barons' wings.

Henry wore himself out at fifty-six and died in 1189, an embittered man because his sons and his queen were all in revolt against him. The new king, Richard (nicknamed *Coeur-de-Lion*), was a magnificent warrior, but he was an ineffective ruler whose absence on the Third Crusade and various French campaigns kept him away from England for all but six months of his ten-year reign. Indeed, it is the highest tribute to the administrative foundations laid by Henry II that the realm could continue along peacefully while this son, "the least English of all the kings of England," sought military glory abroad.

When Richard died in 1199 from a wound inflicted during a petty siege in France, his younger brother John ascended the throne. John was a conscientious ruler, but a hateful man and an unsuccessful warrior. Consequently his reputation has been permanently blackened. Consideration of this most interesting king must, however, be postponed until the discussion of the transition from medieval to modern times in the next Unit.

The Collapse of a Perfect Pyramid   When the Carolingian Empire was formally split up in 843 by the Treaty of Verdun (see page 307), one of the three remnants emerged as the East Frankish kingdom. Rebellions by the German nobles, along with the Magyar invasions, resulted in a weak kingdom loosely ruled by the duke of Saxony.

Finally in 962 the Saxon king Otto I was crowned emperor by the pope —but emperor of what? The central German states formed the nucleus of Otto's power, and later emperors were to extend their control over northern Italy until in the 13th century their lands came to be termed the "Holy Roman Empire." Since, however, such a label is awkward and misleading, we would do better to call this shifting group of territories the German Empire.

Otto I forced the pope to recognize the emperor's right to approve papal elections, even though this so-called empire did not result in permanent Saxon control of Italy, nor did it extend farther west than Lothair's old middle kingdom. But in the 11th century the German kings, unfortunately for their power, became involved in a long and bitter dispute with the Church. Although Henry III was able to keep churchmen co-operating

---

[1]Winston S. Churchill, *A History of the English-Speaking Peoples,* Vol. I, *The Birth of Britain* (McClelland and Stewart, Toronto, 1956), pp. 222-223.

1066

1200

4
1346

3
1300

1415
5

Y. MOULD.

with him, Henry IV, as will be seen in the next chapter, broke decisively with Rome.

While the contest between Empire and Papacy was going on, the German nobility were able to take advantage of the monarch's preoccupation to increase their collective strength against him. The struggle between those who wished to be independent in church and feudal matters under the papacy (*Guelfs*) and those who wanted to control the Church and build up a strong imperial government (*Ghibellines*) rent both Germany and Italy. At length the German bishops so wanted peace that they were willing to support the candidacy of a man who looked as if he might be able to end the struggle. This man was Frederick I, named, because of his red beard, Barbarossa. Although his mother had been a Guelf, Barbarossa belonged to the Hohenstaufen family whose members were the core of the Ghibellines. Thus there appeared to be a good prospect for conciliation, and Barbarossa was unanimously elected king of Germany.

Frederick Barbarossa

Barbarossa proved to be the last of the great medieval emperors. Like Henry II of England, he set about restoring the rights of the monarchy—

Here we may trace the development of armour in the Middle Ages. Norman armour at the time of the English Conquest (1) consisted mainly of the knee-length chain mail shirt or *hauberk*, split front and back to make riding easier. Mail was sometimes also used to protect the legs (although this warrior has quilting and cross garters instead) and the head, which had a further safeguard in the simple helmet with a long *nasal* to protect the nose. The large, elongated shields, decorated with simple non-heraldic patterns, were ideal for horsemen, and the swordbelt could be worn beneath the hauberk when a slit was provided for the hilt. Mail was the basic armour for the next 200 years. In 1200 (2) mail still covered the body, including the hands, but the helmet was blunter in design with a protective band framing the face, and the shield became smaller. Mail stockings were strapped to either the whole leg or the front of the leg. The knight seen here is wearing a *surcoat*, a garment now adopted to protect armour from both sun and dampness, emblazoned with one of the many versions of the Cross worn by crusaders.

By 1300 (3) mail was giving additional protection to vulnerable places. There were solid or plate armour knee caps (*polayns*) and small defensive shoulder shields (*ailettes*), while the head was completely enclosed, first in mail, then in a type of helmet called the *great helm* which was rounded so as to ward off a direct blow. The surcoat was cut away at the front, and the sword was worn almost directly in the middle of the body—held in this position by a most ingenious arrangement of straps. Heraldry had become an intricate science of symbols and markings, and each knight had his own special crest. Where does the knight's insignia appear here? By the time of Crécy (4), plate covered the mail over much of the body. Because more fighting was now done on foot, the surcoat had shortened to the less cumbersome *gipon*. Plate covered the front of arms, shoulders, chest, thighs, legs, and feet, and the sword reverted to its old position at the hip. About this time the helmet began to acquire a movable visor, which could be lifted for better sight and breathing. The war-horse, here sporting a surcoat with its owner's coat of arms, was specially bred to carry the great weight of man and armour. This weight was increased over the next century as the horse acquired very extensive armour of its own. In another half century (5) plate armour was virtually complete, with scant mail armour left. Plate armour was hinged outside and buckled inside so that straps could not be cut away. The *great basinet* helmet took on several different shapes, but had one great disadvantage: its full weight rested squarely on the shoulders, and prevented the head from turning independently of the body. The surcoat was no longer in use, though it returned later in the century as the *tabard*. Note the heavy fighting mace.

Because a fine suit of mail such as (2) would cost about $12,000 at today's values, it would be hopelessly beyond the reach of all but the very wealthy. Common soldiers, archers, and pikemen were protected only by hardened leather or quilting, with perhaps a metal headpiece. It is an eloquent commentary on the value of medieval armour that in the celebrated Bayeux Tapestry depicting the Norman Conquest, unarmoured soldiers may be seen stripping mail shirts from the fallen while the battle rages on!

rights shattered by the preceding century of conflict with the Church—and extending his rule over wider territories. The power of the Guelfs in Germany was finally broken, but Barbarossa had trouble trying to build up a compact royal domain which he could use as a base for his power. Apparently he thought that Italy might serve as this foundation for his strength, and when the Roman aristocracy offered him the imperial throne in 1155 he spoke bluntly to them:

You have related the ancient renown of your city, and have extolled the ancient state of your sacred republic. Agreed! Agreed! In the words of your celebrated author, "there was once virtue in this republic". "Once", I say. Would that we could truthfully say "now"! But Rome has experienced the vicissitudes of time. ... First, as is known to all, the vigour of your nobility was transplanted to the royal city of the East.... Then came the Franks ... who took away by force the remnants of your freedom. ... We have turned over in our minds the deeds of modern emperors, considered how our sacred predecessors, Charles and Otto, wrested your City with the lands of Italy from the Greeks and Lombards, and brought it within the frontiers of the Frankish realm, not as a gift from alien hands but as a conquest won by their own valour.... I am the lawful possessor. ...

Quite clearly Emperor Barbarossa aimed at restoring the Carolingian and Ottonian Empires.

Nevertheless, six separate expeditions into Italy did not finally re-establish the imperial power there. In fact, had Barbarossa not become so involved in Italy he might have been able to consolidate his power in Germany and check the disintegration of the Empire. Unfortunately, however, Frederick was obsessed with his own personal version of feudalism, a version which involved a perfect pyramid with himself at the top and all the princes of Germany as his vassals. In order to put his theories into practice, he attempted (and with considerable success) to make all German bishops and princes his vassals. But instead of increasing his own power by adding all confiscated lands to his royal domain, as English and French monarchs were doing, he parcelled them out again to other princes. When, for instance, he seized the lands of his powerful cousin Henry the Lion, duke of Saxony and Bavaria (Henry had refused aid to an Italian expedition), Barbarossa did not take them over himself but divided them among his vassals. The result was that the princes' power increased while the emperor's decreased. Even so, as long as Barbarossa lived he was able to control Germany.

The great emperor was drowned in 1190 while participating in the Third Crusade. By then the German emperor, while technically overlord of the powerful German feudal princes, had few resources of his own inside Germany. If those outside Germany should ever be lost, the emperor would be practically powerless. When this actually came to pass under Frederick II the monarchy lost almost entirely its control over the German princes, and within a few years the shadowy German Empire had crumbled into a conflicting maze of independent states.

Three Feudal Monarchies The history of the three feudal monarchies that dominated medieval Europe until the 13th century shows that feudalism could lead either to a strongly centralized state under a vigorous king-overlord (as in France and England), or to a decentralized and localized state in which the vassals were victorious (as in Germany). A strong monarchy developed more rapidly in England than in France, and because the German kings did not succeed in organizing a compact block of royal estates on which to base a hereditary monarchy, Germany lagged perhaps 500 years behind France. Of the three feudal monarchies that attempted to build up their power the English was the most successful, its strength born of a line of brilliant kings who made feudalism serve them first and their barons second.

5/SUMMARY AND CONCLUSION

Charlemagne's empire, with its primitive central government ruling over a mainly agricultural society, did not long outlive its founder. New invaders —Vikings, Magyars, and Moslems—attacked it from without, and it was broken up from within by the competing successors of Charlemagne, who divided it into three parts.

With the old empire gone, loyalty to local strong men became important for protection, a loyalty that resulted in a different organization of society under feudalism, or the "feudal system." The man who was wealthy enough to own a horse and buy equipment and armour became a knight. If he had only enough land to support himself he would need protection, and accordingly would surrender his land to some great lord in return for this protection. The lord would then give back the land in the form of a lifetime benefice, and would receive in exchange the vassal knight's military service.

On the basis of this relationship between landholding and military service a social pyramid was erected, each man in it owing service to his direct overlord. In time this relationship became a hereditary one, with the lifetime benefice descending to the heir of a deceased vassal. This was the general pattern by the 10th century.

Knights were free to fight only because their estates were worked by peasants. Although a manor with its three-field system of cultivation provided plenty of services due to a lord, the justice and government owed by a lord to his serfs depended mostly upon his own whims. The serf, on the other hand, owed tax upon tax, and any complaint was likely to be met with an exhortation from either his lord or his Church to accept the station of life in which God had placed him.

Meanwhile, great political states were in the making. The French monarchy of the Capetians became hereditary, and by the end of the 12th century had established the kingdom which was to become modern France. The Norman conquerors of England founded the strongest and most efficient of all the feudal monarchies. However, the Saxon kings, though attempting to revive the memory of Rome in the German Empire, were not able to build up a hereditary monarchy which could overrule rebellious princes.

The formative period of the Middle Ages from the 9th to the 13th century thus saw a new organization of society linked with a new tradition of monarchy, as the quotation at the head of this chapter suggests. This evolution of feudalism was matched by the growing power of the Church and by new traditions in learning. We must examine these in the next chapter.

SOURCE READINGS

(a)

CHARLEMAGNE (768-814) in his Decree concerning Villas laid down detailed instructions as to what he expected from the stewards on his estates. The Emperor took a keen personal interest in checking the efficiency of these superintendents of his estates, which were combined farms and factories.

34. They must provide with the greatest care, that whatever is prepared or made with the hands, that is, lard, smoked meat, salt meat, partially salted meat, wine, vinegar, mulberry wine, cooked wine, garns, mustard, cheese, butter, malt, beer, mead, honey, wax, flour, all should be prepared and made with the greatest cleanliness.

40. That each steward on each of our domains shall always have, for the sake of ornament, swans, peacocks, pheasants, ducks, pigeons, partridges, turtle-doves.

42. That in each of our estates, the chambers shall be provided with counterpanes, cushions, pillows, bedclothes, coverings for the tables and benches; vessels of brass, lead, iron and wood; andirons, chains, pot-hooks, adzes, axes, augers, cutlasses and all other kinds of tools, so that it shall never be necessary to go elsewhere for them, or to borrow them. And the weapons, which are carried against the enemy, shall be well cared for, so as to keep them in good condition; and when they are brought back they shall be placed in the chamber.

43. For our women's work they are to give at the proper time, as has been ordered, the materials, that is the linen, wool, woad, vermilion, madder, wool-combs, teasels, soap, grease, vessels and the other objects which are necessary.

62. That each steward shall make an annual statement of all our income: an account of our lands cultivated by the oxen which our ploughmen drive and of our lands which the tenants of farms ought to plough; an account of the pigs, of the rents, of the obligations and fines; of the game taken in our forests without our permission; of the various compositions; of the mills, of the forest, of the fields, of the bridges, and ships; of the free-men and the hundreds who are under obligations to our treasury; of markets, vineyards, and those who owe wine to us; . . . of the fish-ponds; of the hides, skins, and horns; of the honey, wax; of the fat, tallow and soap . . . of the hens and eggs . . . the number of fishermen, smiths, sword-makers, and shoe-makers . . . of the forges and mines . . . of the colts and fillies; they shall make all these known to us, set forth separately and in order, at Christmas, in order that we may know what and how much of each thing we have.

S. C. Easton and H. Wieruszowski, *The Era of Charlemagne: Frankish State and Society* (Anvil Books, Van Nostrand, 1961), pp. 133-134, 132-133.

(b)

This performance of HOMAGE, FEALTY, and INVESTITURE took place in 1127 at the court of William, count of Flanders, on the death of his father Charles.

Through the whole remaining part of the day those who had been previously enfeoffed by the most pious count Charles, did homage to the new count, taking up now again their fiefs and offices and whatever they had before rightfully and legitimately obtained. On Thursday, the seventh of April, homages were again made to the count, being completed in the following order of faith and security:

First they did their homage thus. The count asked if he was willing to become completely his man, and the other replied, "I am willing"; and with clasped hands, between the hands of the count, they were bound together by a kiss. Secondly, he who had done homage gave his fealty to the representative of the count in these words, "I promise on my faith that I will in future be faithful to Count William, and will observe my homage to him completely, against all persons, in good faith, and without deceit." Thirdly, he took his oath to this upon the relics of the saints. Afterwards, with a little rod which the count held in his hand, he gave investitures to all who by this agreement had given their security and homage and accompanying oath.

T. C. Mendenhall, B. D. Henning, and A. S. Foord, *Ideas and Institutions in European History, 800-1715: Select Problems in Historical Interpretation* (Henry Holt, 1948), p. 11.

(c)

RATHERIUS (890-974) of Liège was a 10th century bishop of Verona, Italy, a gifted writer and earnest pastor. He was deposed and imprisoned for several years by King Hugh of Italy, and while in a dungeon wrote a series of sermons which included the following extract. Note the surprising suggestion in the sixth paragraph that within five generations a family might rise from serfdom to officialdom.

Do you wish to be a Christian, a good Christian among the many [who call themselves] Christians . . .? Be a worker, not only a fair but also a laborious one, content with what you have, cheating no one, hurting nobody, insulting nobody, besmirching no man. Fear God, pray to the saints, haunt the church, honour the priests, offer to God the tithe and the first fruit of your labour, give alms according to your means, cherish your wife . . . raise your children in the fear of God, visit the sick, bury the dead. . . .

Are you a soldier . . .? If you cannot gain a [sufficient] salary in the service, seek for food through the work of your hands, and shun pillage, beware of murder, avoid sacrilege. . . .

Are you a physician . . .? While you cure others of bodily affections, do enough to cure yourself in the health of your behaviour. . . . Charitably assist the poor without charge, for the love of God, and faithfully administer medicine to the rich for the compensation you receive. . . .

Are you a merchant . . .? Do consider what a dangerous profession for your soul you have chosen. . . . That you are enthralled by vice is shown by the word by which you hear yourself called by many, *lover of money*. . . . You have fraudulently hoarded, you have loved usury and surplus. . . . If those from whom you have taken something are present . . . even if you cannot, or do not wish to make a fourfold restitution, at least return the equivalent. . . .

Are you a procurator, a collector . . . an important personage . . . a customs officer, or a man entrusted with any other public function? Beware above all of cruelty, then of cheating, hence of avidity and grabbing, as well as of drunkenness.

Lastly, [beware of] covering up some theft or of protecting a crime which you ought to prosecute, whether you have been enticed by some bribe or deceived by pity. . . .

Are you a master, or, as it is more commonly said by many, a lord . . . ? Whoever you are, who wrongly boast of proud blood, remember that the entire race of men on the earth springs from similar origins. . . . Let us consider a son of a top-rank official, whose grandfather be known to have been a judge, the great-grandfather a tribune or a minor official, and the great-great-grandfather a knight. Who will remember, after all this, whether the father of the knight was a petty merchant or a painter, a bath attendant or a fowler, a fishmonger or a potter, a tailor or a sausage-maker, a muleteer or a driver of animals, or, lastly, a horseman or a peasant, a freeman or a serf . . . ?

Are you a serf? Do not be sad; if you have faithfully served your lord, you will be a freedman of the Lord of all men.

R. S. Lopez, *The Tenth Century: How Dark the Dark Ages?* (Source Problems in World Civilization, Rinehart, 1959), p. 34.

(d)

The two paragraphs that follow are a modern prose translation of a few lines from an anonymous Middle English poem written about 1394. Entitled *Pierce The Ploughman's Crede*, it imitates the theme of William Langland's *The Vision of William concerning Piers the Plowman* (see page 464).

As I went by the way, weeping for sorrow, I saw a poor man hanging on to the plough. His coat was of a coarse stuff which was called cary; his hood was full of holes and his hair stuck out of it. As he trod the soil his toes peered out of his worn shoes with their thick soles; his hose hung about his hocks on all sides, and he was all bedaubed with mud as he followed the plough. He had two mittens, made scantily of rough stuff, with worn-out fingers and thick with muck. This man bemired himself in the mud almost to the ankle, and drove four heifers before him that had become feeble, so that men might count their every rib so sorry looking they were.

His wife walked beside him with a long goad in a shortened cote-hardy looped up full high and wrapped in a winnowing-sheet to protect her from the weather. She went barefoot on the ice so that the blood flowed. And at the end of the row lay a little crumb-bowl, and therein a little child covered with rags, and two two-year olds were on the other side, and they all sang one song that was pitiful to hear: they all cried the same cry—a miserable note. The poor man sighed sorely, and said "Children be still!"

H. S. Bennett, *Life on the English Manor: A Study of Peasant Conditions, 1150-1400* (Cambridge University Press, 1948), pp. 185-186, slightly adapted.

FURTHER READING

1. THE CAROLINGIAN EMPIRE AND FEUDALISM

    ARBMAN, H., *The Vikings* (Thames and Hudson, 1961)

    BLOCH, M., *Feudal Society* (Routledge, 1961)

  * BRONSTED, J., *The Vikings* (Penguin, 1960)

    COULBORN, R., editor, *Feudalism in History* (Princeton, 1956)

† DAVIS, W. S., *Life on a Mediaeval Barony* (Harper, 1923)

  * EASTON, S. C. and WIERUSZOWSKI, H., *The Era of Charlemagne* (Anvil, 1961)

    FICHTENAU, H., *The Carolingian Empire* (Blackwell, 1958)

  * GANSHOF, F. L., *Feudalism* (Torchbook, 1961)

  * HAVIGHURST, F., editor, *The Pirenne Thesis: Analysis, Criticism, and Revision* (Heath, 1958)

  * PIRENNE, H., *Mohammed and Charlemagne* (Meridian, 1957)

† * STEPHENSON, C., *Mediaeval Feudalism* (Cornell, 1956)

    WHITE, L., *Medieval Technology and Social Change* (Oxford, 1962)

2. TRADE AND MANORIALISM

* ADELSON, H. L., *Medieval Commerce* (Anchor, 1962)
† * BENNETT, H. S., *Life on the English Manor: A Study of Peasant Conditions, 1150-1400* (Cambridge, 1960)
CLAPHAM, J. H. and POWER, E., editors, *The Cambridge Economic History of Europe*, Vol. I, *The Agrarian Life of the Middle Ages* (Cambridge, 1953)
† * COULTON, G. G., *Medieval Village, Manor, and Monastery* (Torchbook, 1960)
LATOUCHE, R., *The Birth of Western Economy* (Methuen, 1961)
LEWIS, A. R., *Naval Power and Trade in the Mediterranean, 500-1100* (Princeton, 1951)
———, *The Northern Seas, 300-1100* (Princeton, 1958)
† NEILSON, N., *Medieval Agrarian Economy* (Holt, 1936)
† * PAINTER, S., *Medieval Society* (Cornell, 1951)
* PIRENNE, H., *Economic and Social History of Medieval Europe* (Harvest, 1956)
† * POWER, E., *Medieval People* (Anchor, 1954)

3. THE FEUDAL MONARCHIES

BARRACLOUGH, G., *The Origins of Modern Germany* (Blackwell, 1947)
† * CAM, H., *England before Elizabeth* (Torchbook, 1960)
† * CHURCHILL, W. S., *A History of the English-Speaking Peoples*, Vol. I, *The Birth of Britain* (Bantam, 1963)
† EVANS, J., *Life in Medieval France* (Phaidon, 1957)
FAWTIER, R., *The Capetian Kings of France, 987-1328* (Macmillan, 1960)
† * KELLY, A., *Eleanor of Aquitaine and the Four Kings* (Vintage, 1957)
† * PAINTER, S., *The Rise of the Feudal Monarchies* (Cornell, 1951)
SAYLES, G. O., *The Medieval Foundations of England* (Methuen, 1952)
* STENTON, D. M., *English Society in the Early Middle Ages* (Penguin, 1962)
* WHITELOCK, D., *The Beginnings of English Society* (Penguin, 1952)

11. The Church and Learning

Our books have informed us that the pre-eminence in chivalry and learning once belonged to Greece. Then chivalry passed to Rome, together with that highest learning which now has come to France. God grant that it may be cherished here, that the honour which has taken refuge with us may never depart from France. God had awarded it as another's share, but of Greeks and Romans no more is heard; their fame is passed, and their glowing ash is dead.
CHRÉTIEN DE TROYES, IN THE YEAR 1170

Even though the early Middle Ages witnessed the rise of political states, society continued to be more international than national. And above all states stood the Church. Where lines of authority intersected, the tempers of laymen and churchmen might rub raw.

Of deeper and more lasting importance than the role of the Church in political controversy was her function as an agent of culture. The medieval Church faithfully shielded the light of learning after the fall of Rome and during what are sometimes called the "Dark Ages".

1/THE MEDIEVAL CHURCH

The Cluniac Order

During the formative centuries of the Middle Ages the Church was increasing in prestige and influence. Part of the reason for this was that the 10th century Church experienced a new reforming and purifying movement from within. This movement began when duke William of Aquitaine founded the monastery of Cluny as a new kind of establishment which owed no lord feudal services in return for its monastic lands. Consequently Cluny inherited no feudal entanglements, and under a series of extremely able abbots the Benedictine Rule, which was being loosely enforced in other monasteries, was tightened up. There was, however, one important innovation: the Cluniac Order was to have only one abbot, and any other monasteries which, they hoped, might later follow their Rule, were to be run by priors under the abbot of Cluny. The Cluniac movement was a great success, with many Benedictine monasteries accepting its Rule and its example of firm discipline. In fact within a century regular clergy also had caught the Cluniac reforming zeal and were looking to the papacy for leadership.

The German emperor, Henry III (1039-56) gladly supported the papal beginnings at reform, particularly the attempts to prohibit certain pressure groups from forcing the election of their own candidate as pope. To this end the College of Cardinals was founded to conduct papal elections without being dictated to by the populace of Rome or the feudal kings and emperors. Henry III certainly had no intention of having his own power limited by this move, but succeeding emperors were to be curbed by it. Various other reforms also were carried out, including the enforcement of clerical *celibacy* (the non-marriage of clergy) and the prohibition of *simony* (the purchase of church offices). The Church also sought to control the appointment of its personnel, who, too often, had been chosen by kings and emperors and were later merely ratified by Rome. One way to exert this control was to insist that a new bishop could be invested with his power only in a ceremony at Rome.

The Road to Canossa Hildebrand, who became Pope Gregory VII in 1073, was one of the Church's most vigorous reformers. As pope he determined to reform and rule, to purify the Church, and to make himself the overlord, under God, of both kings and emperors. The high claims that Pope Gregory put forward were clearly expressed in a statement of principles drawn up about 1075:

1. That the Roman church was founded by God alone.
2. That only the Roman pontiff is rightly called universal.
3. That he alone can depose or reinstate bishops.

4. That his legate, even though of inferior rank, has precedence over all bishops in a council; and he can give sentence of deposition against them.

9. That all princes shall kiss the feet of the pope and the pope only.

11. That his name is unique in the world.

12. That he is allowed to depose emperors.

18. That his sentence can be annulled by no one and he alone can annul the sentences of all others.

19. That he should be judged by no one.

22. That the Roman church has never erred nor, as scripture testifies, will ever err.

24. That by his command and licence subjects can accuse their rulers.

26. That one shall not be considered a Catholic who does not agree with the Roman church.

27. That he can absolve subjects from their oaths of fidelity to iniquitous rulers.

The Investiture Controversy

It was on this matter of choosing bishops that the battle was joined. Pope Gregory was determined that he (through his clergy)—and not the lay ruler—should invest new church officials with the symbols of their office, such as the bishop's ring and staff. The quarrel over investiture soon turned into a full-blown attempt by the papacy to take over the control of the German church from the emperor Henry IV, and the sharp letters exchanged between pope and emperor may be read at the end of this chapter. Henry persuaded the German bishops to renounce their obedience to the pope, who in turn deposed Henry, releasing all his subjects from their allegiance to the emperor and cutting him off by excommunication from the Church. This early victory of Gregory's culminated in 1077 when Henry journeyed to Italy, and, according to legend, stood for three days barefoot in the snow outside the castle at Canossa where the pope was lodged until Gregory finally forgave him. But the great pope himself was eventually to be driven from Rome. His famous last words were typical: "I have loved justice and hated iniquity; therefore I die in exile."

The Book of Kells is an illuminated copy of the Gospels in Latin, produced in the monastery of Kells in Ireland about A.D. 800. It is said to be the finest extant example of early Christian art of its kind, and among its rich illustrations is this page bearing the symbols of the four Gospel writers: the man (Matthew), the lion (Mark), the ox (Luke), and the eagle (John). Such incredibly delicate illumination was done by the monks "for the love of God".

It was only a generation later in 1122 that the Concordat of Worms settled the Investiture Controversy by providing for a double ceremony of investiture—by the secular ruler for lands and temporal privileges, by the ecclesiastical superior for spiritual authority. But this compromise did not end the contest between Empire and Papacy: it merely shifted it to other grounds.

"Dieu le veut!" While Gregory VII was involved in his bitter quarrel with Henry IV, the Byzantine emperor begged the pope for help against the Moslem Turks. Gregory was in no position to launch a crusade against the Moslems, but his successor, Pope Urban II (1088-99), not only acted on the idea but enlarged upon it as well. Urban's fervent desire was to launch a great expedition and wrench from the infidels that Holy Land of Palestine where Jesus Christ was born, lived, and was crucified. Accordingly, on November 27, 1095, Pope Urban preached the crusade in a highly effective oration (some extracts from which can be read on pages 355-356) before the nobles and clergy assembled at the Council of Clermont. Aroused and inspired by Urban's sermon, the great crowd responded with a shout, *"Dieu le veut!"*—"God wills it!"

The First Crusade The motives for this First Crusade (1096-99) have been the subject of much debate. Doubtless Urban was sincere in his desire to recover the Holy Land for Christendom, but he must also have known that a successful crusade would greatly increase the prestige and influence of the Church. Moreover, in his anxiety to curb feudal warfare he had reissued the Peace and Truce of God (a ban on war against women, children, or clergymen, and on fighting over week-ends), though he must have realized that to try to restrict the fighting of feudal nobles without providing some outlet for their warlike spirits was a vain hope. Now, however, there could be unlimited warfare—against the enemies of Christ. Finally, Urban must have cherished the idea of healing the recent schism between the Eastern and Western churches, a split that had occurred in 1054 when the Patriarch of Constantinople had rejected certain Roman Catholic doctrines as well as the claim of papal supremacy.

Thus, despite its appearance of spontaneity, the First Crusade was launched after careful thought and plenty of advance publicity. In the last analysis, the Crusades were a manifestation of a general 11th century European expansionism: England and Sicily had both been conquered by Normans; in Spain there was a steady southward push against the Moslems; the Saxons had moved eastward across the Elbe; and Christianity had been taken to Greenland and Iceland.

The Sign of the Cross The Crusade quickly got out of hand. The emotional enthusiasm of wandering preachers like Peter the Hermit

For over one hundred and fifty years this great Crusader castle "stuck like a bone," wrote a Moslem, "in the very throat of the Saracens." Krak des Chevaliers was built by Frankish knights of the 12th and 13th centuries in what is today northern Syria, with the purpose of protecting the Latin Kingdom of Jerusalem. Majestically executed both inside and out, Krak was one of the first castles to be concentrically designed—a design which influenced the plan of Edward I's concentric castles in Wales. Today the imposing fortress which once rang with the commands of mailed knights stands empty and silent, its towers keeping an eerie vigil over the stark, torrid wastelands where centuries before it was the chief Christian bastion of the Holy Land.

swept many simple people off their feet, and eagerly they flocked to take up the cross. Here is how one chronicler of the time describes the earliest crusaders.

The French at that time suffered from famine; bad harvests year after year had raised the price of corn [grain] to a great height, and avaricious merchants according to their wont took advantage of the general misery. . . . Suddenly, the cry of Crusade, sounding everywhere at once, broke the locks and bars that closed the granaries. Provisions that had been beyond price when every one stayed where they were, sold for nothing when every one was stirring and anxious to depart. Famine disappeared and was replaced by abundance. As every one hastened to take the road of God each hurried to change into money everything that was not of use for the journey, and the price was fixed not by the seller but by the buyer. It was strange and marvellous to see every one buying dear and selling cheap: everything that was needed for the journey was very costly, but the rest was sold for nothing. . . . It touched the heart to see these poor crusaders shoeing their oxen as if they had been horses, and harnessing them to two-wheeled carts on which they put their small belongings and their little children. At each castle and each city they passed on the road they stretched out their hands and asked if they had not yet reached that Jerusalem which all were seeking.

The remnant of this ignorant and unorganized rabble who did manage to reach Constantinople waited there for the main body of crusaders, the feudal knights and men-at-arms under the command of the barons of France. By the autumn of 1096 this force had started east, and by the next spring had crossed into Asia. Mutual distrust began to eat into the feudal chieftains from Europe, and, moreover, the Byzantine emperor and his troops soon abandoned the westerners in Palestine and went home. Constantinople thereby handed over all leadership of the Crusade to Rome.

In June 1098 Antioch finally fell to the crusaders. But even in victory there was dissension: the northern French attributed the fall to their superior generalship, and the southern French to the discovery by one of their followers of a sacred relic—the very lance that had pierced Christ's side as he hung on the cross. So bitter did the quarrel become that the question of the authenticity of the relic and the honesty of its discoverer, Peter Bartholomew, were put to the test.

The leaders and the people to the number of fifty thousand came together; the priests were there also with bare feet, clothed in ecclesiastical garments. The invocation was made: "If Omnipotent God has spoken to this man face to face, and the blessed Andrew has shown him our Lord's Lance while he was keeping his vigil, let him go through the fire unharmed. But if it is false, let him be burned, together with the Lance which he is to carry in his hand." And all responded on bended knees, "Amen." The fire was growing so hot that the flames shot up thirty cubits into the air, and scarcely anyone dared approach it. Then Peter Bartholomew, clothed only in his tunic, and kneeling before the bishop of Albar, called God to witness that he had seen him face to face on the cross. . . . Then, when the bishop had placed the Lance in his hand, he kneeled and made the sign of the cross, and entered the fire with the Lance, firm and unterrified. For an instant's time he paused in the midst of the flames, and then by the grace of God passed through. . . . But when Peter emerged from the fire so that neither his tunic was burned nor even the thin cloth with which the Lance was wrapped up had shown any sign of damage, the whole people . . . threw themselves upon him and dragged him to the ground and trampled on him, each one wishing to get a piece of his garment, and each thinking him near some one else. . . . Peter had died on the spot, as we believe, had not Raymond Pelet, a brave and noble soldier, broken through the wild crowd with a band of friends and rescued him at the peril of their lives.

Despite such disputes the crusaders did at last capture Jerusalem in 1099 with a force of 2000-3000 knights and 8000-12,000 infantry. Apparently no action against the Moslems was too revolting so long as it contributed to the glory of Christian arms, and a participant has left us the following description of how the Holy City was taken.

On the top of Solomon's Temple, to which they [the Moslems] had climbed in fleeing, many were shot to death with arrows and cast down headlong from the roof. Within the Temple about ten thousand were beheaded. If you had been there, your feet would have been stained up to the ankles with the blood of the slain. What more shall I tell? Not one of them was allowed to live. They did not spare the women and children.

After the Moslem conquest of Spain in the 8th century a small pocket of Christians in the north-west set about trying to reconquer land from the Moors, and eventually established the kingdom of Leon. By the 10th century they had expanded south-east, building castles to consolidate their hard-won holdings until the new territory was known as the "land of castles", Castile.

In the 11th century Alfonso VI (1072-1109), king of Castile and Leon, took up the "perpetual crusade"—as the Christian reconquest of Spain was called—aided by a Castilian noble, Rodrigo Diaz de Vivar (1043-1099). Diaz is better known as *El Cid Campeador* (from the Arabic *Seid,* "master", and the Spanish *Campeador,* "challenger"), and in the *Poema del Cid,* written about 1150, he is glorified as the noble Christian hero of Spain. Actually the Cid was a ruthless freebooter who was banished from Castile and thereafter offered his services, at the head of 300 free lances, to Christians and Moslems alike, plundering churches and mosques with equal indifference. He is pictured here in the throes of his greatest victory—the capture of Valencia in 1094 when he defeated an army of Almoravides, fanatical Berber Moors from the Sahara. The Cid wields a heavy broadsword against his Moslem opponent, who is about to be aided by the black-hooded Almoravide riding up in the background. For the last five years of his life the Cid ruled Valencia virtually as an independent king.

It is not hard to see why many modern historians regard the Crusades as a blot on the history of the West. "In the long sequence of interaction and fusion between Orient and Occident out of which our civilization has grown, the Crusades were a tragic and destructive episode. The historian as he gazes back across the centuries at their gallant story must find his admiration overcast by sorrow at the witness that it bears to the limitations of human nature. There was so much courage and so little honour, so much devotion and so little understanding. High ideals were besmirched by cruelty and greed, enterprise and endurance by a blind and narrow self-righteousness; and the Holy War itself was nothing more than a long act of intolerance in the name of God...."[1]

In 1147-49, 1189-92, and 1202-04 other crusades followed. They were failures; but worse than that, they revealed the depth to which unrestricted crusading warfare could sink. And this under the sign of the Cross! The Fourth Crusade never did reach the Holy Land: it veered from its course and ended with the capture and sack of Constantinople in 1204. After this tragic climax the movement fell off ingloriously, though other crusades followed over the next two centuries.

The crusading movement was one sign of the Church's new power; for with the Investiture Controversy the Church had become largely free of secular control and was now competing with kings and emperors. The perversion of the Crusades, however, provoked criticism, and a sample of this criticism is voiced by a 13th century French satirist in an imaginary argument between a crusader and a stay-at-home knight.

Am I to leave my wife and children, all my goods and inheritance, to go and conquer a foreign land which will give me nothing in return? I can worship God just as well in Paris as in Jerusalem. One doesn't have to cross the sea to get to Paradise. Those rich lords and prelates who have grabbed for themselves all the treasure on earth may well need to go on Crusade. But I live at peace with my neighbours, I am not bored with them yet and so I have no desire to go looking for a war at the other end of the world. If you like heroic deeds, you can go along and cover yourself with glory: tell the Sultan from me that if he feels like attacking me I know very well how to defend myself. But so long as he leaves me alone, I shall not bother my head about him. All you people, great and small, who go on pilgrimage to the Promised Land, ought to become very holy there: so how does it happen that the ones who come back are mostly bandits? If it were just a question of crossing a stream I would readily jump over—I might even wade through it. But the waters between here and Acre are broad and deep. God is everywhere: to you He may only be in Jerusalem, but to me He is here in France as well.

What, finally, did the Crusades accomplish? Historians used to say that practically everything that happened in the 13th century was caused by them, but now they are much more cautious. Today we know, for example, that most of the science and learning of the East was not brought back to Europe by the crusaders. The most important function of the Crusades seems to have been their role as catalytic agents, speeding up many movements already under way.

[1] S. Runciman, *A History of the Crusades,* Vol. III, *The Kingdom of Acre and the Later Crusades* (Cambridge University Press, 1954), p. 480.

The chief results of the Crusades as seen by modern historians are as follows[2]:

---

1. First-hand geographical acquaintance of Westerners with the East
2. Western demand for Eastern luxury goods—spices, fabrics, perfumes, etc.
3. Commercial profits for Venice, Genoa, Pisa
4. Revival of a money economy in the West and development of international banking
5. Charters of liberties sold by lords to serfs and towns to raise ready cash for crusading
6. Western imitation of Eastern castles and systems of fortifications
7. Western imitation of Eastern surplices over armour, and imitation of heraldic emblems
8. Elimination of many troublesome nobles
9. Increased power and prestige for the papacy
10. Development of a system of *indulgences* (an indulgence was granted in exchange for a money contribution to the Church, which enabled a man to provide for a substitute to do his penance of, for instance, going on a crusade)
11. Special taxes for crusaders, which increased the scope of ecclesiastical and secular taxing power
12. Criticism of these taxes, and, therefore, criticism of the perversion of the Crusades and of the Church

---

By the opening of the 13th century the Church had reached a new height of power. Under Pope Innocent III (1198-1216), the greatest of all medieval popes and the originator of the Fourth Crusade, the Church claimed a new realm in the East when it set out to rebuild the Kingdom of Jerusalem. (It had been established as a feudal kingdom in 1099 and reconquered by the Moslems in 1187.) Innocent III, a contemporary of Philip Augustus of France and John of England, thus stands at the close of an era of consolidation for the Church and, at the same time, at the opening of the era of papal monarchy.

2/MEDIEVAL LEARNING

Now we come to the last of the three great themes of 12th century medieval civilization, the content and structure of medieval learning. It will be considered much more briefly than the organization of society or of the Church, not because it is less important, but because a later chapter entitled "The Thirteenth-Century Synthesis" takes a closer look at the intellectual and cultural expression of the medieval point of view.

[2]Adapted from the excellent summary in J. L. LaMonte, *The World of the Middle Ages: A Reorientation of Medieval History* (Appleton-Century-Crofts, 1949), pp. 510-513.

St. Augustine

Beacons in the Gloom   In the tottering Roman Empire of the 4th and 5th centuries, there lived in the North African city of Hippo a most remarkable man. His name was Augustine (354-430). Augustine turned from the pagan philosophers to Christianity and became bishop of Hippo, where he earned such a reputation for his theology that he is today the most famous of what are called the Church Fathers.

Faced as he was with the political disintegration of his world (even as he died the Vandals were at the gates of the city), St. Augustine searched for some explanation for the catastrophe. The pagans blamed Rome's fall on Christianity, but Augustine answered the accusation in his profound book *The City of God*. The truth is rather, says Augustine, that Rome fell because of her adherence to false gods. God has a plan for mankind, and all history is the story of the conflict between the City of God (composed of good and faithful Christians) and the City of Man (weak Christians and pagans). In so reasoning, Augustine summed up the thought of other Church Fathers of the West and East and furnished an intellectual opposition to paganism. His use and refutation of pagan philosophy provides yet another example of the fusion of Roman and Christian elements in a new medieval civilization.

We have already observed the monasteries of the Benedictine Rule acting as preservers and transmitters of learning during the collapse of organized society in western Europe. In a day when most people were indifferent to learning, monks spent long hours in their *scriptoria* (writing rooms) copying ancient manuscripts. Scholars such as Bishop Gregory of Tours (see pages 299-300) were exceptional for their time, as was the Northumbrian monk, the Venerable Bede (673-735). Bede wrote Latin, could read Greek, and, as well as compiling an *Ecclesiastical History of the English People*, drew on classical scholarship in his commentaries on the Bible and his corrections of St. Jerome's 4th century Latin translation of the Bible, the so-called *Vulgate*.

Bede

It is, in fact, from Bede that we get our most vivid portrayal of how the Benedictines brought Christianity to northern England in the 7th century. The missionaries, he tells us, were invited to preach, and after they had had their say an aged councillor addressed his Northumbrian king thus:

So seems the life of man, O King, as a sparrow's flight through the hall when a man is sitting at meat in wintertide, with a warm fire burning on the hearth but the chill rain storm without. The sparrow flies in at one door and tarries for a moment in the light and heat of the hearth-fire, and then, flying forth from the other, vanishes into the wintry darkness whence it came. So tarries for a moment the life of man in our sight, but what goes before, or what comes after, we know not. If this new teaching tells us aught certainly of these, let us follow it.

It was not only the English who received conversion in this way. Irish monks, who had continued their work on their island after the legions left Britain in the 5th century, were dedicated missionaries too, and had carefully preserved both Christianity and Roman learning in Ireland. Men such

In their *scriptoria* or writing rooms, the monks copied and illuminated manuscripts under the vow of silence. But sometimes they wrote each other notes on the margins of the manuscripts—notes we can still read today! Often it took years to complete a single book. When we realize how valuable the finished copy would be we can appreciate the extent of the misfortune that befell a certain monastery after a hard winter. A hungry bear broke into the library and devoured some leather parchments, including, alas, the monastery's only Bible.

as St. Columban (560-615) founded new monasteries in Gaul and Italy, and carried their excellent Irish tradition of scholarship over to the continent.

St. Columban

The names of St. Jerome, St. Augustine, St. Columban, and Bede are only four among many that might have been chosen to illustrate the intellectual life of the years between the fall of Rome and the era of Charlemagne. Yet these four should suffice to remind us that it is something of a misnomer to call these centuries the "Dark Ages", if that sober label denotes centuries when intellectual accomplishments died. To be sure, the standard of achievement was far below what it had been in the days of Greece or Rome. Gregory of Tours, for example, wrote his 6th century Frankish history in ungrammatical Latin unsure of both genders and cases, and true to his day he accepted all sorts of superstitions and legends. Even the 9th century historical methods of some authors were not too exacting—as witness the bishop of Ravenna who wrote the biographies of his predecessors as follows:

> Where I have not found any history of any of these bishops, and have not been able by conversation with aged men, or inspection of the monuments, or from any other authentic source, to obtain information concerning them, in such a case, in order that there might not be a break in the series, I have composed the life myself, with the help of God and the prayers of the brethren.

The Carolingian Renaissance

Under Charlemagne a significant renaissance (rebirth) of learning occurred. Since the emperor had a consuming interest in education, he imported leading scholars to his palace school at Aachen (Aix-la-Chapelle). In fact it was largely the ambition of Charlemagne alone that was responsible for the Carolingian Renaissance. The Merovingians had palace schools for the nobles' sons; Charlemagne's school affected his whole empire.

Charlemagne was anxious that the children of both freemen and nobles attend his school; but as it turned out, the lower classes attended in greater numbers because many of the nobles wished to avoid contact with those whom they considered their social inferiors. The education provided was intended primarily to produce future ministers of God, as Charlemagne's edict clearly states.

> Be it known, therefore . . . that we, together with our faithful, have considered it to be useful that the bishoprics and monasteries . . . ought to be zealous in the culture of letters, teaching those who by the gift of God are able to learn, according to the capacity of each individual. . . . For when in the years just passed letters were often written to us from several monasteries in which it was stated that the brethren who dwelt there offered up in our behalf sacred and pious prayers, we have recognized in most of these letters both correct thought and uncouth expressions: because what pious devotion dictated faithfully to the mind, the tongue, uneducated on account of the neglect of study, was not able to express in a letter without error. Whence it happened that we began to fear lest perchance, as the skill in writing was less, so also the wisdom for understanding the Holy Scriptures might be much less than it rightly ought to be. And we all know well that, although errors of speech are dangerous, far more dangerous are errors of the understanding.

Alcuin

The foremost teacher in this great educational experiment was Alcuin of York (735-804). Alcuin wrote Bible commentaries, collected and supervised the copying of manuscripts, and ran the palace school. One of the most important results of this revival of learning was a clearer form of handwriting called Caroline minuscule, which replaced the cramped and crabbed Merovingian script. Caroline minuscule is the ancestor of the modern roman type which you are reading at this moment. So successful was Charlemagne's work that a century after he had sent to England for Alcuin to educate Gaul, King Alfred the Great sent to Gaul for teachers to re-educate England.

Gerbert

The most learned man of the 10th century was Gerbert, who stands out like a beacon in the gloom of the crumbling Carolingian Empire.

This modified Caroline minuscule is taken from the *Evangelienbuch (Gospel Book)* written between 863 and 871 by the monk Otfrid of Weissenburg.

Vnſer druhtin nrchiumin·

Gerbert, who died as Pope Sylvester II in 1003, was educated in France and Spain and became especially noteworthy for his mathematical skill. He was a great teacher, inventing various mechanical aids with which to instruct his students. Among his famous inventions were a sphere representing the earth and surrounded by metal bands depicting the orbits of the planets, a hydraulic organ, a water clock, and an abacus for solving numerical problems. He also wrote textbooks and introduced his students to the great authors of Rome. One of the interesting problems connected with Gerbert is the source of his mathematical knowledge. While historians cannot agree, it seems likely that during his study in Spain he somehow came into contact with Moslem mathematics, though he himself knew no Greek or Arabic.

Light from the Infidels Actually the greatest cultural developments between the 7th and 12th centuries we owe to the Moslems; for in their conquests they absorbed and passed on the classical heritage of the East, and in this way it came to the West. Moreover, the Moslems added to the old cultures to produce a new one of their own, which was transmitted to us in the uniform Arabic language of the Koran.

Moslem contributions in commerce, industry, geography, and astronomy can best be seen simply by listing twenty of the Arabic words now used in English:

muslin	admiral	monsoon	calibre	zenith
damask	cable	traffic	magazine	nadir
tabby	sloop	tariff	check	zero
bazaar	barque	risk	average	cipher

The Moslems also made important contributions in mathematics, chemistry, and medicine. Their mathematicians did the Western world a great service in displacing the clumsy Roman numerals by adopting from India the Hindu numerals which we call "Arabic". Try, for example, to divide MDCLXVIII by CXVII without transposing into Arabic numerals! The famous book of numbers, called for short *Al Gebra*, that is, "The Book," introduced the science of algebra in the 9th century. The Moslems also invented the zero, "one of the greatest inventions of the human mind."

Moslem chemistry (*al chemie* in Arabic) was rather primitive, much time and thought being spent on alchemy, the attempt to transmute base metals into gold. Still, a great deal was learned from the experiments performed in this search. As has been said, "the failure of the alchemist was the success of the chemist."

The most interesting Moslem accomplishments were in the field of medicine. The Moslems translated the medical works of the Greek physicians Hippocrates and Galen, and wrote treatises of their own on ophthalmology. In fact their doctors laid the foundations of many modern medical practices —the first clinical account of smallpox was written by a Moslem—and their skill contrasted sharply with Christian superstitions of the time, as you will see on page 356). It is remarkable that at a time when westerners believed in demon possession and thought that plague was some form of divine punishment, Moslem physicians had already discovered how diseases spread. One Arab physician writes:

The existence of contagion is established by experience, study, and the evidence of the senses, by trustworthy reports on transmission by garments, vessels, ear-rings; by the spread of it by persons from one house, by infection of a healthy sea-port by an arrival from an infected land . . . by the immunity of isolated individuals and . . . nomadic Beduin tribes of Africa. . . . It must be a principle that a proof taken from the Traditions has to undergo modification when in manifest contradiction with the evidence of the perception of the senses.

Another reiterates:

The result of my long experience is that if a person comes into contact with a patient, he is immediately attacked by the disease with the same symptoms.

This is a far cry from superstition, magic, and witchcraft.

3/ THE RENAISSANCE OF THE TWELFTH CENTURY

The Arabic learning and the cultural legacies of Greece and Rome which the Moslem soon transmitted deeply affected Western Europe, with the result that in the 12th century the West began to develop its own vigorous intellectual life. This movement is usually called the "Renaissance of the Twelfth Century," a name given it in a famous book by an American medievalist. The 12th century Renaissance affected far more people than did the Carolingian Renaissance of the 9th century, and because it drew on Eastern sources of inspiration it was international in scope. It is, of course, impossible to list in detail all the intellectual accomplishments of the 12th century, and as usual only a selection among great names can be made.

A New Literature and an Old The fact that medieval Latin was not written according to the stylistic and grammatical standards of the age of Cicero does not mean that it was "bad" Latin. Actually medieval Latin was a new language, not a corruption of an old one.

If one is to understand what had been happening to Latin, he should try to say in words known to Shakespeare that "A dive-bomber dropped incendiary bombs on an ammunition dump, creating an explosion which destroyed all telephonic communications, railroad installations, and motor roads, wrecking a large number of tanks, refrigerator cars, and parked aircraft." Obviously the Bard would be hopelessly confused amidst such a vocabulary. It was equally impossible to say in Ciceronian phrase that "The penitent sinner confessed his heresy to the priest, and

THE CHURCH AND LEARNING

was given absolution, whereupon, proceeding to the high altar of the cathedral, he received the eucharist and was relieved of the excommunication which had been placed upon him by the bishop." Yet the Middle Ages needed these words as much as we need the technical vocabulary of our own time, and like ourselves, as new conditions arose, they adapted new words to express new ideas.[3]

In the 12th century, then, rules of grammar were simplified. Latin poetry also departed from classical models, and much of medieval verse celebrated the pleasures of youth:

> Cast aside dull books and thought;
> Sweet is folly, sweet is play;
> Take the pleasure Spring hath brought
> In youth's opening holiday!
> Right it is old age should ponder
> On grave matters fraught with care;
> Tender youth is free to wander,
> Free to frolic light as air.
> Like a dream our prime is flown,
> Prisoned in a study;
> Sport and folly are youth's own,
> Tender youth and ruddy.

There was of course more serious poetry too, such as that written in the 12th century by Bernard of Cluny in a long poem entitled "On Contempt for the World," part of which you may have sung in the hymn "Jerusalem the Golden":

> Jerusalem the Golden,
> With milk and honey blest,
> Beneath thy contemplation
> Sink heart and voice oppressed.
> I know not, O I know not,
> What social joys are there,
> What radiancy of glory,
> What light beyond compare!

The Greek and Roman classics may have been neglected during certain periods of the Middle Ages, but they were never wholly ignored, and in the 12th century they were studied avidly. Most of the Greek classics came to Western Europe via the Moslems who had translated them into Arabic. Then in Spain these Arabic translations were for the most part retranslated into Spanish or Hebrew and thence into Latin. Although a text was apt to become quite garbled in passing from Greek to Arabic to Spanish (or Hebrew) to Latin, this is how most of the Greek scientific classics came to the West. It was in such a roundabout way that Aristotle was rediscovered.

From Doubt to Truth When the works of Aristotle became known in the West, many men were disturbed by apparent contradictions between his views and the Bible's. Medieval churchmen always claimed that there

---
[3]LaMonte, *The World of the Middle Ages*, p. 555.

could be no conflict between reason and faith, and therefore, by an ingenious series of arguments, many scholars set about reconciling Aristotle with Christianity. Their point of view is well summed up in the words of St. Anselm (1033-1109): "Nor do I seek to know that I may believe, but believe that I may know."

Peter Abelard   The most celebrated philosopher of that day, and one of the greatest teachers of all time, was the Frenchman Peter Abelard (1079-1142). Abelard gave more prominence to reason than did Anselm, saying: "I understand in order that I may believe." Abelard's far-ranging mind soon got him into trouble, particularly when he collected a number of passages from Scripture and from the Church Fathers that seemed contradictory, and published them under the title *Sic et Non* ("Yes and No"), an extract from which is in the Source Reading. Abelard was not trying to tear down religion, but he did insist that "by doubting we are led to inquiry, and from inquiry we perceive the truth"—an approach that drew from his long-standing antagonist St. Bernard of Clairvaux (see page 392) the cutting retort that Abelard "sees nothing through a glass darkly but stares at everything face to face." Nevertheless, it is to Peter Abelard that the schools of Paris, from which the University of Paris ultimately sprang, owe their great development.

The revival of Roman law in the 12th century was almost as spectacular as the stimulation of science by the Moslems. The rediscovery of Justinian's *Corpus Juris Civilis* increased legal knowledge without benefit of new Arabic or Greek sources, and law henceforth became a subject of instruction in the schools, with various monarchs highly approving its revival. Phrases from the *Digest* such as "What the prince desires has the force of law" and "The king is above law" were extremely useful to monarchs anxious to be more absolute in their rule than feudal practices warranted.

A Climate Congenial to Learning   Universities were born eight centuries ago, and in some ways their organization has not changed greatly since the Middle Ages.

In the Middle Ages there was, of course, learning that was not "book learning." The fledgling knight was trained as a squire in some baronial castle, just as the apprentice craftsman was trained in some master's house or shop. Any book learning was limited almost entirely to clergymen. After the downfall of the Roman schools, which had existed to train lawyers and civil servants, the schools of Europe from the 6th to the 12th centuries existed mainly to train men for service in the Church. At first monasteries were the chief educational centres; then the cathedral schools were promoted by the Carolingian Renaissance. Finally in the 12th century the universities rose to prominence, though the first "university", the medical school at Salerno in southern Italy, actually goes back to the middle of the 11th century. A more characteristic university was the University of

The medieval university often had no regular buildings, classes being held in corners of churches or in a room rented by the master, with the students sitting literally at his feet. Because the price of books was prohibitive, many a student formed his own "texts" from his lecture notes. Debates known as *disputations* sometimes went on for hours between masters or, as in the illustration, between advanced students, and were a popular test of learning. How were medieval universities financed?

Bologna, established early in the 12th century as a centre for legal studies. The University of Paris developed at approximately the same time, and Oxford a little later.

Since it is impossible to describe all the courses offered, let us concentrate on the so-called "arts" course of a university such as Oxford or Paris. The seven subjects of the curriculum were divided into two groups. The first group included the literary arts: grammar, rhetoric, and dialectic (what we would call logic or the art of reasoning); these three were known as the *trivium*. The second group included the mathematical sciences: arithmetic, geometry, astronomy, and music; these four were known as the *quadrivium*. The seven liberal arts have been truly called "vocational subjects for the clergy." The first three were essential to the priest's preparation of his sermons; arithmetic he needed for keeping parish accounts, geometry (which included surveying) for settling property disputes, astronomy for

The Seven Liberal Arts

setting movable dates (such as that of Easter) in the church calendar, and music for chanting the hymns. These seven liberal arts made up the curriculum of the medieval university, which gave a full training in arts. It was "practical" education for that time, and the degree of Master of Arts (M.A.) was a licence to teach in a university, or a prerequisite for further study in medicine or theology.

The modern university campus with its great lecture halls, laboratories, and athletic facilities bears little physical resemblance to its medieval ancestor. Students of those days—who enrolled at fourteen or fifteen—attended lectures given by the professors in their own rooms or in buildings they rented for the purpose. With the students seated before him, the professor lectured at break-neck speed (see page 358) on some standard text such as the works of Aristotle.

Here is an arts student's daily timetable[4]:

4 a.m.	Rising
5-6 a.m.	First lesson
6 a.m.	Breakfast, consisting only of a small piece of dry bread. Rest then, but no recreation
8-10 a.m.	Principal forenoon lesson
10-11 a.m.	Discussion and argument on the preceding lecture
11 a.m.	Dinner, accompanied by reading of the Bible or the lighter passages in the Lives of the Saints. The Chaplain intoned prayers, and the Principal of the College admonished, praised, blamed, and announced the punishments for the day.
12-2 p.m.	Interrogation on the morning lessons
2-3 p.m.	Rest period, so-called, while someone read aloud from a Latin poet or orator
3-5 p.m.	The principal afternoon lecture
5-6 p.m.	More discussion and more argument on the theme
6 p.m.	Supper
6.30 p.m.	General questions on all the day's lectures
7.30 p.m.	Evening Prayers and Benediction
8 p.m.	And so to bed, in winter; in summer, 9 p.m.

After six years of such austere routine the Paris arts student became a Bachelor of Arts (B.A.), and a Master after up to ten years' further study —in all a possible total of sixteen years of college education.

Of course not all medieval university students spent their whole time with books and study, any more than all modern university students do.

[4]Adapted from N. Schachner, *Mediaeval Universities* (George Allen and Unwin, 1938), p. 320.

Freshmen were hazed unmercifully, many students wasted their parents' money (and wrote ingenious letters home for more), and even the teachers could be unacademic, one German master being dismissed for stabbing a colleague to death in a faculty meeting. But the publicity that attended the exploits of wilder students should not make us think they were all playboys. The average student, then as now, was law-abiding and fairly inconspicuous, and when Robert de Sorbonne preached a sermon to the students at Paris upbraiding them because some "are better acquainted with the rules of dice than with those of logic," we must remember that he was, after all, preaching at a minority.

Perhaps the greatest contribution of the medieval universities was that they provided a climate congenial to learning. Then, too, they trained men for the learned professions—teachers, lawyers, doctors, and theologians. If much that they studied seems out of date to us today, and if their methods seem crude, we might well wonder what the historian of the future will say in writing of certain 20th century universities in which it is possible to learn how to ride a horse, how to play golf, or how to apply make-up!

There were, of course, developments in the 12 century other than the literary, philosophic, scientific, and educational ones. The magnificent 12th century artistic developments, for example, have been passed over completely here since they are to be given a fuller and continuous treatment in a later chapter. Chivalry also is to be discussed further. Even so, enough has been said to prove that the 12th century was one of the great creative periods in human history. It saw the flowering of medieval civilization into an international culture.

4/SUMMARY AND CONCLUSION

Alongside the feudal organization of society marched a Church of increasing power. By the time of Pope Gregory VII the Church had not only carried out internal reforms but was also challenging the power of lay monarchs to perform the investitures of their own clergy. Moreover, in the course of the 12th century the Crusading movement further added to the prestige of the papacy, and extended its political influence as far east as Constantinople.

Western European learning was concerned to justify the truths of Christianity, and though the intellectual life of the Middle Ages was no match for the brilliant accomplishments of Greece and Rome, neither were these centuries truly "Dark Ages." Actually the 9th century Carolingian Renaissance greatly stimulated learning, although after the collapse of Charlemagne's empire new waves of barbarians indifferent to learning surged across Europe.

By the time of Gerbert, Europe was beginning to inherit the classical learning which the Moslem translations re-emphasized. And by the 12th century a new international renaissance resulted in brilliant developments in philosophy, science, law, and education.

In the formative period between the 9th and 13th centuries, the Christian, Germanic, and Roman elements, whose merging in 800 began the Middle Ages, continued to dominate medieval life and thought. But the overall picture was of a new culture which, while international, was beginning by the end of the 12th century to be associated with particular nation states. This new national consciousness was expressed by such authors as the 12th century French poet, Chrétien de Troyes.

How would the international medieval culture be affected by rising nation states? Would the organization of society alter, or would feudalism, with its highly stratified structure congenial to both nobles and church, persist? The 13th and 14th centuries will show us what happened during "the medieval transition."

SOURCE READINGS

(a)

Emperor HENRY IV (1056-1106) wrote in the following manner to Pope GREGORY VII (1073-85), some time in the year 1076.

> Henry king not by usurpation but by the holy ordination of God to Hildebrand at present not the apostle but a false monk. . . . Our Lord Jesus Christ called us to the kingship, but he did not call you to the priesthood. For you ascended by these steps: you have obtained money, which is abhorrent to the monastic vow, by wiles, by money you have obtained favour, by favour you have obtained a sword, and by the sword you have obtained the chair of peace and from the chair of peace you disturb the peace in that you have armed subjects against their rulers, have taught that our bishops called to their office by God are to be spurned, and have usurped for laymen authority over their priests so that these laymen depose or condemn those whom they have received as teachers from the hand of God by the imposition of episcopal hands. On me also who although I am unworthy to be among the anointed have been anointed in the kingship you have lain hands.... the true pope St. Peter announces "Fear God, honour the king." You, however, who fear not God, dishonour me his appointed one. Wherefore St. Paul, when he does not spare an angel from heaven if he preaches otherwise, has not excepted you preaching otherwise on earth. For he says: If anyone, either I or an angel from the sky should preach other than is preached to you, he is accursed. You, therefore, damned by this curse and by the judgment of us and all our bishops descend and leave the purchased apostolic seat: let another ascend the throne of St. Peter who does not excuse violence by religion but teaches the true doctrine of St. Peter. I, Henry, king by God's grace with all our bishops say to you: Descend, descend, damned throughout the ages.

Pope Gregory countered by excommunicating Henry in the same year, thus depriving him of all Christian privileges. Gregory couched his answer to the emperor in the form of a letter to St. Peter:

> O blessed Peter, prince of the Apostles, mercifully incline thine ear, we pray, and hear me, thy servant, whom thou hast cherished from infancy and hast delivered until now from the hand of the wicked who have hated, and still hate me for my

loyalty to thee. Thou art my witness, as are also my Lady, the Mother of God, and the blessed Paul, thy brother among all the saints, that thy Holy Roman Church, forced me against my will to be its ruler. I had no thought of ascending thy throne as a robber, nay, rather would I have chosen to end my life as a pilgrim than to seize upon thy place for earthly glory and by devices of this world. Therefore, by thy favour, not by any works of mine, I believe that it is and has been thy will, that the Christian people especially committed to thee should render obedience to me, thy especially constituted representative. To me is given by thy grace the power of binding and loosing in Heaven and upon earth.

Wherefore, relying upon this commission, and for the honour and defence of thy Church, in the name of Almighty God, Father, Son and Holy Spirit, through thy power and authority, I deprive King Henry, son of the Emperor Henry, who has rebelled against thy Church with unheard-of audacity, of the government over the whole kingdom of Germany and Italy, and I release all Christian men from the allegiance which they have sworn or may swear to him, and I forbid anyone to serve him as king. For it is fitting that he who seeks to diminish the glory of thy Church should lose the glory which he seems to have. . . .

K. M. Setton and H. R. Winkler, editors, *Great Problems in European Civilization* (Prentice-Hall, 1954), pp. 141-142.

(b)

Pope URBAN II (1088-1099) had in his youth been a French noble, but he renounced the world to become a monk at Cluny. He preached the Crusade at the Council of Clermont in southern France on November 27, 1095, in what has been called "the most effective oration recorded in history." Four contemporary writers reported the address, and ROBERT THE MONK, whose version is translated below, claimed to have actually been present when the pope spoke.

"Oh, race of Franks, race from across the mountains, race chosen and beloved by God. . . . To you our discourse is addressed, and for you our exhortation is intended. We wish you to know what a grievous cause has led us to your country, what peril, threatening you and all the faithful, has brought us.

"From the confines of Jerusalem and the city of Constantinople a horrible tale has gone forth and very frequently has been brought to our ears; namely, that a race from the kingdom of the Persians, an accursed race, a race utterly alienated from God, a generation, forsooth, which has neither directed its heart nor entrusted its spirit to God, has invaded the lands of those Christians and has depopulated them by the sword, pillage, and fire; it has led away a part of the captives into its own country, and a part it has destroyed by cruel tortures; it has either entirely destroyed the churches of God or appropriated them for the rites of its own religion. . . .

"Let the deeds of your ancestors move you and incite your minds to manly achievements; likewise, the glory and greatness of King Charles the Great, and his son Louis, and of your other kings, who have destroyed the kingdoms of the pagans, and have extended in these lands the territory of the Holy Church. . . .

"However, if you are hindered by love of children, parents, and wives, remember what the Lord says in the Gospel, 'He that loveth father or mother, more than me, is not worthy of me.' 'Every one that hath forsaken houses, or brethren, or sisters, or father, or mother, or wife, or children, or lands for my name's sake shall receive an hundred-fold and shall inherit everlasting life.' Let none of your possessions detain you, no solicitude for your family affairs, since this land which you inhabit, shut in on all sides by the sea and surrounded by mountain peaks, is too narrow for your large population; nor does it abound in wealth; and it furnishes scarcely food enough for its cultivators. Hence it is that you murder and devour one another, that you wage war, and that frequently you perish by mutual wounds. Let therefore hatred depart from among you, let your quarrels end, let wars cease, and let all dissensions and controversies slumber. Enter upon the road to the Holy Sepulchre; wrest that land from the wicked race, and subject it to yourselves. That land which, as the Scripture says, 'floweth with milk and honey'. . . ."

When Pope Urban had said these and very many similar things in his urbane discourse, he so influenced to one purpose the desires of all who were present that they cried out, "God wills it! God wills it!" When the venerable Roman pontiff heard that, with eyes uplifted to heaven he gave thanks to God and, with his hand commanding silence, said:

"Most beloved brethren. . . . Unless the Lord God had been present in your minds, all of you would not have uttered the same cry. . . . Let this then be your battle-cry in combat. . . . When an armed attack is made upon the enemy, let this one cry be raised by all the soldiers of God: 'God wills it! God wills it!' . . .

"Whoever, therefore, shall determine upon this holy pilgrimage and shall make his vow to God to that effect and shall offer himself to Him as a living sacrifice, holy, acceptable unto God, shall wear the sign of the cross of the Lord on his forehead, or on his breast. When, having truly fulfilled his vow, he wishes to return, let him place the cross on his back between his shoulders. Such, indeed, by two-fold action will fulfil the precept of the Lord, as He commands in the Gospel, 'He that doth not take his cross and follow after me, is not worthy of me.'"

Setton and Winkler, *Great Problems in European Civilization*, pp. 106-108.

(c)

USAMA was a Syrian prince whose memoirs included stories based on the reports of his Arabic physician THABIT. About 1140 Thabit witnessed the two cases of surgery that are described here.

They brought to me a knight with an abscess in his leg, and a woman troubled with fever. I applied to the knight a little cataplasm; his abscess opened and took a favourable turn. As for the woman, I forbade her to eat certain foods, and I lowered her temperature. I was there when a Frankish doctor arrived, who said, "This man can't cure them!" Then, addressing the knight, he asked, "Which do you prefer, to live with a single leg or to die with both of your legs?" "I prefer," replied the knight, "to live with a single leg." "Then bring," said the doctor, "a strong knight with a sharp axe." The knight and axe were not slow in coming. I was present. The doctor stretched the leg of the patient on a block of wood, and then said to the knight, "Cut off his leg with the axe, detach it with a single blow." Under my eyes, the knight gave a violent blow, but it did not cut the leg off. He gave the unfortunate man a second blow, which caused the marrow to flow from the bone, and the patient died immediately.

As for the woman, the doctor examined her and said, "She is a woman with a devil in her head, by which she is possessed. Shave her hair." They did so, and she began to eat again, like her compatriots, garlic and mustard. Her fever grew worse. The doctor then said, "The devil has gone into her head." Seizing the razor he cut into her head in the form of a cross and excoriated the skin in the middle so deeply that the bones were uncovered. Then he rubbed her head with salt. The woman, in her turn, expired immediately. After asking them if my services were still needed, and after receiving a negative answer, I returned, having learned from their medicine matters of which I had previously been ignorant.

C. H. Haskins, *The Renaissance of the Twelfth Century* (Harvard University Press, 1927 reprinted by Meridian Books, 1957), pp. 326-327.

(d)

PETER ABELARD (1079-1142) in the *Sic et Non* examined 158 questions or problems on which various authorities—the Apostles, Church Fathers, Church councils, papal decrees—expressed apparently contradictory opinions. Abelard's aim was not to undermine the Church or its doctrines. Rather he wished to show that these contradictions could be reconciled, and that it was better for scholars to discuss the pros and cons of any question than to accept a statement as authoritative merely because it came from a revered source. For Abelard there was no conflict between reason and faith, and he believed, as he says in the preface to *Sic et Non* quoted below, that truth is best served by persistent questioning.

There are many seeming contradictions and even obscurities in the innumerable writings of the church fathers. Our respect for their authority should not stand in the way of an effort on our part to come at the truth. The obscurity and contradictions in ancient writings may be explained upon many grounds, and may be discussed without impugning the good faith and insight of the fathers. . . .

Not infrequently, apocryphal works are attributed to the saints. Then, even the best authors often introduce the erroneous views of others and leave the reader to distinguish between the true and the false. Sometimes, as Augustine confesses in his own case, the fathers ventured to rely upon the opinions of others.

Doubtless, the fathers might err; even Peter, the prince of the apostles, fell into error; what wonder that the saints do not always show themselves inspired? The fathers did not themselves believe that they, or their companions, were always right. Augustine found himself mistaken in some cases and did not hesitate to retract his errors. He warns his admirers not to look upon his letters as they would upon the Scriptures, but to accept only those things which, upon examination, they find to be true.

All writings belonging to this class are to be read with full freedom to criticize, and with no obligation to accept unquestioningly; otherwise the way would be blocked to all discussion, and posterity be deprived of the excellent intellectual exercise of debating difficult questions of language and presentation. But an explicit exception must be made in the case of the Old and New Testaments. In the Scriptures, when anything strikes us as absurd, we may not say that the writer erred, but that the scribe made a blunder in copying the manuscripts, or that there is an error in interpretation, or that the passage is not understood. The fathers make a very careful distinction between the Scriptures and later works. They advocate a discriminating, not to say suspicious, use of the writings of their own contemporaries.

In view of these considerations, I have ventured to bring together various dicta of the holy fathers, as they came to mind, and to formulate certain questions which were suggested by the seeming contradictions in the statements. These questions ought to serve to excite tender readers to a zealous inquiry into truth and so sharpen their wits. The master key of knowledge is, indeed, a persistent and frequent questioning. . . . By doubting we are led to inquiry, and from inquiry we perceive the truth.

The following are examples of the questions Abelard raised in the *Yes and No*:

Should human faith be based upon reason, or no?
Is God one, or no?
Does the first Psalm refer to Christ, or no?

Is sin pleasing to God, or no?
Is God the author of evil, or no?
Is God all-powerful, or no?
Was the first man persuaded to sin by the devil, or no?
Do we sometimes sin unwillingly, or no?
Does God punish the same sin both here and in the future, or no?
Is it worse to sin openly than secretly, or no?

Knoles and Snyder, *Readings in Western Civilization*, pp. 270-271, slightly adapted.

(e)

The following statute of December 10, 1355, lays down the rules for lecturing in the Arts Faculty of the UNIVERSITY OF PARIS.

In the name of the Lord, amen. Two methods of lecturing on books in the liberal arts having been tried, the former masters of philosophy uttering their words rapidly so that the mind of the hearer can take them in but the hand cannot keep up with them, the latter speaking slowly until their listeners can catch up with them with the pen; having compared these by diligent examination, the former method is found the better. Wherefore, the consensus of opinion warns us that we imitate it in our lectures. We, therefore, all and each, masters of the faculty of arts . . . have decreed in this wise, that all lecturers, whether masters or scholars of the same faculty, whenever and wherever they chance to lecture on any text ordinarily or cursorily in the same faculty, or to dispute any question concerning it, or anything else by way of exposition, shall observe the former method of lecturing to the best of their ability, so speaking forsooth as if no one was taking notes before them, in the way that sermons and recommendations are made in the university and which the lectures in other faculties follow. Moreover, transgressors of this statute, if the lecturers are masters or scholars, we now deprive henceforth for a year from lecturing, honours, offices, and other advantages of our faculty. Which if anyone violates, for the first relapse we double the penalty, for the second we quadruple it, and so on. Moreover, listeners who oppose the execution of this our statute by clamour, hissing, noise, throwing stones by themselves or by their servants and accomplices, or in any other way, we deprive of and cut off from our society for a year, and for each relapse we increase the penalty double and quadruple as above.

Lynn Thorndike, *University Records and Life in the Middle Ages* (Columbia University Press, 1944), p. 237.

FURTHER READING

1. THE CHURCH AND THE CRUSADES

    ATIYA, A. S., *Crusade, Commerce and Culture* (Indiana, 1962)
    COULTON, G. G., *Crusades, Commerce and Adventure* (Nelson, 1930)
† * NEWHALL, R. A., *The Crusades* (Holt, Rinehart and Winston, 1963)
    RUNCIMAN, S., *A History of the Crusades*, 3 vols. (Cambridge, 1951-54)
    SETTON, K. M. and others, editors, *A History of the Crusades*, 2 vols. to date (Pennsylvania, 1955-62)
    ULLMANN, W., *The Growth of Papal Government in the Middle Ages* (Methuen, 1962)
† TREECE, H., *The Crusades* (Bodley Head, 1962)

2. MEDIEVAL LEARNING TO THE 12TH CENTURY RENAISSANCE

ARNOLD, T. W. and GUILLAUME, A., editors, *The Legacy of Islam* (Oxford, 1931)
† ARTZ, F. B., *The Mind of the Middle Ages* (Knopf, 1958)
BOLGAR, R. R., *The Classical Heritage and Its Beneficiaries* (Cambridge, 1954)
CLAGETT, M. and others, editors, *Twelfth-Century Europe and the Foundations of Modern Society* (Wisconsin, 1961)
* CROMBIE, A. C., *Medieval and Early Modern Science*, 2 vols. (Anchor, 1959)
CRUMP, C. G. and JACOB, E. F., editors, *The Legacy of the Middle Ages* (Oxford, 1926)
DALY, L. J., *The Medieval University, 1200-1400* (Sheed and Ward, 1961)
† DUCKETT, E. S., *Alcuin, Friend of Charlemagne* (Macmillan, 1951)
* HASKINS, C. H., *The Renaissance of the Twelfth Century* (Meridian, 1957)
† * ──────, *The Rise of Universities* (Cornell, 1957)
† * HOLMES, U. T., *Daily Living in the Twelfth Century* (Wisconsin, 1962)
* KNOWLES, D., *The Evolution of Medieval Thought* (Longmans, 1962)
LAISTNER, M. L. W., *Thought and Letters in Western Europe, A.D. 500-900* (Methuen, 1957)
* LEFF, G., *Medieval Thought from Augustine to Ockham* (Penguin, 1958)
SCHACHNER, N., *Mediaeval Universities* (Allen and Unwin, 1938)
† TAYLOR, H. O., *The Medieval Mind*, 2 vols. (Harvard, 1949)
* WADDELL, H., *The Wandering Scholars* (Anchor, 1955)

UNIT V
THE MEDIEVAL TRANSITION

12. Trade and Towns

City air makes a free man.
—GERMAN PROVERB

By the end of the 12th century a characteristic medieval civilization had grown up. The birth of the Middle Ages had taken more than three centuries (476-800), the adolescence another four (800-1204). The mature civilization lasted a century (1204-1302), until old age overtook it by 1400. But the Middle Ages were slow to die. From the end of the 13th century there was a changing climate in which medieval civilization withered away "through a long autumn period—an autumn which had its bright sunny days as well as its frosts and rains."

The two centuries that you are about to study formed a bridge between medieval and modern times. In this age of transition city life flourished in Europe, the universal Church—rising to its heights in the 13th century—began to lose both power and prestige, nation states were strengthened, and parliaments sprang to life. Out of the kaleidoscope of the Middle Ages came today's civilization.

1/THE REVIVAL OF TRADE

In order to understand the revival of trade and the rise of towns we must retrace our steps a bit. After the fall of the Roman Empire in the West trade declined, but did not disappear, and the manorial system came to provide reasonably self-sufficient communities for most people. The difficulties of travel for the few enterprising pedlars can scarcely be imagined in these days of superhighways. By chance an account has survived of a journey made in the 9th century from Rheims to Chartres by a canon of Rheims. After sloshing through rain and floods and getting lost in the forest, the weary traveller came at last to a bridge.

> When I reached the bridge it was scarcely light enough to see. Carefully examining the structure I was once more overwhelmed with new misfortunes. For it had so many holes and such great gaps in it that the citizens of the town could scarcely cross it even by daylight in the course of their necessary business. But my quick-witted guide, who was pretty well experienced in travelling, searched about on every side for a skiff. Finding none, he came back to the dangerous task of trying to cross over the bridge. With the aid of heaven he managed to get the horses over safely. Where there were holes he would sometimes lay his shield down for the horses to step on, sometimes place boards across that were lying around, and now bending over, now standing up, first running ahead, then coming back, he finally got safely across with me and the horses. . . .

Medieval society conceived of the ideal community as being divided into three orders or estates: those who prayed (clergy), those who fought (knights), and those who worked (serfs). These orders are well portrayed in this late 13th century illuminated manuscript. Note the absence of representatives of a middle class (merchants and townsmen).

Rivers have traditionally been great highways and were surpassed as routes of travel only in the 19th century. In the Middle Ages heavy goods were often shipped on river barges or boats such as the one seen here. On the shore road, a heavy-wheeled country cart and a pack-horse are followed by a tinker with a square sack on his back, a box of tools, and a pair of bellows for blowing up his charcoal fire, while another traveller carries his children in a kind of shoulder cradle. They are making their way to the town across the river, which has the typical medieval thick walls and flanking towers and is on a main trade route. The illustration is modelled after the famous medieval town of Carcassonne in southern France.

Is it any wonder that, when the dangers of lurking robbers were added to such hazards, trade did not flourish?

By the 11th century all this was beginning to change. The Crusades further stimulated the reopening of ancient trade routes between East and West, with the Italian cities first feeling the surge of new markets. Then land and river routes carried the riches of the East inland as well. But the roads were still terrible and the ships small and dangerous, so that merchants moved in caravans or fleets for protection. How, then, did a merchant get his start?

The Making of a Merchant We have an account of how one peasant's son became a great merchant. Godric was an Englishman born in Lincolnshire about the end of the 11th century, and when he was forced to leave his parents' small holding and strike out on his own he became a beachcomber, gathering wreckage cast up by the sea and salvaging what he could to make a living. In this way he apparently accumulated enough of value to become a pedlar, going about with a pack on his back. After a while he fell in with a band of travelling merchants, and went with them from town to town. Godric prospered, and he and his associates eventually collected sufficient capital to engage in the shipping trade along the Channel and North Sea coasts. Goods in demand overseas were shipped abroad by Godric & Co., and return merchandise was sold where it was scarcest and would realize the highest prices. In this way Godric became a very rich man.

Local merchants could buy goods from foreign merchants for retail selling at the great international fairs, where wholesale trading was carried on. Since these fairs were spaced throughout the year so that there was usually an important one being held somewhere in Europe, the merchants travelled from one to another. During a fair, each day was set aside for dealing in a different commodity—for example, wool would be sold one day, and perhaps hides or wine the next—and on the last day foreign money would be exchanged and accounts settled.

Medieval fairs

Of course business was not the only attraction at the fair, any more than it is today. Jugglers, dancing bears, races, and other forms of entertainment made medieval midways as raucous as our modern ones. Fascinating as the great fairs were, however, it must be remembered that the bulk of business was done in local town markets, where elaborate regulations protected consumers and prevented unscrupulous merchants from *forestalling* (acquiring goods before they appeared in the market), *engrossing* (cornering the market), or *regrating* (buying simply to sell at a profit).

2/THE RISE OF TOWNS

Historians are by no means agreed as to which came first, the trade or the town; but does it not seem reasonable to suggest that they grew up together? Just as no trade route could exist entirely without towns, so no great town arose where there was no trade route. Thus towns usually developed where roads crossed or where goods were unloaded for transshipment, which might be where there was a junction of navigable rivers,

Moreover, kings, counts, and bishops wanted the Eastern luxury goods the merchants could provide—spices, drugs, dyes, perfumes, gems, rugs, arms and armour, and a score of other exotic articles. These were especially coveted by the local prince who lived in a castle or fortified place (= *burg* in German), and here merchants from near and far congregated. But when they discovered that the lord of the castle wanted a toll to allow them to enter the gate of the burg, they set up their stands outside the walls. The cluster of stalls gradually developed into a well established market with a permanent settlement of traders. Here is a chronicler's picture of the birth of one such town.

In order to satisfy the needs of the castle folk, there began to throng before his gate near the castle bridge traders and merchants selling costly goods, then inn-keepers to feed and house those doing business with the prince, who was often to be seen there; they built houses and set up inns where those who could not be put up at the castle were accommodated. The phrase they used was: "Let's go to the Bridge." The houses increased to such an extent that there soon grew up a large town which in the common speech of the lower classes is still called "Bridge", for Bruges means "bridge" in their patois.

Such a settlement outside the walls was called a *faubourg,* or, as we would say, a *suburb*, that is, a place in the shadow of the city walls. Since the new communities were without walls, the inhabitants soon banded together to purchase from the lord of the castle the right to build their own protective wall and to have their own government. The inhabitants were now called *burgers* (in German), *bourgeois* (in French), or *burgesses* (in English), terms that simply meant dwellers in the newly fortified suburb. And just as a modern metropolis tends to spread ever wider, absorbing its suburbs, so the medieval town was constantly enclosing its faubourgs within new walls.

It was by some such process as this that practically all the cities that grew up in Europe before modern industrial times were established.

The Route to Freedom The men who settled in the new suburbs were often the sons of runaway serfs. As early as the 10th century the Spanish province of Catalonia had granted asylum to slaves in order to attract settlers, probably to ensure an extra supply of fighting men to push back the Moslems.

When my grandfather Wilfred, count and marquis [of Barcelona], first constructed and built this castle of Cardona with its ground, he ordered that . . . if a man slave or a woman slave, or any man with a wife or fiancée of another, or a cunning thief, or any forger or criminal should come here, he was to remain safe among all the other inhabitants without any doubt. . . .

Whether slaves or serfs the settlers were landless, and wanted freedom to come and go as their business demanded while at the same time possessing a safe place for their family and their stock of goods. By the 12th century the townsmen were asking their lords for certain privileges: they wanted to be freemen and to own property. To obtain such privileges the townsmen often purchased a charter from the local lord. (An example of such a charter appears on page 379.)

Because the charter usually granted freedom to anyone who lived in the town for a year and a day, a runaway serf who could escape detection for this period would henceforth be free. Some unscrupulous lords even had no compunction about drawing off their neighbours' serfs to towns that they themselves had chartered. Moreover, serfdom and the business pursuits of the townsmen were incompatible, since no townsman could perform the long list of services due a lord. Thus the charter usually stated that townsmen could rent their land and buildings for money; also, the land could be leased or sold freely or could be willed away. Men of the town were usually granted exemption from feudal courts in favour of their own town courts, and the townsman could sell freely in the market of his own town. In exchange for all these rights the lord received money—perhaps to pay for a crusade, perhaps to renovate his castle.

Often a town charter stated specifically that the townsmen could form what was called a *merchant guild,* if one was not already in existence. These guilds were established by the merchants not only to gain a monopoly of local trade but also to provide what we would call social services—the care of widows, orphans, and the sick, even the provision of decent burial for the dead.

Guilds

Alongside the merchant guilds there grew up *craft guilds* to represent the separate professions, the butchers, the bakers, the candle-stick makers, and others. The craft guild, too, took care of its members, living or dead. In addition, it set rigid specifications for the goods produced by its members and laid down regulations governing entrance to the guild.

The prospective guildsman had to serve an apprenticeship of from three to twelve years, at the end of which period he became a journeyman, working as a day (French *journée*) labourer for various masters. When he had acquired enough money to set up a shop of his own he could apply for an examination by a board of his guild, which would decide on his skill. If he was found acceptable he became a master with his own shop and apprentices—although sometimes he was also required to produce a "masterpiece" before he could become a master and a member of the guild. A few of the regulations governing one such guild, the London Hatters, are given in the Source Reading at the end of this chapter.

The Communes

A New and Bad Name Some towns, such as Genoa and Pisa, possessed a measure of political independence that went far beyond the liberties granted in the usual type of town charter. These independent towns were called *communes*. Venice, built on islands off the northeastern Italian coast, was one of the earliest communes, with its elected chief magistrate, the Doge, heading what was really an independent city-state.

In northern Europe the extent of the towns depended on the powers of the local princes. Some communes were suppressed; one in France was defeated twelve times in its bid for independence. Naturally the nobles and clergy hated the communes for destroying what they regarded as their legitimate control over serfs, and one churchman defined "commune" with unmistakable scorn:

Now Commune is a new and bad name of an arrangement for all the poorest classes to pay their usual due of servitude to their lords once only in the year, and to make good any breach of the laws they have committed by the payment fixed by law, and to be entirely free from all other exactions usually imposed on serfs.

Perhaps we can understand the reason for such antagonism if we set beside his definition of commune an account by the same churchman of the struggle between the bishop and citizens of Laon.

On the fifth day of Easter week . . . there arose a disorderly noise throughout the city, men shouting "Commune!" . . . Citizens now entered the bishop's court with swords, battle-axes, bows, hatchets, clubs, and spears, a very great company. . . . The nobles rallied from all sides to the bishop. . . . He, with some helpers, fought them off with stones and arrows. . . . He hid himself in a cask . . . and piteously implored them, promising that he would cease to be their bishop, would give them unlimited riches, and would leave the country. And as they with hardened hearts jeered at him, one named Bernard, lifting his battle-axe, brutally dashed out the brains of that sacred, though sinner's, head; and he, slipping between the hands of those who held him, was dead before he reached the ground, stricken by another blow under the eye-sockets and across the nose. There brought to his end, his legs were cut off, and many another wound inflicted. Thibaut, seeing a ring on the Bishop's finger, and not being able to draw it off, cut off the finger.

The Hanseatic League

In northern Italy, Flanders, the Rhine valley, and along the Baltic shores the numerous towns organized leagues for the protection of their liberties. The most famous of these was the 13th century Hanseatic League (*hansa* = an association of merchants), made up of a number of German cities headed by Lübeck. The Hanseatic League enrolled as many as 200 towns, villages, and districts in a loose federation. Though it was a mercantile association, the Hansa became embroiled in politics from time to time, even defeating the king of Denmark in two naval wars when the League's Baltic monopoly was threatened. The Hansa's main business was, of course, the protection of the members' trading privileges, as one of its 13th century decrees clearly shows:

Inside the medieval walled town space was at a premium, so that numerous houses were often crowded against the wall. Guards on the ramparts and in the watch-tower gave warning of an approaching enemy. Town houses varied from simple mud cottages with thatched roofs to timber structures with elaborate framework patterns and tiled roofs. In some cases the lower storeys were made of stone, and the overhanging upper ones of wood, while windows were small and glass was used very sparingly. Cobble-stone pavement, such as is shown here, might well provide more jolts than the rough, unpaved country roads.

We wish to inform you of the decision taken after discussion in aid of all merchants who enjoy and are ruled by the law of Lübeck. First that each city shall do its utmost to defend the sea against pirates and other ill-doers so that merchant mariners may freely carry on their business. . . . No merchant shall ransom another who has been captured, nor receive payment on account of the latter's debts, under penalty of losing his property in his own city and in all cities where the law of Lübeck prevails. . . . If any lord besieges a city, no other city shall help him, unless he be its lord. . . . If there be war in the land, no city shall on that account injure a burgess from the other cities, in person or goods, but shall assist him in good faith. . . . These decisions shall hold good for one year; and what shall be done thereafter shall be intimated by letters from city to city.

Inside the Walls Suppose someone from the present age was able to transport himself back in time to a typical medieval town. What would his impressions be?

For one thing, he would probably be able to smell the town over the horizon even before he reached it. Then the first thing he would see outside the town's main gate would likely be the public gallows, replete with a corpse or two, while the weathering heads of other criminals crowned iron spikes over the gate. The town wall itself would be perhaps 20 feet high and 10 feet thick. From outside the walls might be seen a jumble of roofs, and (if the town had a bishop) the great bulk of the cathedral overtopping all else.

Inside the town walls the streets were narrow and crooked, far different from the grid pattern of a modern Canadian city. Since space within the walls was limited, the buildings were jammed together and the upper storeys often projected over the street. There were no sidewalks, and the cobblestones sloped to the centre to make a drainway for refuse. Garbage was simply thrown into the streets for pigs to eat and rain to wash away. Hence the modern custom—which is not really modern at all—that when a man is accompanying a woman he should walk on the outside. On his manly shoulders would fall the garbage thrown from a projecting upper storey, while his fair companion could walk protected beneath the overhang.

In the centre of the town was the public square or market-place, usually graced by a fountain. The important buildings of the town surrounded the square: the cathedral, the town hall, the halls of the various guilds, and, nearby, the houses of the wealthiest townsmen.

At night after the curfew (= *couvre feu,* cover the fire), anyone who ventured forth had to do so by lantern or torch light and be on his guard against hoodlums lurking in dark alleys. Most honest men were content to stay at home in their modest houses with thick wooden shutters across the windows. Some of the more luxurious houses had doors that locked rather than bolted, but since locks turned clumsily the keys were often large and heavy—though not all as heavy as this one: "He held a key and struck him in the face with it, so rudely that down he fell."

All who lived in the medieval town, rich and poor alike, had one dread: fire. Firefighting equipment was primitive, consisting of bucket brigades and long hooks to pull down a burning building when it threatened another. Since houses were cold, the usual custom was to sleep on a straw mattress near the fire, and the Coroners' Rolls attest to the frequency of resulting tragedies. Anyone convicted of arson was burned to death.

Yet despite their filth and danger, the towns must have been interesting places in which to live. For one thing there was nearly always something going on—a market, a fair, a religious procession, a play on the cathedral steps, even a hanging. And best of all, town life was *free*; even the lowliest serf could lose his bonds if only he could stay in a town for 366 days.

Although we cannot know for certain the size of these towns, most of them probably had between 5000 and 10,000 inhabitants, while the larger ones had enough people to be classed by modern standards as cities. In the Middle Ages, however, a "city" was where a bishop lived; a large town without a bishop could not qualify as a city, yet a small village with a cathedral church was technically a "city". Accordingly we shall speak of them all as towns. The following population figures are reasonable guesses for some of the larger towns in the first quarter of the 14th century.

	(Population in thousands)		
Milan	200—	Naples	100—
Venice	200—	Paris	80+
Florence	100+	London	40—

3/THE COMMERCIAL REVOLUTION

With new towns springing up, the lord of the manor found it difficult to hold his serfs on the land. For a lump sum of cash or for yearly payments in place of manorial obligations, many lords let serfs settle in towns. Other serfs preferred to stay on the lord's land, but paid a money rental and performed no manorial services. As free peasants they could sell their produce in the neighbouring town and lease land from the profits. The process of exchanging labour services for money payments was called *commutation*.

Between the middle of the 10th and the end of the 13th centuries the population of Europe rose sharply; in fact it did not again go up at a corresponding rate until after 1800. This increase in population, which coincided with the rise of towns, brought an increased demand for agricultural produce, which in turn necessitated improvements in agricultural techniques. A more widespread use of the heavy wheeled plough and the waterwheel, and the invention of the windmill greatly increased the farmer's capacity for the production and milling of grain. More land was brought under cultivation as the great forests were cut back, land which the lords worked, not by serfs, but by hired labour or by renting out new farms.

From the 10th century onwards an ageless institution began to disappear from Christian Europe as slavery tended to decline. The Church had strongly opposed slavery. Now the cumulative effects of the newly available horse-, water-, and wind-power supported this opposition by diminishing the need for slave labour.

. . . from the twelfth and even from the eleventh century there was a rapid replacement of human by non-human energy wherever great quantities of power were needed or where the required motion was so simple and monotonous that a man could be replaced by a mechanism. The chief glory of the later Middle Ages was not its cathedrals or its epics or its scholasticism: it was the building for the first time in history of a complex civilization which rested not on the backs of sweating slaves or coolies but primarily on non-human power.[1]

[1] Lynn White, Jr., "Technology and Invention in the Middle Ages", as cited in A. F. Havighurst, editor, *The Pirenne Thesis: Analysis, Criticism, and Revision* (D. C. Heath, 1958), p. 83.

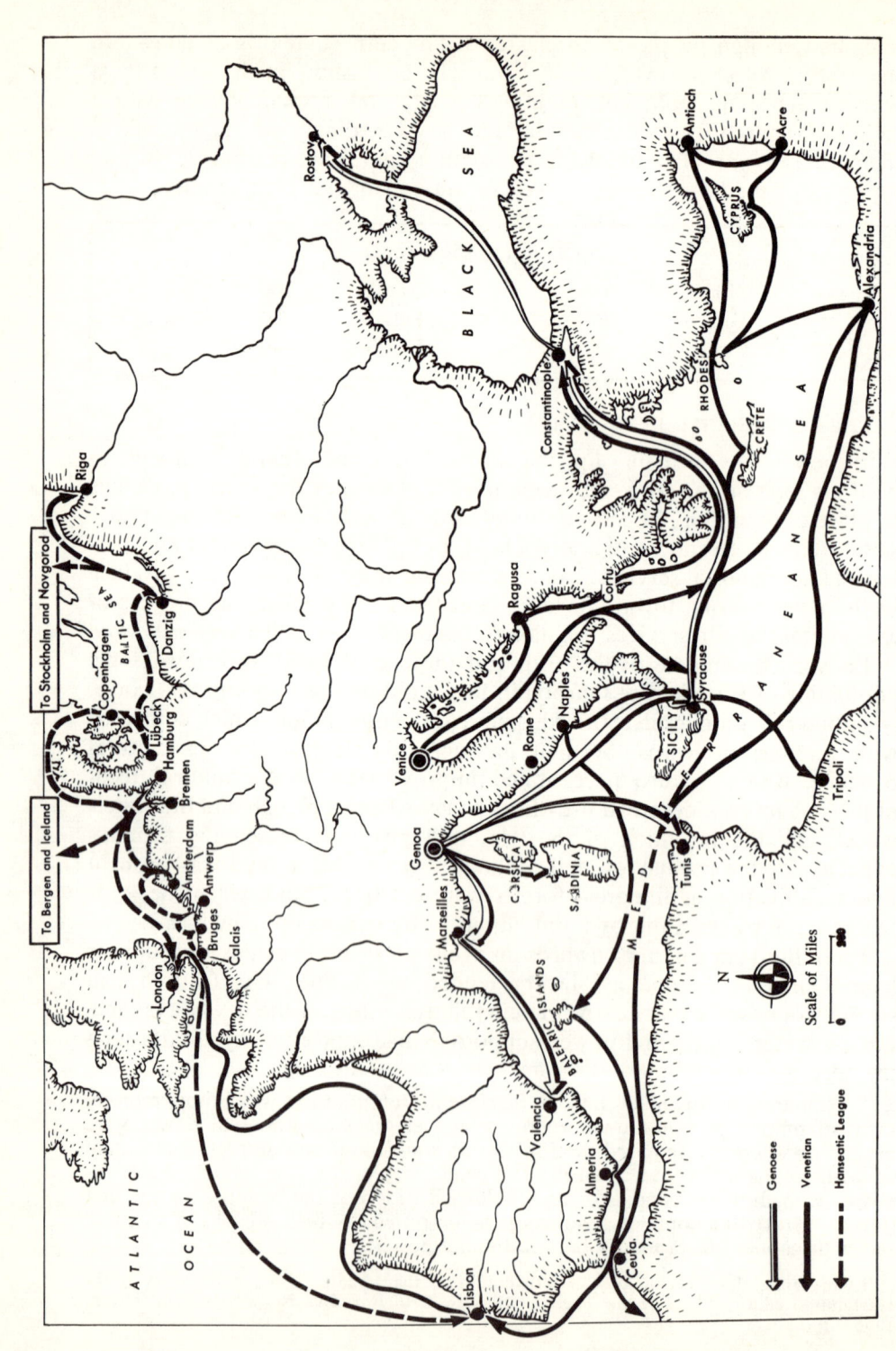

MEDIEVAL COMMERCE: PRINCIPAL SEA ROUTES BY THE 14TH CENTURY

Feudalism was also transformed by the effects of the money economy. Originally there had been only two classes, nobles and peasants. Now the townsmen formed a third class midway between the two, that is, a middle class. The members of this class should not be thought of as being equals, for there was a world of difference between the rich merchant and the poor apprentice. Nevertheless, they had one thing in common—freedom. All lived beyond the arbitrary will of any lord, free to impose their own restrictions under their own town government.

There was no love lost between nobles and townsmen. The noble was likely to resent the fact that the merchant had more ready cash available, whereas the merchant might envy the noble's lands and titles, even if their cash value was slight. Moreover, where the monarchy was strong there grew up an alliance of town and crown. Kings ceased to depend on their feudal and manorial incomes and developed instead a system of regular taxes paid by all subjects. The townsmen paid a large share of these taxes and consequently demanded, and got, a share in the government.

Taming the Knights The 12th and 13th centuries witnessed a change in the knight's way of life. Money made his life softer and more luxurious, and new methods of warfare diminished his military value. To capture the new stone castles that had replaced wooden ones, the knight was less vital than the engineer and the sapper. By the 12th century it became the custom to travel unarmed instead of always riding laden down with heavy armour. Then, should enemies meet, each knight would, as a chivalrous gentleman, give the other time to arm properly. Tournaments became more popular and less violent. They were now great social affairs which even the ladies attended. But more important was the convention that knights defeated in tournaments had to pay a ransom and lost their horse and armour to the victor. Private profit became a greater incentive than public glory.

The status of women was continually improving. Whereas at the opening of the 12th century a knight who tired of his wife could abandon her for another, by the end of the century this action required the sanction of an ecclesiastical court. The spirit of ancient feudalism was dying, and chivalrous manners sought to compensate for a power which the knight no longer wielded.

The changed attitude toward the knight is well illustrated by the *fabliaux,* short humorous verse stories which became popular in the 13th century. The heroes of these stories were not, as of old, knights or nobles: they were shrewd, hard-headed townsmen and merchants who knew what they wanted and had no compunction about how they got it. One of the most famous fabliaux is the *Romance of Reynard,* whose hero, the sly fox, makes fools out of the Lion King and his vassals. All merchants, of course, were not deceitful robbers such as Reynard, but this type of story did indicate an increasingly worldly attitude symbolized by the townsman.

"Romance of Reynard"

The Business World   Like the manor and the fief, commerce and industry, too, felt the impact of town life and an increasing population. What we call capitalism began to grow up during the later Middle Ages, the original capitalists being Italians, in particular Venetians. *Capital*, that is, the money necessary for carrying on business, was acquired through savings and earnings and then invested to bring a profit.

All the following financial practices and instruments (some of which were adapted from the Moslems) were made popular in the West by the Italian bankers:

1. standardized gold coinage	— Florentine *florins* (1252), and Venetian *ducats* (1284)
2. joint-stock company	— a group sells shares in order to build a ship, float a loan, etc.
3. banks	— for deposit and loan
4. maritime insurance	— loan by a shipowner to a merchant, to be returned (with a bonus) by the merchant on the safe arrival of his goods
5. letter of credit	— medieval traveller's cheque
6. bill of exchange	— medieval bank cheque guaranteeing payment in another city of a set sum in that city's currency for goods bought elsewhere
7. life rent	— an annuity; a town paid a life income to one who had loaned money to the municipal corporation
8. underwriting	— each member of a group agreed to guarantee a certain amount of credit to back a particular venture
9. government bonds	— first issued by the Venetian Republic (1157)

Marco Polo

An Astonishing Tale   One 13th century Italian businessman is worth special notice. He is Marco Polo. Marco's father lived in Constantinople, where he traded on behalf of Venice, and in 1271 he took his son on the long three-and-a-half-year overland expedition to the court of the great Mongol ruler of China, Kublai Khan (1257-94). The young Marco stayed in China for 17 years, during which time he won the favour of the Great Khan. After he returned to the West he was captured in a war with Genoa, and while in prison dictated his amazing travel tales of China. Marco Polo's narrative eclipsed all earlier accounts of the East in its accuracy and graphic description.

In Unit I we traced the history of China up to about 500 B.C. In the late 3rd century B.C. a new dynasty, the Ch'in, overthrew the Chou dynasty and unified China, building a Great Wall some 1400 miles long to keep out northern barbarians. But the Ch'in dynasty ended very soon after its mightiest figure, Chih Huang-ti ("First Emperor") died in 210 B.C. Now for the next four centuries (206 B.C.-A.D. 220) it was the Han dynasty that ruled, expanding the empire even farther and adopting Confucianism as the official state religion. The Han dynasty was contemporary with the Roman Empire, and presided over a greater mass of people than Rome ever ruled.

The Han Dynasty

The period of the Han dynasty was one of internal order and prosperity. Irrigation projects and flood control measures were undertaken along the Yellow River, and crops such as alfalfa, grapes, oranges, lemons, and tea were introduced. Trade expanded into South China and Indo-China, and the first contacts were made with Japan. In addition, the Han rulers extended their influence westward to the frontiers of Parthia. By the 2nd century A.D. Chinese records even tell of envoys from "An-tun" (Marcus Aurelius Antoninus, the Roman emperor) in southern Chinese ports.

Then in the 3rd century Han China collapsed internally, for somewhat the same reasons that Rome collapsed in the West. Early in the 4th century a group of Huns ruled north China, though the Chinese civilization tended to absorb its invaders more completely than did the Roman Empire. Nevertheless the unity of China had been broken, and feeble dynasties now divided up the country. It was during this period of internal division that Buddhism spread to China.

Only in the 7th century did a new clan of rulers, the T'ang dynasty (618-907), reunite the country and bring to China one of its most brilliant periods of history. Early in the T'ang period, when the Eastern Turks from Mongolia invaded the empire, the emperor's son addressed his officers as follows:

The T'ang Dynasty

Comrades, the whole plain is nothing but a sea. Night is about to fall, and it will be one of the darkest. The Turks are nothing to fear except when they can fire their arrows. Let us rush them with swords and pikes in hand, and we will overwhelm them before they can get ready for defense.

The Turks were defeated, and the T'angs went on to reorganize the administration, insisting that public officials be selected through civil service examinations based on the Confucian classics. Their dynasty remained in power until the 10th century.

As the T'ang rulers settled into a comfortable sense of invincibility, however, they deteriorated, and in the 8th century the high point of empire passed. Nevertheless there was a steady growth in other ways: The population passed the 100,000,000 mark; industrial techniques for producing silk and lacquer wares were perfected; gunpowder was invented; the abacus was introduced for mathematical calculations; cotton became a staple crop; and ocean trade with Europe prospered.

THE MONGOL EMPIRE (ABOUT 1250)

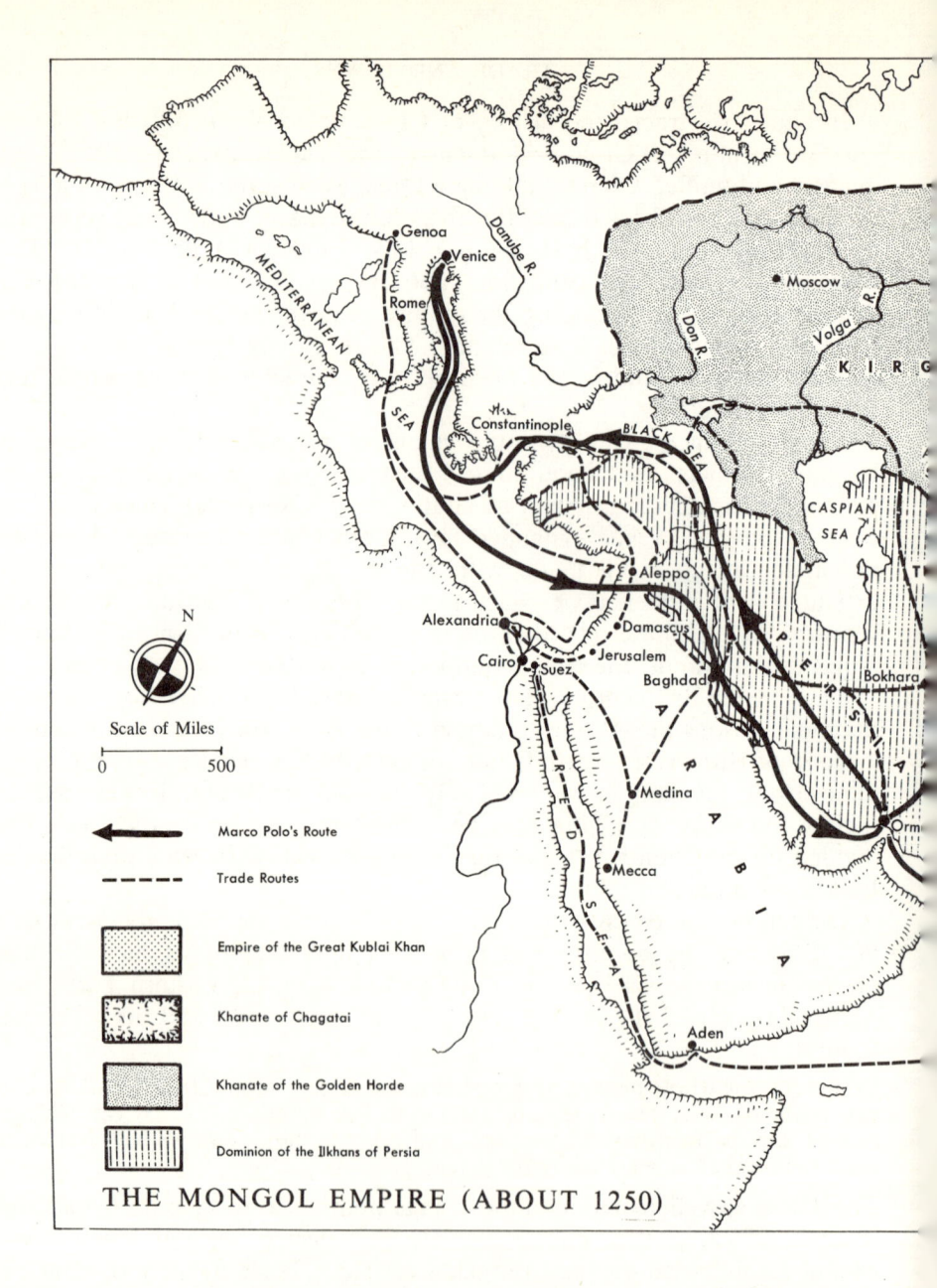

After the era of the T'angs, savage new invaders from Mongolia known as the Tartars swept into North China, murdering, plundering, destroying—and yet in the end partially adopting Chinese civilization. As the ancient proverb has it, China is a sea that salts all the rivers flowing into it. Finally in the 13th century the mighty Mongol emperor Genghiz Khan (1206-1227) not only conquered several of the northern provinces of China but also—far to the west—founded a Tartar state close to the modern city of Volgograd (Stalingrad). A few years later these Tartars were so strong that even the powerful Moslem empire asked for French and English aid to stem their expansion. They were curtly rebuffed. "Let us leave these dogs to devour one another," said one English bishop.

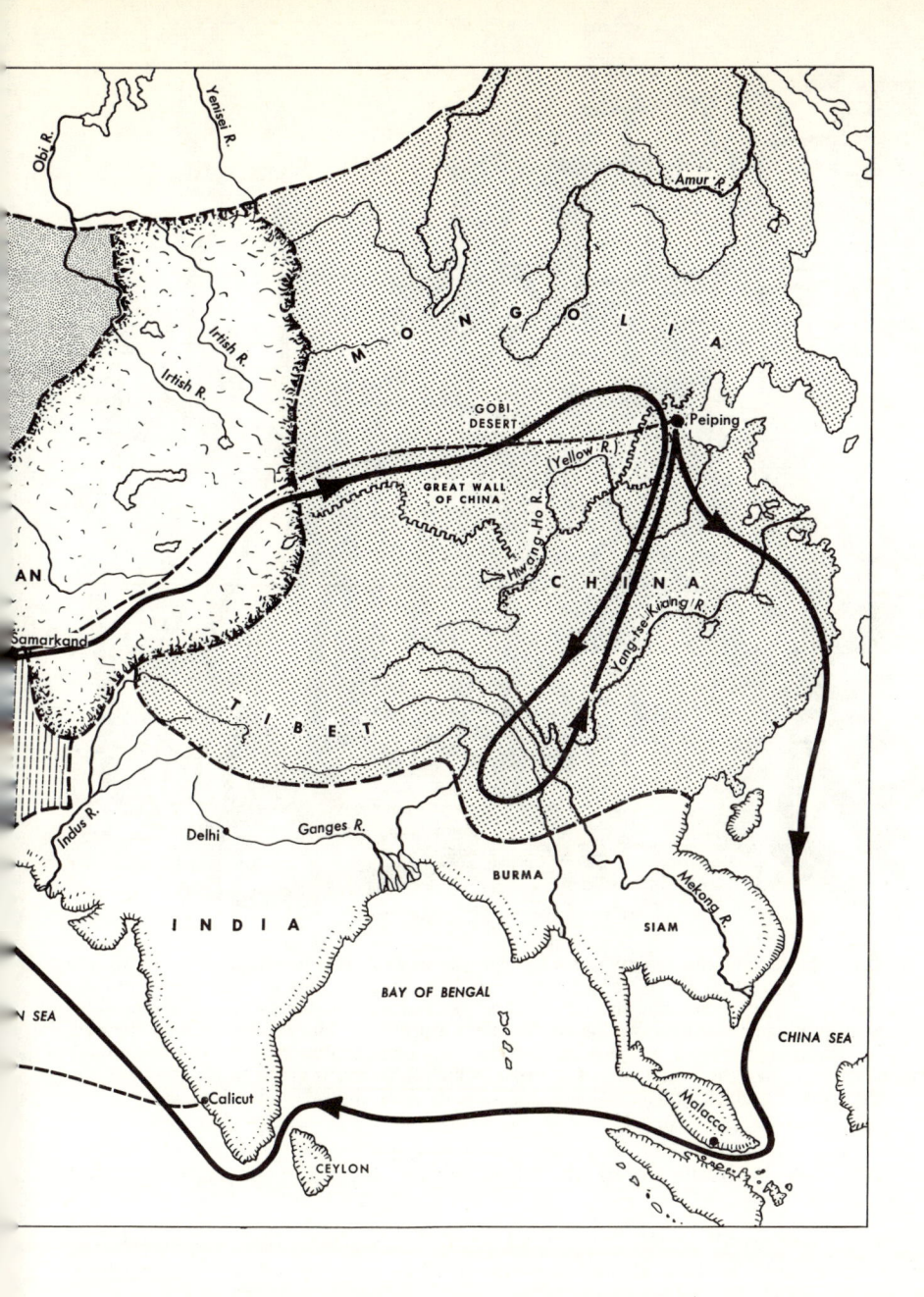

It was the court of Genghiz Khan's grandson, Kublai, that the Polos visited after the Mongol threat to Europe had receded. This great Khan finally united the various Chinese empires, which had shifted shape like an amoeba from dynasty to dynasty. He also sent one of the largest overseas expeditions in world history—140,000 men—to invade Japan. But fate

In 1275 an impressionable twenty-one-year-old came with his father and uncle to the glittering court of Kublai Khan. Marco Polo was to serve the mighty emperor faithfully for the next seventeen years, and has left us this impressive (if flattering) description of his remarkable employer: "Kublai . . . is of middle stature. . . . His limbs are well formed, and in his whole figure there is a just proportion. His complexion is fair, and occasionally suffused with red, like the bright tint of the rose, which adds much grace to his countenance. His eyes are black and handsome, his nose is well shaped and prominent."

intervened. A typhoon aided the determined Japanese in repulsing the invaders.

Marco Polo swept aside the veil of Oriental mystery that shrouded this ancient empire of Kublai Khan, and revealed to an astonished Europe such wonders as a black stone that burned (coal), and money made of paper:

In this city of Kanbalu [Peking] is the mint of the Great Khan, who may truly be said to possess the secret of the alchemists, as he has the art of producing money by the following process.

He causes the bark to be stripped from those mulberry-trees the leaves of which are used for feeding silk-worms, and takes from it that thin inner rind which lies between the coarser bark and the wood of the tree. This being steeped, and afterwards pounded in a mortar, until reduced to a pulp, is made into paper. . . . When ready for use, he has it cut into pieces of money of different sizes, nearly square, but somewhat longer than they are wide. . . . The coinage of this paper money is authenticated with as much form and ceremony as if it were actually of pure gold or silver; for to each note a number of officers, specially appointed, not only subscribe their names, but affix their seals also. When this has been regularly done by the whole of them, the principal officer, appointed by his Majesty, having dipped into vermilion the royal seal committed to his custody, stamps with it the piece of paper, so that the form of the seal tinged with the vermilion remains impressed upon it. In this way it receives full authenticity as current money, and the act of counterfeiting it is punished as a capital offence.

When thus coined in large quantities, this paper currency is circulated in every part of the Great Khan's dominions; nor dares any person, at the peril of his life, refuse to accept it in payment. All his subjects receive it without hesitation, because, wherever their business may call them, they can dispose of it again in the purchase of merchandise they may require; such as pearls, jewels, gold, or silver. With it, in short, every article may be procured.

Stagnation and Depression in Europe Marco Polo's tales of the East, fascinating as they were, fell on sceptical ears (he was nicknamed "Marco Millions"), and did not greatly influence the people of his own time. Actually it was the 15th century before the great trading ventures to the Orient began: 13th century Europe, though it could have benefited from new markets, did not open direct trade with the Far East. For now, Europeans contented themselves with initiating certain international banking practices (see page 372) and promoting the large-scale production of woollen cloth in Flanders and Florence, where great weaving industries spun the wool sheared from the backs of English and Spanish sheep. In fact woollen cloth was the only Western product that the East imported in any quantity. Such mining as there was in iron and coal was not extensive, and at that the best mines were usually under the control of the kings whose sovereignty included the direct possession of the soil.

Apparently 13th century Europe lacked the capital, energy, and technical skill necessary to follow the trail blazed by Marco Polo. Indeed, the overall picture in Europe was of an economy that was becoming stagnant. By the end of the 13th century the steady expansion of population and production had halted. Price fluctuations ruined those nobles who had commuted manorial dues for a fixed money payment. In the towns, guilds became more restrictive in their membership, and an ever-increasing group of day labourers found themselves at loggerheads with the rich merchants. Civil wars often tore rich and poor townsmen apart, and organized strikes by workers in industry became common. Most discontented of all were the peasants, who either found themselves working for fixed wages despite a rise in prices or threatened with a revival of serfdom. So desperate did they become that on three occasions major peasant revolts broke out, in Flanders in 1323-28, northern France in 1358, and England in 1381.

As if these economic problems were not serious enough, in 1348-50 a great plague swept westward across Europe, with lesser waves following in 1361-62 and 1369. This was the so-called Black Death or bubonic plague.

The Black Death

Between one- and two-thirds of the town population of Europe died, though the mortality was not as high in the country. We now know that fleas infected from plague-stricken rats carried the disease, and the frightful lack of public sanitation certainly promoted its easy and rapid spread. The labour shortage caused by the Black Death in England raised wages at a time of falling prices (see page 380). Hence production fell. But as food stocks became depleted, prices bounced back up to unprecedented levels. So severe were the economic crises of the 14th century that they were comparable in their gravity to our own Great Depression of the 1930's. After the Black Death there was little increase in European population for two centuries. Indeed, some of the land that ambitious towns had enclosed within their walls in the 14th and 15th centuries remained open fields right up to the 19th century.

Almost the only bright spot in the commercial life of the 14th and 15th centuries was the increasing refinement of the big business techniques of Italian merchants and bankers. These techniques were to be taken over by the monarchs of France and England in the interests of building up strong national states. Long before such monarchies gained strength, however, the Commercial Revolution—the term for all the economic developments we have been discussing here—had worked its changes.

4/SUMMARY AND CONCLUSION

After the collapse of the Western Roman Empire trade declined and cities disappeared. But in the 11th century, and with the impetus of the Crusades in the 12th, trade was stimulated and merchants travelled more readily through Europe to fairs, while towns sprang up in profusion between 1050 and 1200.

Town life changed Europe. For one thing, townsmen usually bought a charter freeing them from manorial obligations. For another, townsmen wanted their own government, and sometimes even set up small independent republics or banded their towns together in leagues to protect their liberties.

Though the towns were not large by modern standards, the greatest being not as populous as Winnipeg or Hamilton, and though by modern standards of sanitation medieval towns were filthy, still they attracted settlers. Their great advantage was that they afforded freedom; so serfs flocked to them. In the changed commercial society that grew up the knight lost much of his influence, and the townsmen formed a great new middle class whose members, be they rich merchants or poor day labourers, did not fit into the feudal hierarchy of society.

Modern capitalism had its birth in the improved business techniques of the Italian merchants and bankers, and large-scale industries became common. Thus even though depressions in the 14th and early 15th centuries slowed up the development of big business, by the 13th century society had already been transformed.

This Commercial Revolution was, of course, very gradual, beginning in separate areas at various times between the 11th and 13th centuries. But after it had begun nothing was ever the same again. European society was ceasing to be medieval.

SOURCE READINGS

(a)

The CHARTER TO THE BURGESSES OF IPSWICH was granted by King John of England in 1200, and illustrates the privileges that townsmen wanted. In this selection technical terms are italicized and defined.

John, by the grace of God king, etc. Know that we have granted and by our present charter have confirmed to our burgesses of Ipswich our borough of Ipswich, with all its appurtenances and with all its liberties and free customs, to be held of us and our heirs by them and their heirs in hereditary right, paying to our exchequer every year at Michaelmas term, by the hand of the reeve of Ipswich, the just and accustomed *farm* [fixed annual payment]. . . . We have also granted that all burgesses of Ipswich are to be quit of toll, *stallage* [rent paid for stalls in a fair or market], *lastage* [sales tax on goods sold by measure], *pontage* [toll paid for repair of a bridge or for passing over, or under, it], and all other customs throughout all our land and throughout the ports of the sea. We have granted to them that, with the exception of our officials, none of them shall be *impleaded* [prosecuted in a lawsuit] in any plea outside the borough of Ipswich, save only in pleas concerning foreign tenures; and that they shall have their guild merchant and their *hanse* [guildhall]; that no one shall be lodged or shall take anything by force within the borough of Ipswich; that they shall justly have their lands and their pledges and all their debts, by whomsoever owed; that, with regard to their lands and tenures inside the borough, justice shall be assured them according to the custom of the borough of Ipswich and of our free boroughs; that, with regard to their debts established at Ipswich and their pledges made in the same place, the pleas shall be held at Ipswich; and that none of them shall be adjudged in mercy with respect to his chattels except according to the law of our free boroughs. We also forbid any one in all our land, on pain of £10 forfeiture to us, to exact toll, stallage, or any other custom from the men of Ipswich. Wherefore we will and straitly command that the aforesaid burgesses shall have and hold the aforesaid liberties and free customs well and in peace, as they have been and are best and most freely enjoyed by the other burgesses of our free boroughs in England. . . .

C. Stephenson and F. G. Marcham, editors, *Sources of English Constitutional History: A Selection of Documents from A.D. 600 to the Present* (Harper and Brothers, 1937), pp. 96-97.

(b)

The ARTICLES OF THE GUILD OF LONDON HATTERS of 1347 illustrate the controls laid down by the members of a particular craft guild. Some of the articles follow.

The points of the Articles touching the trade of Hat-makers, accepted by Thomas Leggy, Mayor, and the Aldermen of the City of London, at the suit, and at the request, of the folks of the said trade:
In the first place, that six men of the most lawful and most befitting of the said trade shall be assigned and sworn to rule and watch the trade. . . .

Also, that no one shall make or sell any manner of hats within the franchise of the city aforesaid, if he be not free of the same city; on pain of forfeiting to the Chamber the hats which he shall have made and offered for sale.

Also, that no one shall be made apprentice in the said trade for a less term than seven years, and that, without fraud or collusion. And he who shall receive any apprentice in any other manner, shall lose his freedom, until he shall have bought it back again.

Also, that the Wardens of the said trade shall make their searches for all manner of hats that are for sale within the said franchise, so often as need shall be. . . .

Also, whereas some workmen in the said trade have made hats that are not befitting, in deceit of the common people, from which great scandal, shame, and loss have often arisen to the good folks of the said trade, they pray that no workman in the said trade shall do any work by night touching the same, but only in clear daylight; that so, the aforesaid Wardens may openly inspect their work. And he who shall do otherwise, and shall be convicted thereof before the Mayor and Aldermen, shall pay to the Chamber of the Guildhall, the first time forty pence, the second time half a mark, and the third time he shall lose his freedom.

Also, that no one of the said trade shall receive the apprentice or serving-man of another, until he has fully completed his term, or his master has given him a proper dismissal. . . .

Introduction to Contemporary Civilization in the West: A Source Book (Columbia University Press, 2nd edition, 1954), I, 29-30.

(c)

This account of the effects of the Black Death on England was written by a contemporary chronicler, HENRY KNIGHTON. Knighton was an Augustinian monk at St. Mary-of-the-Meadows in Leicester.

The dreadful pestilence penetrated the seacoast by Southampton and came to Bristol, and there almost the whole population of the town perished, as if it had been seized by sudden death; for few kept their beds more than two or three days, or even half a day. Then this cruel death spread everywhere around, following the course of the sun. And there died at Leicester in the small parish of St. Leonard more than three hundred and eighty persons, in the parish of Holy Cross, four hundred, in the parish of St. Margaret's, Leicester, seven hundred; and so in every parish, a great multitude. Then the bishop of London sent word throughout his whole diocese giving general power to each and every priest, regular as well as secular, to hear confessions and to give absolution to all persons with full episcopal authority, except only in case of debt. . . . Likewise the Pope granted full remission of all sins to anyone receiving absolution when in danger of death, and granted that this power should last until Easter next following, and that every one might choose whatever confessor he pleased.

In the same year there was a great murrain [infectious disease] of sheep everywhere in the kingdom, so that in one place in a single pasture more than five thousand sheep died; and they putrefied so that neither bird nor beast would touch them. Everything was low in price because of the fear of death, for very few people took any care of riches or property of any kind. A man could have a horse that had been worth 40*s.* for half a mark [6*s.* 8*d.*]. . . . Sheep and cattle ran at large through the fields and among the crops, and there was none to drive them off or herd them; for lack of care they perished in ditches and hedges in incalculable numbers throughout all districts, and none knew what to do. . . .

In the following autumn a reaper was not to be had for a lower wage than 8*d.*, with his meals; a mower for not less than 10*d.*, with meals. Wherefore many crops wasted in the fields for lack of harvesters. But in the year of the pestilence, as has been said above, there was so great an abundance of every kind of grain that almost no one cared for it.

The Scots, hearing of the dreadful plague among the English, suspected that it had come about through the vengeance of God, and, according to the common report, they were accustomed to swear "be the foul deth of Engelond." Believing that the wrath of God had befallen the English, they assembled in Selkirk forest with the intention of invading the kingdom, when the fierce mortality overtook them, and in a short time about five thousand perished. . . .

E. Rickert, C. C. Olson, and M. M. Crow, editors, *Chaucer's World* (Columbia University Press, 1948), pp. 355-356.

FURTHER READING

1. GENERAL: THE LATER MIDDLE AGES

† * CHEYNEY, E. P., *The Dawn of a New Era, 1250-1453* (Torchbook, 1962)
* HEER, F., *The Medieval World: Europe, 1100-1350* (Mentor, 1963)
PREVITÉ-ORTON, C. W., *A History of Europe from 1198 to 1378* (Methuen, 1951)
WAUGH, W. T., *A History of Europe from 1378 to 1494* (Methuen, 1949)

2. TRADE AND TOWNS

BALDWIN, S., *Business in the Middle Ages* (Holt, 1937)
CLARKE, M. V., *The Medieval City State* (Methuen, 1926)
* MUNDY, J. H. and RIESENBERG, P., *The Medieval Town* (Anvil, 1958)
PACKARD, L. B., *The Commercial Revolution, 1400-1776* (Holt, 1927)
† * PIRENNE, H., *Medieval Cities: Their Origin and the Revival of Trade* (Anchor, 1956)
POSTAN, M. M. and RICH, E. E., editors, *The Cambridge Economic History of Europe*, Vol. II, *Trade and Industry in the Middle Ages* (Cambridge, 1952)
POSTAN, M. M., RICH, E. E., and MILLER, E., editors, *The Cambridge Economic History of Europe*, Vol. III, *Economic Organization and Policies in the Middle Ages* (Cambridge, 1963)
† REYNOLDS, R. L., *Europe Emerges: Transition toward an Industrial World-Wide Society, 600-1750* (Wisconsin, 1961)
SALUSBURY, G. T., *Street Life in Medieval England* (Pen-in-Hand, 1948)

13. The Thirteenth-Century Synthesis

. . . containing in brief whatever, from unnumbered books, I have been able to gather, worthy of consideration, admiration, or imitation as to things which have been made or done or said in the visible or invisible world from the beginning until the end, and even of things to come.
VINCENT OF BEAUVAIS (d. 1264) IN HIS
Speculum Maius (Greater Mirror)

Even though the 13th century lies in the midst of the economic changes that destroyed the Middle Ages, it is often taken to represent various medieval developments at their height. In fact some historians have praised the 13th as "the greatest of centuries." Why are such high claims made for it? Are they justified?

1/ THE THIRTEENTH-CENTURY MONARCHIES

Out of the many rulers of 13th century Europe, it is possible to select a dozen who are roughly representative of all the rest:

PAPACY	FRANCE	ENGLAND	GERMAN EMPIRE
Innocent III (1198-1216)	Philip Augustus (1180-1223)	John (1199-1216)	Frederick II (1215-1250)
Gregory IX (1227-1241)	Louis IX (1226-1270)	Henry III (1216-1272)	Conrad IV (1250-1254)
Boniface VIII (1294-1303)	Philip IV, the Fair (1285-1314)	Edward I (1272-1307)	Rudolf of Habsburg (1273-1291)

These rulers are not all necessarily of equal importance; some of them cannot even be called "great". They have been chosen because they illustrate an important 13th century development—a development that saw the papacy attempt to become "a great super-state", and the feudal monarchies continue the political direction of the 12th century.

Innocent III

The Most Efficient Monarchy Pope Innocent III turned out to be the greatest medieval pope. Having served at the papal court under five popes, he himself came to the papal throne at the age of thirty-seven, and proved to be not only a churchman but a politician and a diplomat as well. Like Gregory VII, Innocent had a very exalted idea of the spiritual domination of all human institutions by the Church of Rome—an attitude that naturally brought him into conflict with the rulers of France, England, and Germany. Innocent, however, made a practice of winning his disputes.

Under Innocent III the Church became the greatest state in Europe. An army of officials was needed to handle the mass of correspondence dealing with everything from personal moral problems to international affairs, and to administer the wide-flung concerns of the papacy. Often a great council was held in Rome, at which the clergy assembled (some 1300 came to the Lateran Council of 1215) to work out church policies. Thus the papacy had come to be the most efficient of medieval monarchies.

Gregory IX

Pope Gregory IX is notable chiefly for continuing Innocent's quarrels with the German Empire and for establishing the court of the Inquisition. *Heresy* (the holding of religious beliefs that the Church defined as untrue) had flourished during the 12th century, especially in the new towns. The Inquisition was created to ensure that heresy would not spread, but would be stamped out. In other words the Inquisition was one means by which the Church sought to obtain obedience and devotion to its teachings.

The Inquisition

The Inquisition followed a set procedure. Inquisitors would preach a sermon in the town square; then heretics received a period of grace during which to discover the error of their ways and be received back into the true faith. At the end of this period the judges conducted an examination of witnesses on oath, that is, a sworn *inquest,* in which persons still accused of heresy would be summoned and questioned. If they admitted their heresy and desired reconciliation with the Church, an appropriate penance was meted out. Most trials ended at this point. But if the heretic refused to recant his heretical beliefs he was handed over to the state to be burned at the stake.

It must be remembered, however, that the Inquisition tried very hard to win back souls for the Church. Execution was only the last resort. The cancer had to be cut out before it spread any further, for those who persisted in their heresy might endanger the faith of others. What shocks us today are the methods used—the anonymous accusation, the secret trial, the questioning under torture, and the invitation to prove one's repentance by denouncing one's past associates. But for the 13th century, heresy was the most dangerous crime imaginable, so vile and insidious that any means to suppress it seemed fully justifiable.

By the end of the 13th century and the pontificate of Boniface VIII, the prestige and power of the papal monarchy had slipped badly. The national monarchies of France and England had become stronger in the course of the century, and Pope Boniface fought stubbornly in an attempt to maintain the high papal claims of Innocent III. His bitter quarrel with France culminated in the bull (Latin *bulla* = seal) *Unam Sanctam,* issued in 1302. This papal pronouncement, since modified, declared among other things that

Boniface VIII

there is one Holy Catholic and Apostolic Church . . . and outside of this there is neither salvation nor remission of sins. . . . We, moreover, proclaim, declare and pronounce that it is altogether necessary to salvation for every human being to be subject to the Roman Pontiff.

But the very next year the Pope was arrested by agents of the French king, and though they soon released him, the broken man died a month later.

To such a level had the papacy fallen in one century.

Leaders of Europe It was Philip Augustus (who has already been mentioned as the founder of France) who pushed John of England out of Normandy and all his other lands north of the River Loire. A few years later, in 1214 at the Battle of Bouvines—"one of the decisive battles of the world"—Philip's greatest hour came.

On that hot July day the dust rose in such clouds that the armies could scarcely see each other. And so the Battle of Bouvines turned into a confused series of mêlées—virtually tournaments, a form of warfare at which the French knights excelled. After about three hours Philip Augustus was completely victorious over the Anglo-Flemish-German coalition. The results were manifold: Flanders came under French rule for a century; King John of England had, once and for all, to give up his plans to recover his former continental dominions; and the German Emperor, Otto IV (1198-1215),

Bouvines

The castle had developed slowly from a wooden structure to the great stone keep, then to the 13th century curtain wall type we see here under siege. This one has round flanking towers so that the defenders can fire along the face of the wall, and overhanging wooden hoardings so they can shower stones, boiling water, or lime on the attackers below. Because these hoardings could be damaged or destroyed, they were later replaced by permanent stone arrangements called *machicolations*. The siege tower or *belfry* was a great movable structure usually built on the scene and covered with hides to protect it from fire. When it was moved close to the castle, troops could climb up inside it and cross over a bridge on to the walls. At the right is a battering-ram, operating under the protection of a shelter called a *cat*. This ram was probably a large tree trunk with an iron head, and swung from the beams of the cat on ropes or chains. It could open a breach in the wall through which troops might pour. In the foreground, from left to right, we see the largest of medieval siege engines, the *trebuchet,* a kind of giant counterpoised beam which could hurl huge stones out of a large sling as the counterweight was dropped. Next to it is the *mangonel*, which sometimes worked from a large bow as seen here, or sometimes on the torsion principle. Both these weapons could throw more imaginative things over castle walls, such as a dreaded inflammable liquid known as "Greek fire," or a dead horse. The archer in the foreground is firing from behind a protective shield called a *mantelet*. Which Roman siege engines do the trebuchet and mangonel resemble?

lost his crown to a rival, Barbarossa's young grandson, Frederick II, who was the king of Sicily (see page 390).

Most important, however, was the fact that with this victory the French monarchy was raised to the heights and Philip Augustus became the most powerful secular ruler in Europe. As he came back to Paris from the battle all the houses along his route were hung with bunting, all the streets strewn with flowers. Peasants in the fields dropped their work and rushed to the roadside to see the triumphal procession pass. And in Paris, the citizens and university students had a non-stop celebration that lasted for a week!

By territorial conquests and by administrative reforms Philip Augustus had bequeathed a new monarchy to France. What would his successors do with it?

The French monarchy was at its strongest by the time Philip Augustus' grandson, Louis IX, or St. Louis as he is commonly known (he was canonized in 1297), came to the throne. Bolstered by this strength, Louis did not have to wage incessant warfare against his vassals and could lead a saintly life in which the pursuit of justice and the maintenance of peace were his overruling ambitions. An intimate picture has come down to us of this good man meting out justice under an oak tree:

Saint Louis

Many a time it happened that in summer time he would go and sit down in the wood at Vincennes, with his back to an oak, and make us take our seats around him. And all those who had complaints to make came to him without hindrance from ushers or other folk. Then he asked them with his own lips: "Is there any one here who has a cause?" Those who had a cause stood up when he would say to them; "Silence all, and you shall be dispatched one after the other." Then he would call Monseigneur de Fontaines or Monseigneur Geoffrey de Villette and would say to one of them: "Dispose of this case for me." When he saw anything to amend in the words of those who spake for others, he would correct it with his own lips. Sometimes in summer I have seen him, in order to administer justice to the people, come into the garden of Paris dressed in a camlet coat, a surcoat of woollen stuff without sleeves, a mantle of black taffety round his neck, his hair well combed and without coif, a hat with white peacock's feathers on his head. Carpets were spread out for us to sit down upon around him, and all the people who had business to dispatch stood about in front of him. Then he would have it dispatched in the same manner as I have already described in the wood of Vincennes.

Nevertheless Louis was no pacifist, and the fact that he settled various territorial disputes on his borders by negotiation and treaty merely freed his resources for use in two crusades against the Moslems. Louis was deeply religious. He supported the activities of the Inquisition, and eventually died in North Africa in the course of his second crusade.

At home, Louis had further developed the institutions he inherited from his grandfather and father. Henceforth all royal vassals possessed the right to come to the king's feudal court. Revenue and justice were dealt with by two subdivisions of the court, the *chambre des comptes* and the *parlement* at Paris, while in local administration the bailiffs (*baillis*) and seneschals (*sénéchaux*) corresponded to the English sheriffs.

Philip the Fair Under St. Louis' grandson, Philip IV, royal government exploited its power to the full. In order to finance Philip's wars with Edward I of England and with the count of Flanders, taxes were greatly increased. Philip even taxed the clergy, a step that brought him into conflict with Pope Boniface VIII.

When a king taxed his nobles, the established custom was that it should be done with their consent, and that this consent should be obtained from them sitting as a body. Two orders of society, clergy and nobility, made up the two "estates" of these assemblies. A third estate was composed of the deputies of the towns, which had at first been consulted individually by the king's agents. These three estates met for all of France in 1302 when Philip IV summoned them to one central session, the Estates-General. The Estates-General did not, however, concern itself with financial matters: these were handled locally. Unlike England, France was too big a nation to have a single assembly represent the whole country. Hence local assemblies were more active than any national body.

The 13th century saw a strong centralized monarchy develop in France, a monarchy that was able to go on from strength to strength without its nobles ever combining against it. It asserted high claims for itself—and not only as far as France was concerned. One of Philip the Fair's ministers, doubtless reflecting his master's pretensions, wrote:

It would be salutary for the whole world to submit to France, for the French make better use of the power of rational judgment than do any other people.

Kings Submit to the Law Meanwhile three kings ruled England in the course of the 13th century. The first of these, John, knew England well and was an able administrator. Nor can there be any doubt that he was intelligent and shrewd (and clean—surviving household accounts tell us how many baths King John took in a certain year, and we may suspect that he was cleaner than most medieval kings).

Yet it was John's very efficiency that won him the hatred of his subjects. Good administrator that he was, he taxed with a heavy hand and ground down his barons mercilessly. He also alienated his clergy by quarrelling with Rome over who should be Archbishop of Canterbury, and, when he

The castle of the Counts of Flanders, at Ghent in Belgium, was erected in 1180 on earlier foundations dating back to the 9th century and was modelled after Crusader fortresses in Syria. The castle pictured here is a 20th century restoration of the 12th century original. Note the moat (formed by the River Lieve), the massive walls protected by a watch-tower, the count's house to the left, and the high, square keep or donjon in which a last stand could be made if the outer defences were breached. By 1194 a trading quarter spreading over 200 acres had grown up across the river from the castle. In that year it was enclosed with a wall, and it was thus that the town of Ghent was born.

did not get his own way, seizing Church property. To these two sources of rancour was added the unpardonable sin for a medieval king: failure in war. His loss of Normandy in 1204 began a succession of defeats in Europe, which culminated in a final disaster at the Battle of Bouvines. John's defeats earned him the nickname "Softsword"; yet it is only fair to note that his barons gave him but half-hearted support on his continental campaigns. Whatever else he was, John was not lazy or cowardly in making war.

But in the end, weakened by foreign defeats and faced by seething rebellion on the part of a minority of his baronage, the wily king saw fit to capitulate. Two generations of Angevin heavy-handedness had exhausted the goodwill of the barons. In a very real sense, John had to pay the bill for the excessively high financial demands made by his father, Henry II, and his Lionhearted brother. Yet as far as tyranny was concerned, John was little worse than his predecessors: like them, he was merely trying to develop the royal power to the full. Caught with a nearly static royal revenue in a time of increasing costs and rising baronial incomes, John was doing his best to even out the situation by exacting more money from his barons in order to balance his budget.

Magna Carta   All this is clear to the modern historian. King John's barons could only see that he was going too far too fast. On June 15, 1215, they forced him to set his seal to their demands, which were edited and published four days later as the famous Magna Carta (*magna* because it was so much longer than previous charters). Magna Carta did not give rights to the "people" of England, nor was it a great monument to national liberty. Rather it was a feudal document reciting the grievances of the barons and defining the rights and obligations of John as their feudal overlord. There was virtually nothing new in it. What the Great Charter did was to insist that the king was and ought to be below the law—something English barons were to remind English kings of many times over.

On John's application, Pope Innocent III annulled Magna Carta as an unwarranted interference with royal rights. Civil war followed, but the situation eased with John's sudden death in 1216 after a bout of dysentery.

Henry III   Henry III was the reverse of his father John—a good man but a weak king. He was no leader. A series of campaigns in France proved fruitless, and finally in 1259 Henry agreed to abandon claims to almost all the old Angevin possessions. The only one he was to retain was Gascony, a fragment of Aquitaine, which was henceforth to be held as a fief from the French king. The wars pulled Henry deeper and deeper into debt, and when he had to throw himself on the not too tender mercies of his barons they, of course, demanded reforms and a greater voice in administering the state. Now more was wanted than Magna Carta's limitations on the king: the goal was complete control of the government. The result was that between 1258 and 1264 the barons really ruled in the name of the king. Finally in 1265 at the battle of Evesham, Henry's son and heir, Edward, defeated the rebellious barons.

Edward I   For a king to be considered great in the Middle Ages he had to be a successful warrior and a faithful son of the Church. Both of these requirements were fulfilled by Edward I, a tall (his nickname was "Longshanks"), sinewy man who was happiest when hunting or making war and who was a frequenter of tournaments ("the best lance in the world," one poet described him). A pale replica of Saint Louis, Edward nevertheless holds a highplace in English history. He has been called the English Justinian, an apt description because, like Justinian, he probably accomplished most in his legal work. The old laws were codified and systematized so that the common law was henceforth channelled in its growth. There is a further parallel between the reigns of Edward and Justinian. Edward, too, left a government crushed under debts, yet committed to foreign wars which were to end in disaster.

For like Justinian, Edward followed an imperialistic policy. In Wales alone he was successful, though even there it took a number of stiff campaigns before the country was finally brought under the English crown. But in Scotland Edward waged annual forays without being able to conquer that kingdom, and in France there was an inconclusive war with Philip IV over the legal status of Gascony. These constant wars so depleted Edward's treasury that his subjects were eventually able to force him to meet with them in national assemblies. How this came about is well shown by the events of the year 1297.

In 1297 a number of English barons refused to accompany the king on a military expedition to the continent, and in addition opposed further taxation. When Edward went overseas anyway the barons rebelled, refusing him any more money until he met their demands—which included the reissuing of Magna Carta. Finally, in November 1297, Edward was forced to agree that no extraordinary and direct taxation might henceforth be collected "except by the common assent of the whole kingdom and for the common benefit of the same kingdom." Just how the "common assent of the whole kingdom" was to be obtained was not made very clear. But it soon became evident that it was most convenient for the king to get the assent of all the groups affected by such taxes at one and the same time by calling them (or their representatives) together in a national meeting. Edward's use of the assembly known as parliament is so important a topic that consideration of it must be postponed until the next chapter.

Wonder of the World One of the most colourful personalities of the entire Middle Ages was Frederick II, the German emperor. His nickname *Stupor Mundi,* meaning literally "wonder of the world", was well deserved when we consider his intellectual interests (he sent questionnaires to other courts asking for all manner of information), his book on falconry, his scientific experiments, and his (rumoured) harem of Moslem beauties. Frederick's truly amazing career has been summed up in one sentence by a modern biographer: "Statesman and philosopher, politician and soldier, general and jurist, poet and diplomat, architect, zoologist, mathematician, the master of six or it might be nine languages, who collected ancient works of art, directed a school of sculpture, made independent researches in natural science, and organized states, this supremely versatile man was the Genius of the Renaissance on the throne of the Emperors."

Frederick II

How had this brilliant man come, at the age of twenty, to sit on the uneasy throne of the German Empire? His father, Henry VI (1190-97), was Barbarossa's son, and besides inheriting his empire gained the kingdom of Sicily through marriage. When Henry died, his infant son Frederick became king of Sicily; but the German Guelf nobles elected Otto of Brunswick, son of Barbarossa's old rival, Henry the Lion, as emperor. Supporters

THE GERMAN EMPIRE IN 1250

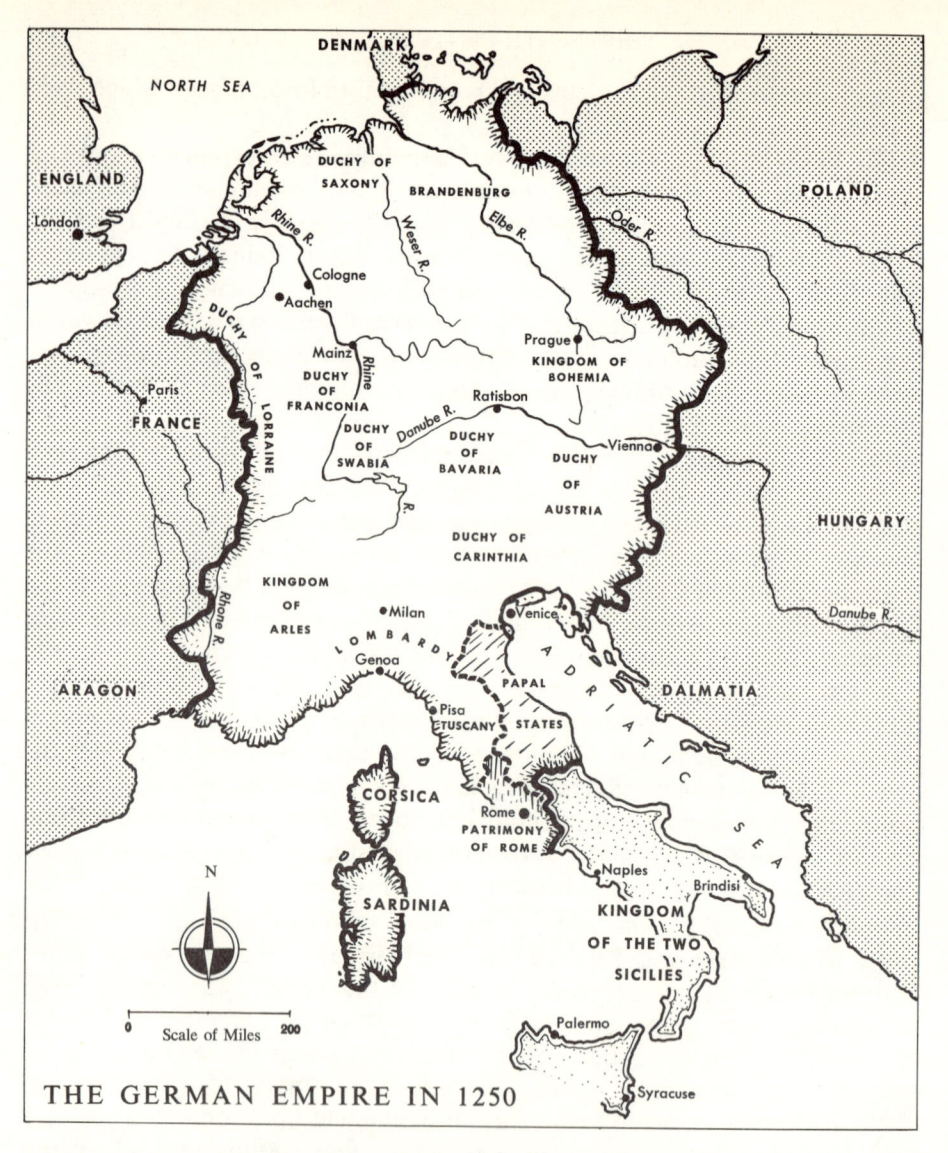

of Otto IV and Frederick sparred for years in a restless Germany until Otto's crushing defeat at Bouvines in 1214 finally resulted in Frederick's coronation as German emperor the next year.

Impressive though his accomplishments were, to the Church of his day Frederick II was a veritable anti-Christ. It is probably not true that he said, as is alleged, that "The world has been fooled by three great impostors: Moses, Christ, and Mohammed." But if he did not say it he might well have, for such a cryptic statement expressed his complete cynicism in religious matters. When in 1229, after much urging by various popes, he finally did lead a crusade which accomplished the restoration of Jerusalem to the Christians, he managed this feat while under a ban of excommunication (for not having started earlier), and won back the Holy City without shedding any blood. Rather he negotiated a treaty with the sultan of Egypt!

As far as the Empire was concerned, Frederick was unable to retain the support of his German princes. Nor was he able to consolidate the Empire's policy in Italy. His failure to do both only hastened a disintegration already begun by the breakdown of central authority in Germany. When he died in 1250 at the age of fifty-six, loyalties to local princes were already flourishing.

Frederick II's son, Conrad IV, was the last of the Hohenstaufens, the family of Frederick Barbarossa. Conrad had been a kind of figurehead while his father was in Italy. Now as emperor he came to Italy himself, but he accomplished little and died in 1254.

For the next nineteen years there was no central authority in the Empire. Finally in 1273 Count Rudolf of Habsburg was chosen emperor, largely because he appeared to be too weak to bother the great German princes. But strangely enough Rudolf of Habsburg won Austria, a conquest that transformed his family into one of the most powerful of all German dynasties. The German crown, however, was still weak, and the Empire only a loose alliance of semi-independent princes.

Thus the 13th century continued the political direction taken by the 12th. The papacy had become a great monarchy, but its power was already declining by the end of the 13th century. The French and English monarchies had become stronger and had clashed, whereas the German monarchy had become weaker and feudal decentralization had not been arrested.

At about the same time as leaders in Rome and in the national monarchies were struggling with problems of Church and State, three men of a very different stamp were making their mark, men of faith whose inspiration was transforming monasticism. It is now time to turn to them.

2/THREE MONASTIC REFORMERS

From time to time sincere men had tried to institute reforms in the Church, and various Rules, such as the Benedictine and Cluniac, had been devised to govern life in the monasteries. By the mid-11th century, however, even the Cluniac movement had lost its drive, and the time was ripe for new monastic orders.

One such order was founded in 1098 at Cîteaux (or Cistercium) in eastern France. The Cistercians strove to restore literal observance of St. Benedict's Rule, each abbey being independent with its own abbot, but all being required to follow the customs and discipline of Cîteaux. Because they made a practice of establishing their monasteries in out-of-the-way places, the Cistercians soon found themselves experts in introducing agriculture in undeveloped areas and became especially adept in the production of wool.

The Cistercian Order

Go Forth into the World In 1113 a young Burgundian noble named Bernard (1090-1153) became a Cistercian monk. Soon after, he set out with a dozen companions to found another monastery at Clairvaux, one that proved to have such an attraction for new monks that by the end of Bernard's life Clairvaux missionaries had founded over 160 daughter houses. Nor was St. Bernard's influence confined to the cloister. He was the friend of kings and popes, their adviser and their critic.

Clairvaux was a far from lax place, for Bernard was an uncompromising ascetic who railed against luxury, gracious living—even education. He rose to great heights of rhetoric in denouncing the female dress of those women of his day who "borrow . . . appearance from the furs of animals and the work of [silk] worms . . . women burdened rather than adorned with ornaments of gold, silver, and precious stones, and all the raiment of a court . . . dragging long trains of most precious material behind them, stirring up clouds of dust as they go."

In keeping with Cistercian tradition, this great monastery of Fountains Abbey was established in 1132 in a wild and desolate part of Yorkshire, a place "fit rather to be the lair of wild beasts than the home of human beings." This famous institution, which once sent missionaries to Denmark and Norway, is now in ruins. The whole plan was centred around the beautiful Gothic church (1) and chapel (2). Next in importance was the cloister (3), which was placed on the south side to take advantage of the sun's warmth since it was here that the monks did much of their work. Note also the infirmary (4), infirmary chapel (5), kitchen (6), monks' dormitory (7), abbot's house (8), *hospitium* or guest-house (9), and lay brothers' infirmary (10). In the warming-room (11) a fire burned from November till Easter, and monks were permitted to warm themselves at certain hours. The chapter-house (12) was the meeting place where once a day the abbot might preach a sermon or read a portion of some book. The refectory (13) or dining hall was, of course, next to the kitchen. A mill house to grind the grain was close by, as was a bakery to make bread. Gardens may be seen to the right of the chapel. Notice how part of the buildings are situated directly over the River Skell. If this is a typical Cistercian monastery, what would be its main product?

Nor did Bernard approve of the architecture of his day, which aimed to make palaces of beauty out of the monasteries. "What do you suppose," he asks, "is the object of all this?"

Is it the repentance of the contrite, or the admiration of the beholders? Oh! vanity of vanities! but not more vain than foolish. The church's walls are resplendent, but the poor are not there . . . the curious find wherewith to amuse themselves, the wretched find in them no stay for their misery. What has all this imagery to do with monks, with professors of poverty, with men of spiritual minds? I will not speak of the immense height of the churches, of their immoderate length, of their superfluous breadth, costly polishing, and strange designs, which while they attract the eyes of the worshipper hinder the soul's devotion, and somehow remind me of the ancient Jewish ritual. Let all this pass. We will suppose that it is done, as we are told, for the glory of God. But, as a monk myself, I ask other monks: "Tell me, O ye Professors of Poverty, what does gold in a holy place?" . . . Again, in the cloisters, what is the meaning of those ridiculous monsters, of that deformed beauty, that beautiful deformity, before the very eyes of the brethren reading there? . . . In fact, such an endless variety of forms appears everywhere that it is more pleasant to read in the stonework than in books, and to spend the day in admiring these oddities than in meditating in the Law of God. Before God, if they blush not at the impropriety, why do they not flinch before the expense?

The main purpose of the monasteries was to save the souls of the monks, and any social services they provided, such as tending the sick or helping the poor and needy, were only undertaken within the monastery itself or at its gate. What was needed was a monastic order that instead of retiring from the world would go out into it, and in the 13th century two such orders were founded. They are known as "mendicant" orders, from the Latin verb *mendicare* meaning "to beg." The friars, as the mendicants were called, were sent forth into the world to live in poverty amongst the common people, serving them and getting food by begging.

St. Francis of Assisi

The first order was founded by St. Francis (1181-1226), the son of a well-to-do cloth merchant of the Italian town of Assisi. As a youth St. Francis was devoted to luxury and pleasure, but he experienced a conversion and dedicated his life to the service of the poor. He has been described as "probably the most Christ-like man that has ever lived." Francis drew up a brief and simple Rule for his followers (some excerpts from which are on pages 407-408), and the great Pope Innocent III authorized the new religious order, whose members, the Friars Minor ("Little Brothers"), came in time to be known as Franciscans.

The Dominican Order

A second mendicant order founded about this time was associated with a cultured Spaniard named Dominic (1170-1221), who had been involved in a mission in France to win back heretics to the faith. St. Dominic's concern was to found an order that would combat ignorance by preaching the Church's teachings. As he told his followers: "Henceforth the world is your home: go forth into the whole world, teach and preach." Dominic's new order was named the Friars Preachers, though it too came to be called after its founder and was known as the Dominican Order. Because of its emphasis on doctrinal learning and preaching, it was natural that the Dominican Order should play a prominent part in the affairs of the Inquisition.

Occasionally the friars incurred the wrath of the local clergy, on whose parishes and jurisdiction they seemed to trespass. (That none too scrupulous associate of Robin Hood, the corpulent Friar Tuck, is a caricature of a certain type of friar.) Yet in reviving the influence of the Church in the new towns, in adjusting Church doctrine to Moslem learning, and in combatting heresy the friars made an inestimable contribution to the strengthening of the medieval Church.

Finally, it was the Church that gave a unity to Europe by embracing all good citizens who were not heretics. For it was not simply one church among many: it was *the* Church, and its teachings thoroughly coloured all those aspects of 13th century culture which we must now explore.

3/LEARNING

As a result of Moslem stimulus, natural science began its great development in the 13th century. For a full appreciation of this new enlightened approach, it must be contrasted with an example of the earlier attitude. A popular handbook of the early Middle Ages felt it should describe animals for the moral they pointed up, the following scientific gem occurring under "Lion":

> The Lion has three characteristics; as he walks or runs he brushes his footprints with his tail, so that hunters may not track him. This signifies the secrecy of the Incarnation. . . . Secondly, the Lion sleeps with his eyes open; so slept the body of Christ upon the Cross. . . . Thirdly, the lioness brings forth her cub dead; on the third day the father comes and roars in its face, and wakes it to life. This signifies our Lord's resurrection on the third day.

It is obvious that in the early Middle Ages science was dominated by religious considerations. But by the 13th century the attitude was much different.

Nature for Its Own Sake   The work of Albertus Magnus (1193-1280). an outstanding biologist as well as a theologian, exemplifies the new outlook. Albert the Great disagreed with and criticized some of Aristotle's scientific writings, and from his own examinations described plants and animals. Albert's powers of observation can be seen in his famous description of sheep-dogs.

Albertus Magnus

> The dogs . . . that follow sheep for the sake of guarding them differ in size, but all are habitually trained to run down wolves. The female, especially, is distinguished in this chase. I have seen one of them teach her young to pursue a wolf by running ahead of them to incite them on their course. When the wolf threatened to escape, she would hold it until the young dogs had caught up; then she would let it go again. And if the wolf bit the young dogs, she would not immediately come to their aid; for she wanted them to be provoked against the wolf. These dogs vary in size, some being very large, bigger and stronger than wolves, and some being smaller. All belonging to this breed, however, are larger and fiercer than other dogs.

Now admittedly it is a great deal easier to observe sheep-dogs than lions, but even so note the difference in the observer's attitude. Our early medieval author writes a nonsensical passage on lions in order to prove a religious doctrine, whereas our 13th century author is only interested in describing sheep-dogs as they actually behaved according to his own observations. Likewise Emperor Frederick II's *The Art of Hunting with Birds* gave elaborate descriptions of the anatomy of birds, based on the writings of Aristotle, on Moslem writings, and on his own observations.

Noteworthy scientific works were also produced by three Englishmen: Adelard of Bath, Robert Grosseteste, and Roger Bacon. Adelard of Bath (died 1145) was tutor to the future Henry II of England, and a pioneer student of Moslem science and philosophy who had travelled widely in Europe and the Middle East. He wrote a number of scientific treatises, including a translation from Arabic of Euclid's *Elements,* and a work called *Questions on Nature,* a composition arranged in the form of a dialogue with his nephew in which he passed off his own ideas as the new knowledge he had acquired from "his Arabs". The nephew questioned the uncle on seventy-six matters, such as:

Adelard of Bath

3. How different plants grow in the same region.
7. Why certain beasts chew the cud, and certain others not at all.
11. Why certain animals have a stomach, and others do not.
13. Whether beasts have souls.
15. Why men are not born with horns or other weapons.
31. Why we smell, taste, and touch.
38. Why men cannot walk when they are born, as animals do.
50. How the earth moves.
61. Whether, if one atom is set in motion, all are set in motion, since whatever is moved moves something.
75. What food the stars eat, if they are animals.

When the uncle could not produce natural explanations for some questions, the nephew suggested it might be "better to attribute all the operations of the universe to God." Adelard's reply shows how far he had come from the moralizing attitude that marked the early Middle Ages.

I do not detract from God. Everything that is, is from him and because of him. But [nature] is not confused and without system and so far as human knowledge has progressed it should be given a hearing. Only when it fails utterly should there be recourse to God.

Adelard is one of the first men of the Middle Ages to recommend studying nature for its own sake in order to learn what laws govern it.

Letabitur iustus in domino

This 13th century Gothic script from a medieval Psalter shows how letter forms became narrower and higher, with the pointed arch replacing the curve at the top and bottom of letters—perhaps a reflection of Gothic design. Compare specifically the letters *e, t, u, r, n,* and *m* with the corresponding ones in the Caroline minuscule on page 347.

Robert Grosseteste

Robert Grosseteste (1168-1253), chancellor of Oxford University and bishop of Lincoln, has been called by his modern biographer "the first to set out a systematic and coherent theory of experimental investigation . . . the first to use and exemplify such a theory in the details of original research into concrete problems." Grosseteste employed controlled experiments, and tried to explain the nature of rainbows by studying refraction in lenses. In his account of mirrors he even added some remarks on perspective lenses, with which, he said, "we may make things a long distance off appear as if placed very close, and large near things appear very small. . . . we may make small things placed at a distance appear any size we want, so that it may be possible for us to read the smallest letters at an incredible distance, or to count sand, or grains, or seeds, or any sort of minute objects." Others must have experimented with lenses, for we know that spectacles were in use before the end of the 13th century.

Roger Bacon

One of Grosseteste's pupils was the Oxford Franciscan, Roger Bacon (1214-94). Bacon was, however, little more than a theorist who often criticized his predecessors without fully understanding them. If he was original in anything it was in his appreciation of the technical skill of engineers, architects, and craftsmen. Although some of his accounts of practical experiments are valuable (see pages 408-409), Albert the Great and Robert Grosseteste deserve to rank far above that propagandist for science, the voluble Roger Bacon.

To Consolidate All Knowledge In the Middle Ages philosophy and theology were essentially one subject, and the student studied them in a *summa* (= sum total), that is, a book dealing exhaustively with a particular topic, usually by the method of question and answer. Thus the textbooks of the 13th century consisted of the Bible, the writings of the Church Fathers such as St. Augustine, and the various *summae,* which incorporated commentaries on the Greek and Roman classics along with their 12th century translations. Something of this movement to consolidate all knowledge was observed in the chapter dealing with the Renaissance of the 12th century, and in particular in the sections dealing with Anselm and Abelard. The trend culminated in the work of the 13th century Dominicans, Albert the Great and his pupil Thomas Aquinas (1225-74), who wrote monumental books in an attempt to sum up all knowledge and reconcile it with Christian doctrine.

Impressive as was the progress in natural science, the teaching of the Church was much more the concern of the 13th century. In fact Albert the Great's fame as a scientist is far surpassed by his fame as the teacher of Thomas Aquinas, whose *Summa Theologica* (*Summary of Theology*) is the greatest intellectual achievement of the 13th century. Thomas was born in southern Italy and eventually joined the Dominican Order, coming under the influence of Albert at Cologne and Paris. He himself taught in Italy and at Paris, and wrote prodigiously. He died in 1274 before he was fifty.

St. Thomas Aquinas

The *Summa Theologica* was a staggering undertaking, which in a modern English translation runs to twenty volumes. In it St. Thomas poses some 631 questions on all conceivable aspects of life and Christianity; then an answer is given to each question; a number of opinions contradicting each answer are presented; and finally St. Thomas disposes of the contrary opinions by stating his own. In all, St. Thomas answers some 10,000 objections to his conclusions.

No man before or since has subjected his own philosophy to such a comprehensive and searching logical examination. For St. Thomas, there could be no conflict between faith and reason. If anyone thought there was a conflict it was only because of faulty reasoning. "The highest knowledge we can have of God in this life," he wrote, "is to know that He is above all we can think concerning Him." Or as he also wrote, "The wise man creates order." Thus St. Thomas created a great structure of learning, elaborate, consistent with both Christianity and Aristotelianism according to his own reasoning, and authoritative. And although some of what he wrote was condemned as heretical only three years after his death, by 1323 the Church had canonized him. Today he remains the official philosopher of the Roman Catholic Church.

Certainly in learning the 13th was no mean century.

4/LAW AND POLITICAL THOUGHT

Roman Law

During the 12th century Renaissance there had been a great revival of the teaching of Roman law, particularly at the Italian university of Bologna. The revived Roman law of Justinian proved particularly attractive to the lay monarchs of the 13th century, since phrases from the Digest (see page 284) fitted in well with their lofty claims and their struggles with the papacy. Moreover, the customary law which varied from feudal state to feudal state was slowly being replaced by the principles of a uniform Roman law. England, however, was little affected by Roman law, owing to the strong tradition of English common law built up by Henry II and his successors.

Canon Law

The Church had its own law, administered in ecclesiastical courts presided over by bishops. This canon law was based on the Bible, decrees of Church councils, and papal decrees. Naturally national monarchs did not take kindly to the loss of legal jurisdiction over their clerical subjects, who in this way escaped prosecution in the secular courts. Hence canon law proved to be yet another bone of contention in the great medieval quarrel between Church and State.

Spiritual vs. Temporal Much of the political thought of the 13th century dealt with the question of what ought to be the relationship between spiritual and temporal authorities. The Investiture Controversy was one aspect of this question. The papalists emphasized the ideal of *limited* monarchy, and argued among other things that Christ in appointing Peter the custodian of the Church had given Peter's successors, the popes, power to—as Pope Boniface VIII put it in *Unam Sanctam*—"establish the earthly power and judge it, if it be not good. Thus, in the case of the Church and the power of the Church, the prophecy of Jeremiah is fulfilled: 'See, I have this day set thee over the nations and over the kingdoms.' "

Of course the lay monarchs would accept no such political theory, and in their turn used many arguments. They were especially fond of quoting Christ's statement, "Render therefore unto Caesar the things which are Caesar's; and unto God the things that are God's" (*Matthew* 22: 21), and maintaining that it proved conclusively that Christ Himself recognized the authority of secular rulers. This was the verse cited again and again by Philip IV in his running fight with Boniface VIII. As the future was to bear out, however, the papacy's political power was already waning by the end of the 13th century.

5/LITERATURE, MUSIC, AND ART

Vernacular literature

By the 13th century, literature written in the vernacular—the native European languages as opposed to the international Latin of the Roman Empire and early Middle Ages—was common. This is not to say that Latin literature did not continue to be produced. But alongside it there was a luxuriant growth of German, Scandinavian, Irish, and French writings. Autobiographies, travel books, poetry (such as the *fabliaux*), and drama associated with the Church all enjoyed a great vogue from the 13th century onwards.

Especially popular was a long French allegorical poem, the *Roman de la Rose* (*Romance of the Rose*). The Rose typifies the lady-love of the poet, and through the poem wander a host of other characters—Idleness, Danger, Evil-Tongue, Fear, Shame, Reason, Youth, and so on—whose intrigues help or hinder the lover's efforts to pluck the Rose. As a satire on courtly manners the *Roman de la Rose* expresses the approach of townsmen to outmoded ideals of chivalry. It also served as a model for the later morality plays such as *Everyman*.

Music Music was, of course, primarily religious. By the 11th century many-part music began to rival one-part music, and secular music had begun to rival church music. The earliest instrumental music that has survived in written form comes from the 13th century, and was composed for great occasions such as coronations, tournaments, or weddings. It was to be performed by ten to twenty musicians on such instruments as viols, harps, zithers, recorders (a kind of flute), oboes, trumpets, and drums.

At the beginning of the 14th century an English monk wrote a famous round called "Sumer is icumen in," which was designed to be sung by different voices picking up the tune one after another just as in modern rounds such as "Three Blind Mice." Compare the first line of "Sumer is icumen in" as it appears below in the medieval manuscript with a modern piece of musical notation.

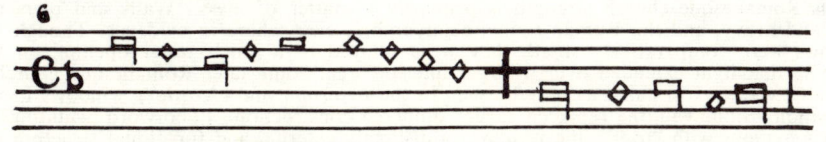

Su-mer is i-cum-en in,   Lhu-de sing cuc-cu,

Per-spi-ce Chris-ti-co-la,   que dig-na-ti-o,

Complicated variations in music began to develop in later medieval music. However, like most new styles, they met with opposition, as is evident from the comment of the medieval musician who wrote scornfully of those whose voices "throw sounds about at random, like awkward people throwing stones at a mark, without hitting it once in a hundred times," or the stern churchman who wrote that the simple souls at church were distracted "by the riot of the wantoning voice, by its eager ostentation, and by its womanish affectations in the mincing of notes and sentences."

Strength and Light The early churches of the Middle Ages were basilicas, simple rectangular buildings with wooden roofs. Since the roofs often caught fire and burned, destroying the whole church, builders turned

In this Romanesque church strength is primarily a matter of mass. Walls and piers are thick and heavy, and the Roman semicircular arch is everywhere in evidence. Though the windows were comparatively small, they were splayed to a much wider opening on the inside so that light extended over a larger interior area. Some later Romanesque churches combined windows in pairs, as may be seen in the *triforium* and *clerestory* windows in the illustration. Stone vaulting replaced earlier timbered construction. (The word "vaulting" is not synonymous with "roof": the roof was a distinctly external and functional structure for the protection of the interior and the vaulted "ceiling." The apparent Romanesque strength and solidity is often at odds with the poor workmanship involved in its construction. Massive looking walls might be just a masonry shell with rubble filling and very poor bonding, and joints were sometimes wide and coarse.

Romanesque architecture

their attention to the erection of a church with a non-inflammable roof, one made of stone. The arched stone "ceiling" that evolved was called a vault, and the new style of church architecture that made extensive use of it was called *Romanesque* (that is, Roman-like).

These long barrel vaults were very heavy, requiring massive stone pillars and thick walls to support them. The resulting building had only a few narrow windows in order not to weaken the thick walls, and consequently the interior was gloomy with the emphasis on solidity and horizontal lines. In an attempt to make the church appear less ponderous, sculptors often decorated the exterior facing with elaborately fashioned figures representing

Bible stories and enhanced the bulky interior pillars with carved capitals. The carving was stylized, that is, not lifelike but rather designed to bring out a particular sort of symbolism, the human figures often being elongated to harmonize with the columns.

The Romanesque style was dominant in the 11th and 12th centuries, just when stone castles were beginning to replace wooden ones. About the middle of the 12th century, however, several innovations led to the development of a new style of architecture called *Gothic*. Created in France, the Gothic style was to achieve greater elaboration and national variations as it spread to England, Germany, and Spain.

Gothic architecture

Some churches started out in one style and ended up in another. In the mid-12th century, for instance, the abbey church of St. Denis just outside Paris began as Romanesque and finished as the first great Gothic structure. How did Gothic differ from Romanesque?

Since the gloomy Romanesque style was not suited to the shorter days and long winters of northern Europe, a series of important discoveries led to a more suitable form. The chief of these inventions was the ribbed vault which you see in the illustration on page 404. Diagonal ribs concentrated the weight of the vault at certain points, so that the vault itself need not be nearly so heavy. Moreover, by using pointed arches greater height was possible, and the thrust of the vault downward and outward could be supported on slender columns if exterior "flying buttresses" outside the walls transferred the pressure to massive buttresses rising from the ground. Thus the scaffolding, in the form of the flying buttresses, was actually harmoniously incorporated into the finished building. Consequently the walls of the church did not have to be heavy, and could, in fact, be punctured with large windows to let light flood the interior. The whole impression given by a Gothic cathedral is one of vertical lines constantly striving upward, soaring, as it were, to heaven.

Gothic decoration was brilliantly conceived. The large pointed windows were filled with stained glass, while on the façade and surrounding the doorways lifelike sculptures gilded with fresh paint (far different from the grimy gray stone of today) clustered in profusion.

Suger, the abbot of St. Denis (1081-1151), was the moving spirit behind the creation of the Gothic style, and himself aptly described the purpose and impact of the new architecture.

The plan was to build a new church and to integrate the old with the new, except for the gracious and laudable extension of a cluster of chapels by which the entire sanctuary should be suffused with a marvellous and perpetual light from the most holy windows, adding lustre to the beauty within. . . .

When the house of God, many-coloured as the radiance of precious stones, called me from the cares of the world . . . I seemed to find myself, as it were, in some strange part of the universe, which was neither wholly of the baseness of the earth, nor wholly of the serenity of heaven; but by the grace of God, I seem lifted in a mystic manner from this lower, toward that upper sphere.

The Romanesque and Gothic cathedrals pictured here illustrate the contrasting elements of the two styles of architecture. Their chief differences may be summarized as follows:

ROMANESQUE (11th-12th centuries)	GOTHIC (12th-15th centuries)
1. barrel vaulting	1. ribbed vaulting
2. round arches	2. higher pointed arches
3. massive pillars	3. slender columns
4. narrow windows	4. large stained glass windows
5. thick walls	5. thinner walls
6. gloomy interior	6. light-flooded interior
7. horizontal lines	7. vertical lines
8. stylized sculpture	8. more lifelike sculpture

Defying Gravity to Glorify God "Our civilization has not produced a greater miracle than the Gothic cathedral," writes a modern historian. And it must be admitted that the present age of structural steel has not been able to reproduce comparable churches. If you live in a large city you will probably be able to have a look at a modern cathedral-type Gothic church —St. Paul's in Toronto is one, St. John the Divine in New York City another.

Yet the effect of our churches, even though they are stylistically correct, cannot compare with that of the medieval cathedrals. Our stained glass, experts concede, cannot match the medieval product. Our churches are not usually set off by a square, nor do they tower over their surroundings as medieval cathedrals do. Indeed today's cathedrals are often dwarfed by huge office buildings of plainer style. It requires considerable effort on our part to realize how and why these medieval cathedrals were built, one of the tallest stretching to the height of a modern fifteen-storey building, not counting additional towers and spires. In truth, as has been said, the Gothic cathedral rose "defying gravity to glorify God" in a miracle of medieval engineering.

Something as complex as a cathedral was not built by amateur architects or dabbling masons. Cathedrals required expert plans and swarms of workmen, both specialists and unskilled labourers. The extract quoted in the Source Reading testifies to the popular enthusiasm with which building materials were assembled for one great cathedral. But there is a difference between hauling stone and timber to the building site and cutting and shaping it to erect an arch over a hundred feet high. The whole town may have turned out to help, but there never was such a thing as a cathedral-raising bee.

The construction was supervised by a master mason, or a master builder as he was sometimes called, who was a highly skilled and well paid professional. The assisting masons, too, had to be experts, since even a small error in one section of the vaulting could ruin everything. Medieval drawings and models show us how carefully construction was planned. Proficient carpenters were needed to erect the scaffolding on which the masons raised the arches, and to construct a protective wood covering shingled over with lead, thin stone, or tile, under which the vaults could be completed. In addition, the carpenters fashioned the choir stalls and other furniture. Since smiths, plumbers, and glaziers were also required for any major construction job such as a cathedral, great numbers of wage-earning craftsmen must have travelled about from one large building project to another.

Church building was usually not a year-round activity. The work went ahead during the summer or as money became available, though a skeleton staff was maintained during the off seasons. It is little wonder, then, that a cathedral like Notre-Dame in Paris, begun in 1163, was not finally finished until some time after 1300, or that many cathedrals never were completed up to the last tower or buttress. Yet the productivity of these times is astonishing even by modern standards. In France alone, between 1170 and 1270 some 80 cathedrals and 500 churches of cathedral proportions were begun.

Why were all these cathedrals built? The most obvious motive was, of course, to glorify God—but this is hardly the whole story. The cathedral was the heart and centre of the medieval city, and each town was eager to have the greatest church. When, for instance, the citizens of Amiens completed a magnificent cathedral with an interior height in the nave of 141 feet, the men of Beauvais determined that their cathedral should soar even higher. So they built the nave to the incredible height of 154 feet. But they had, alas, succeeded in glorifying God without defying gravity, for the vaulting collapsed twice. And though rebuilding and reinforcing finally did produce the highest choir in Christendom (157 feet), the Beauvais cathedral was never finished.

As the focus of the town, the medieval church corresponded to the modern community centre. Municipal meetings and ceremonies were held in the cathedral square; the Gothic façade provided a dramatic backdrop for medieval miracle plays performed on the church steps; and the town fair often spilled over into the cathedral square.

Amiens Cathedral, built in 1220-1402, is considered the finest example of Gothic architecture in France. This cut-away drawing illustrates some of the main parts of Gothic cathedrals: (1) central pointed vault, (2) nave arcade, (3) triforium galleries, (4) clerestory, (5) pointed arch, (6) exterior pier or buttress, (7) crockets and finials, (8) flying buttress, (9) timber-trussed roof. The peak of the roof at Amiens is approximately 200 feet from the ground, the interior height of the nave being 141 feet from floor to vaulting. A multitude of stained glass windows bathe the interior with colour, producing an effect of lightness and richness which is accentuated by the ribbed vaulting and ornamental carving.

The carvings and glass of the great buildings reminded the high- and low-born alike of the eternal truths of Christianity. "The pictures and ornaments in a church are the text and scriptures of the laity," wrote a 13th century bishop, and a Bible in glass and stone is exactly what the cathedral was for the illiterate masses. Into it medieval men poured all their learning and all their art. Those who could read had the *Summa* of Thomas Aquinas. Those who could not, had a *summa* in the cathedral.

The skeleton of the Gothic cathedral is both functional and beautiful, and the whole structure is a superb lesson in the play of pressure and support. In this analytical diagram of Amiens Cathedral the arrows show how the weight and thrust of the vaulting and roof are gathered on the flying buttresses and pier buttresses, which are in a sense the backbone of the building. Even the pinnacles of the pier buttresses exert a stabilizing downward thrust. Why did the use of buttresses permit interior effects impossible to achieve in Romanesque building?

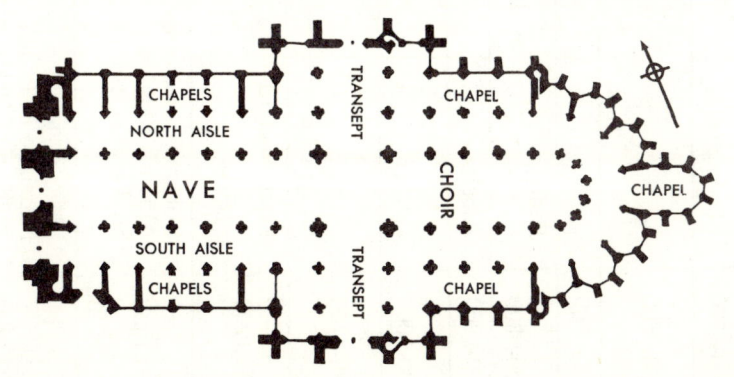

This ground plan of Amiens shows the typical arrangement of the Gothic cathedral.

A MEDIEVAL CULTURAL CALENDAR[1]

	PHILOSOPHY	ARCHITECTURE AND ART	SCIENCE	LITERATURE
300- 400	St. Jerome	Basilica churches		St. Jerome's *Vulgate*
400- 500	St. Augustine			St. Augustine's *City of God*
500- 600		Hagia Sophia; mosaics		St. Benedict's *Rule* Justinian's *Corpus Juris Civilis* Gregory of Tours' *History*
600- 700		Irish illuminated manuscripts		The Koran
700- 800	Bede	Moslem mosques	Bede Moslem translations	Bede
800- 900	Alcuin	Caroline minuscule	Moslem translations	Alcuin
900-1000	Gerbert	Theophilus (Roger) and stained glass	Gerbert	
1000-1100	St. Anselm	Suger of St. Denis Romanesque cathedrals, sculpture, stained glass	Moslem scientists	
1100-1200	Peter Abelard	Romanesque to Gothic	Moslem scientists Adelard of Bath	*Song of Roland* Bernard of Cluny Chrétien de Troyes
1200-1300	Robert Grosseteste Albertus Magnus St. Thomas Aquinas	Gothic cathedrals, sculpture, stained glass	Robert Grosseteste Albertus Magnus Roger Bacon Emperor Frederick II	Vincent of Beauvais *Romance of Reynard* *Roman de la Rose*
1300-1400	Marsiglio of Padua William of Ockham John Wycliffe	Gothic cathedrals (See also the Renaissance Cultural Calendar on p. 453.)		Dante Froissart (See also the Renaissance Cultural Calendar on p. 453.)

[1] Again the student should be cautioned that this is a highly selective chart, as are all the Cultural Calendars in this book. A blank space does not mean that nothing of cultural note happened.

6/SUMMARY AND CONCLUSION

The 13th century was a time of great men and great learning. Among rulers, none put forth higher claims than Pope Innocent III; but by the end of the century the papacy's power had greatly declined, while the French and English monarchies were attaining to new heights of influence. The German emperors, on the other hand, were not able to reverse an earlier trend: competing feudal princes still successfully resisted all attempts to create a strong monarchy.

Though the Church declined in political power, its monastic reformers grew ever more active. Some even abandoned the cloisters to go out into the world to save sinners and help improve the miserable lot of the common man. Nowhere was this church-inspired society, which in the Middle Ages was one and international, more creative than in learning and the arts, a creativity symbolized in the *Summa Theologica* and the vast Gothic cathedrals.

This chapter is called "The Thirteenth-Century Synthesis." A synthesis is a coherent whole made up of many parts. A great deal, though not all, of what has been discussed here bore a close relationship to the Christian faith. For religion inspired the man of the Middle Ages as nationalism does us, or as the city-state commanded the loyalty of the ancient Greek. The 13th century witnessed the last great attempt to organize all ideas—political, social, and economic—under the leadership of a Universal Church. The attempt failed.

It was bound to fail, for a civilization that thinks it knows all the answers is already suffering from hardening of the intellect. While a brilliant attempt at synthesis was made, there were certain things, such as the changes of the Commercial Revolution, that would not quite fit into the pattern. It took only a few extra strains—ambitious national monarchs dominating a weakened papacy, long expensive wars increasing the political power of the middle class—to bring down the whole structure of medieval life.

SOURCE READINGS

(a)

ST. FRANCIS (1181-1226) had been content to base his Rule on a few simple precepts from the Gospels. After his death, however, more detailed regulations, still based on St. Francis' essential beliefs, were drawn up. This Franciscan Rule, a few extracts from which follow, was approved by Pope Honorius III in 1223.

In the Lord's name thus begins the way of life of the brothers minor. The rule and life of the brothers minor is this, to observe the holy Gospel of our Lord Jesus Christ, living in obedience, without property, and in chastity. . . .

. . . If any wish to take up this life and they come to our brothers, the latter shall send them to their provincial ministers. . . . And the ministers shall carefully examine them concerning the Catholic faith and the sacraments of the Church. . . . the ministers shall bid them, in the words of the holy Gospel, to go and sell all their goods and carefully distribute all to the poor. . . . And when the year of probation is over, they shall be received into obedience, promising ever to observe this life and rule. . . . And those who have promised obedience shall have one gown with a hood and, if they wish, one without a hood. And those who really need them can wear shoes. And all the brothers shall wear poor clothing, which they may repair with sackcloth and other scraps, with God's blessing. And I warn and exhort them not to despise or judge men whom they see clothed in soft and coloured raiment or enjoying rich food and drink; but each shall rather judge and despise himself.

I strictly command all the brothers never to receive coin or money either directly or through an intermediary. The ministers and guardians shall make provision, through spiritual friends, for the needs of the infirm and for other brothers who need clothing, according to the locality, season or cold climate. . . .

The brothers shall possess nothing, neither a house, nor a place, nor anything. But, as pilgrims and strangers in this world, serving God in poverty and humility, they shall confidently seek alms, and not be ashamed, for the Lord made Himself poor in this world for us. . . .

. . . And [the brothers] shall not attempt to teach the illiterate, but shall strive for that which they should desire above all things, to have the spirit of the Lord and its holy working, ever to pray to Him with a pure heart, to have humility and patience in persecution and infirmity, and to love those who persecute and censure and revile us, for the Lord says "Love your enemies. . . ."

Select Documents of European History, 800-1492, edited and translated by R. G. D. Laffan (Henry Holt, no date), pp. 106-108.

(b)

When it was noised about that ROGER BACON (1214-94) was writing a great book on philosophy, Pope Clement IV heard of the project and asked Bacon to send him a copy. But the best Roger could do was to give the pope an outline of what he intended to write. This discussion of the rainbow, quoted from Bacon's outline, the *Opus Maius (Major Work)*, is probably derived from his teacher, Robert Grosseteste.

I now wish to unfold the principles of experimental science, since without experience nothing can be sufficiently known. For there are two modes of acquiring knowledge, namely, by reasoning and experience. Reasoning draws a conclusion and makes us grant the conclusion, but does not make the conclusion certain. . . . if a man who has never seen fire should prove by adequate reasoning that fire burns and injures things and destroys them, his mind would not be satisfied thereby, nor would he avoid fire, until he placed his hand or some combustible substance in the fire, so that he might prove by experience that which reasoning taught. . . .

He therefore who wishes to rejoice without doubt in regard to the truths underlying phenomena must know how to devote himself to experiment. For authors write many statements, and people believe them through reasoning which they formulate without experience. Their reasoning is wholly false. For it is generally believed that the diamond cannot be broken except by goat's blood, and philosophers and theologians misuse this idea. But fracture by means of blood of this kind has never been verified, although the effort has been made; and without that blood it can be broken easily. For I have seen this with my own eyes, and this is necessary, because gems cannot be carved except by fragments of this stone. . . .

But if we give our attention to particular and complete experiments and such as are attested wholly by the proper method, we must employ the principles of this science which is called experimental. I give as an example the rainbow and phenomena connected with it. . . .

Let the experimenter first, then, examine visible objects, in order that he may find colours arranged as in the phenomena mentioned above and also the same figure. For let him take hexagonal stones from Ireland or from India . . . and let him hold these in a solar ray falling through the window, so that he may find all the colours of the rainbow, arranged as in it, in the shadow near the ray. . . . Let the experimenter proceed further, and he will find this same peculiarity in crystalline stones correctly shaped, and in other transparent stones. . . . And further let him observe rowers, and in the drops falling from the raised oars he finds the same colours when the solar rays penetrate drops of this kind. The same phenomenon is seen in water falling from the wheels of a mill; and likewise when one sees on a summer's morning the drops of dew on the grass in meadow or field, he will observe the colours. Likewise when it is raining, if he stands in a dark place and the rays beyond it pass through the falling rain, the colours will appear in the shadow near by; and frequently at night colours appear around a candle. Moreover, if a man in summer, when he rises from sleep and has his eyes only partly open, suddenly looks at a hole through which a ray of the sun enters, he will see colours. Moreover, if seated beyond the sun he draws his cap beyond his eyes, he will see colours; and similarly if he closes an eye the same thing happens under the shade of the eyebrows; and again the same phenomenon appears through a glass vessel filled with water and placed in the sun's rays. Or similarly if one having water in his mouth sprinkles it vigorously into the rays and stands at the side of the rays. So, too, if rays in the required position pass through an oil lamp hanging in the air so that the light falls on the surface of the oil, colours will be produced. Thus in an infinite number of ways colours of this kind appear, which the diligent experimenter knows how to discover. . . .

J. B. Ross and M. M. McLaughlin, editors, *The Portable Medieval Reader* (Viking Press, 1949), pp. 626-627, 629-632.

(c)

The abbot HAIMON, of St. Pierre-sur-Dives in Normandy, wrote to an English prior this account of the religious enthusiasm attending the building of the cathedral at CHARTRES (55 miles southwest of Paris), in the year 1145.

. . . For who ever saw, who ever heard, in all the generations past, that kings, princes, mighty men of this world, puffed up with honours and riches, men and women of noble birth, should bind bridles upon their proud and swollen necks and submit them to wagons which, after the fashion of brute beasts, they dragged with their loads of wine, corn, oil, lime, stones, beams, and other things necessary to sustain life or to build churches, even to Christ's abode? Moreover, as they draw the wagons we may see this miracle that, although sometimes a thousand men and women, or even more, are bound in the traces (so vast indeed is the mass, so great is the engine, and so heavy the load laid upon it), yet they go forward in such silence that no voice, no murmur, is heard; and, unless we saw it with our eyes, no man would dream that so great a multitude is there. When, again, they pause on the way, then no other voice is heard but confession of guilt, with supplication and pure prayer to God that He may vouchsafe pardon for their sins; and, while the priests there preach peace, hatred is soothed, discord is driven away, debts are forgiven, and unity is restored betwixt man and man. If, however, anyone be so sunk in evil that he will not forgive those who have sinned against him, nor obey the pious admonition of the priests, then is his offering forthwith cast down from the wagon as an unclean thing; and he himself, with much shame and ignominy, is separated from the unity of the sacred people. . . .

... Nor can we wonder that the older and more aged undertook this burdensome labour for the multitude of their sins; but what urged boys and children to this work? Who brought them to that good Teacher who hath perfected His praise in the mouths and works of children? Hath perfected, I say, that by all means the work begun among the elders may be proved to have been completed by the children; for you might see them, with their own little kings and leaders, bound to their laden wagons, and not dragging with bowed backs like their elders but walking erect as though they bore no burden, and (more wonderful still) surpassing them in nimbleness and speed. Thus went they in a fashion far more glorious, holy, and religious, than any words of ours could express.

When they were come to the church, then the wagons were arrayed around it like a spiritual camp; and all that night following, this army of the Lord kept their watches with psalms and hymns; then waxen tapers and lights were kindled in each wagon, then the sick and infirm were set apart, then the relics of the saints were brought to their relief, then mystical processions were made by priests and clergy, and followed with all devotion by the people, who earnestly implored the Lord's mercy and that of His blessed Mother for their restoration to health. ...

Life in the Middle Ages, selected, translated, and annotated by G. G. Coulton (Cambridge University Press, 1929), II, 19-21.

(d)

THEOPHILUS was a Greek who, after travelling extensively, became a Benedictine under the name of Roger, and lived in a monastery near Paderborn, Germany. His manual of art techniques, from which this extract is taken, was written late in the 10th century.

When you wish to compose glass windows, first make for yourself a flat wooden table, of such breadth and length that you can work upon it two portions of the same window; and taking chalk, and scraping it with a knife over all the table, sprinkle water everywhere, and rub it with a cloth over the whole. And when it is dry, take the dimensions of one portion of the window in length and breadth, marking it upon the table with rule and compass with the lead or tin; and if you wish to have a border on it, portray it with the breadth which may please you, and in the pattern you may wish. Which done, draw out whatever figures you will, first with the lead or tin, then with a red or black colour, making all outlines with study, because it will be necessary, when you have painted the glass, that you join together the shadows and lights according to the [drawing on the] table. Then arranging the different tints of draperies, note down the colour of each one in its place; and of any other thing which you may wish to paint you will mark the colour with a letter. After this take a leaden cup, and put chalk, ground with water, into it: make two or three pencils for yourself from hair, either from the tail of the marten, or badger, or squirrel, or cat, or the mane of the ass, and take a piece of glass of whatever kind you like, which is in every way larger than the place upon which it is superposed, and fixing it in the ground of this place, so that you can perceive the drawing upon the table through the glass, so portray with the chalk the outlines upon the glass. And if the glass should be so thick that you cannot perceive the lines which are upon the table, taking white glass, draw upon it, and when it is dry place the thick glass upon the white, raising it against the light, and as you look through it, so portray it. In the same manner you will mark out all kinds of glass, whether for the face, or in draperies, in hands, in feet, in the border, or in whatever place you intend to place colours.

Lopez, *The Tenth Century: How Dark the Dark Ages?,* pp. 44-45.

FURTHER READING

1. **THE 13TH CENTURY CHURCH AND THE MONARCHIES**
   BALDWIN, M. W., *The Medieval Papacy in Action* (Macmillan, 1940)
   CANNON, W. R., *History of Christianity in the Middle Ages from the Fall of Rome to the Fall of Constantinople* (Abingdon, 1960)
   * CORBETT, J. A., *The Papacy* (Anvil, 1956)
   DANIEL-ROPS, H., *Cathedral and Crusade: Studies of the Medieval Church, 1050-1350* (Dent, 1957)
   FLICK, A. C., *The Decline of the Medieval Church*, 2 vols. (Knopf, 1930)
   † MASSON, G., *Frederick II of Hohenstaufen* (Secker and Warburg, 1957)
   MOORMAN, J. R. H., *Church Life in England in the Thirteenth Century* (Cambridge, 1944)
   PACKARD, S. R., *Europe and the Church under Innocent III* (Holt, 1927)
   PETIT-DUTAILLIS, C., *The Feudal Monarchy in France and England* (Routledge, 1936)
   * RUNCIMAN, S., *The Sicilian Vespers: A History of the Mediterranean World in the Later Thirteenth Century* (Penguin, 1960)
   † WARREN, W. L., *King John* (Eyre and Spottiswoode, 1961)

2. **LEARNING, LAW, AND POLITICAL THOUGHT**
   * COPLESTON, F. C., *Medieval Philosophy* (Torchbook, 1961)
   CROMBIE, A. C., *Robert Grosseteste and the Origins of Experimental Science, 1100-1700* (Oxford, 1953)
   † * DAWSON, C., *Medieval Essays* (Image, 1959)
   EASTON, S. C., *Roger Bacon and His Search for a Universal Science* (Columbia, 1952)
   * FREMANTLE, A., *The Age of Belief: The Medieval Philosophers* (Mentor, 1955)
   HASKINS, C. H., *Studies in the History of Mediaeval Science* (Harvard, 1927)
   ———, *Studies in Mediaeval Culture* (Oxford, 1929)
   † HAWKINS, D. J. B., *A Sketch of Mediaeval Philosophy* (Sheed and Ward, 1947)
   * MORRALL, J. B., *Political Thought in Medieval Times* (Torchbook, 1962)

3. **LITERATURE, MUSIC, AND ART**
   * ADAMS, H., *Mont-Saint-Michel and Chartres* (Anchor, 1959)
   BRANNER, R., *Gothic Architecture* (Braziller, 1961)
   EVANS, J., *Art in Medieval France* (Oxford, 1948)
   HARVEY, J., *The Gothic World* (Batsford, 1950)
   * HUIZINGA, J., *The Waning of the Middle Ages* (Anchor, 1954)
   MOREY, C. R., *Medieval Art* (Norton, 1942)
   * PEVSNER, N., *An Outline of European Architecture* (Penguin, 1963)
   REESE, G., *Music in the Middle Ages* (Norton, 1940)
   SIMSON, O. VON, *The Gothic Cathedral* (Pantheon, 1962)
   † * TEMKO, A., *Notre-Dame of Paris* (Compass, 1959)

14. Popes, Kings, and Parliaments

A most just law, established by the careful providence of sacred princes, exhorts and decrees that what concerns all should be approved by all....
PARLIAMENTARY WRITS OF 1295 TO
EDWARD I'S CLERGY

After the glories of the 13th century, the next hundred years must have seemed an anticlimax to both the Church and the national monarchies. Not only the popes but also the French, English, and German kings became weaker. In fact, the only political progress in the 14th century was the evolution of English representative government. This chapter, then, concentrates on one of our most important inheritances from the Middle Ages —parliamentary government.

1/THE DECLINE OF THE MEDIEVAL CHURCH

The Babylonian Captivity We have seen the way in which King Philip the Fair of France humiliated the papacy in the person of Pope Boniface VIII. In 1305 the French archbishop of Bordeaux became Pope Clement V, and, on the pretext of disturbed political conditions in the Papal States, took up what was assumed to be only temporary residence at Avignon on the east bank of the Rhone River. Actually he never did move to Rome, nor did his six French successors; and while Avignon was not technically in France it was surrounded by French lands, so that the papacy could hardly escape French pressure. For this reason historians have used the term "Babylonian Captivity" (see page 62) to describe the period (1309-76) when the French popes reigned from Avignon.

At Avignon the popes concentrated on devising means to increase their financial resources, and the resulting luxury was severely censured by some contemporaries. One of these critics, himself a former papal office-holder, bitterly contrasted the early Church with the court at Avignon.

> Here reign the successors of the poor fishermen of Galilee; they have strangely forgotten their origin. I am astounded, as I recall their predecessors, to see these men loaded with gold and clad in purple, boasting of the spoils of princes and nations; to see luxurious palaces and heights crowned with fortifications, instead of a boat turned downwards for shelter.

The individuals who held the office of pope during this period may not have been any worse than other monarchs of the century, but neither were they much better. The English and the Germans denounced the papacy for allying itself with France—and hence against them; and even the Italians (who missed the political and economic advantages of the papacy in their midst) denounced Avignon as the "Babylon of the West."

The Church and Its Critics The popes knew that their position at Avignon was disadvantageous, but it was not until the last years of his pontificate that Pope Gregory XI, the seventh Avignonese pope, moved to Rome. Probably Gregory intended only a visit to Rome, but when he died there before he could return to Avignon the Romans forced the College of Cardinals (a majority of whom were French) to choose an Italian, Urban VI (1378-89), as his successor. Only four months later, however, the French cardinals declared Urban's election invalid and elected a French pope, Clement VII, who again established the papal residence at Avignon. Now the Church had two popes, each excommunicating the other, each claiming to be the only true successor to Peter.

The Great Schism This unhappy period of the Church's history is called the "Great Schism" and lasted for a generation. Indeed, before it was settled in 1417 there were at one time three popes. It is not hard to see how the Babylonian Captivity and the Great Schism would destroy much of the prestige and influence of the popes, and many sincere Christians in the 14th century were harshly critical of both the popes and their office. Four of these critics are worth mentioning, two Italians and two Englishmen.

Dante Alighieri (1265-1321) was a cultured Italian who was exiled from Florence after supporting a municipal revolution in which Boniface VIII played a part. Dante may have been unsuccessful in politics, but he turned out to be a literary genius. His most famous poem, the *Divine Comedy*, relates his adventures as he progressed through Hell, Purgatory, and Paradise. In the poem the emperor Henry VII (whose ambition was to join a unified Italy to the German Empire) is in Heaven, while Pope Boniface VIII is confidently awaited in Hell. This is not, of course, the main message of the great poem, but is mentioned here simply as an illustration of 14th century papal criticism.

Dante

Marsiglio of Padua (1270-1343), in a remarkable book called *Defensor Pacis* (Defender of the Peace), wrote that since the Church was the entire body of believing Christians it possessed sovereign authority. The pope and all the clergy were, therefore, not the rulers but the servants of the people, whose will was expressed through a general council of the Church. Marsiglio also had his views on secular government. Here he favoured a limited monarchy in which the "better elements" of the citizenry could, if need be, depose a king who defied the laws.

Marsiglio

In England, William of Ockham (1290-1350), an Oxford Franciscan whose theology was called into question by Avignon, also joined in an attack on the power of the popes. He stated that the pope possessed no temporal authority and that he might even err in spiritual matters.

Finally, John Wycliffe (1320-1384), a doctor of divinity at Oxford and one of England's greatest 14th century theologians, came into prominence in the quarrel over whether or not, as a result of King John's surrender of England to Innocent III in 1213, the king of England owed a yearly tribute to the pope. When Edward III of England, now at war with France, refused the tribute to the Avignonese papacy, Wycliffe provided the intellectual justification for the refusal, and went on to assert that the king could even deprive the papacy of its property. Wycliffe's arguments appealed to English nationalism, and he began to attract large and enthusiastic audiences.

John Wycliffe

Many of Wycliffe's opinions are now commonplaces of Protestant thought. But in the 14th century his ideas were rankest heresy, and Pope Gregory XI condemned some of them as erroneous and reminiscent of those of Marsiglio "of damned memory." Eventually Wycliffe was forbidden to preach or lecture, with the result that he spent his last years in retirement working on an English translation of the Bible.

It was perhaps symptomatic of the Church's decline that, despite the heavy fire it was under in the 14th century, no new reforming orders were created. In fact a new Cluniac or a new Franciscan order might have been frowned upon as dangerous and heretical. But the seeds of unrest had been sown. A Church that would not reform itself might, in time, have reform thrust upon it.

2/FRANCE, ENGLAND, AND THE EMPIRE

With the Church in a state of spiritual bankruptcy, the leadership of society passed to the secular monarchs. The new rulers of Europe tried to assume in the 14th century the power the papacy had wielded in the 13th, but they found themselves under severe handicaps. The economic depression complicated by the ravages of the Black Death made it impossible for them to fill the political vacuum left by the papacy's collapse.

Before looking briefly at the political history of France, England, and the German Empire, let us list their rulers.

FRANCE	ENGLAND	GERMAN EMPIRE
Louis X (1314-16)	Edward II (1307-27)	Henry VII (1308-13)
Philip V (1316-22)		
Charles IV (1322-28)		
Philip VI (1328-50)		Louis IV (1314-47)
John II (1350-64)	Edward III (1327-77)	
Charles V (1364-80)	Richard II (1377-99)	Charles IV (1347-78)
Charles VI (1380-1422)	Henry IV (1399-1413)	Wenceslas IV (1378-1400)

Not all of these rulers are equally important, but the table will allow you to keep track of their chronological sequence.

The War that Lasted a Century In 1337 King Edward III of England laid claim to the throne of France. He based his claim on the fact that he was the son of Isabelle, the daughter of Philip IV of France, and this hereditary right was the ostensible cause for a war that broke out between the two countries and lasted in an on-again off-again fashion until the middle of the 15th century. This was the famous Hundred Years' War.

Actually almost every king of England since William the Conqueror had been at war with France at some time; so it was not very difficult to find excuses for renewing hostilities between the two traditional enemies. Gascony was a festering sore in France's side; French and English sailors cheerfully practised piracy on each other in the Channel; and French interference in Flanders was always a possible threat to England's best market for her wool.

Although France was the larger country, with greater resources and a population of 16 million, England's 3 million inhabitants held a distinct military advantage. The French army, consisting mainly of heavy-armed feudal knights, found itself up against an efficient army of paid mercenaries ready for indefinite service and reinforced by archers with the devastating English long-bow. And while the cross-bowmen of the French army wielded a formidable weapon, a good long-bowman could shoot a dozen arrows a minute to every two bolts of the cross-bowman. The artillery of the day was most primitive, being more frightening on account of its noise than deadly on account of its accuracy.

Compared to the total war of our 20th century, the Hundred Years' War seems a rather strange sort of adventure. Armies wandered around after each other in France and the occasional castle was captured, but decisive engagements were few. Finally in 1346 the French and English armies could no longer avoid a full-scale battle. At Crécy (see page 426), wave after wave of French knights broke against a hill defended by English long-bows, until at last the outnumbered English men-at-arms swept aside what was left of the French army. It was a great victory for England, and Edward III went on to capture Calais—an excellent invasion port for future English expeditions. In 1356 Edward's son, the Black Prince (so-called from the colour of his armour), used the tactics of Crécy to win another smashing victory over the French at Poitiers.

Crécy and Poitiers

By now France was exhausted and the war was unpopular. In 1357, in order to get more money for the war, the king was forced by the French Estates-General of the North (only rarely did both North and South meet as in 1302) to grant a *Grande Ordonnance* embodying reforms which, among other things, gave control of the collection of taxes to the Estates and made royal officers responsible to it. If this document had remained in force it might have made France a constitutional monarchy like England; but disorder, including a peasant revolt, allowed the French monarchy to reassert its authority. Peace with England was negotiated in 1360, and each side had a brief breathing spell.

In 1369, however, the Hundred Years' War was renewed, the English exhausting their resources in valiant but useless sorties through the French countryside, and the French grimly holding on, recapturing castles and cutting the English supply lines instead of risking pitched battles. By 1378 both Edward III and the Black Prince were dead, and two years later so was Charles V of France. Although the war still dragged on rather casually until 1396, when a twenty-year truce was concluded, after 1380 there were so many domestic troubles in each kingdom that the war was neglected. In France, civil wars over the succession to the throne produced a chaotic situation, and in England the unpopular Richard II was deposed and murdered by his rebellious barons—much as his great-grandfather Edward II had been over seventy years earlier.

Thus we see that in the 14th century neither the French nor the English monarchs were very successful rulers. The French kings were weaker in 1400 than they had been in 1300, and were in continual hot water with their subjects. English kings were equally weak, but there was this difference: the parliament had, as we shall see at the end of this chapter, assumed a position of dominance which the French Estates-General never succeeded in winning.

In the meantime, what was happening in the German Empire?

The Age of the Princes The 14th century in Germany has been called the "Age of the Princes." The emperor Henry VII was the last ruler to try to control Italy as well as Germany, and after 1313 the urban groups of Italy and the stronger nobles of Germany split up the power.

An event of some importance did, however, occur in the reign of Charles IV. In a document called the Golden Bull, Charles attempted to define the power of the greater princes by fixing the number eligible to elect the emperor at seven—three bishops and four lay princes. Charles hoped by this means to end disputed elections and civil wars. "The Golden Bull," writes one historian, "was promulgated in 1356; in 1493 Maximilian I ascended the throne. Between these two dates there is scarcely an event worth chronicling in imperial affairs." During the 14th and 15th centuries Germany can hardly be either defined or described, so many and so bewildering were the wars, alliances, and conquests among more than a hundred principalities. Chaos reigned: emperors did not.

The 14th century feudal monarchies of France, England, and Germany, then, failed in their exercise of power just as dismally as the papacy had failed during the preceding century. Of course these were not the only feudal monarchies; but these were the greatest, and therefore the ones from which most might have been expected. In trying to achieve power too quickly they only succeeded in weakening themselves.

3/THE EVOLUTION OF ENGLISH REPRESENTATIVE GOVERNMENT

Of all the ways in which the monarchs of the 14th century lost power, the most remarkable was through the development of parliament in England. The story of this evolution from Anglo-Saxon times to the end of the 14th century both reviews and summarizes the history of the Middle Ages.

In the beginning the word "parliament" was a sort of slang expression, a word that meant simply a "parleying" or a talking together. The first recorded usage is in the 12th century *Chanson de Roland* when an emir says to a dying king, "I cannot hold long *parlement* with you." And soon after their marriage King Louis IX of France and Queen Marguerite met on the back stairs of the palace to hold their "parlement" free from the jealous eyes (and ears) of the king's mother.

As time went on the term acquired a wider significance. By the 12th century Italian cities held general assemblies which they called *parlamenti*, and a contemporary of King John referred to the assembly at Runnymede as a "parliament" in which John "gave his charter to the barons." What

Westminster Hall in London dates from 1097. It was built by William Rufus, son of William the Conqueror, and though it is 240 feet long and 68 feet wide he is said to have remarked that it was "a mere bedchamber compared with what I had intended to build." Floods from the nearby Thames so damaged the hall over the centuries that Richard II determined to reconstruct it in the perpendicular style. It was then that the columns were removed and the richly carved hammerbeam roof of oak was installed, "the greatest triumph of English medieval carpentry . . . the largest timber roof in Europe unsupported by pillars." Westminster Hall has been the scene of such historic events as the deposition of Richard II (1399), the trials of Charles I (1649) and Warren Hastings (1788-95), the Coronation Banquet of George IV (1821), and the lying-in-state of George VI (1952).

we must try to discover is how a simple parley was turned into a parliament in which a national and an elective assembly were finally united.

A National Assembly When Duke William of Normandy conquered England in 1066 he found there a machinery of government both central and local. The Anglo-Saxon kings had governed their realm through a household, a kind of embryo civil service, supplemented by an aristocratic council called the *witan*. If the king was wise he remembered to consult this witan in order to get its backing, and the witan accordingly provided a time and place for men from all over the kingdom to meet for discussion.

This machinery of central government stood ready for Norman adoption and adaptation, and William the Conqueror took it over. He was elected king by the witan and was consecrated by an Anglo-Saxon primate. William brought Norman feudalism to England. He owned all the land and parcelled it out to his barons, specifying that he was to receive in return the services of 5000 knights. But William did not merely try to reproduce the feudal conditions of Normandy. He now exacted allegiance not just from his tenants-in-chief, the barons of whom he was the direct overlord, but from *all* his vassals.

At a large meeting of tenants-in-chief and sub-tenants held on Salisbury Plain in 1086, all swore homage and allegiance to William as overlord and king. Moreover, William turned the king's council and court (*curia regis*) into a kind of feudal court to which his representatives might come either as summoned by the king or on their own initiative as part of their privileges as vassals. On special occasions the council was enlarged to include all vassals (as at Salisbury), though ordinarily it included only the greater barons. Thus William the Conqueror welded Norman and Anglo-Saxon institutions together in such a way that the English government of the Middle Ages became incomparably stable.

An Elective Assembly   Anglo-Saxon local government had been based on units called shires, or, as we would say, counties. The shire-reeve (= sheriff) supervised the courts held in the shire, or in the *hundreds* and *townships* (smaller subdivisions of the shire). Again, the Normans took over this local government and later added to it the administration of the mercantile towns called *boroughs*. In shire and hundred courts, the reeve and four men represented each township along with twelve burgesses representing each borough.

In speaking for the whole shire Henry II's juries were virtually miniature parliaments. Under Richard I, the record of a case transferred to the central court had to be brought to Westminster by four knights of the shire. It was this time-tested system of shire representation on a national scale that King John adapted in 1213 when he summoned to his council four knights from each shire.

Henry III also summoned shire representatives to his council when he needed money for the French wars. These royal *writs of summons,* that is, the written orders from the central government to the sheriff, specified that two knights were to be selected by the shire court to represent the whole shire. In 1258, as a result of a baronial uprising, the king was forced by the Provisions of Oxford to summon three parliaments each year whether he wanted them or not, and to submit to supervision of the policy of his permanent council. Moreover, these Provisions specified the election in each county of "four discreet and lawful knights" to hear grievances. Since the practice of presenting written complaints to royal judges in royal courts was already well known, it was now easy to extend this practice to that of presenting petitions in parliament. Henry III, however, had no intention of keeping within the bounds of the Provisions of Oxford, and when he openly flouted them civil war broke out in 1261.

Simon de Montfort   When Henry III was defeated at the battle of Lewes in 1264, the supreme power came to rest in the hands of the victor—Henry's brother-in-law, Simon de Montfort. After being deserted by many of his baronial allies, this rebel baron issued writs in the king's name directing the sheriffs to select two knights from every shire to come to a parliament at London on January 20, 1265. But Simon did not stop there. Royal letters were also dispatched requiring the citizens of each borough to "send two of the more discreet, lawful, and upright citizens or burgesses."

It is this provision to include representatives of the English burgher class that has sometimes earned for Simon de Montfort the proud title of "founder of the House of Commons." For the first time the burgesses and knights were summoned to a parliament simultaneously. But Simon de Montfort was no great democrat. The loss of his baronial adherents by 1264 meant that he had to broaden the base of his support by summoning knights and burgesses along with the lords. Doubtless Simon had no intention of founding a fully representative parliament, yet even so he did create the precedent. True, Henry III did not follow it; but his son Edward I continued, from time to time, to call burgesses to parliament. Out of Simon's need for personal support had come a decisive step.

Between 1258 and 1272 there were at least 31 parliaments convoked in England. But despite the innovations of Simon de Montfort the position of the representatives of shires and towns was precarious. The main business of parliaments was carried on in parleys involving king, council, judges, barons, and bishops, whereas the representatives played a distinctly minor role. In fact their main function seems to have been to agree to bind their shires or towns to pay the taxes upon which the king and his great men had already agreed. It is not during the reign of Henry III but rather during the reigns of the first three Edwards that parliament enjoyed its greatest development.

Representation for Taxation Edward I called parliaments, not because he was interested in furthering something called the English Constitution, but to secure his own advantage. During his reign, only three of the sixteen parliaments to which the commons (as the knights and burgesses eventually came to be called) were summoned were not asked for money. This indicates a strong financial motive for summoning the commons—nor can it be mere chance that the appearance of the knights and burgesses in the first "general parliament" summoned by Edward I (April, 1275) coincided with the first specific tax on wool.

These representatives were not considered to be acting only on their own behalf. The writs clearly specified that when the representatives came to parliament they should have authority to bind their shires and boroughs then and there to whatever the parliament would agree upon. Such a formula continued in official use for over 500 years. Indeed, it looks as if Edward insisted that knights and burgesses have this binding authority to act on behalf of their communities so that taxes could be levied in parliament with the assurance that they would be collected later.

These same knights and burgesses also brought information to the king on the condition of his realm. The king's dispensation of justice in dealing with their petitions (there were 450 of them in the 1305 Parliament) could make his whole kingdom aware of the strength of the royal arm.

Under Edward I the knights and burgesses gained experience and were re-elected to successive parliaments. But the growth of popular representation was a very slow process. True, in summoning the bishops to parliament in 1295, Edward could quote in the writs a phrase from Justinian's Code "that what concerns all should be approved by all;" but we may suspect that this was little more than a rhetorical flourish. Between 1290 and 1310 representatives of the shires and boroughs came to only 13 out of 34 parliaments.

Edward I presides over this opening session of a late 13th century parliament, flanked by the members of his council including the Archbishops of Canterbury and York. In the centre of the floor may be seen the chief justices, along with various other judges and lawyers. They are seated on woolsacks, that is, sacks stuffed with sheep's wool and covered with red cloth, homely symbols of an important source of wealth in a nation which exported some 30,000 sacks of wool each year. (Incidentally, the Lord Chancellor still sits on a woolsack in the House of Lords, and until a few years ago judges of the Supreme Court of Canada sat on woolsacks in the Senate.) About to take their seats along the sides of the hall are the lords of church and state, and the knights of the shire; while at the far end of the hall—and still apart from the knights—stand the burgesses from the towns.

In the reign of Edward II, historians have seen a turning-point in the fortunes of the commons. Once again in this reign, as in that of Henry III and of John, a baronial revolt attempted to establish a commission of barons to control the king. The Ordinances of 1311 were no more successful in doing this than had been the Provisions of Oxford of 1258 or Magna Carta of 1215. But the Ordinances marked an advance over previous attempts to control the king, and this advance can be seen in their emphasis on the role of parliament. The Ordainers of 1311 did not, it is true, stress the functions of the commons; but they did insist that war might not be waged without baronial assent *in parliament,* that royal grants might not be made without baronial assent *in parliament,* and that even royal ministers might not be appointed without baronial assent *in parliament.* The Ordainers also specified that a "committee" of parliament be elected to hear complaints against ministers who violated the Ordinances. The important thing is that in 1311 the barons chose to make parliament the vehicle of their reforms. From 1311 to 1327 representatives of the commons came to 17 out of 19 parliaments, and after 1327 the commons were invariably present in parliament.

The following diagram illustrates this evolution of parliament into a national elective assembly.

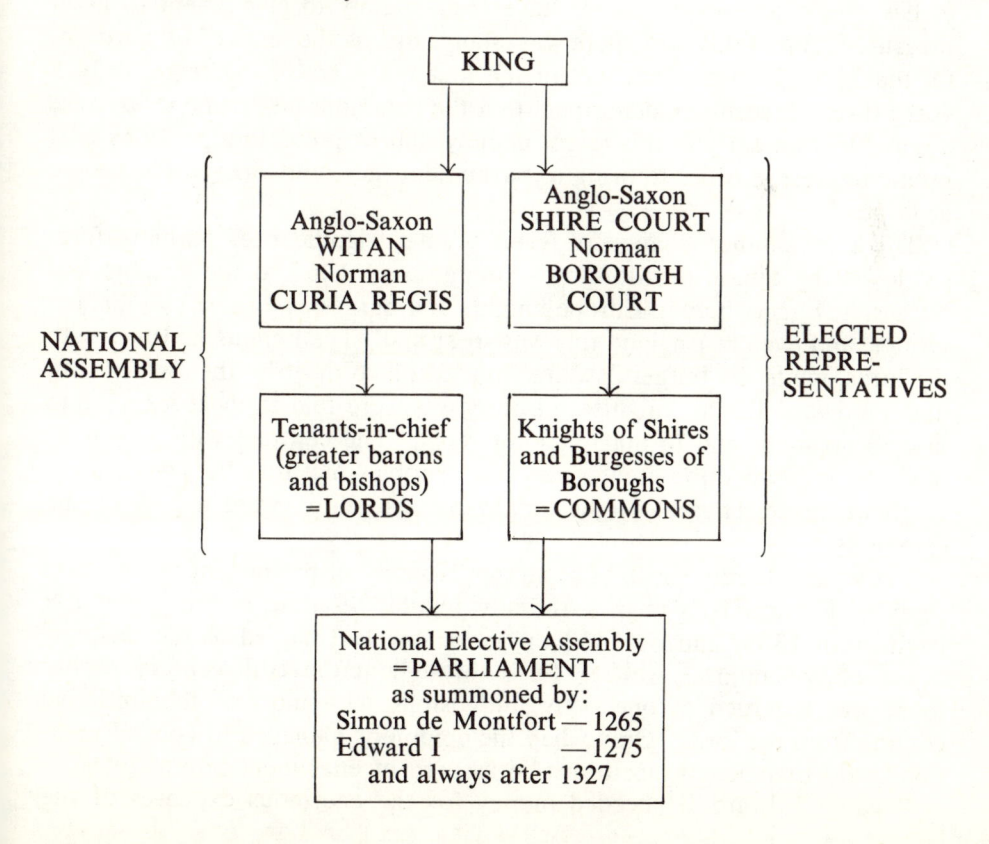

```
                           ┌─────────┐
                           │  KING   │
                           └────┬────┘
               ┌────────────────┴────────────────┐
               ▼                                 ▼
      ┌─────────────────┐              ┌─────────────────┐
      │  Anglo-Saxon    │              │   Anglo-Saxon   │
      │     WITAN       │              │  SHIRE COURT    │
      │    Norman       │              │     Norman      │
      │  CURIA REGIS    │              │    BOROUGH      │
      │                 │              │     COURT       │
      └────────┬────────┘              └────────┬────────┘
               │                                │
               ▼                                ▼
      ┌─────────────────┐              ┌─────────────────┐
      │ Tenants-in-chief│              │Knights of Shires│
      │ (greater barons │              │ and Burgesses of│
      │   and bishops)  │              │    Boroughs     │
      │    =LORDS       │              │   =COMMONS      │
      └────────┬────────┘              └────────┬────────┘
               │                                │
               └────────────────┬───────────────┘
                                ▼
                   ┌──────────────────────────┐
                   │ National Elective Assembly│
                   │      =PARLIAMENT          │
                   │    as summoned by:        │
                   │ Simon de Montfort — 1265  │
                   │ Edward I          — 1275  │
                   │   and always after 1327   │
                   └──────────────────────────┘
```

NATIONAL ASSEMBLY { ... } ELECTED REPRESENTATIVES

As the 14th century progressed, the nature of parliament continued to change significantly. Up until the reign of Edward I it had been mainly a factory for the dispensation of justice; now it shifted its concentration to the intricacies of politics. In addition, parliament came to be separated distinctly into king, lords, and commons. How did such a transformation come about?

Under Edward III, the changes in emphasis of parliamentary business begun under Edward II were consolidated. Whereas previously the official element, the king's professional counsellors, ministers, and judges, had predominated and done nearly all the work, the political repercussions of Edward II's reign tended to push them into the background. The king's enlarged council came to be more a baronial body and less a professional one, the nobles now regarding summons to the council as a right rather than a duty. Consequently the functions of parliament changed, and private petitions and judicial work dropped into a position of secondary importance while politics became the main concern. It was also in Edward III's reign that a group of about 50 nobles first began to constitute a rudimentary "house of lords."

Meanwhile, how were the commons developing? They originally seem to have been summoned, as we have seen, mainly to give assent to fiscal measures. But in this, too, there was change during the reign of Edward III. Of the 21 parliaments that met in the first ten years of the reign, only 5 voted taxes. It seems evident, then, that the commons also came to exercise a new function early in this reign, namely that of presenting petitions of a public interest in order to bring to parliament the complaints of the people at large.

There is another change of even greater significance, perhaps foreshadowed by Simon de Montfort's summons of 1265 to the knights and burgesses. Everywhere in Europe knights and burgesses formed two distinct groups, and even in England this was so at first. The knights had met with the lords, while the burgesses were shunted off to meet by themselves. But under Edward III the knights and burgesses were told to meet *together* to discuss taxation, and to make a joint reply. The contrast with France is obvious: in France the greater and lesser nobles met together, whereas in England the lesser nobles (the knights of the shire) were a part of the commons.

Yet we can hardly speak of separate "Houses" of parliament in the early years of Edward III's reign. Actually regular separate meetings did not begin until 1339, and by 1352 the commons had moved to the chapterhouse of Westminster Abbey. Even throughout the 14th century parliament was regarded as one body, indivisible, with most of the initiative coming from the lords. Only when the commons managed to control taxation could their lengthy petitions be assured of enactment into statutes.

Because Edward III needed money for the enormous expenses of the

Hundred Years' War he summoned parliaments with great regularity—48 in 50 years. The commons, as a result, came to assert certain claims. These claims were as follows:

1. Parliamentary consent for all taxation
2. Agreement of both lords and commons on legislation
3. Right to inquire into and amend abuses of administration
4. Right to impeach (i.e. put on trial) the king's ministers for maladministration
5. Right to have a chairman, and to have secrecy of debate in their separate meetings

We are very fortunate in possessing a comprehensive account of the Parliament of 1376, in the form of an eyewitness description of a commons' debate which illustrates all these claims. This account will be found on pages 427-428.

After 1376 no contemporary could underestimate the leading part that the commons could take—provided they had the backing of the lords. It is almost impossible to estimate the extent to which they were able to act by and for themselves.

During Richard II's reign the commons reasserted the claims made under Edward III, with the lords, as usual, being responsible for any successful commons' actions. That, at least, is the clear implication given by a satirist who described a parliament of Richard II in 1398, and whose description may be read in the Source Reading.

The Seeds of Democracy Though the commons were hardly independent by 1400, they at least were present in parliament and had won a number of fundamental rights. The growth of parliament must have been a baffling phenomenon to medieval Englishmen, since there was, in fact, nothing quite like it in the rest of Europe.

What accounts for this success of the English parliament? In parliament's formative period—the 13th and 14th centuries—there was maintained a rough balance of power among king, lords and commons. The king could not become a tyrant; the lords could not become an oligarchy; and the commons could not turn the kingdom into a democracy. For the political power of the king or the lords was balanced by the economic power of the knights and burgesses. On occasion, two of these groups might co-operate against a third, but no alliance could destroy any one of the three. Each needed the other two. Moreover, all three could meet in parliament reasonably conveniently because England was a small country.

Contrast the situation in France. There the kingdom had taken a much longer time to become unified, and no corporate feeling grew up between

French lords and townsmen. The French Estates-General was, as we have seen, first convened by Philip the Fair in 1302, but it never gained enough experience to exert significant pressure on the king. The nobility had only opposed the king on a local level and was apparently unable to transfer this technique to the national scene, while the towns were jealous of one another and did not co-operate. The Spanish story is the same. Moreover, since both of these countries were much larger than England, national assemblies were expensive and difficult to arrange.

What was the situation in the assemblies of other smaller countries? In Italy and the various German states, city and state assemblies existed but were not at all national in scope. Leagues of Italian or German cities did not respect any national authority. In other words the rulers were pushed around at will by both nobles and townsmen.

Thus in neither France nor Spain, Italy nor Germany, did there develop national assemblies in which a balance was maintained amongst king, nobles, and townsmen. Only in England did this occur, thanks to a happy conjunction of geography and political and economic development. The outcome was an English parliament in which there was, by the 14th century, popular representation. This elective representative element, an essential element in all modern governments that claim to be governments by the people, had its origins in the day-to-day government of medieval England.

4/SUMMARY AND CONCLUSION

In the 14th century both the Church and the national monarchies declined. A French papacy at Avignon, followed by a divided papacy, provoked the harsh criticism of some of the great thinkers of the time. And for these critics the Church had no satisfactory answer.

In 1337 the French and English kingdoms began a war which continued at intervals for over a century, with the result that by 1400 both countries were exhausted and internally divided, while their monarchies became progressively weaker. At the same time the German emperors were deferring completely to their princes, a policy resulting in such confusion that even the modern historian is left panting for breath.

In one political development only was the 14th century a time of progress, and that was in the evolution of English representative government. A successful blend of Anglo-Saxon and Norman institutions provided the starting-point for the development of a system of representative government. What began as a common sense measure of strong kings, when Edward I insisted that electors had to be bound by the consent that their representatives in parliament gave on their behalf, became, through the centuries, an unbreakable rule. Government *for* the people thus became government *by* the people, or as it has been aptly called, "self-government at the king's command."

The rule that electors must be bound by the undertakings of their representatives in parliament could, in time, be applied so that *only* with the

TIME CHART FOR THE MIDDLE AGES
(476 - 1400)

	WEST	EAST
476 500	Odoacer rules Italy on behalf of Emperor at Constantinople Clovis in Frankish kingdom; Theodoric in Italy	
600	Founding of Benedictine Order Justinian reconquers Italy, N. Africa, part of Spain Pope Gregory the Great	Justinian at Constantinople (Byzantium) Mohammed in Arabia
700	Mayors of the palace rule Frankish kingdom Moslems cross from N. Africa to Europe	Moslem expansion in Middle East
800	Charles Martel turns back Moslems at Tours; rise of feudalism Pepin the Short becomes king of the Franks; Papal States founded Charlemagne crowned emperor in the West	Leo the Isaurian saves Constantinople from the Moslems; outlaws use of icons
900	Carolingian Renaissance Collapse of Carolingian Empire and barbarian invasions Alfred the Great in England Beginning of Medieval Agricultural Revolution	
1000	Otto the Great in Germany and Italy Hugh Capet in France Cluniac reform	
1100	Rise of Towns; feudalism declining Norman Conquest of England; Henry IV and Gregory VII quarrel over investiture First Crusade	Schism between Papacy and Eastern Orthodox Church at Constantinople Latin Kingdom of Jerusalem founded
1200	Renaissance of the 12th century; rise of Universities Henry II in England; Frederick Barbarossa in Germany John of England quarrels with Pope Innocent III	Second Crusade fails Third Crusade fails Fourth Crusade captures Constantinople
1300	Battle of Bouvines leaves Philip Augustus supreme in Europe Frederick II in Italy and Germany Simon de Montfort summons burgesses along with knights to English parliament Pope Boniface VIII; Philip the Fair convenes French Estates-General	Kublai Khan rules Mongol Empire; Marco Polo travels to China
1400	Papacy at Avignon Outbreak of Hundred Years' War; the Black Death Growth of English parliament under Edward III The Great Schism in the Church Death of Geoffrey Chaucer	The Black Death Ottoman Turks push into Europe

consent of elected representatives could any direct tax be imposed. It took until 1688 to make this rule ironclad in England, but the medieval tradition paved the way.

If parliamentary development in England by the end of the 14th century was the prologue to modern times, it was also the epilogue of the Middle Ages. After 1400 the power of a feudal nobility and a Universal Church had to be shared with parliaments, and in six centuries (800-1400) medieval civilization had completed its growth.

These centuries were unusually creative ones, and everywhere today we find their bequests to us. For "people who deposit money in a bank, who elect representatives to a national assembly, who rely on the precedents of English common law, who receive degrees from universities and believe that science is an important part of education, who worship in Gothic churches, and who read books written in modern European languages . . . would find their lives rather limited and unsatisfactory if they could do none of these things, and yet the basic idea of every one of these activities was worked out in the Middle Ages and not in ancient Greece or Rome."[1]

SOURCE READINGS

(a)

SIR JOHN FROISSART (1338-1410) was a Fleming, a clerk, traveller, poet, and chronicler who wrote an account entitled *The Chronicles of France, England, Scotland and Spain,* covering the years from 1326 to 1399 and consisting almost entirely of accounts of 14th century wars. This extract is only a part of his description of the Battle of Crécy (August 26, 1346). Crécy shines as "one of the most complete victories in the annals of English warfare", though, ironically enough, it was not fought in the way English knights preferred, by chivalrous combat. Edward III's 9,000 men were vastly outnumbered by the French, and won simply by making the enemy come to them—in fifteen or sixteen uphill charges—to be killed. The fact is that Edward III owed his victory to his numerical inferiority: he dared not risk a hand-to-hand fight between knights in the open.

There is no man, unless he had been present, that can imagine or describe truly the confusion of that day, especially the bad management and disorder of the French, whose troops were out of number. . . . The English, who, as I have said, were drawn up in three divisions, and seated on the ground, on seeing their enemies advance, rose up undauntedly and fell into their ranks. . . .

You must know that the French troops did not advance in any regular order, and that as soon as their king came in sight of the English his blood began to boil, and he cried out to his marshals, "Order the Genoese forward and begin the battle in the name of God and St. Denis." There were about 15,000 Genoese cross-bow men; but they were quite fatigued, having marched on foot that day six leagues, completely armed and carrying their cross-bows, and accordingly they told the

---

[1] J. R. Strayer, *Western Europe in the Middle Ages, A Short History* (Appleton-Century-Crofts, 1955), p. 10.

constable they were not in a condition to do any great thing in battle. The Earl of Alençon hearing this, said, "This is what one gets by employing such scoundrels, who fall off when there is any need for them." During this time a heavy rain fell, accompanied by thunder and a very terrible eclipse of the sun; and, before this rain, a great flight of crows hovered in the air over all the battalions, making a loud noise; shortly afterwards it cleared up, and the sun shone very bright; but the French had it in their faces and the English on their backs. When the Genoese were somewhat in order they approached the English and set up a loud shout, in order to frighten them; but the English remained quite quiet and did not seem to attend to it. They then set up a second shout, and advanced a little forward; the English never moved. Still they hooted a third time, advancing with their cross-bows presented, and began to shoot. The English archers then advanced one step forward, and shot their arrows with such force and quickness, that it seemed as if it snowed. When the Genoese felt these arrows, which pierced through their armour, some of them cut the strings of their cross-bows, others flung them to the ground, and all turned about and retreated quite discomfited.

The French had a large body of men-at-arms on horseback to support the Genoese, and the king, seeing them thus fall back, cried out, "Kill me those scoundrels, for they stop our road without any reason." The English continued shooting, and some of their arrows falling among the horsemen, drove them upon the Genoese, so that they were in such confusion, they could never rally again.

W. O. Hassall, editor, *They Saw It Happen: An Anthology of Eye-witnesses' Accounts of Events in British History, 55 B.C.-A.D. 1485* (Blackwell, 1957), pp. 152-153.

(b)

The ANONIMALLE CHRONICLE was written at the Benedictine Abbey of St. Mary's, York, about the year 1400. The account of the Parliament of 1376 was taken over by the St. Mary's chronicler from someone else who was an eyewitness of events in parliament, and perhaps later in the chapter-house, to which the knights and burgesses retired on the second day.

. . . And they began to talk about their business, the matters before the parliament, saying that it would be well at the outset for them to be sworn to each other to keep counsel regarding what was spoken . . . among them. . . . And to do this all . . . agreed, and they took a good oath to be loyal to each other. Then one of them said that, if any of us knew of anything to say for the benefit of the king and the kingdom, it would be well for him to set forth among us what he knew and then, one after the other, [each of the rest could say] what lay next his heart.

Thereupon a knight of the south country rose and went to the reading desk in the centre of the chapter house so that all might hear and, pounding on the said desk, began to speak in this fashion: ". . . My lords, you have heard the grievous matters before the parliament—how our lord the king has asked of the clergy and the commons a tenth and a fifteenth and customs on wool and other merchandise for a year or two. And in my opinion it is much to grant. . . . Also, as I have heard, there are divers people who . . . have in their hands goods and treasure of our lord the king amounting to a great sum of gold and silver; and they have falsely concealed the said goods. . . . For the present I will say no more. . . ." And he went back to his seat among his companions.

Thereupon another knight arose and went to the reading desk and said: "My lords, our companion has spoken to good purpose, and now, as God will give me grace, I will tell you one thing for the benefit of the kingdom. . . ."

And the third man rose . . . and said: "My lords, our companions have spoken very well. . . . But it is my opinion that . . . it would be well . . . to pray our lord the king and his wise council in the parliament that they may . . . assign to us certain bishops and certain earls, barons . . . such as we may name, to counsel and aid us. . . ." And to this all agreed.

About the same time a knight from the march of Wales . . . named Sir Peter de la Mare, began to speak . . . and he said: "My lords, you have well heard what our companions have had to say and what they have known and how they have expressed their views; and, in my opinion, they have spoken loyally and to good purpose." And he rehearsed, word for word, all the things that they had said, doing so very skilfully. . . . [and] because the said Sir Peter de la Mare had spoken so well and had so skilfully rehearsed the arguments and views of his companions, and had informed them of much that they did not know, they begged him on their part to assume the duty of expressing their will in the great parliament before the said lords, as to what they had decided to do. . . . And the said Sir Peter, out of reverence to God and his good companions and for the benefit of the kingdom, assumed that duty. . . .

Stephenson and Marcham, *Sources of English Constitutional History*, pp. 220-222.

In the next two weeks the commons travelled back and forth to parliament with their complaints; and then one day

. . . the lords assembled in their parliament and sent for the commons, to hear what they had to say; and the commons came before the lords in parliament, openly and in a body. And the aforesaid Sir Peter [de la Mare] commenced to speak. "Lords, we have come here before you, at your direction to show you what is grieving our hearts; and we say that we have declared, to you and to all the council of parliament, several trespasses and extortions made by different people, and we have had no redress; nor are there any around the king who will tell him the truth or give him loyal and profitable counsel, but always they mock and scoff and work for their own profit. Wherefore we declare to you that we will say nothing further until those who are about the king, who are traitors and evil counsellors, be removed and ejected from his presence, and until the present chancellor and treasurer be removed from their offices, for they are not worth a straw . . . and until our lord the king shall assign to be members of his council, three bishops, three earls, and three barons, such as will not shirk telling the truth and will improve matters. No great business shall be carried out or finished without them, or any wards or marriages bestowed without their advice, and they shall redress that which has been ill done and used before this time, in deceit of the king. . . . And that they shall hear and amend by their good counsel and advice, the trespasses which have been made, as we have shown before." The lords replied that this would be an advantageous step, and that they would willingly inform the king of their agreement and counsel and of their purpose; and they went away for this day, without doing more. . . .

B. Wilkinson, *Constitutional History of Medieval England, 1216-1399; with Select Documents*: Vol. II, *Politics and the Constitution, 1307-1399* (Longmans, Green, 1952), p. 222.

(c)

RICHARD THE REDELESS is the title of a poem written in 1399 in criticism of Richard II. The following modernized selection is an attack on the parliament of 1398, which was dominated by Richard.

When the day for action arrived, the lords were assembled and also the knights of the shire. According to the usual form, the cause of meeting was first declared and then the king's will. A clerk began with a dignified speech setting forth the main points before them all, and asking above all for money, with flattery to the great men to avert complaints. When the speech was ended, the knights of the commons were commanded to meet on the morrow, before dinner, with the citizens of the shires to go through the articles which they had just heard and grant all that had been asked them. But to save appearances, and in accordance with custom, some of them falsely argued at some length, and said: "We are servants and we draw a salary, we are sent from the shires to make known their grievances, to discuss matters on their behalf and to stick to that, and only make grants of their money to the great men in a regular way, unless there is war. If we are false to the people who pay our wages, we are not earning them."

Some members sat there like a nought in arithmetic, that marks a place but has no value in itself. Some had taken bribes, so that the shire they represented had no advantage from their presence. . . . Some were tattlers, who went to the king and warned him against men who were really good friends of his and deserved no blame either from king, council or commons, if one listened carefully to the very end of their speeches. Some members slumbered and slept and said little. Some stammered and mumbled and did not know what they meant to say. Some were paid dependents, and were afraid to take any step without their masters' orders. Some were so pompous and dull-witted that they got hopelessly involved before they reached the end of their speeches, and no one, whether he sat on the bench or whether he was a burgess, could have made out what they wanted to say, there was so little sense in it. . . .

Some had been got at beforehand by the council and knew well enough how things would have to end, or the assembly would be sorry for it. Some went with the majority, whichever way they went, while some would not commit themselves. Some were quite openly more concerned about the money the king owed them than about the interests of the commons who paid their salaries, and these were promised their reward; if they would vote the taxes, their debts would be paid them. And some were so afraid of great men that they forsook righteousness.

H. M. Cam, *Liberties and Communities in Medieval England* (Cambridge University Press, 1944), pp. 230-231.

FURTHER READING

1. THE 14TH CENTURY CHURCH AND THE MONARCHIES
    BOASE, T. S. R., *Boniface the Eighth, 1294-1303* (Constable, 1933)
    GREEN, V. H. H., *The Later Plantagenets* (Arnold, 1955)
    † * MYERS, A. R., *England in the Late Middle Ages* (Penguin, 1956)
    PERROY, E., *The Hundred Years War* (Eyre and Spottiswoode, 1951)
    ULLMANN, W., *The Origins of the Great Schism* (Burns Oates, 1948)

2. PARLIAMENT
    CAM, H. M., *Liberties and Communities in Medieval England* (Cambridge, 1944)
    ——————, *Law-Finders and Law-Makers in Medieval England* (Barnes and Noble, 1963)
    † HASKINS, G. L., *The Growth of English Representative Government* (Pennsylvania, 1948)
    LYON, B. D., *A Constitutional and Legal History of Medieval England* (Harper, 1960)
    † * MACKENZIE, K., *The English Parliament* (Penguin, 1951)
    STEPHENSON, C., *Mediaeval Institutions: Selected Essays* (Cornell, 1954)
    † THOMPSON, F., *A Short History of Parliament, 1295-1642* (Minnesota, 1953)
    TASWELL-LANGMEAD, T. P., *English Constitutional History from the Teutonic Conquest to the Present Time* (Houghton Mifflin, 1960)
    WHITE, A. B., *Self-Government at the King's Command* (Minnesota, 1933)
    WILKINSON, B., *Constitutional History of Medieval England, 1216-1399*, 3 vols. (Longmans, 1952-58)

UNIT VI
MAKING THE MODERN WORLD

15. Europe and Her Wider World

*... for to him is nothing more joyous than to know
his subjects to live peaceably under his laws
and to increase in wealth and prosperity.*
HENRY VII OF ENGLAND (1489)

*For here [Cuba] must be a large population, and very
valuable productions, which I hope to discover before I
return to Castille. I say that if Christendom will find profit
among these people, how much more will Spain....
[no] stranger should trade here, or put his foot in the
country, except Catholic Christians, for this was
the beginning and end of the undertaking; namely
the increase and glory of the Christian religion....*
CHRISTOPHER COLUMBUS (1492)

Between 1400 and 1600 the outlook of Europeans changed completely—although once again there was no abrupt break with the past. The marks of this new civilization can readily be detected. The rivalry of city-states and nation-states became sharper and transferred itself overseas as Europeans vied with each other to build empires there. At the same time a new attitude towards man's place in the world emerged, as he concentrated on acquiring a culture centred around himself and struck out along fresh paths of philosophy and religion.

Just as the centuries from the 9th to the 13th witnessed the formation of medieval civilization, so those from the 14th to the 17th saw the making of the modern world.

1/CITY-STATES AND NATION-STATES

From early in the 14th century the word "Europe" frequently replaces "Christendom" in the writings of men such as Petrarch. The change is significant. For while the idea of one Christian continent did not die with the Middle Ages, the new term came to stand for a civilization that was increasingly worldly and increasingly divided politically. That potent force called *nationalism*, a force which we have already seen building up in the feudal monarchies of the Medieval Transition, now quickly came to replace the old ideals of an overruling Christian or secular empire.

In 1454 one of Europe's best informed men, the papal legate in Germany, who was to become Pope Pius II (1458-64), wrote as follows to a friend in Rome:

I prefer to be silent, and I could wish that my opinion may prove entirely wrong and that I may be called a liar rather than a true prophet.... For I have no hope that what I should like to see will be realized; I cannot persuade myself that there is anything good in prospect.... Christianity has no head whom all will obey. Neither the pope nor the emperor is accorded his rights. There is no reverence and no obedience; we look on pope and emperor as figure heads and empty titles. Every city has its king and there are as many princes as there are households.

Michelangelo's David *appears as a god-like, clear-eyed Greek youth rather than the simple Jewish lad who challenged a giant with a sling-shot. Frowning and tensely watchful as he balances before taking aim, the colossal* David *epitomizes the self-confidence of the Renaissance.*

Such pessimism was well borne out by the conduct of the 15th and 16th century monarchs, the chief of whom are listed in the following table.

FRANCE	ENGLAND	SPAIN
Charles VII (1422-61)	Henry VI (1422-61)	
Louis XI (1461-83)	Richard III (1483-85)	Ferdinand and Isabella (1479-1516)
Charles VIII (1483-98)	Henry VII (1485-1509)	
Louis XII (1498-1515)	Henry VIII (1509-47)	Charles V (1516-58)
Francis I (1515-47)	Edward VI (1547-53)	
	Mary (1553-58)	
Henry IV (1589-1610)	Elizabeth I (1558-1603)	Philip II (1558-98)

Italy's Big Three In early modern times Italy witnessed the chaotic competition of great cities. The map on page 435 shows how the peninsula was divided. In the south was the Kingdom of the Two Sicilies, in the middle the states of the Church, and in the north a group of city-states. Of these city-states the three most important were Venice, Milan, and Florence.

Venice

Venice was Italy's chief maritime city. "Scattered," as an Ostrogothic king of Italy wrote, "like sea-birds' nests over the face of the waters," the Venetian islands resisted all conquerors, and by virtue of being directly on the trade route from the East to northern Europe became the great emporium of the medieval world. Thus Venice was an Italian rarity— a Latin city with a Byzantine culture.

The weary traveller must have sailed up the Grand Canal in those days with much the same feeling of wonder as the one who today enters New York harbour for the first time. Byzantine opulence glittered down on him from all sides. Listen to a 15th century Doge of Venice address his fellow citizens:

You Venetians are the only people to whom land and sea are alike open. You are the canal of all riches; you provision the whole earth; all the universe is interested in your prosperity: all the gold of the world comes home to Venice.

The prosperous Venetians enjoyed stable government under a Doge elected by the Great Council (= assembly) of businessmen. In this Venice was fortunate. Other city-states were not nearly as stable.

Milan

The northern city of Milan, in the plain of Lombardy, was aggressive in politics. She had led the neighbouring cities in their successful opposi-

This striking statue portrays one of the greatest soldiers of Renaissance Italy, Bartolommeo Colleoni (1400-75), at once a prince living in a luxurious court and a campaigner accustomed to the hardships of military life. The tenseness of Colleoni's body, the grim face and piercing eyes, all show the impetuosity and energy of the *condottiere*. It is interesting to note that while the model for the statue was produced by a Florentine, Andrea del Verrocchio (1435-88), the actual casting of the colossal bronze was undertaken by a later artist after Verrocchio's death. It now stands, perhaps the most imposing equestrian statue in the world, in Venice.

tion to Frederick Barbarossa and had become the ruler of the Po Valley. As Milanese territory increased the government grew more and more despotic, until finally Milan was under a military dictator.

It was customary for Italian towns to carry on wars by means of hired mercenaries under the leadership of soldiers of fortune (*condottieri*). One such adventurer, General Francesco Sforza, was so successful at playing off Milan and Venice that he rose to be Duke of Milan (1450-66). Some idea of the methods used in the jockeying for power among Italian despots may be gained from the politician Machiavelli (1469-1527), who wrote a treatise called *The Prince* as a handbook for successful rulers. Extracts from this treatise may be found on pages 445-446.

Florence The most turbulent of the Italian city-states was Florence, situated on the river Arno about 150 miles north of Rome. The Florentines seem to have been a politically vibrant people, for Florence early developed representative institutions, and frequent clashes between competing political parties occurred. Florentine wealth was based on wool manufacture; in fact so skilful were Florentine weavers that rough woollen cloth was imported from Flanders to be rewoven and re-exported. Florence ultimately became the financial capital of Europe for three centuries.

By the middle of the 15th century one of the Florentine bankers, Cosimo de' Medici, took over control of the city. We would call him a political "boss". The powerful Medici family provided enlightened, if absolute, rulers, their most distinguished member being the grandson of Cosimo, Lorenzo the Magnificent (1469-92). *Il Magnifico,* as he was called, was a great despot and a great patron of the arts—"as good a judge of cattle as of statues . . . as ready to discourse on Plato as to plan a campaign or plot the death of a dangerous citizen." Naturally there were those who hated him and resented his high-handed financial and political dealings, and who plotted a ruthless revenge.

Assassination at Easter When Pope Sixtus IV, fearing the political ambitions of the Medici, transferred the management of the papal revenues to a rival Florentine banking house run by the Pazzi family, Lorenzo systematically set about ruining his competitors. The enraged Pope was privy to a Pazzi plot to overthrow the Medici, and plans were carefully laid.

Now there were obvious advantages in those days to assassinations in church: the marked family was usually all together, and the members were usually unarmed and unguarded. So it was that on Easter Sunday, 1478, the Pazzi family conspired to murder Il Magnifico and his younger brother Giuliano in the cathedral in Florence.

The signal for the assassins' daggers was to be the elevation of the host—the most sacred moment of the mass. The two professionals quickly dispatched Giuliano, but the two priests assigned to Lorenzo, being amateurs, (the paid assassin, squeamish about doing the deed in church, had backed out at the last minute), bungled their part, and Lorenzo was able to fight his way to safety. The congregation fled shrieking in terror, one Pazzi managed to get out of town, and the mob, loyal to Lorenzo, busied itself by hanging an archbishop, his attendants, the other Pazzi, anyone, in fact, whom it could remotely associate with the crime, while the frightened officials attempted to appease the rioters by tossing some of the guilty down to them from the windows of the town hall.

Such was the horror of one Easter Sunday in 15th century Florence.

Europe's Battleground South of these city-states and stretching across the centre of the peninsula sprawled the Papal States. With the re-

CITIES OF RENAISSANCE ITALY

establishment of a strong papacy at Rome in the 15th century the popes became involved in Italian affairs, and, as we shall see in discussing the Reformation, the Renaissance really captured Rome.

The whole southern half of the Italian peninsula belonged to the Kingdom of Naples. This kingdom had a chequered history, having been ruled in turn by a branch of the Hungarian royal family, by the Spanish rulers of Aragon and Sicily (who thus reunited the Kingdom of the Two Sicilies), and finally by Charles VIII of France, who conquered Naples in 1495. By the end of the 15th century Italy was about to become the battleground for the greater European states.

A Divine Mission Outside Italy, the competition between nation-states was just as fierce. In 1415 Henry V of England (1413-22) broke the long truce to reopen the war with France, and in October won the great battle of Agincourt. He also managed to gain the friendship of the Burgundians, whose lands, independent of France, formed almost an eastern semicircle around Paris. When both Henry V and Charles VI of France died, the crowns of France and England passed by treaty to the infant Henry VI; but most of France south of the Loire River acknowledged the son of Charles VI as their king. The Dauphin, so called from his territory of the *Dauphiné* which was traditionally regarded as the property of the king's eldest son, was a sickly, weak-willed youth. Yet he found a champion—a seventeen-year-old peasant girl, the daughter of Jacques Darc.[1]

The rest of the story is well known. Jeanne persuaded the Dauphin that she had a divine mission to drive the English out of France and have him crowned as king, and this is exactly what she set about doing. She so heartened the dispirited French that they finally raised the English siege of Orléans, and Charles was crowned at Rheims. But Jeanne herself was captured by a Burgundian soldier in a siege in 1430 and handed over to the gleeful English, who put her on trial for witchcraft at Rouen in the spring of 1431. The kind of questions the court asked Jeanne and her replies may be seen in the extracts from the trial transcript in the Source Reading. But cleverness and innocence were not enough, for while the court was properly constituted it was devoted to the English cause. Jeanne went to the stake a martyr to English hatred and French indifference.

Charles VII, who raised not a finger to rescue his saviour, continued to reign triumphantly—in fact henceforth French kings enjoyed a rule not effectively challenged until the Revolution of 1789. The war ended in 1453 with only Calais left to the English.

Three years later the Rouen judgment on Jeanne was reversed.

For France and Personal Glory Louis XI, the son of Charles VII, was so adept at spinning webs of intrigue that he earned the nickname of the "Universal Spider." Modern historians have exhausted their armoury of adjectives in trying to pin down this elusive monarch's character.

Louis XI

... He was neurotic, diseased, suspicious, cruel, treacherous and false, without a moral sense. His religion was a strict and superstitious formalism, in which he believed he could bribe the celestial hierarchy as he did men in his dealings. He took a base delight in lying, tyranny, and revenge. Yet he was endowed with great gifts, which stamped his personality on the government and development of France. He was restlessly industrious, always on the move, and kept his hand on all the acts of government. No king knew his subjects better. He was, with all his loquacity and contempt of his species, a true master of men. He was shrewd, alert, energetic, and persevering. His aims were grandly conceived, original, and usually well judged. At the root of his policy lay a conviction of his duties as king, to whom absolute rule was given by right divine, and he claimed at the end of his life that he had fulfilled his task with all diligence to the benefit of France.[2]

Louis perfected a system of heavy taxes, among them a direct tax (*taille*) paid by non-nobles, sales taxes (*aides*), and a salt tax (*gabelle*). In addition he brought the powerful duchy of Burgundy under the crown.

[1] The name was spelled d'Arc after Charles VII ennobled the family in 1429.
[2] C. W. Previté-Orton, *The Shorter Cambridge Medieval History* (Cambridge University Press, 1952), II, 1026.

In 1428 the English armies in France were successful in pushing their conquests south to the River Loire, even laying siege to Orléans, the pivot of the Loire line. The Dauphin became so discouraged at these reverses that he contemplated flight to Scotland. At this crucial moment Jeanne d'Arc persuaded him to let her attempt to relieve Orléans, which, by April 1429, had become a symbol of French resistance.

She is seen here wearing her "white" (i.e. polished steel) armour at the head of an army as she advanced on Orléans from the north—where, as yet, the besiegers had failed to erect fortifications. It is recorded of the seventeen-year-old Jeanne that she was sturdily built, with a bright smiling face and dark hair cut short like a page boy's. Believing that they were led by an angel from heaven, the heartened French raised the siege in May, driving off the demoralized English who were convinced that their attackers were led by a devil from hell. A modern commentator has said of Jeanne that she provided "probably the greatest example of the importance of morale in the history of warfare."

Under Louis' son, Charles VIII, France at last became a united territorial state with the acquisition of Brittany in 1491. But Charles VIII wanted more than national unity: he wanted new territory for France and personal glory for himself. An invasion of Italy would satisfy both desires, and it would also strike a blow against the Habsburgs, since they, as German Emperors (see page 391), were a power in Italy. In 1495 Charles was able to take Naples without firing a shot. But now suspicion of French ambi-

Charles VIII

tions was aroused, and Venice, Milan, and the papacy called on Spain and the German Empire for support. And so, to maintain their influence there, not only Charles VIII but also his two successors, Louis XII and Francis I, found themselves making forays into Italy for half a century. In 1559 France was finally forced to abandon her claims on Naples and Milan.

A Shrewd Lancastrian The English defeat in the Hundred Years' War was followed by a long struggle between the ruling house of Lancaster (whose emblem was a red rose) and the rival house of York (represented by a white rose). The Yorkists won these Wars of the Roses, but when Richard III seized the throne from his infant nephew a rebellion broke out against the usurper. Henry Tudor, a Lancastrian heir with a shrewdness akin to that of Louis XI, led the uprising and won the crown at Bosworth Field in 1485. As Henry VII he founded England's most famous and forceful dynasty, that of the Tudors.

Under Henry VII the well-to-do English merchant class worked with the monarchy to re-establish law, order, and prosperity. Known as "the Solomon of his age", Henry started his reign in debt and ended it with treasure worth perhaps £2,000,000. Because Henry was an adept financier, he only had to summon parliament seven times in all his twenty-four years of reign. But even though he met parliament seldom, Henry wisely realized that he could not dispense with it entirely.

Henry VII was also an accomplished diplomat. He established friendly relations with the French, German, and Spanish royal houses. So important, in fact, did he consider England's alliance with Spain that when his eldest son Arthur, who was married to Catherine of Aragon (the daughter of Ferdinand and Isabella), died within a few months of the marriage, King Henry got a papal dispensation for his second son, Prince Henry, to marry the widowed Catherine.

The Most Absolute Monarchy The unification of Spain was a slow process, accompanying as it did five centuries of crusades south by Spanish Christians in their bitter struggle to clear the Moslems from the Iberian peninsula. But when Ferdinand of Aragon and Isabella of Castile married, they combined the two largest kingdoms and drove the Moslems from Granada in 1492—an eventful year for Spain. The Spanish monarchy reigned with a firm hand: Moslems and Jews were expelled, and the Spanish Inquisition was founded. Ferdinand and Isabella ruled so strongly that their successors were able to consolidate Europe's most absolute monarchy.

Here the history of disunited Germany merges with that of Spain. For the Habsburg Charles V (grandson of the German emperor Maximilian I [1493-1519]), having inherited Spain in 1516, also came to rule over the Low Countries, the German Empire, parts of Italy, and Hungary. When the nineteen-year-old Charles became emperor in 1519 he reigned over the largest territory to be under one man since Charlemagne's death. France was encircled. Neither her 16 million nor England's 4 million inhabitants were any match for Spain's 8 million, augmented by the German states' 15 to 20 million and a large portion of Italy's 12 million.

2/OVERSEAS EXPANSION AND EUROPEAN CAPITALISM

While city-states and nation-states were consolidating their power in Europe, they were also casting hopeful glances overseas. From the time of the Crusades onward Europe had been expanding. In the 13th century the west sent missionaries and merchants to the Mongols—though in fact the European eastern horizon was soon to contract. Only a century later the Ottoman Turks crossed the Dardanelles from Asia Minor to the Gallipoli peninsula, and pushed westward through Macedonia and northward through Bulgaria to the Danube. Finally, in 1453, Sultan Mohammed II conquered the second Rome, Constantinople herself, and the dying Byzantine empire was at last taken over by the Ottomans.

Gospel, Glory, and Gold But if eastern Europe's frontiers were contracting, hitherto unimaginable vistas beckoned in the west. Amongst the reasons for this new expansion were technological innovations in the field of navigation. As early as the end of the 13th century, Moslem and Christian sailors on the Mediterranean were using a primitive form of magnetic compass. In addition, their ships were being fitted out with a large new rudder fixed to the stern post. This rudder, well under water and therefore not affected by the sweep of the waves, allowed ships to sail close to the wind. The consequences of these two inventions were tremendous: shipping made more progress over the next two hundred years than it had made during the previous 4000. When accurate maps were added to rudder and compass it became possible to navigate ships with greater precision away from the sight of land. Europe began to look outwards to the open sea.

Thus Europeans were contemplating unknown horizons even before the Turks captured Constantinople in 1453. The Turks did not, as was once believed, cut off trade to the West, though Italian merchants in Constantinople did lose some of their trading privileges there. Venice still controlled the spice trade to the Middle East after 1453; indeed, the trade increased in volume until the end of the 15th century. Galled by the Venetian monopoly, other states continued avidly to seek alternative routes to the East.

The impelling motives that now made Europeans brave vast oceans in tiny ships have been tersely summed up as "Gospel, Glory, and Gold". The quickening of religious faith in the 15th and 16th centuries inspired men to go out to convert the heathen; the Renaissance monarchs and the new strong nation-states sought to gain prestige; and gold was to accelerate the changes of the Commercial Revolution in an expanding European economy.

Gold was not, of course, the only commodity sought by Europeans. Silk, rugs, jewellery, sugar, and above all spices (pepper, cloves, and cinnamon) had come from the East. Why not now go direct to India and China, the sources of supply, and cut out the Italian or Moslem middlemen?

Henry the Navigator

One of the great driving forces behind the exploration movement was Prince Henry the Navigator (1394-1460), son of the King of Portugal. Prince Henry was not, strictly speaking, an explorer. He was, however, intensely interested in exploration (see pages 447-448). For this reason he commissioned the drawing of better maps and charts, and sent out numerous expeditions both to the oceanic islands off Portugal and along the African coast past Cape Verde. The work went on after Prince Henry's death, reaching a high point in 1488 when Bartholomew Diaz rounded the southernmost tip of Africa, the Cape of Good Hope.

In the next half century Europe's knowledge of the world increased phenomenally. The following table indicates the scope of exploration in that time.[1]

1488	—Bartholomew Diaz (Portuguese) rounds Cape of Good Hope.
1492	—Christopher Columbus (Genoese, agent of Spain) discovers West Indies.
1497-99	—Vasco da Gama (Portuguese) reaches India (1498).
1497, 1498	—John Cabot (Venetian, agent of England) discovers Cape Breton Island, Newfoundland, Labrador.
1500, 1501, 1502	—Gaspar and Miguel Corte Real (Portuguese) explore coasts of Greenland, Labrador, Newfoundland.
1499-1500	—Vincente Yañez Pinzon (Spanish) discovers north-east coast of South America.
1500	—Pedro Alvares Cabral (Portuguese) reaches Brazil on way to India.
1509-11	—Portuguese under Alfonso de Albuquerque conquer Ormuz, Goa, Malacca.
1513	—Ponce de Léon (Spanish) discovers Florida.
1513	—Vasco Nuñez de Balboa (Spanish) crosses the Isthmus of Panama and discovers the Pacific.
1514	—Portuguese reach China.
1519-22	—Hernando Cortés (Spanish) explores and conquers Mexico.
1519	—Alvarez de Pineda (Spanish) sails along coast from Florida to Mexico.
1519-22	—Ferdinand Magellan (Portuguese, agent of Spain) killed in Philippines, but his ship *Victoria* completes circumnavigation of the globe. Of the 5 ships and 243 men that began the voyage, 1 ship and 18 men got back.
1534, 1535-36	—Jacques Cartier (French) lands on Gaspé peninsula, and explores St. Lawrence, spending winter in Canada.

[1]Adapted from the table in S. B. Clough and C. W. Cole, *Economic History of Europe* (D. C. Heath, revised edition, 1946), p. 109.

These explorations resulted in the establishment of great New World empires—Spanish, Portuguese, and French. English and Dutch colonies developed later with English settlement at Jamestown, Virginia, in 1607 and Plymouth, New England, in 1620, and Dutch settlement at New Amsterdam in 1612. The competition for empire even spread to Africa and Asia.

New territories brought an interchange of products between America (so called for an Italian navigator, Amerigo Vespucci, whose claim to have discovered the "new world" was accepted by early map-makers) and Europe. Here is a partial list of these products.

EUROPEAN IMPORTS FROM AMERICA

hot peppers   maize
peas   string beans
strawberries   Canada geese
tomatoes   turkeys
potatoes   tobacco

AMERICAN IMPORTS FROM EUROPE

oranges   rye   plums
lemons   coffee   goats
sugar-cane   carrots   sheep
wheat   apples   horses
barley   pears   pigs

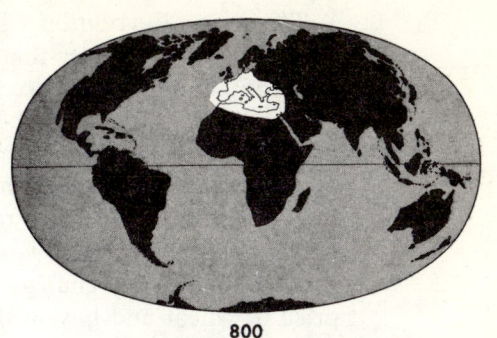

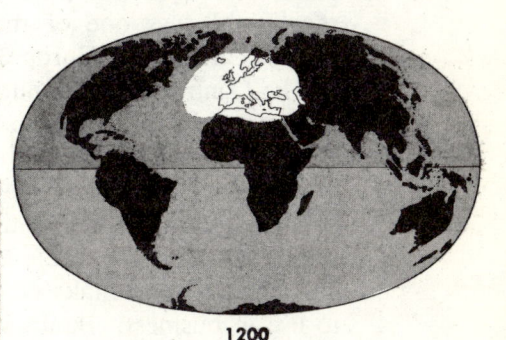

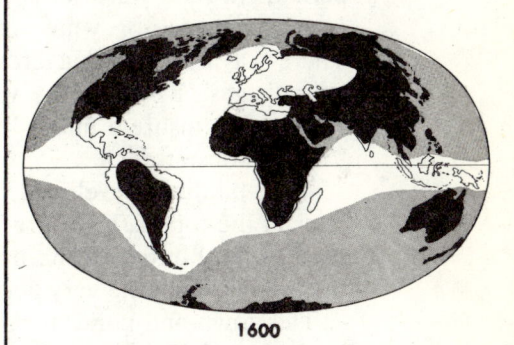

800

1200

1600

THE WIDENING OF THE EUROPEANS' WORLD

Spices previously imported from the Middle East now came directly by water to Atlantic ports. As a result, in 1504 the price of spices in Lisbon, for example, was one-fifth of that in Venice. The Italian cities were sliding into a long decline; the future lay with the Atlantic ports.

It was only a matter of time until a new "product" assumed great importance—slaves. The Portuguese and Spanish enslaved negroes with the noble motive of converting them to Christianity, and, incidentally, of supplying cheap labour for the colonies.

The Price Revolution Europe's main import, however, was the rich river of gold and silver that surged in from the Americas. Between 1500 and 1650 the stock of precious metals in Europe probably tripled. This tremendous influx made possible a great rise in the amount of coined money in circulation, which in turn caused an average 2% or 3% rise in prices almost every year, until by 1650 prices had doubled or tripled over the figures in 1500. The result of this situation in which wages did not keep up with prices was what we call *inflation*. Average figures give only a rough idea of the changes brought about by inflation, but with the 1650 price of wheat and hay in the Paris market fifteen times that of the 1500 price, one does not have to be an economist to realize that a "price revolution" was one of the major consequences of the importation of precious metals into Europe.

The Commercial Revolution, which was discussed in an earlier chapter, did not end in 1500. Accelerated by overseas expansion, it lasted on into the 18th century. Businessmen of the 16th century devised improved methods of handling money; for instance, it became the common practice to endorse bills of exchange so that they could be passed from hand to hand, and promissory notes payable to the bearer were developed as well. Such changes served to make credit instruments more negotiable, that is, easier to use in business. Banks were established under government regulation; permanent fairs were set up for year-round business; and *bourses* (French *bourse*=purse), or what we would call stock exchanges, were created. There was, too, a great growth in the writing of insurance contracts for sea voyages or other risky ventures. It was even possible in 16th century Antwerp to insure a man's life without his knowing it—an open invitation for fraud or murder.

As capitalism developed, the prosperous merchants formed a growing class at the top of the bourgeoisie, while the labouring class of journeymen sank to the bottom of the middle class. A great gap had opened between the very rich and the very poor.

The town-and-guild framework of medieval times was also affected by 15th and 16th century changes. During the Middle Ages England had exported raw wool to Flanders and imported finished cloth, but by the 15th century certain Englishmen were already doing their own spinning, weaving, and dyeing. To avoid restrictions imposed by towns and guilds they "put out" this work to country people, lending them looms and equipment. This method of manufacturing was called the "putting-out" or *domestic* system, and remained the typical method of production for certain goods, such as cloth and hardware, until the factory system was introduced in the 18th century.

Naturally it took a substantial amount of money to set up such a system, and those who supplied it became powerful financiers. Capitalists were also essential in co-ordinating the activities of hundreds of handicraftsmen in the first mass production, which was undertaken to manufacture muskets for the new national armies.

New Mental Horizons One of the new industries was particularly expensive to set up, and particularly far-reaching in its influence. This was printing.

Long before the middle of the 15th century printing from movable type had been known in China; but whether Europeans learned it from the Chinese or invented it for themselves is uncertain. In the 12th and 13th centuries wooden blocks had been used in Europe to print designs on cloth, and by the 15th century block printing was done on paper. It took only a few changes to adapt these techniques to a more refined system of printing, and about 1450 a German from Mainz, Johann Gutenberg (1398-1468), first printed from movable type.

Gutenberg

Paper was another Chinese invention that had been carried to Europe by the Moslems. Some paper was being produced in Italy and southern France in the 13th century, but the invention of the spinning wheel about 1300 revolutionized the paper-making industry by greatly reducing the price of cloth and thereby making available much more abundant quantities of linen rags for use in the manufacture of paper. Without such a cheap supply of paper the large-scale printing of books might not have been economically feasible.

Printing was an epoch-making invention. It spread very rapidly: by the end of the 15th century there were 73 presses in Italy, 51 in Germany, 39 in France, 24 in Spain, 15 in the Low Countries, and 8 in Switzerland. Not everyone, however, welcomed the new invention; manuscript collectors and many scholars denounced printed books as vulgar. Nevertheless it soon became obvious that the new presses were the greatest boon to scholarship yet conceived. The press of Aldus Manutius (established in Venice in 1490), for example, became famed for its printing of Greek classics, and the printer believed his work to be so vital that his motto was, "If you have to speak to Aldus be quick about it, because time presses." The Aldine Press was one of the most potent forces for promoting that revival of learning which we associate with the Italian Renaissance.

These lines from the Fust and Schoeffer Psalter (1457) were probably designed by Gutenberg and hence are some of the earliest letters moulded into movable type. Based on 15th century manuscript styles, they are superb examples of severe Gothic letters and illustrate Gutenberg's concept of fine printing, which was to make it look as much like the monks' artistic script as possible.

Nō sic impij nō sic:
sed tanq̃ puluis

The printing press affected another segment of the population—the masses of the people, whose horizons were now extended in an entirely new way. Before the end of the 16th century some educators were arguing for more pictures in children's schoolbooks, so afraid had they become that the printed word would destroy the visual image! It is almost impossible for us now to imagine what the world was like before books standardized knowledge. Just as this same page can be read in Newfoundland or British Columbia, men of the 15th and 16th centuries could now read the same Greek classic from one end of Europe to the other. Businessmen could keep stricter accounts of time and money, and political appeals could make a hitherto undreamed of impact on thousands of people. The mental horizons of Europe were expanding along with her geographical ones. "As the Apostles of Christ formerly went through the world announcing the good news," wrote a German scholar, "so in our days the disciples of the new art [of printing] spread themselves through all countries, and their books are as the heralds of the Gospel and the preachers of truth and science." So widespread was the demand for literature that by 1500 Europe boasted nine million books, thirty thousand different titles, and over a thousand printers.

A Favourable Balance of Trade The ambitious monarchs of Europe were naturally eager to make their countries self-sustaining. Hence colonies were much sought after, not only as a source of precious metals but of other raw materials as well. An elaborate theory was worked out to cover all aspects of national economic policy, a theory called *mercantilism*. It varied from nation to nation, but in general it encouraged the import of gold and silver and such raw materials as the mother country did not possess, while equally stressing the export of manufactured goods. On the other hand mercantilism discouraged the import of foreign manufactures or the export of raw materials, the production of luxury goods, and any idleness or unemployment. To carry out such a complex economic policy and assure a "favourable balance of trade," the state had to regulate both trade and industry.

Mercantilism

So it was that as well as selling offices, and devaluing the currency, and quarrelling with parliaments—all accepted monarchal ways of raising money in the 17th century—kings backed mercantilist policies. Such policies tended to emphasize the importance of middle-class merchants as the necessary support for rulers.

3/SUMMARY AND CONCLUSION

So far we have traced the political, economic, and social foundations that underlie the modern world. On these foundations Europeans erected a new structure of learning and religion.

In the 14th and 15th centuries the city-states and nation-states of Europe competed fiercely for political and economic advantage, until finally Italy became Europe's battleground and the high seas her ultimate horizons. As the European world widened physically it broadened mentally. Capitalism flourished, and the new wealth was responsible for the setting up, among other industries, of the printing trade. Nationalism had made itself felt in politics and economics. Soon it would be evident in learning and religion as well. The next chapter describes how this process began in Italy.

SOURCE READINGS

(a)

NICCOLO MACHIAVELLI (1469-1527) was a Florentine diplomat who was exiled from his city by the Medici. During his forced retirement from politics he wrote a *History of Florence*, and a small book called *The Prince*. Though *The Prince* was circulated in manuscript form, it was not printed until 1532, five years after its author's death. Some of the most famous passages are quoted in this selection.

I conclude, therefore, with regard to being feared and loved, that men love at their own free will, but fear at the will of the prince, and that a wise prince must rely on what is in his power and not on what is in the power of others, and he must not contrive to avoid incurring hatred, as has been explained.

How laudable it is for a prince to keep good faith and live with integrity, and not with astuteness, everyone knows. Still the experience of our times shows those princes to have done great things who have had little regard for good faith, and have been able by astuteness to confuse men's brains, and who have ultimately overcome those who have made loyalty their foundation.

You must know, then, that there are two methods of fighting, the one by law, the other by force: the first method is that of men, the second of beasts; but as the first method is often insufficient, one must have recourse to the second. It is therefore necessary for a prince to know well how to use both the beast and the man. . . .

A prince being thus obliged to know well how to act as a beast must imitate the fox and the lion, for the lion cannot protect himself from traps, and the fox cannot defend himself from wolves. One must therefore be a fox to recognize traps, and a lion to frighten wolves. Those that wish to be only lions do not understand this. Therefore, a prudent ruler ought not to keep faith when by so doing it would be against his interest, and when the reasons which made him bind himself no longer exist. If men were all good, this precept would not be a good one; but as they are bad, and would not observe their faith with you, so you are not bound to keep faith with them. Nor have legitimate grounds ever failed a prince who wished to show colourable excuse for the non-fulfilment of his promise. Of this one could furnish an infinite number of modern examples, and show how many times peace has been broken, and how many promises rendered worthless, by the faithlessness of princes, and those that have been best able to imitate the fox have succeeded best. But it is necessary to be able to disguise this character well, and to be a great feigner and dissembler; and men are so simple and so ready to obey present necessities, that one who deceives will always find those who allow themselves to be deceived. . . .

It is not, therefore, necessary for a prince to have all the above-named qualities, but it is very necessary to seem to have them. I would even be bold to say that to possess them and always to observe them is dangerous, but to appear to possess them, is useful. Thus it is well to seem merciful, faithful, humane, sincere, religious, and also to be so; but you must have the mind so disposed that when it is needful to be otherwise you may be able to change to the opposite qualities. And it must be understood that a prince, and especially a new prince, cannot observe all those things which are considered good in men, being often obliged, in order to maintain the state, to act against faith, against charity, against humanity, and against religion. And, therefore, he must have a mind disposed to adapt itself according to the wind, and as the variations of fortune dictate, and, as I said before, not deviate from what is good, if possible, but be able to do evil if constrained. . . .

. . . A certain prince of the present time, whom it is well not to name, never does anything but preach peace and good faith, but he is really a great enemy to both, and either of them, had he observed them, would have lost him state or reputation on many occasions.

European Inheritance, II, 168-170.

(b)

JEANNE d'ARC (1412-31) was tried in 1431 at Rouen Castle in English-occupied France. The Bishop of Beauvais, who had represented the English in engineering the ransom of Jeanne from her Burgundian captors, presided at the court and was assisted by representatives of the University of Paris and the Inquisition. Jeanne was condemned and burned at the stake in Rouen on May 30, 1431. The following questions, and Jeanne's answers (which are marked by a dash), have been telescoped together from separate court sessions lasting from February 24 to March 17.

"Do you know if you are in the grace of God?"
—"If I am not, may God place me there; if I am, may God so keep me. I should be the saddest in the world if I knew that I were not in the grace of God. But if I were in a state of sin, do you think the Voice would come to me? I would that every one could hear the Voice as I hear it. I think I was about thirteen when it came to me the first time."
"This Voice that speaks to you, is it that of an Angel, or of a Saint, or from God direct?"
—"It is the Voice of Saint Catherine and of Saint Margaret. Their faces are adorned with beautiful crowns, very rich and precious. To tell you this I have leave from Our Lord. . . ."
"How can you make sure of distinguishing such things as you are free to tell, from those which are forbidden?"
—"On some points I have asked leave, and on others I have obtained it. I would rather have been torn asunder by four horses than have come into France without God's leave."
"Which did you care for most, your banner or your sword?"
—"Better, forty times better, my banner than my sword!"
"Who caused you to get this painting [of the world with an angel on each side] done up on your banner?"
—"I have told you often enough, that I had nothing done but by the command of God. It was I, myself, who bore this banner, when I attacked the enemy, so that I might kill no one. I have never killed any one."
"Does not Saint Margaret speak English?"
—"Why should she speak English, when she is not on the English side?"
"In what likeness did St. Michael appear to you?"

—"I did not see a crown: I know nothing of his dress."
"Was he naked?"
—"Do you think God has not the wherewithal to clothe him?"
"Have you spoken to Saint Catherine since yesterday?"
—"I have heard her since yesterday, and she has several times told me to reply boldly to the Judges on what they shall ask me touching my case."
"Will you refer yourself to the decision of the Church?"
—"I refer myself to God Who sent me, to Our Lady, and to all the Saints in Paradise. And in my opinion it is all one, God and the Church; and one should make no difficulty about it. Why do you make a difficulty?"
"Do you know if Saint Catherine and Saint Margaret hate the English?"
—"They love what God loves; they hate what God hates."
"Does God hate the English?"
—"Of the love or hate God may have for the English, or of what He will do for their souls, I know nothing; but I know quite well that they will be put out of France, except those who shall die there, and that God will send victory to the French against the English."

Jeanne d'Arc, Maid of Orleans, edited by T. D. Murray (William Heinemann, new and revised edition, 1907), pp. 24, 28, 30, 34, 42, 44, 64, 79, 80.

(c)

PRINCE HENRY THE NAVIGATOR (1394-1460) was not much interested in politics, marriage, or court life, but he was fascinated by geographical exploration. At the age of twenty-one he was sent by his father, King John I of Portugal, to attack Ceuta, and he captured it. Appointed governor of the Algarves, a province in southern Portugal, Henry established a little court at Sagres from which he sent out his expeditions. This account of his work is from the *Chronicle of the Discovery and Conquest of Guinea* of GOMES EANES DE ZURARA (1410-74).

You must take good note that the magnanimity of this prince constrained him always to begin, and lead to a good conclusion, high exploits; and for this reason, after the taking of Ceuta, he had always at sea ships armed against the Infidels. And because he desired to know what lands there were beyond the Canary Isles and a cape which was called Bojador, for up to that time no one knew, whether by writing or the memory of any man, what there might be beyond this cape.

. . . And the Infante [son of the Portuguese king] Dom Henrique desired to know the truth of this; for it seemed to him that if he or some other lord did not essay to discover this, no sailor or merchant would undertake this effort. . . .

. . . if in these territories there should be any population of Christians, or any harbours where men could enter without peril, they could bring back to the realm many merchandises at little cost, by reason that there would be no other persons on these coasts who would negotiate with them; and . . . in like manner one could carry to these regions merchandise of the realm, of which the traffic would be of great profit to the natives. . . .

[Also it] was his great desire to increase the holy faith in Our Lord Jesus Christ, and to lead to this faith all souls desirous of being saved, recognizing that the whole mystery of the Incarnation, the death, and the passion of Our Lord Jesus Christ took place to this end: namely, that lost souls should be saved; and the Infante was fain, by his efforts and his expenditure, to lead these souls into the true path, understanding that man could render the Lord no greater service....

[The Infante Dom Henrique, having completed his preparations in his town of Sagres, began to send his caravels and his men along the western coast of Africa, with the mission of rounding Cape Bojador, and bringing back word to him as to what they found beyond this limit, which no man until then had passed.]

... However, although many set out—and they were men who had won fair renown by their exploits in the trade of arms—none dared go beyond this cape....

... The Infante always welcomed with great patience the captains of the ships which he had sent to seek out these countries, never showing them any resentment, listening graciously to the tale of their adventures, and rewarding them as those who were serving him well. And immediately he sent them back again to make the same voyage, them or others of his household, upon his armed ships, insisting more and more strongly upon the mission to be accomplished, and promising each time greater rewards to those who should bring him the intelligence he desired.

And at last, after twelve years of effort, the Infante had a barque fitted out, appointing the captain his squire Gil Eannes, whom he afterwards knighted and rewarded largely. This captain made the same voyage that the others had made, and overcome by the same dread did not pass beyond the Canary Isles, where he took captives, and returned to the kingdom. And this took place in the year 1433 of Jesus Christ. But the following year the Infante again had the same barque fitted out, and sending for Gil Eannes, and speaking with him alone, he recommended him strongly to do all that was possible to go beyond the cape; and that even if he did no more on this voyage, that would seem to him sufficient....

The Infante was possessed of great authority; his remonstrances, even the lightest, were for wise men of great weight. And this was proved on this occasion, for having heard these words Gil Eannes promised himself resolutely that he would never again appear before his lord without having accomplished the mission with which he had been charged. And it was even so, for on this voyage, disdaining all peril, he passed beyond the cape, where he found matters very different from what he and others had imagined....

And on his return he related to the Infante how the voyage had passed; having lowered a small boat into the sea, he had approached the shore and had landed without finding any person or sign of population. "And because it seemed to me," said Gil Eannes, "that I ought to bring back some token of this country, since I was there, I gathered these plants, which in our kingdom we call roses of St. Mary."

Mendenhall, Henning, and Foord, *Ideas and Institutions in European History, 800-1715*, pp. 182-183.

FURTHER READING

1. GENERAL

    ARTZ, F. B., *From the Renaissance to Romanticism: Trends in Style in Art, Literature, and Music, 1300-1830* (Chicago, 1962)

   † * BRONOWSKI, J. AND MAZLISH, B., *The Western Intellectual Tradition from Leonardo to Hegel* (Torchbook, 1962)

    * CLARKE, G., *Early Modern Europe from about 1450 to about 1720* (Galaxy, 1960)

THE RENAISSANCE 449

† FERGUSON, W. K., *Europe in Transition, 1300-1520* (Houghton Mifflin, 1962)
GRANT, A. J., *A History of Europe from 1494 to 1610* (Methuen, 1952)
GREEN, V. H. H., *Renaissance and Reformation: A Survey of European History between 1450 and 1660* (Arnold, 1952)
LUCAS, H. S., *The Renaissance and the Reformation* (Harper, 1960)
RANDALL, J. H., *The Making of the Modern Mind* (Houghton Mifflin, 1940)
THOMSON, S. H., *Europe in Renaissance and Reformation* (Harcourt, Brace and World, 1963)

2. MONARCHIES AND EMPIRES
ABBOTT, W. C., *The Expansion of Europe: A Social and Political History of the Modern World, 1415-1789* (Holt, 1924)
ELTON, G. R., *England under the Tudors* (Methuen, 1958)
HAMILTON, E. J., *American Treasure and the Price Revolution in Spain, 1501-1650* (Harvard, 1934)
HART, H. H., *Sea Road to the Indies* (Hodge, 1952)
HAY, D., *Europe: The Emergence of an Idea* (Edinburgh, 1957)
MATTINGLY, G., *Renaissance Diplomacy* (Cape, 1955)
† MORISON, S. E., *Admiral of the Ocean Sea: A Life of Christopher Columbus* (Little, Brown, 1942)
* NOWELL, C. E., *The Great Discoveries and the First Colonial Empires* (Cornell, 1954)
† * PARRY, J. H., *Establishment of the European Hegemony, 1415-1715: Trade and Exploration in the Age of the Renaissance* (Torchbook, 1961)
* PENROSE, B., *Travel and Discovery in the Renaissance, 1420-1620* (Atheneum, 1962)
† * SCHEVILL, F., *The Medici* (Torchbook, 1960)
* ———, *Medieval and Renaissance Florence*, 2 vols. (Torchbook, 1963)

16. The Renaissance

O wonder!
How many goodly creatures are there here!
How beauteous mankind is! O brave new world,
That has such people in it!
SHAKESPEARE, *The Tempest*

A French book published in 1855 first applied the label "Renaissance" to that period of European history following the Middle Ages when, so it was alleged, the classical culture of ancient times was reborn in "a new world of light and science." An enthusiastic Florentine expressed the same point of view in the mid-1430's.

Sculpture and architecture, for long years sunk to the merest travesty of art, are only today in process of rescue from obscurity; only now are they being brought to a new pitch of perfection by men of genius and erudition. Of letters and liberal studies at large it were best to be silent altogether. For these, the real guides to distinction in all the arts, the solid foundation of all civilization, have been lost to mankind for 800 years and more. It is but in our own day that men dare boast that they see the dawn of better things. . . . Latin, so long a byword for its uncouthness, has begun to shine forth in its ancient purity, its beauty, its majestic rhythm. Now, indeed, may every thoughtful spirit thank God that it has been permitted to him to be born in this new age, so full of hope and promise, which already rejoices in a greater array of nobly-gifted souls than the world has seen in the thousand years that have preceded it.

Modern historians no longer accept such views. They know how much the Renaissance owes to the Medieval Transition.

1/THE RENAISSANCE IN ITALY

It is, as a matter of fact, the debt that the Renaissance owes to the past that makes it so difficult to be definite about drawing up its chronology. Taken at its widest limits, the Renaissance is often said to stretch from the time of the Italian Dante (1265-1321) to the Englishman Shakespeare (1564-1616). At its beginning it owed much to the Middle Ages; at its end it overlapped the Reformation. There is also a great deal of debate over the characteristics of the Renaissance, particularly since it developed unevenly in Europe, being different things at different times in different places. However, the following tendencies, some of which we have already noted in earlier chapters, seem to be contrasted in the Middle Ages and Renaissance.

First, there was a marked shift away from feudalism to the monarchies of strong nation-states. Such a shift in political power had its counterpart in the composition of society, which now became less aristocratic and more middle-class—as well as in the economic basis of the state, which now depended more on capital than on landed wealth. Then, too, as the monarchs came to rely more and more on the rich merchant class for taxes, the crown found itself allied with the town.

The Renaissance also witnessed a change in religious outlook. The men of the Middle Ages had cherished a backward-looking vision of a revived Roman Empire or a Universal Church, a Church which had dominated both religion and education. Now this idea was breaking down before an increasingly secular philosophy of life that took education out of the hands of the Church and emphasized the cultivation of the individual's tastes. The new secular material that was being introduced into intellectual life was not, however, thought of as being antagonistic to Christianity. It has been well said that the men of the Renaissance devoted themselves to their classical studies "in the secure conviction that they could have all this and Heaven too."

Nevertheless the effect of the new secular material was to make society far less church-centred. Even medieval Latin was being replaced by the vernaculars or by classical Latin. We hear of one churchman who made a habit of washing out his mouth after saying mass, because he considered the medieval Latin of the service so debased! Art, too, while it continued to deal with religious subjects, was no longer completely church art. It also dealt with secular subjects, and when it did take religion as its theme the work was often commissioned by some lay patron.

Tempered by this amalgamation of viewpoints—a blend of old and new—scholars and statesmen of Italy created one of the world's most colourful and impressive cultures.

An Ideal Birth-place Why did the light of the new learning burn first and brightest in Italy? We have already seen the general political, economic, and social changes that affected Europe from the late Middle Ages onwards. These changes were especially marked in the vigorous Italian city-states of the 13th and 14th centuries, where a favourable geographical location midway between Western Europe and the Middle East had much to do with their prosperity. Nowhere else in Europe did capitalism develop so early, and nowhere else until much later did such a relatively small class of capitalists accumulate such wealth. In the busy cities of Italy society became more flexible, its distinctions based not so much on nobility of birth as on the possession of wealth and culture. Wealth conferred leisure, and leisure could be used to acquire culture.

Italy was the land where classical culture had once flourished, where its tangible remains were still plainly to be seen. The upper classes of 14th century Italy were more secular-minded than their medieval predecessors had been. Merchants, bankers, and princes very consciously pursued this world's goods. Even the Church was becoming more secular, particularly in Rome; its very officers—cardinals, papal secretaries, bishops, and abbots—were now recruited from the princely courts or the ranks of the upper classes in the cities. It is, then, hardly to be wondered at that it was in Italy that the Renaissance had its birth.

The Tempietto ("little temple") of San Pietro in Montorio in Rome by Donato Bramante (1444-1514) was built in 1502 to mark the spot on which St. Peter was believed to have been crucified. The severe, unadorned style effected in the Tempietto is one of the most perfect examples of the Renaissance desire to imitate the classic lines of a Greek temple, and its creation was followed by the highest of honours. In 1506 Pope Julius II commissioned Bramante to rebuild St. Peter's, and it was his plan that Michelangelo (see page 463) eventually took over and modified.

During the Renaissance the progress of literature and learning was marked by three stages. In the first there was a vigorous beginning of vernacular literature; next came a scholarly insistence on classical Latin and Greek as the true languages of gentlemen; a third stage saw the triumph of the vernaculars, which now incorporated some of the vocabulary and much of the viewpoint of the classics. We might almost say that a scholarly Renaissance was followed by a popular one. In both southern and northern Europe these three stages may be seen, but let us first deal with the rise of vernacular literature in Italy.

Dante

A Vigorous Vernacular We have already met Dante[1] in connection with 14th century opposition to the political claims of the papacy. But Dante wrote in Latin as well as Italian, and hence belonged to the Renaissance as well as to the Middle Ages. As a student once wrote in his examination, "Dante stood with one foot firmly planted in the Middle Ages, and with the other saluted the rising star of the Renaissance"! Thus his epic poem, the *Divine Comedy*, is medieval enough in its setting—Hell, Purgatory, and Paradise. Yet through these places roam individual personalities, lively figures far different from the stereotypes of a medieval morality play. The Renaissance environment shows, too, in the author's devotion to an earthly city: "Rejoice, Florence, since thou art so great that thou beatest thy wings over sea and land, and thy name is spread through Hell!"

Petrarch

Another prolific Renaissance writer was Francesco Petrarca, better known as Petrarch. Petrarch settled in Avignon after his parents were exiled from Florence, and abandoned a legal career to enter the Church in minor orders. He did not, however, become a priest, but devoted himself to writing 14-line poems in a form now known as the Petrarchian sonnet. These verses in Italian, often developed from folk songs (sonnet = a little song), earned Petrarch a laurel crown as poet laureate from the so-called "Senate of Rome"—by then a sort of municipal council—and for the last twenty-five years of his life he returned to live in his beloved native Italy.

Boccaccio

A third exponent of the vernacular was Giovanni Boccaccio, the son of a Florentine banker. Rather than follow in his father's footsteps he turned to scholarship, and was, for a time, a teacher in the University of Florence. Boccaccio is famous chiefly as a prose writer. His celebrated *Decameron* describes a gay group of seven young ladies and three men who flee the Black Death raging in Florence and take up their abode in a country villa. The *Decameron* consists of the stories they tell to enliven their isolation,

---

[1] For the dates of Renaissance literary and artistic figures see the "Renaissance Cultural Calendar."

[2] Neither musical nor scientific figures are included in this calendar. Considerable as the musical developments were, they are overshadowed by the literary and artistic aspects of the period. Such scientific developments as were accomplished are really only of significance in relationship to the full story of the 17th century Scientific Revolution, which is beyond the scope of this book.

A RENAISSANCE CULTURAL CALENDAR[2]

LITERATURE AND LEARNING	PAINTING	SCULPTURE	ARCHITECTURE
Dante (1265-1321)	Giotto (1266-1336)		Giotto
Petrarch (1304-74) Boccaccio (1313-75) Langland (1332-95) Chaucer (1340-1400)			
	Van Eyck (1380-1441)	Brunelleschi (1377-1446) Ghiberti (1378-1455) Donatello (1386-1466)	Brunelleschi Ghiberti
Valla (1407-57) Villon (1431-63)			
	Botticelli (1444-1510)		
Lefèvre d'Etaples (1450-1536) Leonardo da Vinci (1452-1519) Reuchlin (1455-1522) Pico della Mirandola (1463-94) Erasmus (1466-1536) Machiavelli (1469-1527)	Leonardo	Leonardo	Leonardo
	Dürer (1471-1528)		
Michelangelo (1475-1564)	Michelangelo Titian (1477-1576)	Michelangelo	Michelangelo
Castiglione (1478-1529) More (1478-1535)	Raphael (1483-1520) Holbein (1497-1543)		
Cellini (1500-71)		Cellini	
	Breughel (1525-69)		
Montaigne (1533-92) Shakespeare (1564-1616)			

stories that poke fun at the Church and Christianity, and show the characters gaily sinning and getting away with it.

Something more unmedieval than the *Decameron* would be hard to find. It established the form of the modern short novel.

The Study of Man   Overlapping this first stage of the Italian Renaissance was a second phase, which was also evident in the career of Petrarch. After he had made his great reputation as a poet, Petrarch turned his back on the vernacular and cultivated Latin. He so admired the classics that he wrote letters in Latin to the ancient classicists—Homer, Livy (see page 470), Cicero, Virgil, and other worthies. And though he never learned Greek, he bought all the Greek manuscripts he could, esteeming them, even though they were closed books to him, as "more valuable merchandise than anything offered by the Arabs or the Chinese."

Petrarch managed to interest his friend Boccaccio in helping him search for classical manuscripts, and Boccaccio made a very lucky find: he discovered a copy of Tacitus in the Benedictine monastery at Monte Cassino. He even learned Greek and taught in the University of Florence. Together, Petrarch and Boccaccio became the initiators of a very far-reaching movement known as *humanism*.

Humanism

The widest meaning of the term "humanism" denotes an interest in everything connected with the potentialities of man as a human being, his language, ideas, and art. But the term also has a narrower and more technical meaning when used in connection with the Italian Renaissance. The scholars of the 15th century, who first used the term, meant by it a group of academic subjects—grammar, rhetoric, history, poetry, and moral philosophy—all of which were to be studied in the original ancient Latin and Greek writings. The all-inclusive terms "humanities" and "liberal arts", which we use in our universities today, refer to many of these same interests.

The educational programme of the Italian humanists was well summed up by one of them in the concluding words of a short treatise.

That high standard of education to which I referred at the outset is only to be reached by one who has seen many things and read much. Poet, orator, historian and the rest, all must be studied, each must contribute a share. Our learning thus becomes full, ready, varied and elegant, available for action or for discourse in all subjects. But to enable us to make effectual use of what we know we must add to our knowledge the power of expression. . . . Proficiency in literary form, not accompanied by broad acquaintance with facts and truths, is a barren attainment; whilst information, however vast, which lacks all grace of expression, would seem to be put under a bushel, or partly thrown away. . . . Where, however, this double capacity exists—breadth of learning and grace of style—we allow the highest title to distinction and to abiding fame. . . . My last word must be this. The intelligence that aspires to the best must aim at both. In doing so, all sources of profitable learning will in due proportion claim your study. None have more urgent claim than the subjects and authors which treat of Religion and of our duties in the world; and it is because they assist and illustrate these supreme studies that I press upon your attention the works of the most approved poets, historians and orators of the past.

THE RENAISSANCE

A tremendous stimulus to the humanist cult of the classics was provided at the end of the 14th century when a distinguished Byzantine scholar, Manuel Chrysoloras (1350-1415), was appointed to a university professorship at Florence. Chrysoloras translated Homer, and Plato's *Republic*, and compiled the first Greek grammar to be used in the West. The enthusiasm aroused by his eloquent lectures swept Florence, as one of his hearers testifies:

Chrysoloras

> I was then studying Civil Law, but ... I burned with love of academic studies, and had spent no little pains on dialectic and rhetoric. At the coming of Chrysoloras, I was torn in mind, deeming it shameful to desert the law, and yet a crime to lose such a chance of studying Greek literature; and often with youthful impulse I would say to myself, "Thou, when it is permitted thee to gaze on Homer, Plato and Demosthenes, and the other poets, philosophers, and orators, of whom such glorious things are spread abroad, and speak with them and be instructed in their admirable teaching, wilt thou desert and rob thyself? Wilt thou neglect this opportunity so divinely offered? For seven hundred years, no one in Italy has possessed Greek letters; and yet we confess that all knowledge is derived from them. ... There are doctors of civil law everywhere; and the chance of learning will not fail thee. But if this one and only doctor of Greek letters disappears, no one can be found to teach thee." Overcome at length by these reasons, I gave myself to Chrysoloras, with such zeal to learn, that what through the wakeful day I gathered, I followed after in night, even when asleep.

After Chrysoloras returned to Constantinople in 1407 the Greek revival in Italy progressed rapidly, and a number of Greek manuscripts were subsequently imported from the East.

Humanism also had its more technical side, as may be seen in the career of Lorenzo Valla, who represents classical scholarship at its best. Valla spent most of his life in Naples and Rome, and his most famous accomplishment was his proof that the so-called Donation of Constantine (a document purporting to be a record of the donation by the Emperor Constantine to the pope of temporal rule over Rome and the western half of the Empire) was actually an 8th century forgery. He also wrote a series of notes in Latin pointing out errors, misunderstandings, and questionable translations in the Vulgate version of the Greek New Testament. This was a daring departure, because up until now the Vulgate, the Latin translation of the Bible produced over ten centuries earlier by St. Jerome, had been unquestioningly revered as authoritative. As it turned out, Valla's New Testament notes remained unpublished, but the manuscript had, as we shall see, a profound effect on another humanist in northern Europe.

Valla

The middle of the 15th century saw the completion of the major achievements of Italian humanism. The next generation only continued the trend by building on their predecessors' work, profiting from further manuscripts brought from Constantinople by Greek refugees fleeing the Turks in 1453. The Greek revival in Italy was accelerated by these Greek literary and philosophical works being more readily available, as well as by the new printing presses, which in the second half of the 15th century turned out many scholarly editions of Greek and Latin classics.

Pico della Mirandola

It was inevitable that interest in ancient philosophies should once more be stimulated, and numerous humanist scholars now concentrated on popularizing them. One of the most ambitious of these was Giovanni Pico della Mirandola, who set himself the impossible task of making a coherent system out of Platonic, Aristotelian, Pythagorean, Zoroastrian, Hebrew, and Christian elements! He could read Greek, Latin, Hebrew, and Arabic, and when he was only twenty-three he drew up a list of 900 theses which he was prepared to defend in debate. Part of the Oration which he proposed as an introduction may be read in Source Reading (b) at the end of this chapter.

Even Machiavelli, who wrote as a practising politician, was a humanist. Both his *Discourses* (comments on history and politics) and his *Art of War* (views on military matters) were based on Latin works. Classical authors were his models, and in following them he displayed a humanistic approach to the politics of his day.

Because of their love for the classics, the humanists did much to restore Greek and Latin to their ancient purity, as well as to discover forgotten manuscripts and to develop literary and historical criticism. But strangely enough, the 20th century remembers them not so much for the purity of their Greek and Latin as for their popularization of the vernaculars coupled with their secular ideal of virtue. One of the most widely read vernacular books in the 16th century was *The Book of the Courtier* by Baldassare Castiglione, an analysis, in Italian, of the qualities of the well-rounded gentleman of the Renaissance. Source Reading (c) lists some of these characteristics.

2/ITALIAN ART AND ARCHITECTURE

We have seen literary humanism emphasizing the recovery of the classics, and, at the same time, expressing a more secular attitude toward life. These same two notes were sounded, with exceptions of course, in the art and architecture of the Renaissance. Here again we are dealing with Italy, since it was some time before the Renaissance crossed the Alps.

Giotto

Painting in Three Dimensions Under an artist named Giotto a new trend in painting was inaugurated. Before Giotto's time most painting was stiff and formalized, often looking like a painted mosaic rather than a picture with perspective and gradations in colour and shading. But Giotto began a revolution: he was the first artist since the fall of the Roman Empire to paint in such a way that his figures looked three-dimensional. The new advanced technique was perfected by Botticelli, who emphasized line and pattern, and whose graceful paintings of classical themes portrayed the spiritual beauty of the soul.

Leonardo da Vinci's meticulous anatomical sketches are among his most remarkable creations. Indeed, some of them are so exact that they have been matched point by point with X-rays of the same areas. Leonardo was, in fact, the true originator of scientific illustration, claiming to have dissected more than thirty bodies and producing as a result a wonderful series of drawings. Here, in his studies of the shoulder of an old man, can be seen his stylized representation of muscles as something resembling elastic bands. It should be noted that Leonardo's interest in anatomy stemmed not from any medical curiosity, but rather from his insatiable desire as a painter to know how men's bodies are put together and work, so that he might make the figures on his canvas live and move with the authenticity of life itself.

A contemporary of Botticelli was Leonardo da Vinci. Leonardo was to art what Aristotle was to philosophy: the great Italian master has left us more than 5000 pages of notes on botany, geology, zoology, optics, hydraulic and military engineering, grammar, perspective, colour, anatomy —in fact practically everything. We cannot, of course, consider every side of Leonardo's genius. Let us deal very briefly with him, first as a scientist and then as a painter.

Leonardo da Vinci pointed the way toward the Scientific Revolution of the 17th century. His keen interest in scientific problems is displayed in the 120 notebooks which he filled with his observations and conclusions, all written in Italian and in "mirror-writing" (from right to left on the page). Only a part of these amazing notebooks now survives, dispersed in European libraries and museums, and it is therefore difficult to know exactly what Leonardo's purpose was in compiling them. In any event they remain a remarkable indication of Leonardo's far-reaching interests.

Although the notebooks incorporated uncritical digests of medieval scientific treatises and hence contained much that we would classify as magic and astrology, they included many astute observations. Leonardo described fossil shells found high in the mountains far from the sea (thus anticipating the science of paleontology), and provided painstaking anatomical descriptions. There were even plans for building a flying machine and a submarine. The notebooks, however, were not generally available until long after his day—with the result that Leonardo's scientific formulations had only limited influence. In fact since the bulk of them were not published until the 19th century, they constitute "a backwater in the history of science." But they do indicate the amazing scope of his researches, a scope illustrated by the following extract:

And this old man, a few hours before his death, told me that he had lived a hundred years, and that he did not feel any bodily ailment other than weakness, and thus while sitting upon a bed in the hospital of Santa Maria Nuova at Florence, without any movement or sign of anything amiss, he passed away from this life.

And I made an autopsy in order to ascertain the cause of so peaceful a death, and found that it proceeded from weakness through failure of blood and of the artery that feeds the heart and the other lower members, which I found to be very parched and shrunk and withered; and the result of this autopsy I wrote down very carefully and with great ease, for the body was devoid of either fat or moisture, and these form the chief hindrance to the knowledge of its parts.

The other autopsy was on a child of two years, and here I found everything the contrary to what it was in the case of the old man.

The old who enjoy good health die through lack of sustenance. And this is brought about by the passage to the mesaraic veins becoming continually restricted by the thickening of the skin of these veins; and the process continues until it affects the capillary veins, which are the first to close up altogether; and from this it comes to pass that the old dread the cold more than the young, and that those who are very old have their skin the colour of wood or of dried chestnut, because this skin is almost completely deprived of sustenance.

And this network of veins acts in man as in oranges, in which the peel becomes thicker and the pulp diminishes the more they become old. And if you say that as the blood becomes thicker it ceases to flow through the veins, this is not true, for the blood in the veins does not thicken because it continually dies and is renewed.

If, however, Leonardo's scientific work had little effect on his contemporaries, his painting certainly had. As a matter of fact Leonardo regarded himself primarily as a painter, and wrote this about the purpose of painting:

A good painter has two chief objects to paint—man and the intention of his soul. The former is easy, the latter hard, for it must be expressed by gestures and movements of the limbs. . . . A painting will only be wonderful for the beholder by making that which is not so appear raised and detached from the wall.

Two of Leonardo's paintings are known to everyone. The *Mona Lisa*, now in the Louvre in Paris, is a portrait of an Italian nobleman's wife that is both a likeness and a psychological study. In 1963 this painting, perhaps the most famous in the world, was lent by the French government to the United States for exhibitions in Washington and New York, where crowds thronged to study its enigmatic beauty. Almost as familiar to us are reproductions of Leonardo's *Last Supper*, which was painted on the wall of a refectory in Milan and is today much faded as a result of the ravages of dampness and an imperfect experimental pigment technique adopted by Leonardo. Later restorations have left nothing of the artist's original

The Piazza del Campidoglio in Rome was designed by Michelangelo for the Capitoline Hill, though it was not completed in his lifetime. The bronze equestrian statue of Marcus Aurelius is Roman, and echoes of antiquity may be seen in the use of classical columns adapted to the faces of the buildings, and, the arch, stairs, doorway, and tower. But the rest of the square is in the Renaissance style. The whole group of buildings conveys a sense of mathematical interrelationship and harmony which can also be found in the proportions of the individual buildings themselves.

colours. Yet there can still be seen that dramatic moment in the upper room when Christ has just announced, "One of you will betray me"; and the twelve disciples, grouped in threes about him, have each asked, "Is it I?" Here, incidentally, is an example of Renaissance art that is certainly not worldly or secular—another illustration of the danger of generalizing about so complex an intellectual movement as the Renaissance.

Titian

Two other painters, Titian and Raphael, became famous for their portraits and their frescoes. In some respects Titian was the most remarkable of all Renaissance artists, because he produced an average of one picture a month for 80 years. His portraits of Renaissance popes and rulers form a vivid gallery of 16th century politicians. On the other hand, the classical inspiration of the period is shown in Raphael's *School of Athens*, a vast fresco in which dozens of Greek scholars are grouped around Plato and Aristotle, a picture generally regarded as the greatest painted during the Renaissance.

Raphael

Raphael's career was as brilliant as it was brief. In 1514, loaded with honours and commissions, he was appointed chief architect of St. Peter's in Rome at the age of 31. Within six years he was dead.

Goldsmiths' Apprentices Renaissance sculptors followed the lead of late Gothic masters in making more lifelike copies of their subjects than had early medieval artists. Their devotion to art, as well as the public's adulation of them, is well illustrated by the story of the Baptistry doors in Florence.

Ghiberti In 1401 the citizens of Florence held a competition to replace the east doors of the Baptistry—a small, rather insignificant building opposite the cathedral. The Baptistry already had the finest of doors facing the cathedral, but for the Florentines a century of art had gone by, and they yearned for even better. Lorenzo Ghiberti won the contest, and his runner-up, Filippo Brunelleschi, was to assist him. Brunelleschi, however, magnanimously resigned the task to Ghiberti alone, so that he might complete his creation unhampered by another craftsman.

So Ghiberti began to work. It was a staggering task, and it was years before his bronze doors were set up and gilded in 1424. The citizens were pleased, in fact delighted, with his work, so much so that they decided to commission Ghiberti to make an even finer set for this place of honour opposite the cathedral, and move the doors just completed to another side of the Baptistry. . . . Another set of doors! Ghiberti began again, and laboured on for 27 years until all told he had spent 48 years on Baptistry doors. But finally there they stood, the stories of the Old Testament immobilized in bronze, doors pronounced fit to be the Gates of Paradise, doors which have remained in their place of honour until this day.

Donatello Ghiberti's contemporary, Donatello, also worked in bronze. In fact he was the first artist since Roman times to produce a bronze equestrian statue, that of the *condottiere* Gattemalata. Donatello stressed both lifelikeness and beauty, as can be seen in his handsome young *St. George* who appears as a refined youth rather than a feudal dragon-killer.

Benvenuto Cellini was born a generation after Donatello. He was also a celebrated worker in bronze, though his fame as a goldsmith equals that as a sculptor, and he wrote a disarmingly frank *Autobiography* combining sensitivity and violence, beauty and the beast.

Cellini Cellini's career illustrates one reason for the cultural leadership of Florence. Florence was a city of goldsmithing, a trade that required its craftsmen to master every skill needed in the other arts and in every material used by them. The amazing versatility of Italian Renaissance artists who could work in stone, bronze, silver, or gold, who could draw, paint, and carve was probably due to the fact that men such as Cellini, Donatello, Brunelleschi, Ghiberti, and even Leonardo began their apprenticeship in the goldsmith's shop.

Brunelleschi Both Giotto the painter and Brunelleschi the sculptor also excelled in architecture. Giotto's contribution was a bell-tower (*campanile*) for the cathedral at Florence. Brunelleschi, on the other hand, had not been

Ghiberti's second set of Baptistry doors still face the cathedral in Florence today. Carved in bronze, the graceful Gothic figures in the ten panels portray scenes from the Old Testament, the whole work being gilded to produce an effect of scintillating elegance. Impressed by the exquisite workmanship of the doors, Michelangelo exclaimed, "They are so beautiful that they might fittingly stand at the gates of Paradise." Incidentally, Ghiberti immortalized his own likeness by including his portrait twice in the decorative borders of the doors.

Benvenuto Cellini, the celebrated Florentine goldsmith and sculptor, created this exquisite cup of gold. The swirling grace of the shell and brilliant enamelling of the sphinx handle and dragon and tortoise base reflect the sumptuous taste of the Renaissance.

unmarked by his defeat at the hands of Ghiberti of the Paradise doors, and when a new contest was announced to choose an architect to build a dome for the cathedral, Brunelleschi put heart and soul into winning it.

Now roofing over a building with a dome was not a new problem, as witness the Pantheon in Rome and Hagia Sophia in Constantinople. But what Brunelleschi planned was different. Whereas earlier domes had been partly concealed from the outside, Brunelleschi wanted to emphasize the dome from the exterior, to show it dominating both the cathedral and the surrounding city. To do this he conceived a plan for an octagonal dome which entirely fascinated the Florentines.

But the novelty of the effort had its drawbacks. Florence was full of busy tongues saying just how it should be done, and gradually a lack of confidence infected Brunelleschi's watchers. Did he know what he was doing, they murmured, or would he end up ruining their cathedral and squandering their money? Then they hit upon a plan: appoint a supervisor to watch over this man and make sure he did not lead them into a catastrophe. That supervisor was Brunelleschi's old rival, Ghiberti of the Paradise doors.

Unfortunately Ghiberti was not a big enough man to refuse the job; so Brunelleschi decided upon a course of action. When the dome reached its most difficult point of construction, Brunelleschi was taken with a mysterious and distressing ailment which kept him from his work and left Ghiberti alone with the job. Poor Ghiberti was floored. He was totally incapable of proceeding, as Brunelleschi—and soon all Florence—well knew. Accordingly, he was relieved of his supervisory capacities and Brunelleschi was permitted to complete his work alone, so that it stands today, a larger dome than St. Peter's, and as strong as it is beautiful. Perhaps it could receive no higher award of merit than the fact that Michelangelo gained knowledge and inspiration from it for his dome of St. Peter's in Rome.

Master of the Arts. Now we come to Michelangelo—the very soul of the Renaissance.

Michelangelo Buonarroti of Florence was perhaps the world's greatest artist. He was indebted to the Medici for patronage, and one of his most impressive works was a tomb for this family in which he created not only the sculpture but the whole building as well. Earlier he had produced his magnificent *David,* a majestic statue 18 feet high, carved free-hand from a huge block of marble that had been mauled earlier by another sculptor.

Michelangelo's skill with marble was unmatched. When asked how he did it, the great sculptor is said to have replied that the statue "was already in the marble block.... I had only to cut away the little pieces around it." And Michelangelo the poet wrote:

> The best of artists hath no thought to show
> Which the rough stone in its superfluous shell
> Doth not include: to break the marble spell
> Is all the hand that serves the brain can do.

Sculptor, poet, what more? A brilliant painter too. The pope commissioned him to paint the ceiling of the Sistine Chapel in Rome, an assignment that the artist began unwillingly, hampered not only by his assistants but by the elements themselves which covered his paintings with mould almost as soon as he finished them. But he was not to be released from his mammoth task. The cause of the mould (the lime base had been put on too damp a surface, and the water oozed through it) was discovered, and Michelangelo had to proceed with his work. "I am not a painter," he would protest as he worked fourteen hours a day for four years, lying on his back, exhausted and nerve-wracked on a scaffold that he himself had designed. "Crosswise I strain me like a Syrian bow," he wrote. But at last the epic of the *Creation and Fall of Man* was complete, over 700 square yards of it. Then thirty-four years later another pope gave him a commission, and Michelangelo was to spend four more years decorating the end wall of the Sistine with the *Last Judgment.* During this tremendous effort the artist fell off the scaffolding and broke his leg, an accident that left him lame for the rest of his life.

Sculptor, poet, artist—and architect. At seventy (he lived to be eighty-nine) the great man became chief architect of St. Peter's, conceiving a circular Renaissance plan with a great dome rising 435 feet above the floor. The Renaissance plan differed from the Gothic in that it was designed on circular lines, with a central altar rather than one at the end of a long nave. Platonic thought had emphasized the perfection of the circle, and the idea was revived with enthusiasm in Renaissance philosophy.

Even as the great dome rose over the Roman skyline as if to symbolize the unity and solidity of the Church, that Church was being threatened; and before Michelangelo died Roman Catholics and Protestants were already at each others' throats. How this had come about involves a brief consideration of the northern Renaissance.

3/THE NORTHERN RENAISSANCE

The Middle Ages hung on longer north of the Alps than in Italy; hence the literary renaissance was slower to reach the north, not fully conquering it until a century after Petrarch. A Renaissance in the north had to wait until Italian scholars carried their culture north of the Alps, or northern tourists visited Italy, to return home full of Renaissance ideas. Most of all, it had to wait until books and works of art circulated outside Italy.

But as in the south, so in the north there were forerunners of the academic Renaissance, forerunners who did much to popularize the vernacular. Three only can be mentioned: two Englishmen and a Frenchman. The Englishmen, William Langland and Geoffrey Chaucer, were both younger contemporaries of Petrarch and Boccaccio.

William Langland

William Langland was a poverty-stricken cleric from Shropshire who wrote a long poem, *The Vision of William concerning Piers the Plowman*, in alliterative blank verse. In his archaic English, Langland takes the unmedieval stand of defending the common man by criticizing the existing social order. The characters of his poem remind one of the two-dimensional figures in a medieval fresco—bloodless creatures rather like puppets on a string.

Geoffrey Chaucer

More modern is Geoffrey Chaucer. He came from a family of English merchants and became a senior civil servant. Chaucer's most famous work is the *Canterbury Tales*, which like the *Decameron* is a series of stories held together by a literary device—in this case a pilgrimage to St. Thomas Becket's shrine at Canterbury, during which the pilgrims regale each other with well-known tales. The *Canterbury Tales* also resemble the *Divine Comedy* in their brilliant characterization; the knight, parson, monk, prioress, and especially the inimitable Wife of Bath are no mere stock characters but flesh and blood men and women. Medieval though they may be, their reactions to life are as vital as if they were living today, and the language of the poem is the colloquial English of 14th century London, which, with a little practice, can still be easily read.

In his new use of old materials Chaucer was an unquestioned master. As such he marks the end of one age and the beginning of another.

Across the Channel in France, the vernacular was given vivid expression by the poet François Villon, whose haunting verses exude the atmosphere of both the Middle Ages and the Renaissance. As a convicted thief, guilty of manslaughter (if not murder) in a street brawl, Villon understood and hence was able to express the insecurity felt by those who realized that the medieval world was breaking up. *François Villon*

Humanism Adopts Christianity In northern Europe, as in Italy, academic humanism followed in the steps of its vernacular beginnings. But the North did not display the secular tendencies of the South to the same extent. Instead, this new humanism combined an intense faith in Christianity with all the reverence for antiquity professed by the Italian humanists. Among the many noteworthy northern scholars there are four of particular interest: Johann Reuchlin, Jacques Lefèvre d'Étaples, Thomas More, and Desiderius Erasmus.

Reuchlin, a German scholar, had been induced to study Hebrew by the example of Pico della Mirandola, and in 1506 published the first Hebrew grammar north of the Alps. His opposition to a scheme for seizing and burning Hebrew books in Germany, however, resulted in his being tried for heresy in 1513. Reuchlin answered his critics by claiming that the Bible should be studied in its original languages—in Hebrew as well as in Greek; and as the trial dragged on for six years it came to involve most of the prominent scholars of Europe. Lined up on one side were the humanists, with Reuchlin their champion, and on the other stood the conservative churchmen of Europe. By the time the trial finally ended in Reuchlin's favour, much bitterness had been aroused on both sides. *Johann Reuchlin*

Jacques Lefèvre d'Étaples was the leader of the humanists at the University of Paris. After producing translations of nearly all of Aristotle's work he eventually turned to the study of the texts of the Bible, and in 1512 published an edition of St. Paul's Epistles which presented the Vulgate version along with his own translation from the Greek. Finally, he completed a French translation of both the Old and New Testaments. "We must," he said, "affirm nothing of God but what the Scriptures tell us about him." But Lefèvre's ideas got him into trouble with the faculty of theology at Paris, and he left the city. He never, however, broke with the Church, cherishing the hope that it could be reformed from within. *Lefèvre d'Étaples*

Meanwhile in England there lived the brilliant Sir Thomas More. This scholar, familiar with Greek and classical Latin, was, as well, a successful lawyer and judge, member of Parliament at twenty-six, Under-Sheriff of London, and a distinguished diplomat and royal official. In 1515-16 More wrote a famous book in Latin entitled *Utopia*, a portrayal of an ideal, classless society in which there was no money, a society organized along monastic lines and situated in a place called Utopia (Nowhere). So realistically did More describe Utopia that one English clergyman applied to go *Thomas More* there as a missionary! The book was, of course, a criticism of the European society of that day from a Christian point of view, as the author made perfectly clear:

> Therefore when I consider and weigh in my mind all these commonwealths, which nowadays anywhere do flourish, so god help me, I can perceive nothing but a certain conspiracy of rich men procuring their own commodities under the name and title of the commonwealth. They invent and devise all means and crafts . . . to hire and abuse the work and labour of the poor for as little money as may be. These devices, when the rich men have decreed . . . be made laws.

Erasmus

The most influential northern humanist, and the one who made Christian humanism a truly international movement, was Erasmus of Rotterdam. Born in Holland and educated in a monastic school, he became something of an international citizen, living in France, England, Italy, Germany, and Switzerland, and teaching at Oxford, Cambridge, and Paris. His wide classical learning and his flawless Latin earned Erasmus the title of "Prince of Humanists; and his editions of Greek and Latin classics, now reproduced in quantity by the new printing press and carried by horse or ship from one end of the continent to the other, spread his fame.

In 1505 Erasmus discovered and published the manuscript of Valla's notes on the New Testament. Inspired by these, he continued his humanist predecessor's work and in 1516 produced an edition of the Greek New Testament, combined with a Latin translation and notes. While the humanists applauded, conservative churchmen held up their hands in horror. This was nothing less than an open challenge to the theologian's monopoly to read and interpret Scripture. Now there was a new translation —and one frankly designed to correct any errors that might exist in the Latin Vulgate. Not only that, but in the preface to his book Erasmus made an even more radical proposal.

> I dissent most vehemently from those who would not have the divine Scriptures read by the unlearned, translated into the vulgar tongue, as though Christ taught in such an involved fashion that he would be but barely understood by a few theologians, as though the safeguard of the Christian religion lay only in its being unknown. . . . I would desire that all little women should read the Evangel, should read the Pauline epistles. And would that these might be translated in all languages, so that they might be read and known, not only by Scots and Irishmen, but even by Turks and Saracens. . . . Would that from these the ploughman might sing at the plough, and that with something from these the weaver might keep time to his loom; would that with stories from these the traveller might while away the tedium of his voyage.

Erasmus always remained a faithful son of the Church. Yet by his courageous search for accuracy in the Scriptures and his insistence that all should be able to read them in translation, this intrepid scholar helped to prepare the way for the Reformation.

Montaigne

Popular Culture The impact of the Italian Renaissance on the popular culture of northern Europe can also be seen in the works of two 16th

Hans Holbein the Younger's portrait of Erasmus (painted about 1523) is a proclamation on canvas of the humanists' belief that the pen is mightier than the sword. It is also an unforgettable portrayal of the Prince of Humanists, whose wit was as sharp as his nose.

century authors: the French essayist Michel de Montaigne, and the English poet and dramatist William Shakespeare. Montaigne published a volume of essays in 1580 which demonstrated his extraordinary range of interests in anything even remotely connected with mankind. His essay "Of Cannibals" includes this idyllic picture of life unspoiled by civilization—a fairly novel point of view in the 16th century.

> These nations, then . . . are still very near to their original simplicity. The laws of Nature govern them still, very little debased with any mixture of ours; but they are in such a state of purity that I am sometimes vexed that the knowledge of them did not come earlier, at a time when there were men better able to judge of them than we are. I am sorry that Lycurgus and Plato did not know them. . . . This is a nation, I should say to Plato, wherein there is no manner of traffic, no knowledge of letters, no science of numbers, no name of magistrate or political superiority, no use of service, no riches or poverty, no contracts, no successions, no partitions of property, no employments but those of leisure, no respect of kinship save the common ties, no clothing, no agriculture, no metal, no use of corn or wine. The very words that signify lying, treachery, dissimulation, avarice, envy, detraction, pardon, never heard of.

William Shakespeare is the brightest star of the English Renaissance. Shakespeare made man the centre of his universe, and by his admiration for Greece and Rome, his keen interest in new-found lands, his patriotism, and his belief in strong monarchy he showed that he was a true son of the Renaissance.

Shakespeare

We know that at this time the English took over words from Latin, Greek, and French. In this borrowing, Shakespeare was outstanding: more new words may be found in his works alone than in almost all the rest of the English poets combined. So extensive was the borrowing in fact that one scholar was moved to write that if "English pens maintain that stream we have of late observed to flow from many, we shall within a few years be fain to learn Latin to understand English." This was, to be sure, an exaggeration. Shakespeare, knowing "small Latine and lesse Greeke", depended on vernacular translations for most of the classics.

Just as the classical learning of the Middle Ages had crystallized in Dante, so the humanism of the 15th and 16th centuries produced a Montaigne and a Shakespeare.

In the field of art, northern Europe again remained more medieval than Italy: the nobles of France, for example, continued to collect beautiful books illuminated in the style of medieval manuscripts. By the 15th century the centre of northern painting had shifted to Germany and the Low Countries, where Jan van Eyck, and later Albrecht Dürer, Hans Holbein the Younger, and Pieter Breughel produced a brilliant series of lifelike portraits and village and country scenes. Dürer gained the reputation of being the first artist in history to become a "best seller" when he created copper engravings and wood-cuts which were used to reproduce hundreds of copies.

Incidentally, Jan van Eyck may have been the inventor of a new technique. At any rate he was the first major artist to use this technique—to create oil paintings rather than to fresco on plaster walls. Van Eyck applied his colours in thick, transparent glazes whose clear tones still radiate light over five centuries later.

Chenonceaux, in the Loire valley of France, is a typical 16th century château. The regularity of its architectural design and the formal gardens illustrate the penetration of Italian Renaissance ideas into France. Built as it is astride the River Cher on the pilings of an ancient mill Chenonceaux seems to rise like some fairy-tale castle out of the water that mirrors it.

In sculpture and architecture, too, northern Europe continued to develop medieval tendencies. In England an architectural style called *Perpendicular* developed, a late Gothic variation with stiff, straight lines, vertical window tracery, and fan-vaulting. Two particularly fine examples of Perpendicular are the Chapel of King's College in Cambridge University (1446-1515), and a magnificent chapel that Henry VII had built by an Italian craftsman in Westminster Abbey (1503-19).

It was not until the 16th century that French architecture showed influences other than Gothic. But with Charles VIII and Francis I, Italian ideas began to be imported into France. Francis especially was a great patron of the arts, and commissioned Leonardo da Vinci and Benvenuto Cellini to come to France to work for him. Now elegant country palaces began to be the fashion, as the nation settled down under strong monarchs and the need for protective castles disappeared. The most famous of these châteaux were built in the Loire Valley, where their many windows and patterns of classical ornamentation gave a romantic delicacy to the old square castle plan with rounded towers at the corners.

In northern Europe as in Italy, much of the Middle Ages lingered on underneath an ornate Renaissance façade. It was, in fact, a typically medieval figure—a monk from Germany—who was destined to shatter that perfect world which the men of the Renaissance were certain they were on the way to attaining.

4/SUMMARY AND CONCLUSION

Such a complex period of cultural change as the Renaissance cannot be adequately dealt with in a few pages, nor adequately illustrated by a few drawings and photographs. Indeed, historians are far from being agreed as to either the chronological limits or the correct interpretation of the Renaissance. Certain trends away from medieval ways of doing things can, however, first be detected in the urban civilization of northern Italy, where the economic prosperity stemming from the Commercial Revolution created a class of rich patrons of learning and art—political bosses with culture.

As well as emphasizing a return to the classics, the humanists of the Renaissance promoted new vernaculars. They possessed, too, a supreme self-confidence, a certain individualistic exuberance and optimism. "Man," said one Renaissance writer, "can make whatever he will of himself;" and another said "I have made myself." An increasing worldliness thus accompanied the revived classical learning.

Yet painting, sculpture, and architecture had by no means become suddenly and exclusively secular, as is evident in the work of Leonardo or Michelangelo. What is completely different about the art of the Renaissance is the extent to which the mass of the urban population appreciated artistic activities. Nothing like it had been seen since the days of Periclean Athens. Art was not the hobby of a fortunate minority but a feature of daily life in Medicean Florence.

SOURCE READINGS

(a)

FRANCESCO PETRARCA or PETRARCH (1304-74) has been called "one of the world's most interesting men." As a romantic poet, lover of nature, Italian patriot, and humanist he summed up in his person what the men of the Renaissance exalted as the "universal man". The letter below, addressed to the Roman historian Livy, expresses the nostalgic longing of the early humanists for the recovery of the classical world.

I should wish (if it were permitted from on high) either that I had been born in your age or you in ours; in the latter case our age itself, and in the former I personally should have been the better for it. I should surely have been one of those pilgrims who visited you. For the sake of seeing you I should have gone not merely to Rome, but indeed, from either Gaul or Spain I should have found my way to you as far as India. As it is, I must fain be content with seeing you as reflected in your works—not your whole self, alas, but that portion of you which has not yet perished, notwithstanding the sloth of our age. We know that you wrote one hundred and forty-two books on Roman affairs. With what fervour, with what unflagging zeal must you have laboured; and of that entire number there are now extant scarcely thirty.

. . . It is over these small remains that I toil whenever I wish to forget these regions, these times, and these customs. Often I am filled with bitter indignation against the morals of today, when men value nothing except gold and silver, and desire nothing except sensual, physical pleasures. If these are to be considered the goal of mankind, then not only the dumb beasts of the field, but even insensible and inert matter has a richer, a higher goal than that proposed to itself by thinking man. But of this elsewhere.

It is now fitter that I should give you thanks, for many reasons indeed, but for this in especial: that you so frequently caused me to forget the present evils, and transferred me to happier times. As I read, I seem to be living in the midst of Cornelii Scipiones Africani . . . Fabius Maximus . . . Brutus . . . of Cato. . . . It is with these men that I live at such times and not with the thievish company of today among whom I was born under an evil star. . . .

Farewell forever, matchless historian!

Written in the land of the living, in that part of Italy and in that city in which I am now living and where you were once born and buried, in the vestibule of the Temple of Justina Virgo, and in view of your very tombstone; on the twenty-second of February, in the thirteen hundred and fiftieth year from the birth of Him whom you would have seen, or whose birth you could have heard, had you lived a little longer.

Setton and Winkler, *Great Problems in European Civilization*, pp. 221-222.

(b)

GIOVANNI PICO, COUNT OF MIRANDOLA (1463-94), was the youngest member of a family that claimed descent from the emperor Constantine. At fourteen he was sent to the University of Bologna, and, after travelling widely, came to Rome when he was twenty-three. The speech below, which illustrates the Renaissance exaltation of man, is part of the disputation he prepared in Rome when he offered to defend his 900 theses against anyone in Europe—even volunteering to pay the expenses of scholars from distant points! The disputation was, however, prohibited by Pope Innocent VIII, though later Pope Alexander VI cleared Pico of charges of heresy. He died at the age of thirty-one when he was stricken by a fever.

... At last it seems to me I have come to understand why man is the most fortunate of creatures and consequently worthy of all admiration and what precisely is that rank which is his lot in the universal chain of Being—a rank to be envied not only by brutes but even by the stars and by minds beyond this world. It is a matter past faith and a wondrous one. ...

... God the Father, the supreme Architect, had already built this cosmic home we behold, the most sacred temple of His godhead, by the laws of His mysterious wisdom. ... But, when the work was finished, the Craftsman kept wishing that there were someone to ponder the plan of so great a work, to love its beauty, and to wonder at its vastness. Therefore, when everything was done ... He finally took thought concerning the creation of man. But there was not among His archetypes that from which He could fashion a new offspring, nor was there in His treasure-houses anything which He might bestow on His new son as an inheritance, nor was there in the seats of all the world a place where the latter might sit to contemplate the universe. ...

At last the best of artisans ordained that that creature to whom He had been able to give nothing proper to himself should have joint possession of whatever had been peculiar to each of the different kinds of being. He therefore took man as a creature of indeterminate nature and, assigning him a place in the middle of the world, addressed him thus: "Neither a fixed abode nor a form that is thine alone nor any function peculiar to thyself have we given thee, Adam, to the end that according to thy longing and according to thy judgment thou mayest have and possess what abode, what form, and what functions thou thyself shalt desire. The nature of all other beings is limited and constrained within the bounds of law prescribed by Us. Thou, constrained by no limits, in accordance with thine own free will, in whose hand We have placed thee, shalt ordain for thyself the limits of thy nature. We have set thee at the world's centre that thou mayest from thence more easily observe whatever is in the world. We have made thee neither of heaven nor of earth, neither mortal nor immortal, so that with freedom of choice and with honour, as though the maker and moulder of thyself, thou mayest fashion thyself in whatever shape thou shalt prefer. Thou shalt have the power to degenerate into the lower forms of life, which are brutish. Thou shalt have the power, out of thy soul's judgment, to be reborn into the higher forms, which are divine."

Introduction to Contemporary Civilization in the West, I, 535-537.

(c)

BALDASSARE CASTIGLIONE (1478-1529) *in* The Book of the Courtier *gave a minute analysis of the courtly virtues he found in the entourage of the Duke of Urbino between 1504 and 1508. A few paragraphs from his book follow.*

Not that we would have him look so fierce, or go about blustering, or say that he has taken his cuirass to wife, or threaten with those grim scowls ... because to such men as this, one might justly say that which a brave lady jestingly said in gentle company to one whom I will not name at present; who, being invited by her out of compliment to dance, refused not only that, but to listen to the music, and many other entertainments proposed to him,—saying always that such silly trifles were not his business; so that at last the lady said, "What is your business, then?" He replied with a sour look, "To fight." Then the lady at once said, "Now that you are in no war and out of fighting trim, I should think it were a good thing to have yourself well oiled, and to stow yourself with all your battle harness in a closet until you be needed, lest you grow more rusty than you are;" and so, amid much laughter from the bystanders, she left the discomfited fellow to his silly presumption.

Therefore let the man we are seeking, be very bold, stern, and always among the first, where the enemy are to be seen; and in every other place, gentle, modest, reserved, above all things avoiding ostentation and that impudent self-praise by which men ever excite hatred and disgust in all who hear them. . . .

I think that what is chiefly important and necessary for the Courtier, in order to speak and write well, is knowledge; for he who is ignorant and has nothing in his mind that merits being heard, can neither say it nor write it. . . .

And this I say as well of writing as of speaking: in which however some things are required that are not needful in writing,—such as a good voice, not too thin and soft like a woman's, nor yet so stern and rough as to smack of the rustic's,—but sonorous, clear, sweet and well sounding, with distinct enunciation, and with proper bearing and gestures; which I think consist in certain movements of the whole body, not affected or violent, but tempered by a calm face and with a play of the eyes that shall give an effect of grace, accord with the words, and as far as possible express also, together with the gestures, the speaker's intent and feeling. . . .

In such fashion would I have our Courtier speak and write; and not only choose rich and elegant words from every part of Italy, but I should even praise him for sometimes using some of those French and Spanish terms that are already accepted by our custom.

I would have him more than passably accomplished in letters, at least in those studies that are called the humanities, and conversant not only with the Latin language but with the Greek, for the sake of the many different things that have been admirably written therein. Let him be well versed in the poets, and not less in the orators and historians, and also proficient in writing verse and prose, especially in this vulgar tongue of ours; for besides the enjoyment he will find in it, he will by this means never lack agreeable entertainment with ladies, who are usually fond of such things. . . .

Introduction to Contemporary Civilization in the West, I, 551-553.

FURTHER READING

ADY, C. M., *Lorenzo dei Medici and Renaissance Italy* (English Universities Press, 1955)

* BERENSON, B., *The Italian Painters of the Renaissance* (Meridian, 1957)

CHABOD, F., *Machiavelli and the Renaissance* (Bowes and Bowes, 1958)

† DURANT, W., *The Renaissance: A History of Civilization in Italy from 1304-1576* (Simon and Schuster, 1953)

† * FERGUSON, W. K., *The Renaissance* (Holt, 1940)

——————, *The Renaissance in Historical Thought: Five Centuries of Interpretation* (Houghton Mifflin, 1948)

* GILMORE, M. P., *The World of Humanism, 1453-1517* (Torchbook, 1962)

† GOLDSCHEIDER, L., *Michelangelo: Paintings, Sculpture, Architecture* (Phaidon, 1954)

HALE, J. R., *Machiavelli and Renaissance Italy* (English Universities Press, 1962)

HAY, D., *The Italian Renaissance in its Historical Background* (Cambridge, 1961)
* HAYDN, H. C., *The Counter-Renaissance* (Evergreen, 1961)
* HUIZINGA, J., *Erasmus and the Age of Reformation* (Torchbook, 1957)
† KETCHUM, R. M., editor, *The Horizon Book of the Renaissance* (Doubleday, 1961)
KREY, A. D., *A City that Art Built* (Minnesota, 1936)
* KRISTELLER, P. O., *Renaissance Thought: The Classic, Scholastic, and Humanist Strains* (Torchbook, 1961)
† LUCAS-DUBRETON, J., *Daily Life in Florence in the Time of the Medici* (Allen and Unwin, 1960)
PHILLIPS, M. M., *Erasmus and the Northern Renaissance* (English Universities Press, 1949)
POTTER, G. R., editor, *The New Cambridge Modern History*, Vol. I, *The Renaissance, 1493-1520* (Cambridge, 1961)
† * ROEDER, R., *The Man of the Renaissance* (Meridian, 1958)
SELLERY, G. C., *The Renaissance, Its Nature and Origins* (Wisconsin, 1950)

17. The Reformation

... I believe I see a golden age dawning in the near future.
ERASMUS IN 1518

... the Pope is fallible. ... His authority cannot prevail over the Holy Scriptures.
LUTHER IN 1518

The Reformation is the name commonly given to the religious upheavals of the first half of the 16th century. Chronologically speaking, then, the Reformation overlaps the Renaissance. But because it was essentially a very different sort of movement we must make a detailed separate examination of it, even though its roots are firmly fixed in the Medieval Transition and the Renaissance.

1/THE RENAISSANCE CHURCH AND ITS CRITICS

The economic and political changes that affected Europe from 1300 on also influenced the Church. The return of the papacy to Rome in 1377 had resulted in the Great Schism and in the creation of two lines of popes, the Roman and Avignonese. Now all Europe divided its allegiance. Clement VII was supported by France, Scotland, Navarre, Castile, and Aragon, while Italy was hopelessly split. On the side of Urban VI were England, Gascony, and Flanders—France's enemies—along with Portugal, Bohemia, Hungary, and most of Germany. It was inevitable that such an unhappy situation should produce outspoken criticism.

The Great Schism

More than criticism was needed. What was imperative was a solution, some way to heal the schism. As they pondered the problem, more and more churchmen came to accept the belief that only one solution remained: a General Council. Such a Council would, it was argued, possess authority superior to that of any pope. But who would summon it?

In 1409 a majority of the two colleges of cardinals acted independently of both popes and summoned a General Council to meet in Pisa. When neither Gregory XII nor the Avignon pope, Benedict XIII, would consent to attend, the Council deposed them and elected a new pope, Alexander V. Because no one of the three popes would now defer to the other two, the Council of Pisa failed, and in the end a new council was convoked at Constance in 1414 to set matters right.

At length a new pope, Martin V, replaced all the others in 1417, and the Council of Constance wearily broke up without solving the problem of Church reform, a problem which, it had previously stated, lay within the province of a Church council. It was firmly declared (though on becoming pope two years later Martin V did not accept the assertion) that a Church council

has its authority directly from Christ; and everybody, of whatever rank or dignity, including also the pope, is bound to obey this council in those things that pertain to the faith, to the ending of this Schism, and to a general reform of the Church in its head and members.

HEALING THE GREAT SCHISM

Gregory XI (1370-1378)
—returns to Rome in 1377—

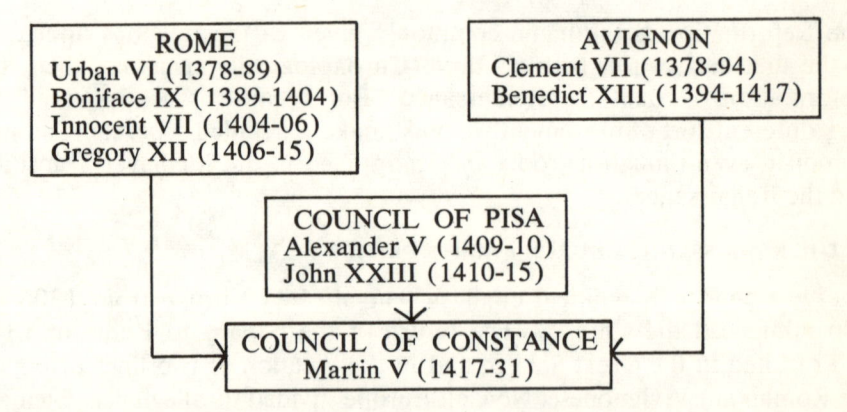

```
ROME                          AVIGNON
Urban VI (1378-89)            Clement VII (1378-94)
Boniface IX (1389-1404)       Benedict XIII (1394-1417)
Innocent VII (1404-06)
Gregory XII (1406-15)

            COUNCIL OF PISA
            Alexander V (1409-10)
            John XXIII (1410-15)

            COUNCIL OF CONSTANCE
            Martin V (1417-31)
```

The Council of Constance did successfully heal the schism. It also condemned for heresy and burned John Huss, the Bohemian professor of philosophy who had translated many of the writing of John Wycliffe into Czech. But the "general reform of the Church in its head and members" fell by the wayside.

Prophet of Doom Although later popes summoned several councils, in the course of the 15th century the Renaissance captured Rome. The popes became Italian princes absorbed in political activities and patronage of the arts, but, able politicians as they may have been, they were "intellectual mediocrities" when compared with a man like Innocent III. Even the lesser clergy became notorious for their worldliness. No one could deny the need for reform, but the Church seemed sluggish, and the popes of this period (some, but not by any means all, of which are listed in the accompanying table) would neither reform themselves nor submit to the correction of Church councils.

CHIEF RENAISSANCE AND REFORMATION POPES
Pius II (1458-64)
Alexander VI (1492-1503)
Julius II (1503-13)
Leo X (1513-21)
Clement VII (1523-34)
Paul III (1534-49)

One strange interlude in these years illustrates the collision between the papacy and those who would reform the Church. Girolamo Savonarola was born in Ferrara in 1452, and early became so disgusted with Renaissance court life that he fled from his family to a Dominican monastery. In time he became a fiery preacher, a prophet of doom, whose reputation was so great that in 1490 Pico della Mirandola persuaded Lorenzo de' Medici to make this zealot prior of the Dominican convent of San Marco in Florence. Two years after the death of Lorenzo it began to look as if Savonarola had been right in claiming that an avenger would come to punish and purify the Church. The French army of Charles VIII invaded Italy; the Medici were expelled from Florence; and the friar became the spokesman of a popular party and the leading political figure in the city. For the next four years Savonarola ruled Florence, denouncing luxury and vice, organizing bands of young people to go about collecting cards, dice, masks, and books, and then burning these "vanities" in huge bonfires in the city square.

Savonarola.

Savonarola was more fanatic than saint, and he overestimated his power. He denounced the pope, Alexander VI, and was excommunicated —whereupon he challenged Alexander's competence to excommunicate him. But the climate was changing in Florence: the aristocratic faction began to regain the upper hand, and Savonarola's political power waned. In short order he was arrested, tried, tortured, made to confess to heresy, hanged, and his body burned in that same square where he had burned the vanities of Florence.

An Unhealthy Situation Evidently there was to be no reform of the Church by a fanatical friar in the medieval tradition. Instead, the Church sought more money and practised simony right and left. Pope Leo X is estimated to have realized an income of more than a million dollars a year on the sale of over two thousand ecclesiastical offices.

There were other financial abuses that cried out for correction. Clergymen had been supported by the revenue of a piece of land called a benefice, and on taking possession of his benefice a clergyman had to pay part of the first year's proceeds to his superior, be he bishop or pope. Often bishops awarded benefices to friends or relatives—sometimes the same benefice being promised to several persons even before it fell vacant—and popes appointed Italians to "reserved" benefices all over Europe, from which revenues were drawn while their holders remained in Italy. Naturally it galled the new strong national monarchs to see wealth being constantly drained off from their realms into Rome.

Indulgences

Among the many other fees which poured into the papal treasury, that charged for papal letters of indulgence led to widespread criticism. An *indulgence* is a remission of all or part of the "temporal" punishment (that is, punishment limited in time) still due to sin after the guilt has been forgiven. Originally indulgences had been granted for making pilgrimages or for going on crusades. But later they were granted in exchange for a money contribution which could be used to build a church, monastery, or hospital, or to send someone else on a crusade. It is easy to see how the granting of indulgences lent itself to commercialization and abuse.

By means of an indulgence a truly repentant sinner, could, after confessing and being absolved of his sins through the sacrament of penance, be relieved of all or part of his temporal punishment. The relief could also be extended to cover punishment still owing at death, and so carried over into Purgatory, a place in which the soul was cleansed and prepared for Heaven. By the 16th century, indulgences could be purchased on behalf of those already dead.

Of course many ordinary people did not fully appreciate the careful qualifications and conditions laid down by the Church for the granting of indulgences. Many were prepared to believe that there were indulgences that could forgive the guilt of their sins, or even release the souls of their dead relatives from the pains of Purgatory without delay. To some it seemed that the Church was selling tickets of admission to Heaven.

Meanwhile there were those who were not slow to take advantage of the Church's failings. The ambitious middle class coveted both Church lands and revenues, and argued that monastic estates ought to be confiscated. Consequently they were more than willing to back lay rulers who wished to seize the Church's wealth.

Linked to such financial complaints, and at least as important, were political considerations—the great conflict between Church and State, accompanied by the declining prestige of the papacy. In 14th century England, Wycliffe's teaching that Church property might be forfeited to the lay ruler coincided with a rising English nationalism. In 15th century Europe the lay monarchs saw control of the Church within their national boundaries as a legitimate exercise of their sovereignty. In fact to many the international Church appeared to be the foe of the strong national state.

Above all there had been, since the Church's beginning, those who argued over what constituted the correct beliefs for a Christian, and from time to time the Church had dealt firmly with her critics in order to combat what she defined as heresy. With the 15th and 16th centuries the studies of the humanists tended to provoke serious questions—questions to which the Church seemed unable to return satisfactory answers. Thus to financial and political grievances were added intellectual ones. They are best summed up in the person of Erasmus.

Like many another humanist, Erasmus came to contrast the life of the Church of his day with the life of the founder of Christianity, and he set about to reform the Church according to "the philosophy of Christ." To this end, in 1516, he published his Greek edition of the New Testament. Nevertheless Erasmus was no dry-as-dust scholar, and this translation was followed by some years of less serious works of criticism in which he employed satire with deadly effect. Erasmus was not afraid to poke fun at the religious superstitions of his day, and denounced bigotry in high and low places. For example, he is probably the anonymous author of a little satire called *Julius Exclusus*, which describes Pope Julius II's astonishment at the cool reception given him by St. Peter at the heavenly gates. In another of his works, Erasmus sarcastically comments on the faking of relics by suggesting that there was enough wood from the true Cross drifting around Europe to build a ship. In yet another, the *Praise of Folly*, written while he was staying with Thomas More in England, he bitingly caricatures some monks who, at the Last Judgment, try to get into heaven by telling what they had done on earth:

Erasmus

One shows a trough full of fish, another so many bushels of prayers, another brags of not having touched a penny for sixty years without at least two pairs of gloves on, another that he has lived for fifty-five years like a sponge, always fixed to the same place. But Christ answers them: "There is only one law which is truly mine, and of that I hear nothing."

Even though he remained a faithful son of Rome and never questioned the pope's authority, Erasmus's programme for the Church to reform itself from within proved damaging. As a 16th century epigram put it, "where Erasmus laid the eggs, Luther hatched the chicks."

The Renaissance Church was in an unhealthy state from which would-be reformers could not seem to rescue it. A fusion of causes—financial, political, and intellectual—brought about a long delayed explosion.

2/MARTIN LUTHER'S REVOLT

The explosion came in Germany. Probably Germans felt the demands of the papacy more because they had no strong national government, as did England and France, to stand up to papal claims. There had long been

Lucas Cranach the Elder (1472-1553), who was court painter to the Elector of Saxony, was, like Holbein, celebrated for his portraits. Unlike Holbein, however, he had no sympathy with the spirit of the Renaissance but was devoted instead to the sombre simplicity of the Reformation. Here he has painted his personal friend, Martin Luther, in a powerful portrait which captures both the sturdy peasant's son with the square, determined jaw and the medieval monk with the deep-set, searching eyes.

complaints about these exactions, and there was growing unrest among the German peasants and the poorer classes in the towns who resented the impositions of the princes. This unrest expressed itself in the revolutionary idea that social inequalities were contrary to God's law. In all of 16th century Europe, the situation was potentially most explosive in Germany. It was the son of one of these German peasants who was to light the fuse.

The Cloistered Monk "If ever a monk got to heaven by his monkery I should certainly have got there," wrote Luther twenty years after he had entered the Augustinian monastery at Erfurt in 1505. What had led this sincere young man into the monastery? And what led him out again in a slashing attack on his Church?

Martin Luther (1483-1546) was the eldest of seven children in a Saxon peasant family. But his father turned to copper mining, becoming a skilled workman who eventually owned his own mines and furnaces. In his affluence, he wished to see his son rise in the world. Accordingly, Martin received an excellent education at the University of Erfurt, which he entered at eighteen, and four years later, at the request of his father, he began the study of law. Then something happened. Luther believed he had received a call from heaven, and he entered the local monastery. In two years time he became a priest, and in 1508 went briefly to the house of the Augustinian Order at Wittenberg, where he was to become a professor of philosophy in the university recently established there by the Elector of Saxony, Frederick the Wise. (The Golden Bull of 1356 had made the duke of Saxony one of the seven electors of the German Empire.) Luther made a trip to Rome in 1510-11, and found the Eternal City depressingly temporal. It was with relief that he returned to his studies, and between 1511 and 1517 he again lectured at Wittenberg.

But if these were years of outward calm there was turmoil within. Luther seems to have been obsessed with the need for the salvation of his soul, the problem that had led him into the monastery in the first place. Finally, in his despair at ever being able to merit salvation, Luther chanced on a Bible verse that seemed to sum up all his ponderings and relieve all his doubts. St. Paul's *Epistle to the Romans* 1: 17 reads, "The just shall live by faith." The verse was, of course, a familiar one; but in 1515 it took on a new significance. "This passage of Paul," wrote Luther, "became to me a gate to heaven." The meaning was clear. Man was saved by his faith in God and by that faith alone. Eventually Luther came to deny that the good works of the Church, be they fasts, pilgrimages, indulgences, or even "monkery", were essential. They were, in fact, quite superfluous. For Luther, justification (= salvation) was by faith *alone*.

It was over the matter of indulgences that his ire was first aroused. Albert of Brandenburg, who was not only the Archbishop of Mainz but held two other bishoprics at the same time, had been forced to pay the pope well for permission to be three bishops in one—so well that he had found it necessary to borrow 30,000 florins from the banking house of Fugger. He therefore proposed in 1515 to repay the loan by acting as the agent of the Medici pope, Leo X, for the granting of indulgences to those who helped finance the rebuilding of St. Peter's basilica at Rome.

The Dominican friar John Tetzel was entrusted with proclaiming the indulgences, and though he was forbidden to enter Saxony many citizens from Wittenberg sallied across the nearby border to hear this high-pressure salesman at work and to buy his wares. The content of Tetzel's remarkable sermons on these occasions was reported as follows by a witness who heard a number of them.

He gained by his preaching in Germany an immense sum of money . . . and especially at the new mining works at St. Annaberg, where I, Frederick Mecum, heard him for two years, a large sum was collected. It is incredible what this ignorant and impudent monk gave out. . . . Item—if they contributed readily, and bought grace and indulgence, all the hills of St. Annaberg would become pure massive silver. Item—so soon as the coin rang in the chest, the soul for whom the money was paid would go straightway to Heaven.

. . . He gave sealed letters stating that even the sins which a man might wish to do hereafter were forgiven. The Pope had more power than all the apostles, all the angels and saints, even than the Virgin Mary Herself. For these were all subject to Christ, but the Pope was equal to Christ. After His ascension into Heaven Christ had nothing more to do with the government of the Church till the last day, but had entrusted all to the Pope as His vicar and viceregent.

Luther reacted in a typically academic way. On All-Saints' Eve (October 31st), 1517, he posted on the door of the Castle Chapel at Wittenberg 95 theses in Latin. These statements, on which he was prepared to debate in the University of Wittenberg, attacked the whole doctrine of indulgences. Here are a few of the theses:

5. The Pope has neither the will nor the power to remit any penalties, except those which he has imposed by his own authority, or by that of the canons.

21. Thus those preachers of indulgences are in error who say that, by the indulgences of the Pope, a man is loosed and saved from all punishment.

24. Hence the greater part of the people must needs be deceived by this indiscriminate and high-sounding promise of release from penalties. . . .

27. They preach mad, who say that the soul flies out of purgatory as soon as the money thrown into the chest rattles.

32. Those who believe that, through letters of pardon, they are made sure of their own salvation, will be eternally damned along with their teachers.

37. Every true Christian, whether living or dead, has a share in all the benefits of Christ and of the Church, given him by God, even without letters of pardon. . . .

43. Christians should be taught that he who gives to a poor man, or lends to a needy man, does better than if he bought pardons. . . .

50. Christians should be taught that, if the Pope were acquainted with the exactions of the preachers of pardons, he would prefer that the Basilica of St. Peter should be burnt to ashes, than that it should be built up with the skin, flesh, and bones of his sheep. . . .

86. Again; why does not the Pope, whose riches are at this day more ample than those of the wealthiest of the wealthy, build the one Basilica of St. Peter with his own money, rather than with that of poor believers. . . ?

94. Christians should be exhorted to strive to follow Christ their head through pains, deaths, and hells.

95. And thus trust to enter heaven through many tribulations, rather than in the security of peace.

The results of Luther's action were sensational. The theses were translated into German, printed, and distributed all over Germany, and much to his surprise Luther found himself engaging in ever-widening circles of debate.

The Wild Boar When Luther posted the 95 theses in 1517 he had no intention of breaking with the Church. But as time went on and he continued to work out from his own interpretation of the Bible the implications of his belief in justification by faith alone, he was driven from one argument against Church tradition to another. He attacked the authority of both pope and council, advocated the reduction of the sacraments of the Church from seven to two (only Baptism and the Lord's Supper having originated with Christ), and proposed an extension of the priesthood to include all who truly believed, not just those ordained as priests by the Church.

Meanwhile Pope Leo X published from his hunting lodge a bull of excommunication against Luther, condemning forty-one specified propositions.

Arise, O Lord, and judge thy cause. A wild boar has invaded thy vineyard. Arise, O Peter, and consider the case of the Holy Roman Church, the mother of all churches, consecrated by thy blood. Arise, O Paul, who by thy teaching and death hast and dost illumine the Church. Arise, all ye saints, and the whole universal Church, whose interpretation of Scripture has been assailed. We can scarcely express our grief over the ancient heresies which have been revived in

Germany. . . . Our pastoral office can no longer tolerate the pestiferous virus of the following forty-one errors. . . . We can no longer suffer the serpent to creep through the field of the Lord. The books of Martin Luther which contain these errors are to be examined and burned. . . . Now therefore we give Martin sixty days in which to submit, dating from the time of the publication of this bull in his district. Anyone who presumes to infringe our excommunication and anathema will stand under the wrath of Almighty God and of the apostles Peter and Paul.

Dated on the 15th day of June, 1520.

While the decree of excommunication was en route, Luther published *The Address to the German Nobility* with its ringing appeal to nationalism. "Rome is the greatest thief and robber that has ever appeared on earth, or ever will. . . . Poor Germans that we are—we have been deceived! We were born to be masters and we have been compelled to bow the head beneath the yoke of our tyrants. . . . It is time the glorious Teutonic people should cease to be the puppet of the Roman pontiff."

Sixty days after he had received the bull of excommunication, Luther and the Wittenbergers solemnly cast it into a great bonfire. The cloistered monk had indeed become a wild boar.

A further decree excommunicating Luther a second time required the co-operation of the secular authorities of the German Empire if it was to take effect. Accordingly in 1521 the Imperial Diet, or parliament, was summoned by the young emperor Charles V to meet in the German town of Worms. Though Luther was given a safe-conduct to and from Worms, it took courage to go; for had not the Bohemian heretic John Huss, trusting in a similar safe-conduct to the Council of Constance, been burned in 1415? And had not Luther been denounced as the Saxon Huss? Nevertheless, to Worms Luther went. It was fortunate that he had a powerful patron, the Elector of Saxony. *Diet of Worms*

So it was that at Worms, faced with a demand to retract his writings and the errors it was alleged they contained, Luther spoke forth in words that have never ceased to inspire Protestants:

Unless I am convicted by the evidence of Scripture or by plain reason—for I do not accept the authority of the Pope or the Councils alone, since it is established that they have often erred and contradicted themselves—I am bound by the Scriptures I have cited and my conscience is captive to the Word of God. I cannot and will not recant anything, for it is neither safe nor right to go against conscience. God help me. Amen.

And he walked out a national hero.

Though condemned by part of the Imperial Diet, Luther had the support of Elector Frederick, who kept him in what might be called protective custody in a lonely castle, the Wartburg. There for a year Luther worked at a German translation of the New Testament, later to be combined with his translation of the Old Testament into a German Bible which did much to fix the style of the German vernacular.

Here is portrayed one of history's most dramatic moments as Martin Luther stands before the throne of the twenty-one-year-old ruler of the Empire, Charles V, at the Diet of Worms in April, 1521. Surrounded by the members of his Imperial Diet, the young emperor listens intently to this monk who, for the second day in a row, has been confronted by a collection of his books and asked whether he wishes to defend them all or to reject a part of what he has written. But far from recanting, Luther went on to defend his teachings. When he concluded with the ringing peroration (delivered both in German and Latin) quoted on page 481, the Spaniards, it is said, broke into hisses and the Germans into applause. As Luther left the hall surrounded by his friends, their hands held high in the old German sign of victory, the emperor's Spanish guards were heard to cry "To the fire with him!" But the rebel escaped. He had already left Worms under the emperor's safe-conduct before a part of the Diet could agree to proclaim him a heretic and an outlaw in May, 1521.

When it was safe to return to Wittenberg, Luther set about organizing his own church. Luther's church practices included no relics, saints, fasts, or monasteries, two sacraments only, and permission for clergy to marry and live in the world. There was to be a simplified ritual in German rather than Latin, with more emphasis on congregational participation in the service. Luther now modified his earlier belief in the "priesthood of all believers," saying that, though every Christian was entitled to preach and administer the sacraments, for the sake of order one or more should be "chosen or accepted to exercise this right in the name and place of all."

This organization of a separate church finally lost Luther the support of Erasmus and other humanists who had cherished the idea of a peaceful reform accomplished by education and without turmoil. Erasmus wrote Luther bitterly that "you treat . . . as if it were your chief aim to prevent the tempest from ever becoming calm, while it is my greatest desire that it should die down." Moreover, the humanists' optimistic belief in the goodness of man and the power of his reason were sadly at variance with Luther's pessimistic insistence that man's "good works" merited him nothing. It was God, said Luther, who "has promised to save me, not according to my working or manner of life, but according to His own grace and mercy." The loss of humanist support tended to make Luther's church more conservative and rigid.

There was another way in which Luther lost some of his early supporters. From the beginning, the new church tended to come largely under the control of state governments because the German princes converted their subjects by setting up Lutheranism as the state religion. Thus however radical Luther's ideas about church reform might appear to be, the churches themselves ended up in the conservative hands of kings and princes. Inevitably this lost him the support of the peasants and poorer classes in the towns. Indeed Luther had no sympathy with a German Peasants' War in 1524-25 against manorialism. The circumstances of his time forced him to throw in his lot with the princes; and when the Imperial Diet of 1529 insisted, at Emperor Charles V's dictation, that Lutheran princes had *not* the right to determine the religion of their subjects, the princes drew up a formal "protest"—hence the later name *Protest-ant* applied to non-Roman Catholics.

Nevertheless, however much support Luther lost by the defection of the humanists, peasants, and poorer townsmen, his ideas did make a strong appeal to the princes and the middle and upper classes. The princes found the Lutheran church a valuable prop to their governments, and the burghers found that it provided them with a more satisfying philosophy of life than had the medieval church. When Luther died in 1546 nearly half of Germany was Lutheran.

Meanwhile the reforming movement was spreading, and the Protestant initiative had passed to other hands.

3/THE SPREAD OF PROTESTANTISM

The abuses denounced by Luther were not, of course, peculiar to Germany, and two Swiss reformers did much to develop Protestant ideas. In fact while Lutheranism's stronghold was Germany and Scandinavia, the rest of Protestant Europe took its lead from Switzerland.

The first important Protestant leader to be produced by Switzerland was Ulrich Zwingli (1484-1531). Zwingli was a more intellectual and less emotional man than his contemporary Luther, and while he established no great church he did deepen Protestant thinking. It was Zwingli who persuaded his congregation in Zurich to remove all images and decorations

Ulrich Zwingli

from their cathedral, to think of the new communion service as a purely symbolic commemoration of Christ's last supper with his disciples (denying that the body and blood of Christ were actually present in the bread and wine), and to adopt a service devoid of ritual, in which the sermon occupied the central place. These new practices became characteristic of the so-called "Reformed Churches", as distinguished from the Lutheran.

Power-House of the Protestant World So far Protestantism had neither a firm, systematic body of beliefs nor a strong church organization. Both were given it by John Calvin (1509-64).

Calvin was the son of a middle-class French family, and pursued legal and classical studies at the universities of Paris, Orléans, and Bourges. In 1536 he was persuaded to prolong a visit to Geneva in order to help the local Protestant leader, and except for a brief exile was to spend the rest of his life there.

Calvin's first great accomplishment was the writing of the *Institutes of the Christian Religion* (see page 497), a carefully worded summary of essential beliefs which formed a manual for Protestantism as the *Summa Theologica* did for Catholicism. Calvin emphasized the Bible as the first and last authority, the foundation of all doctrine. God was, for Calvin, a stern judge who appointed all to their particular "calling" or station in life, and decided who should and who should not be saved.

Calvin considered it to be the duty of both church and state to legislate morality. Hence his second major accomplishment was to set up in Geneva the Protestant City of God. Dancing, card-playing, swearing, drunkenness, attendance at plays, and work on Sunday were only a few of the pro-

The student who made these sketches of John Calvin—perhaps during a lecture—certainly caught a vivid likeness of this stern, uncompromising reformer who ruled Geneva with a rod of iron. It is recorded that Calvin was of medium height with a pallid complexion and clear eye, and that he was sparing in his food and simple in his dress.

hibitions of life in Calvinist Geneva. What the Bible did not expressly permit must be forbidden. Geneva became a Protestant theocracy, a state, that is, in which God ruled. Of course to Roman Catholics the Geneva ruled by John Calvin was "the power-house of the Protestant world, a veritable 16th century Moscow."

To Geneva came refugees who returned to their homelands imbued with Calvinist doctrine. In this way Calvinism spread to France (Huguenots), Scotland (Presbyterians), and Holland (Dutch Reformed Church), and affected Germany (Baptists) and England (Anglicans and Puritans). It was also the religion of the Pilgrim Fathers, the 17th century New England Puritans. No form of Protestantism was more influential and more aggressive than Calvinism.

Defender of the Faith All the various channels of Reformation thought so far described flowed together to sweep Roman Catholicism out of England. The case of the English Reformation is, however, peculiar because, as has been said, it was set in motion by the king for reasons that had almost nothing to do with religion.

At the age of eighteen Henry VIII, a handsome, athletic young man, came to the English throne. He was well educated, and in 1521 wrote a book called *The Defence of the Seven Sacraments*, an indictment of Lutheran theology which he dedicated to the pope. For such piety Pope Leo X designated Henry as "Defender of the Faith," a title by which successive English sovereigns are still proudly described in their coronation rites, though the faith has altered in the interval. What changed Henry from a defender into an attacker of Rome?

Henry longed for a male heir—as what king did not? But his wife, Catherine of Aragon (widow of his brother Arthur, and aunt of Emperor Charles V) had not borne him one, and moreover he had become enamoured of Anne Boleyn, a maid-in-waiting at the court. Henry sought a papal annulment of his marriage on the ground that he ought not to have been permitted to marry his brother's widow; the pope neither could nor would grant his request. Henry made threats; the pope did not budge. Finally in 1531 Henry forced the English clergy to acknowledge him as "their singular protector, only and supreme lord, and, as far as the law of Christ allows, even Supreme Head." And the Archbishop of Canterbury annulled his marriage.

Henry now proceeded to have parliament cut all the ties that bound England to Rome. In 1534 the Act of Supremacy declared "that the king, our sovereign lord, his heirs, and successors, kings of this realm, shall be taken, accepted, and reputed the only supreme head in earth of the Church of England called *Anglicana Ecclesia*. . . ." An end was put to all financial obligations and legal appeals to Rome, and shortly afterwards the English monasteries were dissolved, their property and wealth seized, and their

THE RELIGIOUS DIVISIONS OF EUROPE IN 1600

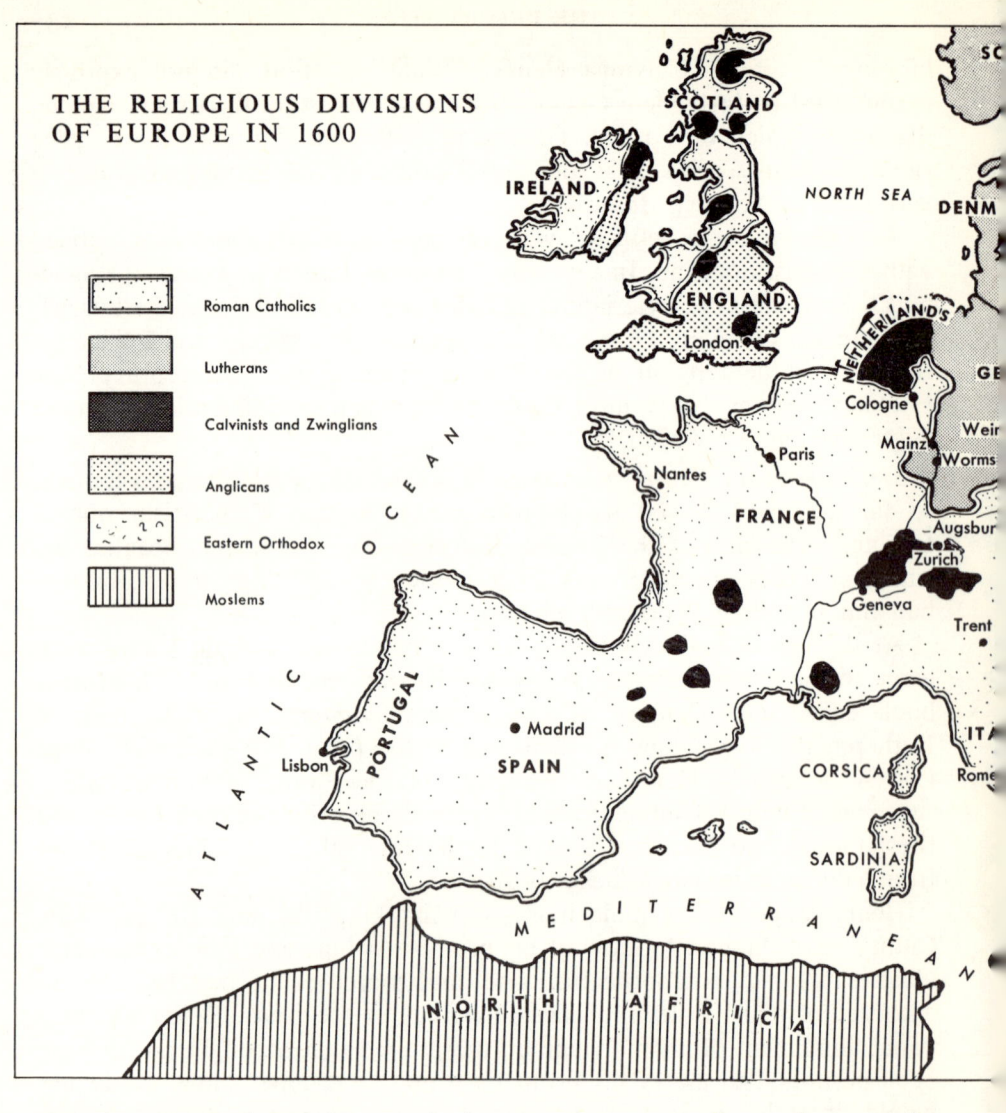

libraries scattered and destroyed. A contemporary antiquary laments the wanton waste:

A grete nombre of them whych purchased . . . those librarye bokes, some used to scoure theyr candlestyckes and some to rubbe theyr bootes. Some they sold to grossers and sopesellers, and some they sent over seas to the bokebynders, not in small nombre, but at tymes whole shyppes full. . . . I know a merchaunte man, whych shall at thys tyme be nameless, that boughte the contents of two noble libraryes for 40 shyllyngs pryce, a shame to be spoken.

Any opposition to the king's innovations was mercilessly crushed. He who resisted them could, by the Treasons Act of 1534, be executed, and among those martyred for refusing to affirm Henry's supreme headship of the church was Thomas More, the humanist, who had been chancellor of

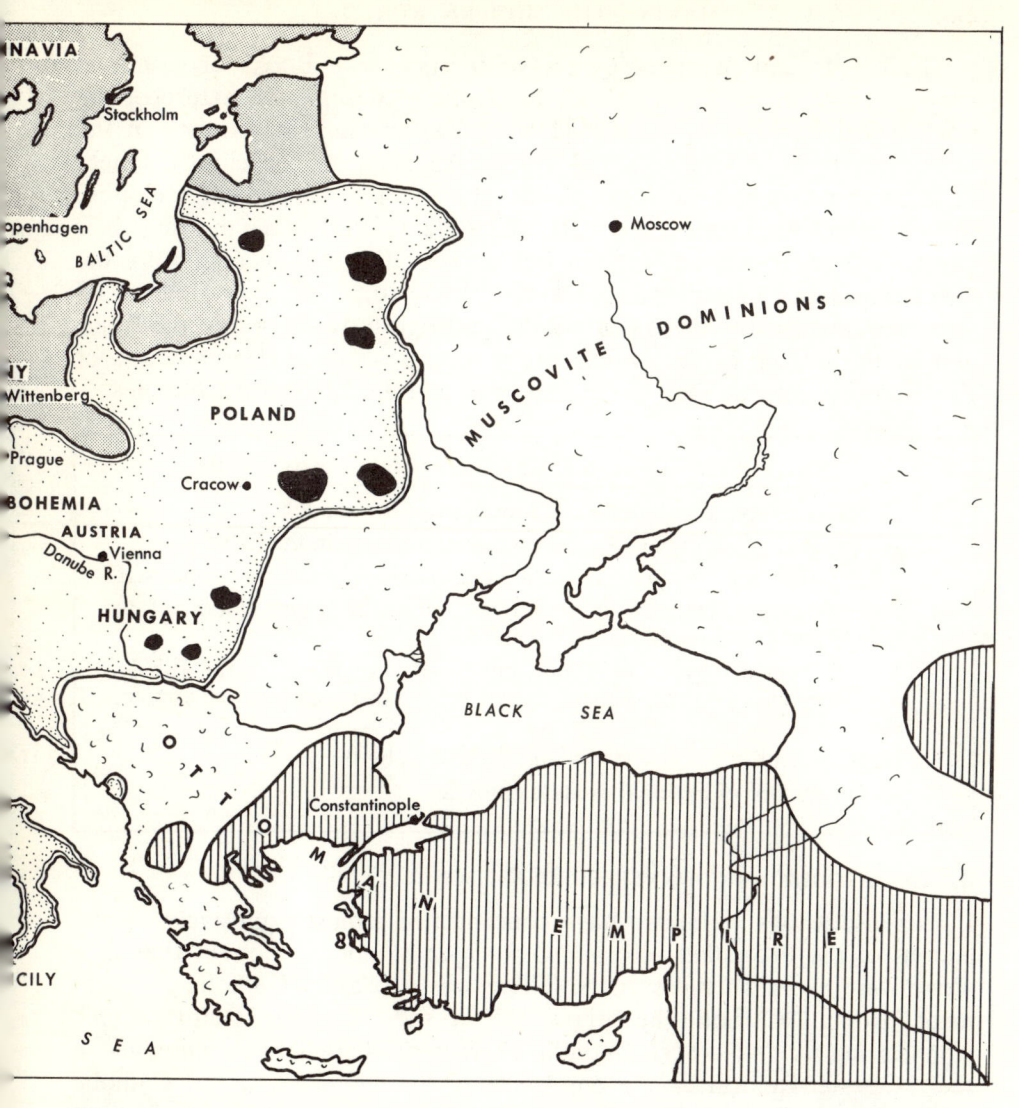

England from 1529 to 1532. Yet actually there was relatively little persecution. In fact most Englishmen readily accepted the break from Rome because it seemed, at least at first, to require no great readjustment in their thinking. Henry was cautious; all he really wanted to do was abolish papal supremacy and close down the monasteries. In short, he was no Protestant. Yet Lutheran and Calvinist doctrines spread in England despite Henry, who was not averse to persecuting such Protestant heresies. For although the king's marital troubles were the immediate cause of England's break with Rome, there were also operating all the grievances that we have noticed in the continental Reformation. Henry VIII merely brought them into focus. Actually it was not until the reign of Henry's son, Edward VI, that Protestant ideas became official beliefs.

The beliefs and practices contained in successive Books of Common Prayer (1549, 1552, 1559) show the strongly conservative nature of the English Reformation as contrasted with the radical Protestantism of the Continent. Both in structure and liturgy Anglicanism consciously sought to preserve many links with the medieval as well as the primitive church. Indeed, many Anglicans today still follow Henry VIII in saying that their church is not "Protestant" at all, but a restoration of the continuity between the Church of earlier days and the 16th century. Yet even Henry VIII approved the distribution to churches of an English translation of the Bible, and by the middle of the 16th century certain common denominators of Protestantism may be seen in the Lutheran and the Reformed Churches.

EARLY PROTESTANT BELIEFS AND PRACTICES

1. Denial of absolute authority of Church, pope, or council
2. Acceptance of authority of God's word as revealed in the Bible
3. Affirmation of individual's right to judge religious questions for himself
4. Assertion that Protestantism was a restoration or reformulation of the beliefs and practices of the early Church
5. Repudiation of church practices and beliefs not found in the Bible—e.g. worship of saints, veneration of relics, fasts, certain sacraments, ritual, etc.
6. Replacement of Latin services and Scriptures by the vernacular
7. Establishment of national or state churches

4/THE CATHOLIC OR COUNTER-REFORMATION

In our discussions so far we have concentrated almost exclusively on the actions of the Protestant reformers, but it must not be thought that they alone were active in these years. There was a parallel Catholic development. Roman Catholic historians call it the "Catholic Reformation", emphasizing its native-born Catholic character, whereas Protestants call it the "Counter-Reformation", emphasizing its character as a Catholic response to the Protestant challenge. Here, as with so much else in history, the phenomenon was neither all one or the other; it was both at once, a Catholic *and* a Counter-Reformation.

The Catholic Reformation saw a revival of piety among both clergy and laity, a revival based on monastic tradition and Christian humanism. The Counter-Reformation saw constitutional reforms and doctrinal definitions, and resulted in coercive measures aimed at curbing Protestantism.

The rapid spread of Protestantism must have been nothing short of terrifying to a Roman Catholic. What could be done to stem this tide of heresy? Were there no new Cluniacs or Franciscans to reform from within?

Soldiers of Christ An early reformer of the Church was the Spanish cardinal, Ximenes de Cisneros (1435-1517), who was for the last twenty-

Ximenes

two years of his life Archbishop of Toledo and primate of Spain. The Spanish experience of over seven centuries of struggle against the Moslems had left the country intensely Catholic. Ximenes had, in fact, taken part in the final campaigns against Moslem Granada in 1492, saying that he enjoyed the smell of gunpowder as much as that of incense.

As the chosen adviser of Ferdinand and Isabella, however, Ximenes was much more than a warrior prelate. He set about reforming the Spanish church and Spanish monasteries. He also founded the University of Alcalá, and—two years before Erasmus published his Greek edition of the New Testament—supervised the production of a scholarly edition of the Old and New Testaments with Hebrew, Latin, and Greek in parallel columns. But with Ximenes the aim was more to reinforce than correct the authority of the traditional Vulgate version. His purpose in introducing the new learning was to serve the old theology, and it was he more than anyone else who was responsible for Spain's resistance to the winds of Protestantism which blew north of the Pyrenees.

Like Ximenes, the Christian humanists—Reuchlin, Lefèvre, More, and Erasmus—also wished to reform rather than to divide, though the Church was not always happy with the results of their scholarship. Then in Rome in 1513 a group of distinguished clergymen organized what they called the "Oratory of Divine Love", with the purpose of starting a wider reform among the clergy. At about the same time, several new religious orders were founded.

Of all the attempts at reform, the most spectacularly successful is associated with a soldier. While Luther hurled defiance at the Diet of Worms, a young Spanish nobleman, Ignatius Loyola (1491-1556), had been fighting in the armies of Charles V against the French. His leg was shattered by a cannon-ball and refused to knit properly, with the result that Loyola, a short man five feet two inches in height, found himself crippled for life. During his convalescence he read a life of Christ and some biographies of saints, and experienced a conversion. "What if you should be a saint like Dominic or Francis?" he asked himself. "What if you should even surpass them in sanctity?" Loyola thus became a soldier of Christ.

Ignatius Loyola

He wrote a book called *Spiritual Exercises* (see page 498), and began to win converts. Then in 1540 he received papal approval for his "Company of Jesus," a new monastic order designed to restore the prestige and power of the Roman Catholic Church. The Jesuit Order, as it is commonly known, became an effective body of "ecclesiastical shock troops" for the papacy. It was organized along military lines with Loyola as its first "general". The Jesuit did not wear a distinctive habit but dressed as his particular job might require, be it priest, teacher, missionary, or secret diplomatic agent. After swearing a special oath of obedience to the pope, he underwent a most rigorous testing-time in what almost amounted to an officer training school. Jesuits were not to spend time and energy on ascetic practices but were to win new converts and reconvert the straying. It is not surprising that this efficient and dedicated body of men became especially noted as teachers (incidentally, they originated the system of marks or grades to regulate work in the classroom), and soon came to control education in most Roman Catholic countries.

These soldiers of Christ were, naturally, the most fervent missionaries the Church had. They won back to the Roman fold most of Bohemia, Poland, Hungary, and southern Germany. In India and China Jesuit missionaries proved outstandingly successful, and every Canadian knows the inspiring story of their missions to the Indian tribes in the days of New France.

Reform from the Top Finally, what could be done to reform the hierarchy of the Church from the papacy down? In the first place high positions could be given only to men of superior moral character who would work reforms; and beginning with Pope Paul III the popes were men of higher calibre. In the second place, the old remedy of a general council could be tried to counter Protestant criticism.

Early in his pontificate Paul III called a general council and followed up this summons by appointing a commission to study and report on reform. (Some extracts from this commission's report may be read in the Source Reading.) It took Paul III some nine years to get the council together, and, once assembled, it met in three separate sessions at intervals between 1545 and 1563 in the German town of Trent near the Italian border. There the clergy hammered out a statement of beliefs under the watchful eye of the papacy.

The Council of Trent — Henceforth no compromise with Protestantism was possible, for at Trent such fundamental Roman Catholic dogmas as the role of the priesthood, the seven sacraments, the importance of both faith and works, the validity of both Scriptures and Church tradition, were all reaffirmed. The Council particularly emphasized that the Church of Rome was the only true interpreter of Scripture. Certain moral and administrative reforms were also provided: bishops, for example, must be properly qualified and must reside in their sees, and in every diocese a seminary was to be set up to train priests. A new edition of the Vulgate was to be prepared and services were to be conducted in Latin, though preaching was permitted in the vernacular. The discipline of monastic orders was to be tightened up, and celibacy of the clergy was to be strictly enforced. Nor was reform the only antidote to Protestantism. Coercion was also revived in the form of the Inquisition and an *Index librorum prohibitorum (Index of Prohibited Books)* listing heretical writings which Roman Catholics were not to read.

Perhaps the greatest gainer of the Council of Trent, some of whose decrees are printed on pages 499-500, was the papacy. It successfully resisted any limitations on its power (though there was a party of bishops who believed that a council ought to be superior to a pope), and the

Since 1563 only two international Roman Catholic Councils have met—one in 1869-70, and the one convened by Pope John XXIII (1958-63) in October, 1962. In this photograph St. Peter's basilica in Rome forms an impressive setting for the opening of Vatican Council II, with its 2400 prelates seated on temporary wooden seats set up in the nave of the church. The interior of the basilica is constructed on a colossal scale. For example, Bernini's (1598-1680) enormous bronze *baldacchino*, supported over the papal altar by four gigantic twisted columns, weighs 93 tons and is two-thirds the height of Niagara Falls.

St. Peter's geometrical symmetry of domes, columns, and round arches makes it the supreme Renaissance adaptation of Roman elements. But it is also a reflection of the Counter-Reformation. The basilica was originally planned by Bramante as a central church—modelled after his Tempietto—based on a Greek cross scheme, and Michelangelo, too, preferred this plan. However, the Council of Trent's emphasis on a return to the usages of the past was reflected in a revival of the basilica type church based on a Latin cross plan, and subsequently a tremendous nave, half as long again as that of a Gothic cathedral, was added by another architect to make St. Peter's the largest church in the world. The mighty basilica stands today a synthesis of classical and Christian traditions that marks the beginning of an exuberant style known as Baroque.

Council voted that none of its acts should be valid unless accepted by the papacy. It is noteworthy that of the 255 prelates who signed the Trent decrees, 189 were Italians.

And so, strengthened by their reformation, the papacy and the Church proceeded to try to exterminate Protestantism. The first half of the 16th century was a period of definition of faiths, the second a savage struggle for power between them. Each was intolerant, each ready to persecute the adherents of the other.

5/COMPETING RELIGIONS COEXIST

A glance at the religious map of Europe on page 486 shows us that while southern Europe, France, Poland, and Hungary were Catholic, there was a Protestant wedge which included Scotland, England, the northern Netherlands, Switzerland, northern and central Germany, and Scandinavia, with a good deal of overlapping along the borderlands. In no country was the Roman Catholic Church stronger than in Spain, the Spain of Charles V whom we earlier saw encircling France. When, in 1554, his son Philip II married Mary Tudor, the daughter of Henry VIII, it seemed as if the ring had closed round the French.

The Humiliation of Spain   If we are to make any sense at all out of the confusing wars, persecutions, and counter-persecutions in the second half of the 16th century, we will do better to stand still at Madrid than to take a grand tour of Europe.

Although Charles V had encircled France, he looked stronger on paper than he really was. He could not count on the Protestant princes of Germany, nor on England, who, though France's traditional enemy, was not anxious to see Spain dominate the continent. In the end Charles's efforts to crush France and politically dominate Europe failed, as did his efforts to restore universal Roman Catholicism, and in 1555 he agreed to the Peace of Augsburg. This settlement recognized the religious stalemate in the German Empire; henceforth each ruler was free to determine the religion of his subjects for them.

Elizabeth I

Could the son succeed where the father had failed? The English had not relished Mary Tudor's Spanish husband, Philip II (who left England about a year after the marriage), and there was opposition to Mary's attempts to reconvert England to Roman Catholicism. With Mary's death, another of Henry VIII's daughters, Elizabeth, came to the throne, and the separation from Rome was renewed. Shocked at this turn of events in England, Philip's ambassador wrote home to Spain: "What can be expected from a country governed by a Queen, and she a young lass who, although sharp, is without prudence? . . . The kingdom is entirely in the hands of young folks, heretics and traitors." But under Elizabeth I England enjoyed twenty-six years (1559-85) of comparative peace. And the kingdom underwent less bloodshed than under its Catholic sovereign. In the five years of Mary's reign 273 Protestants were executed; in the forty-five years of Elizabeth's, 183 Roman Catholics suffered the same fate.

In 1559 parliament repealed Mary's legislation that had put England back under the Church of Rome. Elizabeth now became "supreme governor" of the English church. Yet she said she had no desire to "make windows into men's souls". In other words, so long as Catholics on the one hand did not commit treason, and Puritans (those wishing to purify the Church of England of its alleged Roman Catholic usages and ideas) on the

---

TIME CHART FOR EARLY MODERN TIMES[1]
(1400-1600)

1400	Death of Chaucer; Chrysoloras stimulates cult of classics in Italy
	Battle of Agincourt; Council of Constance
	Healing of the Great Schism with Pope Martin V
1425	
	Trial and death of Joan of Arc
	Prince Henry the Navigator promotes Portuguese exploration
1450	Francesco Sforza in Milan
	France wins Hundred Years' War with England; fall of Constantinople to Ottoman Turks
	Gutenberg in Germany
	Wars of the Roses in England
1475	Lorenzo the Magnificent in Florence
	Louis XI in France
	Conquest of Naples by Charles VIII of France; Ferdinand and Isabella unify Spain
1500	Henry VII in England
	Luther posts his 95 Theses at Wittenberg
	Charles V emperor of German Empire and ruler of Spain; Diet of Worms
1525	Machiavelli in Italy; Erasmus
	Zwingli in Zurich
	Jacques Cartier in North America; Henry VIII becomes Head of the Church of England
	Calvin in Geneva; Jesuit Order founded
1550	Michelangelo in Italy
	Peace of Augsburg; Council of Trent
	Elizabeth I restores the Church of England
1575	
	Revolt in the Netherlands against Philip II of Spain
	Defeat of the Spanish Armada
	Edict of Nantes
1600	Shakespeare in England

---

[1] See also the Renaissance Cultural Calendar on page 453 for the complete list of artists, scholars, and men of letters mentioned in Unit VI.

Contrary to our fondest traditions, a number of ships of the Spanish Armada that fought the English in 1588 were no bigger, and some were even smaller, than their English counterparts. Here to the right we see a smaller Spanish craft, one of their four *galleasses* (propelled by both sail and oars), being battered by a broadside from the larger English galleon in the foreground. The Spaniards still cherished the notion that guns were somehow ignoble, and that sea battles should consist of grappling the enemy's ships, boarding them, and then fighting it out on their decks. The English, on the other hand, trusted to their longer-range guns, their superior seamanship, and above all their agility as contrasted with the near-immobility of the majority of Spanish vessels—the galleons (see extreme right background), converted merchantmen, and the "hulks" or store-ships. During the Channel fighting in 1588 the English—to their great dismay—found that their shot was too light to be of more than nuisance value, and that the Spanish guns were actually firing heavier ammunition. Consequently they abandoned their tactics of keeping the fighting at a distance and moved in for the battle of Gravelines on July 29th. In no time they gained the advantage. Not only had the Spanish ammunition by then been pretty well exhausted, but as the nimbler English ships darted back and forth they found they could, at close range, riddle the hulls of the lumbering Spanish vessels almost at will.

other hand did not infringe on her prerogatives, the queen was satisfied. With Elizabeth, England came first and Protestantism second. And when in 1570 a papal bull deposed and excommunicated Elizabeth and English Catholics had to choose between their religion and their queen, most of them chose their queen.

Philip II   In 1585 an undeclared war broke out between England and Spain over the Netherlands, where Philip II was faced with a revolt. The Spanish king had little understanding or sympathy for this part of his empire, and his repression was such that the Catholic southern Netherlands joined the Calvinist north to throw off his overlordship.

At the same time the Huguenots (the name given to French Protestants of the 16th and 17th centuries) in France had grown to about one-sixteenth of the total population of 16 million, and in the savage religious struggles that ensued there Philip could never be sure of victory for Roman Catholicism and friendship for Spain. Consequently when a Protestant, Henry IV, succeeded to the French throne in 1589, Philip's forces soon invaded France. However, during the war Henry realized that France would never really be reconciled to a Protestant ruler; so he prudently returned to the Catholic fold. "Paris is well worth a mass," he is alleged to have said.

But Henry did not forget his Protestant subjects. In 1598 he issued the Edict of Nantes, which, like the Peace of Augsburg, recognized the religious stalemate. Henceforth French Protestants could hold office and practise their religion, though in Paris and 200 other towns Protestant services were prohibited. When Roman Catholics objected to Henry's toleration he made this eloquent reply: "I must insist on being obeyed. It is time that we all, having had our fill of war, should learn wisdom by what we have suffered."

The Edict of Nantes,

In the same year, 1598, Spanish intervention in France ended and all Spanish conquests were returned. Philip had not only seen Protestantism and Catholicism fight to a draw in France, but he had seen the French monarchy gain a position from which it would go on to become the strongest in 17th century Europe.

Philip's most humiliating failure was, however, already ten years behind him by 1598. We all know the story of how, goaded beyond endurance by Elizabeth's sea-dogs who relieved Spanish galleons of their New World gold and cut off Spain from sea access to the Netherlands, Philip II outfitted a mighty Armada to invade England. When the great contest in the England Channel was all over, the victorious English struck a commemorative medal. *Afflavit Deus et dissipati sunt*, it read: "God blew and they were scattered."

As Philip II breathed his last in his vast palace of El Escorial, his cherished goal—one European Church, one Emperor—was far from being achieved.

Towards Tolerance   By the end of the 16th century, men were beginning to learn tolerance for each other's religions. In the middle of the century an inquisitor could say, "It is no great matter whether they that die on account of religion be guilty or innocent, provided we terrify the people by such examples." But by the end of the century—even though there were to be more religious wars—it was beginning to look as if neither side was strong enough to crush the other. Was it not, then, possible to live and let live?

6/SUMMARY AND CONCLUSION

By the end of the 15th century the Renaissance monarchs had consolidated their power, and France, England, and Spain were strong states. But the Renaissance Church, though it had healed the schism of the papacy, was falling on evil days.

The worldly abuses of the Church disgusted many sincere Christians, whose complaints against the financial extortions, political manipulation, and superstitious practices of Rome grew louder. The open revolt came in Germany with Luther's denial of papal infallibility and his repudiation of certain religious beliefs and practices. Similar denials and repudiations, similarly supported by princes and kings, spread the influence of the reforming Protestants throughout north central Europe and to the British Isles.

The Catholic Reformation consisted of a mighty effort at reconversion along with some genuine reforming measures, and a firm restatement of Roman Catholic doctrines which made any future compromise with Protestantism exceedingly difficult. So ended half a century of definition and redefinition.

In the second half of the 16th century Catholic Spain attempted to unite Europe and bring her back into the fold. That effort failed. The religious stalemate was officially recognized in Germany and France, and Protestantism became the state religion in England. Out of all the blood and persecution, the thought was slowly emerging by 1600 that the religious and national divisions of Europe could, after all, become permanent, or at least that coexistence was possible. The certainty of the 13th century had given way to the uncertainty of the 16th.

SOURCE READINGS

(a)

JOHN CALVIN (1509-64) was a Frenchman who studied scholastic philosophy in Paris and law in Bourges and Orléans. He was introduced to the writings of Luther through Lefèvre d'Étaples and Erasmus, and by 1534 his religious opinions were such that he felt it advisable to leave France. In 1536, at Basel in Switzerland, Calvin published the *Institutes of the Christian Religion* in Latin, and in 1541 he was invited to take charge of the affairs of Geneva. That same year he translated the *Institutes* into French. The following very brief selections give Calvin's convictions concerning the total depravity of man and the absolute sovereignty of God, who predestines man either to salvation or damnation.

. . . Original sin, therefore, appears to be an hereditary pravity and corruption of our nature, diffused through all the parts of the soul. . . . And therefore infants themselves, as they bring their condemnation into the world with them, are rendered obnoxious to punishment by their own sinfulness, not by the sinfulness of another. For though they have not yet produced the fruits of their iniquity, yet they have the seed of it within them; even their whole nature is as it were a seed of sin, and therefore cannot but be odious and abominable to God. Whence it follows, that it is properly accounted sin in the sight of God, because there could be no guilt without crime. . . . this depravity never ceases in us, but is perpetually producing new fruits, those works of the flesh, which we have before described, like the emission of flame and sparks from a heated furnace, or like the streams of water from a never failing spring. . . . For our nature is not only destitute of all good, but is so fertile in all evils that it cannot remain inactive. Those who have called it *concupiscence* have used an expression not improper, if it were only added, which is far from being conceded by most persons, that every thing in man, the understanding and will, the soul and body, is polluted and engrossed by this concupiscence; or, to express it more briefly, that man is of himself nothing else but concupiscence. . . .

. . . Predestination we call the eternal decree of God, by which he has determined in himself, what he would have to become of every individual of mankind. For they are not all created with a similar destiny; but eternal life is foreordained for some, and eternal damnation for others. Every man, therefore, being created for one or the other of these ends, we say, he is predestinated either to life or to death. . . .

. . . In conformity, therefore, to the clear doctrine of the Scripture, we assert, that by an eternal and immutable counsel, God has once for all determined, both whom he would admit to salvation, and whom he would condemn to destruction. We affirm that this counsel, as far as concerns the elect, is founded on his gratuitous mercy, totally irrespective of human merit; but that to those whom he devotes to condemnation, the gate of life is closed by a just and irreprehensible, but incomprehensible judgment. . . .

... Foolish mortals enter into many contentions with God, as though they could arraign him to plead to their accusations. In the first place they inquire, by what right the Lord is angry with his creatures who had not provoked him by any previous offence; for that to devote to destruction whom he pleases, is more like the caprice of a tyrant than the lawful sentence of a judge; that men have reason, therefore, to expostulate with God, if they are predestinated to eternal death without any demerit of their own, merely by his sovereign will. If such thoughts ever enter the minds of pious men, they will be sufficiently enabled to break their violence by this one consideration, how exceedingly presumptuous it is only to inquire into the causes of the Divine will; which is in fact, and is justly entitled to be, the cause of every thing that exists. For if it has any cause, then there must be something antecedent, on which it depends; which it is impious to suppose. For the will of God is the highest rule of justice; so that what he wills must be considered just, for this very reason, because he wills it. When it is inquired, therefore, why the Lord did so, the answer must be, Because he would.

Institutes of the Christian Religion by John Calvin, translated by J. Allen (Presbyterian Board of Education, Philadelphia, 6th American edition, revised and corrected, no date), I, 229-230; II, 145, 149, 165.

(b)

IGNATIUS LOYOLA (1491-1556) wrote the *Spiritual Exercises* over a long period of time, and they were not actually published until 1548. Their purpose was to produce submission to the Church. A few selections follow.

RULES FOR THINKING WITH THE CHURCH
In order to think truly, as we ought, in the church militant, the following rules are to be observed.

1. Laying aside all private judgment, we ought to keep our minds prepared and ready to obey in all things the true Spouse of Christ our Lord, which is our Holy Mother, the Hierarchical Church.
2. The second is to praise confession made to a priest, and the reception of the Most Holy Sacrament, once a year, and what is much better once a month, and much better still every eight days, always with the requisite and due dispositions.
6. The sixth is to praise the relics of saints, showing veneration to the relics, and praying to the saints. . . .
7. The seventh is to praise the precepts with regard to fasts and abstinences. . . .
8. To praise the building and the ornaments of churches; and also images, and to venerate them according to what they represent.
9. Finally, to praise all the precepts of the Church, keeping our minds ready to seek reasons to defend, never to impugn them.
13. To attain the truth in all things, we ought always to hold that we believe what seems to us white to be black, if the Hierarchical Church so defines it; believing that between Christ our Lord the Bridegroom and the Church His Bride there is one and the same Spirit, which governs and directs us to the salvation of our souls; for our Holy Mother the Church is guided and ruled by the same Spirit and Lord that gave the Ten Commandments.

Introduction to Contemporary Civilization in the West, I, 699-701.

(c)

POPE PAUL III (1534-49) was a noted reformer even before his elevation to the papacy. As pope, he early appointed a commission of nine cardinals who, in 1537, submitted to him a *Report of the Cardinals and Other Prelates on the Reform of the Church*—"one of the most remarkable documents of the Renaissance"—from which the following extracts are taken.

Most blessed Father, we cannot express the gratitude which Christendom should feel toward God for having raised you up in these times as a shepherd to His flock, for the spirit of God has determined through you to restore the falling Church to her pristine sublimity. You have instructed us to declare, without regard for the feelings of yourself or anyone else, those abuses and grievous ailments from which the Church suffers and particularly the Roman Curia. The trouble is that your predecessors surrounded themselves with sycophants, dextrous in proving that whatever they liked was licit. From this Trojan horse issued such ills upon the Church that she is like not to recover. . . .

To begin with, laws should be observed. Nothing is more subversive of laws than dispensations. Nor should the vicar of Christ in exercising the power of the keys have an eye to gain. One abuse is the ordination of priests, even mere lads, utterly unqualified reprobates. No one should be ordained save by a bishop who should have a teacher in his see for clerics in minor orders. Benefices should be conferred only on learned, upright men who will be in residence. An Italian should not be appointed to a post in Spain or Britain. A great abuse is the reservation of revenues designed for the indigent but consigned to the wealthy. The permutation of benefices for gain is simony. By no subterfuge should benefices be treated as legacies. The granting of expectations on benefices engenders the wish that somebody will die. . . . The holding of bishoprics and sometimes several by cardinals is incompatible with their office, because they should assist Your Holiness in the governance of the Church universal, whereas a bishop should look after his flock. How is the Holy See to correct the abuses of others if there be so many in the chief members? Nothing is more important, blessed Father, than that bishops should reside in their sees. What more grievous sight could afflict a traveller throughout Christendom than to view so many neglected churches? Equally is it an abuse that the cardinals should reside in their provinces rather than here at Rome. . . . A most pernicious abuse is the teaching of impiety by professors in the universities and especially in Italy. The *Colloquies* of Erasmus should not be used for the instruction of youth. . . . Indulgences should not be allowed in a given territory more than once a year. . . . The cardinals should endeavour to compose feuds between Roman citizens. There are in this city hospitals and widows who should be your especial care.

Blessed Father, these are our modest proposals to be tempered by your goodness and wisdom. You have taken the name of Paul. May you imitate the love of Paul that you may make the Church a washed and spotless dove.

R. H. Bainton, *The Age of the Reformation* (Anvil Books, 1956), pp. 152-153.

(d)

THE COUNCIL OF TRENT met at intervals from 1545 to 1563. It spent nine sessions drawing up its decrees, some examples of which appear here.

Decree on the Sacraments in General

CANON I—If any one saith, that the sacraments of the New Law were not all instituted by Jesus Christ, our Lord; or, that they are more, or less, than seven, to wit, Baptism, Confirmation, the Eucharist, Penance, Extreme Unction, Order, and Matrimony; or even that any one of these seven is not truly and properly a sacrament; let him be anathema.

Decree on Reformation

. . . There is nothing that continually *instructs others unto piety,* and the service of God, more than the life and example of those who have dedicated themselves to the divine ministry. For as they are seen to be raised to a higher position, above the things of this world, others fix their eyes upon them as upon a mirror, and derive from them what they are to imitate. Wherefore clerics called to have the Lord for their portion, ought by all means so to regulate their whole life and conversation, as that in their dress, comportment, gait, discourse, and all things else, nothing appear but what is grave, regulated, and replete with religiousness; avoiding even slight faults, which in them would be most grievous; that so their actions may impress all with veneration. . . .

On the Sacrament of Matrimony

CANON IX—If any one saith, that clerics constituted in sacred orders, or Regulars, who have solemnly professed chastity, are able to contract marriage, and that being contracted it is valid, notwithstanding the ecclesiastical law, or vow; and that the contrary is nothing else than to condemn marriage; and, that all who do not feel that they have the gift of chastity, even though they have made a vow thereof, may contract marriage; let him be anathema. . . .

Decree Concerning Indulgences

Whereas the power of conferring indulgences was granted by Christ to the Church; and she has, even in the most ancient times, used the said power, delivered unto her of God; the sacred holy Synod teaches, and enjoins, that the use of Indulgences, for the Christian people most salutary, and approved of by the authority of sacred Councils, is to be retained in the Church; and It condemns with anathema those who either assert, that they are useless; or who deny that there is in the Church the power of granting them. In granting them, however, It desires that, in accordance with the ancient and approved custom in the Church, moderation be observed; lest, by excessive facility, ecclesiastical discipline be enervated. And being desirous that the abuses which have crept therein, and by occasion of which this honourable name of Indulgences is blasphemed by heretics, be amended and corrected, It ordains generally by this decree, that all evil gains for the obtaining thereof,—whence a most prolific cause of abuses amongst the Christian people has been derived,—be wholly abolished. But as regards the other abuses which have proceeded from superstition, ignorance, irreverence, or from whatever other source, since, by reason of the manifold corruptions in the places and provinces where the said abuses are committed, they cannot conveniently be specially prohibited; It commands all bishops, diligently to collect, each in his own church, all abuses of this nature, and to report them in the first provincial Synod; that, after having been reviewed by the opinions of the other bishops also, they may forthwith be referred to the Sovereign Roman Pontiff, by whose authority and prudence that which may be expedient for the universal Church will be ordained; that thus the gift of holy Indulgences may be dispensed to all the faithful, piously, holily, and incorruptly. . . .

Amen.

Knoles and Snyder, *Readings in Western Civilization,* pp. 399, 402, 403, 404-405.

FURTHER READING

† * BAINTON, R. H., *Here I Stand: A Life of Martin Luther* (Mentor, 1955)
 * _____, *The Reformation of the Sixteenth Century* (Beacon, 1956)
 * _____, *The Age of the Reformation* (Anvil, 1956)
† * BRODRICK, J., *The Origin of the Jesuits* (Image, 1960)
 * BURNS, E. M., *The Counter Reformation* (Anvil, 1964)
   DANIEL-ROPS, H., *The Protestant Reformation* (Dent, 1961)
   _____, *The Catholic Reformation* (Dent, 1962)
† DURANT, W., *The Reformation: A History of European Civilization from Wyclif to Calvin, 1300-1564* (Simon and Schuster, 1957)
   ELTON, G. R., editor, *The New Cambridge Modern History*, Vol. II, *The Reformation, 1520-1559* (Cambridge, 1958)
   FIFE, R. H., *The Revolt of Martin Luther* (Columbia, 1957)
† GRIMM, H. J., *The Reformation Era, 1500-1650* (Macmillan, 1954)
 * HARBISON, E. H., *The Age of Reformation* (Cornell, 1955)
   HOLBORN, H., *A History of Modern Germany*, Vol. I, *The Reformation* (Knopf, 1959)
† * HUGHES, P., *A Popular History of the Reformation* (Image, 1960)
 * HUIZINGA, J., *Erasmus and the Age of Reformation* (Torchbook, 1957)
   JANELLE, P., *The Catholic Reformation* (Bruce, 1949)
   KIDD, B. J., *The Counter-Reformation, 1550-1600* (Society for the Propagation of Christian Knowledge, 1933)
 * MOSSE, G. L., *The Reformation* (Holt, Rinehart and Winston, 1963)
   PARKER, T. M., *The English Reformation to 1558* (Oxford, 1950)
 * POWICKE, F. M., *The Reformation in England* (Oxford, 1961)
 * SMITH, P., *The Reformation in Europe*, 2 vols. (Collier, 1962)

INDEX

NOTE:

The pronouncing guide incorporated in this index is as accurate as it is possible to make it without using the technical phonetic symbols.

AACHEN [aw'kun] (Aix-la-Chapelle), 346
Abacus, 347, 373
Abelard, Peter, 350, 357-358, 397
Abraham, 59-60, 64
Achaean [a-kee'un] Greeks, 82, 84; Achaean League, 194
Achilles, 86, 154
Acre, 342
Acropolis, at Athens, 118-119
Act of Supremacy (1534), 486
Actium, battle of, 219-220
Acts, Book of, 258
Adelard of Bath, 395-396
Adrianople, battle of, 243
Aedileship [eed'yl-ship], 171-172, 174-176, 191, 213
Aegeus [ee'jus], 81
Aegina [ee-jy'nuh], 97, 120
Aeneas [i-nee'us], 165, 255
Aeolian [ee-ohl'ee-un] Greeks, 84
Aequians [ee'kwee-uns], 166
Aeschylus [ee'skil-us], *Agamemnon,* 140
Aetolian [ee-toh'lee-un] League, 194
Africa, 11, 49-50, 185-186, 191, 194, 206, 218, 232, 235, 244, 251, 440-441
Agamemnon, 84
Agincourt, battle of, 435
Agora, at Athens, 115
Agriculture: invention of, 11-13, 42; medieval, 317-318, 369
Agrippina, 233
Ahab [ay'hab], 65-66
Ahaz [ay'haz], 68
Ahriman [aw'ri-mun], 59
Ahuramazda, 59, 73
Aides, 436
Aids: customary, 311; extraordinary, 311
Akhenaton, *see* Amenhotep IV
Akhetaton [aw-keh-taw'tun], 25
Akkad [ak'ad], 15, 31
Alaric [al'ur-ik], 244
Albertus Magnus, 395, 397
Albuquerque, Alfonso de, 440
Alcalá, University of, 489
Alchemy, 3, 347
Alcibiades [al-suh-by'uh-deez], 119-120
Alcuin, 294-295, 346
Alexander the Great, 19, 57-58, 84, 105, 115, 122, 131, 144-151, 152, 155-156, 158-159, 193, 210, 288
Alexander V, Pope, 474
Alexander VI, Pope, 471, 475
Alexandria, 146-147, 153-154, 212, 218-219, 254, 274, 279; Museum at, 153-154
Alfonso VI, 341
Alfred the Great, 306, 324, 346

Algebra, 19, 347
Almoravides [al-mor'uh-vydz], 341
Alphabet: Chinese, 39; Etruscan, 164; invention of, 3, 50; Roman, 254
Amenhotep III [aw-mun-hoh'tep], 25
Amenhotep IV (Akhenaton), 25-26, 28
Amiens, cathedral of, 403-405
Amon [aw'mun], 27; Temple of, at Karnak, 28
Amos, 61, 66-68
Amphipolis [am-fip'oh-lis], 118
Amphora, 91
Analects, 41, 46
Anaxagoras [a-nax-sag'ur-us], 136-137
Anaximenes [a-nax-sim'un-eez], 132
Angevins, 325-327, 387-388
Angles, 273, 324
Anglo-Saxons, 245, 417-418, 421, 424
Anjou, 322
Anselm, St., 350, 397
Antigonus I [an-tig'uh-nus], 151
Antioch, 259, 274-275, 340
Antiochus III [an-tee'uh-kus] of Syria, 194
Antonines, 231, 238, 263
Antoninus Pius, 231, 238
Antony, Mark, 216-221, 224, 255
An-yang, 38-39, 41
Apella [a-pel'a], 92
Apennines, 161-163
Aphrodite [af-ruh-dy'tee], 256
Apollo, 77, 141
Appian, 179
Appian Way, 209
Appius Claudius, 179
Apulia [a-pyew'leeuh], 190
Aqueducts, 218, 227, 250, 252; Pont du Gard, 252
Aquinas [a-kwy'nus], St. Thomas, 397-398, 405; *Summa Theologica,* 397, 405, 407, 484
Aquitaine, 322-323, 388
Arabia, 50, 61, 150, 240, 286-288, 296
Aragon, 435, 473
Aramaeans, 50
Archimedes [ar-ki-mee'deez], 154
Architecture: Assyrian, 53; Baroque, 491; Byzantine, 285-287; early Christian, 263; Egyptian, 22-23, 27-28, 31; Etruscan, 164; Greek, 77, 138-139, 146-147, 451; Hellenistic, 153; medieval, Romanesque, 400-402; medieval, Gothic, 401-405, 409-410, 464, 469; Mesopotamian, 17-19; Minoan, 79; Mycenaean, 85; Persian, 58; Renaissance, 459-460, 462-464, 469
Archon, 93, 98
Areopagus [ar-ee-op'uh-gus], 93, 113

Arginusae [*ar-juh-noo'see*], battle of, 120
Argos, 121
Arianism, 279, 283-284
Aristarchus [*ar-is-tar'kus*], 154
Aristides [*ar-is-ty'deez*], 106
Aristides, Publius Aelius, 240
Aristophanes [*ar-is-tof'uh-neez*], 140-142, 155; *Knights*, 141
Aristotle, 101, 131, 135-137, 143-144, 153, 159; *Ethics*, 136; *History of Animals*, 136; *Politics*, 136
Arius, 279
Armada, Spanish, 494, 496
Armenians, 219
Armour, medieval, 328-329
Army: Carolingian, 294, 304, 307, 309; **Merovingian, 289, 296-297**; Roman, 168-169, 195, 198, 200, 204-206, 226-228, 240-242, 244, 248-249, 250-251; arms and armour of, 204-205; centuries in, 168; **siege engines of, 204-205**
Art: Assyrian, 3, 53, 58; Byzantine, 285; Greek, 138-140; Minoan, 79; Mycenaean, 85; Paleolithic, 8-9, 10-11, 13; Persian, 58; Renaissance, 456-463, 468; Roman, 253
Aryans [*air'ee-unz*], 34-35
Ashur [*aw'shur*], 48, 52-53
Ashurbanipal [*aw-shur-baw'ni-pawl*], 3, 51, 53
Asia, 24, 194, 210, 441
Asia Minor, 78, 82, 84, 87, 89, 103, 108-109, 122, 145-146, 151, 163, 194, 206, 210, 212, 241, 250-251, 256, 260, 287, 340, 439
Asoka [*a-soh'kuh*], 152
Assyria, 16, 18, 25, 48, 51, 62, 66-68, 72, 103; Assyrians, 3, 27, 43, 48, 50-53, 71
Astrology, Chaldean, 56
Astronomy: Egyptian, 28; Mesopotamian, 19; Moslem, 347; Roman, 254
Aswan High Dam, 19
Athanasius, 279
Athena [*a-thee'nuh*], 118, 138-139
Athens, 81, 84, 88, 93-99, 134-137, 141, 150, 153, 156, 174, 260
Aton-worship, 25-26
Attalus II [*at'tuh-lus*] of Pergamum, 115
Attalus III of Pergamum, 194, 201
Attica, 84, 93-94, 97-98, 105, 115-116, 140; standard of currency, 150
Attila [*at'i-luh*], 245
Augsburg, Peace of, 492
Augustine, St., 344-345, 397; *City of God*, 344
Augustus, 216-219, 221, 225-232, 240, 252, 255, 264, 266; *Monumentum Ancyranum*, 266
Austria, 391
Avaris-Tanis, 26
Avignon (papacy at), 412-413, 424, 452, 474

BAALS [*bayls*], 60, 65, 68
Babylon, 16-18, 25, 44, 48, 51, 53, 56-58, 63-64, 69, 103, 149, 151, 412

Babylonian Captivity: of Jews, 62-63; of Church, 412-413
Bacon, Roger, 395, 408-409
Bactria, 147, 151-152
Bailiff, 315, 386
Balboa, Vasco Nuñez de, 440
Balearic Islands, 307
Balkans, 194, 244
Banalities, 319
Banks, 372, 426, 442, 451
Baptistry doors, 460-462
Barbarians, *see* Germanic barbarians
Barnabus, 259
Barracks Emperors, 231, 241
Bartholomew, Peter, 340
Basilica, 263, 399
Bavaria, 294, 330
Bayeux Tapestry, 329
Beauvais, 403, 446
Becket, St. Thomas, 464
Bede, the Venerable, 344-345
Behistun [*bay-hi-stoon'*] inscription, 73
Benedict, St., 275-277, 297; Rule of, 276-277, 297-299, 336, 344, 391; Order of, 391, 410, 427
Benedict XIII, Pope, 474
Benefices, 289, 309-310, 331; clerical, 476, 499
Bernard of Clairvaux, 350, 392-393
Bernard of Cluny, 349
Bernini, Giovanni Lorenzo, 491
Bible, 64, 344-345, 349, 397, 401, 413, 465, 479, 480-481, 484-485, 488
Bill of exchange, 372, 442
Bithynia [*bi-thin'ee-uh*], 261, 267
Black Death, 377-378, 380-381, 414, 452
Black Prince, 415
Boccaccio [*boh-kaw'cheeo*], Giovanni, 452-454; *Decameron*, 452, 454, 464
Boeotia [*bee-oh'shuh*], 84, 115; Boeotian League, 222
Bohemia, 473, 490
Boleyn, Anne, 485
Bologna, 163, 189; University of, 351, 470
Bonds, government, 372
Boniface VIII, Pope, 382-383, 386, 398, 412-413
Boniface IX, Pope, 474
Book of the Dead, 27, 44-45
Boon-work, 319
Boroughs, 379, 418-419, 421
Bosworth Field, battle of, 438
Botticelli [*bot-uh-chel'ee*], Sandro, 453, 456-457
Boulê [*boo'lay*], 98, 113, 123
Bourgeois, 364, 442
Bourges, University of, 484
Bourses, 442
Bouvines, battle of, 383-384, 387
Brahma, 35; Brahmins, 36
Bramante, Donato, 451, 491; Tempietto, 451, 491
Brandenburg, Albert of, 479
Breughel [*broy'gul*], Pieter, 453
Britain, 12, 49, 232, 235, 238, 241, 324, 344, 499
Brittany, 437

Bronze Age, 3, 5, 13-14, 16, 24, 27, 30, 34, 42, 56, 77-78, 82, 99, 162
Bruges, 364
Brundisium, 216
Brunelleschi, Filippo, 453, 460, 462-463
Bucephalus [*byew-sef'uh-lus*], 144-146, 150
Buddha, 35-37, 43, 45; Buddhism, 152, 373
Bulgars, 284-285
Burgers, 364
Burgundy, 245, 322, 436; Burgundians, 245, 277, 279, 307
Byblos, 21, 64
Byzantine (East Roman) Empire, 242-244, 246, 248, 250-251, 264, 273, 283-285, 287-288, 290-291, 295-296, 439
Burgesses, 364, 379, 418-423
Byzantium, 109, 242, 283, 296

CABOT, John, 440
Cabral, Pedro Alvares, 440
Caesar, Julius, 3, 182, 201, 210-216, 217-219, 221, 225, 229-230, 233, 254; *Gallic Wars*, 211-212, 278
Caesarea, 260
Calais, 415, 436
Caligula, 208, 231-232
Calvin, John, 484-485, 497-498; *Institutes of the Christian Religion*, 484, 497-498
Cambyses II [*kam-by'seez*], 56
Campania [*kam-pay'nee-uh*], 190
Canaan, 59-60, 65
Canary Islands, 447-448
Cannae [*kan'ee*], battle of, 188, 190
Canossa, 337
Canute, 324
Cape Breton Island, 440
Cape of Good Hope, 440
Capet [*kap'it*], Hugh, 321-322; Capetians [*ka-pee'shunz*], 322-323, 331
Capitalism, 372, 378, 442-445, 451
Capri (Capraea), 232, 268
Caracalla, 231, 239, 249; Baths of, 252
Carcassonne, 362
Carchemish [*kar'kuh-mish*], 53
Cardinals, College of, 336, 412
Caria, 84
Carloman, 292
Caroline minuscule, 346-347, 396
Carolingian dynasty, 292-297, 303-307, 318, 322; Empire, 327, 330, 346
Cartage, 319
Carter, Howard, 26
Carthage, 182-186, 191-194, 198, 214, 287; Carthaginians, 108, 222
Cartier, Jacques, 440
Castiglione [*kaws-tee-leeoh'nay*], Baldassare, 453, 456, 471-472; *Book of the Courtier*, 456, 471-472
Castile, 341, 431, 473
Castles, 312, 314, 321, 339, 343, 387, 415; castle-guard, 311
Catalonia, 364
Cathedrals: Gothic, 393, 401-404, 407, 426, 464; Romanesque, 400-401
Catherine of Aragon, 438, 485
Catholic or Counter-Reformation, 488-492, 496

Cato the Elder, 198
Cellini [*chuh-lee'nee*], Benvenuto, 453, 460, 462, 469
Censors, 171-172, 175-176, 198, 209, 213
Centuriate Assembly, 171-174, 176, 178, 203-204
Centuries, of Roman populace, 173-174
Chaeronea [*kair-uh-nee'uh*], battle of, 122
Chaldeans [*kal-dee'unz*] (New Babylonians), 27, 51, 53, 55-56
Châlons, battle of, 245
Champagne, 322
Chancellor: Carolingian, 303; English, 420, 428
Chandragupta Maurya [*chun-druh-gup'tuh mour'yuh*], 152
Chanson de Roland, 294
Charlemagne, 273-274, 276, 292, 294-297, 303-305, 307, 309, 331-332, 345-346, 353, 355
Charles the Bald, 307, 310
Charles Martel, 287, 289-291, 296-297, 307, 309
Charles I, of England, 417
Charles IV, of Germany, 414, 416
Charles V, of France, 414
Charles V, of Germany and Spain, 432, 481-483, 489, 492
Charles VI, of France, 414, 435-436
Charles VII, of France, 437
Charles VIII, of France, 432, 435, 437-438, 469
Charters, town, 365, 379
Chartres, cathedral of, 361, 409-410
Châteaux, 468-469
Chaucer, Geoffrey, 274, 453, 464; *Canterbury Tales*, 464
Chenonceaux, 468
Cheops [*kee'ops*] (Khufu), 22
Chih [*chee*], Huang-ti, 373
Ch'in dynasty, 373
China, 37-43, 71, 152, 250, 284, 287, 372-377, 440, 443, 490; Great Wall of, 373
Chios [*ky'os*], 84
Chivalry, 371
Chou [*joh*] dynasty, 40, 373
Chou-k'ou-tien [*joh-goh-tee-en'*], 38
Chrétien de Troyes, 335, 354
Christianity, 57, 72, 151, 234, 252, 256-264, 267-270
Chrysoloras, Manuel, 455
Cicero, Marcus Tullius, 209-210, 211-213, 216-217, 224, 250, 254-255, 348; *Philippics*, 217; *Second Speech against Verres*, 224
Cilicia [*si-lish'uh*], 194, 207
Cimon [*sy'mun*], 113
Cistercians, 391-393
Cîteaux (Cistercium), 391
Citizenship, Roman, 169-170, 201-203, 206, 213, 215, 227, 239, 248-249
City-states, Greek, 88-90, 93, 100, 120, 122, 125-126, 156
Civilization, 4-5, 13
Clairvaux, 392
Claudius, 208, 231-233
Cleanthes [*klee-an'theez*], 153

Cleisthenes [*klys'thuh-neez*], 93, 97-99, 113, 174
Clement IV, Pope, 408
Clement V, Pope, 412
Clement VII, Pope, 412, 473-475
Cleon [*klee'on*] 117-119, 129-130, 141
Cleopatra, 212, 218-221
Clermont, Council of, 338, 355
Clovis [*kloh'vis*], 281-282, 296, 299-300
Cluny, 336, 355; Cluniac Order, 336, 391, '413, 488
Coinage: first Greek use of, 89; gold, in medieval Europe, 372
Colleoni [*kol-lay-oh'nee*], Bartolommeo, 433
Coloni, 314
Colosseum, 208, 236
Columban, St., 345
Columbus, Christopher, 3, 14, 155, 431, 440
Comitatus, 278, 309
Commodus [*kom'uh-dus*], 231, 238
Commons, in English parliament, 418-424, 426-429
Communes, 366
Commutation (of manorial services), 369
Compass, magnetic, 439
Condottiere (*pl.* condottieri), 433, 460
Confucius, 41-43, 46; Confucianism, 373
Conrad IV, 382, 391
Constable, Carolingian, 303
Constance, Council of, 474, 481
Constantine, 231, 241-242, 262, 273-274, 279, 283, 296, 455, 470; sculpture of, 253
Constantinople, 242, 244, 246, 253, 273-274, 277, 282-285, 287-288, 296, 338, 340, 342, 353, 355, 372, 439, 455
Consulship, 171-172, 175-176, 191, 195, 203-204, 206-207, 209-210, 211-213, 226, 266
Corcyra [*kor-sy'ruh*], 116-117
Corinth, 112, 114-116, 119-122, 194, 214
Corinthian Gulf, 86
Corinthian War, 121
Cornelia, 198
Corona civica, 161, 266
Corsica, 185, 194, 294
Cortés [*kor-tez'*], Hernando, 440
Corvée, 319
Cos [*kohs*], 84
Counties, Carolingian, 303
Covenant (Hebrew), 66-67, 69, 74; New Covenant, 69, 74, 257
Cranach [*kraw-nawkh'*], Lucas the Elder, 478
Crassus, Marcus Licinius, 209-212
Crécy, battle of, 415, 426-427
Crete, 21, 77-82, 84, 99, 133, 307
Crimea, 87, 95
Croesus [*kree'sus*], 56, 103
Cross-bow, 415, 426
Crusades, 294, 329, 338-343, 353, 355-356, 362, 378, 439; First, 338-340; Second, 342; Third, 327, 330, 342; Fourth, 274, 342-343; of Frederick II, 390-391; of St. Louis, 386; results of, 342-343

Cultural Calendars: Greek, 157; Medieval, 406; Renaissance, 453
Cuneiform [*kew-nee'uh-form*], 18-19, 50, 53
Curia regis, 417, 421
Curiate Assembly, 165, 173
Cyprus, 50, 80, 151
Cyrene [*sy-ree'nee*], 151
Cyrus I, the Great, 56, 63, 69, 103

DACIA [*day'shuh*], 237
Damascus, 50, 147, 259
Dante Alighieri [*dawn'tay aw-lee-gair'ee*], 413, 452, 468; *Divine Comedy*, 413, 452-453, 464
Dardanelles, 56, 82, 145, 439
Darius I [*duh-ry'us*], 56-58, 73, 97, 107-108, 151
Darius III, 145-147
Dark Ages, 274, 317, 336, 345, 353
Dauphin, *see* Charles VII, of France
David, 60
Dead, Book of the, 44-45
Dead Sea, 64; Dead Sea Scrolls, 64
Deccan, 32
Delos [*dee'los*], Confederacy of, 112-113, 117, 126, 138
Delphi [*del'fee*], 139, 141; oracle at, 87
Demesne [*di-meen'*], 315-316
Demetrius of Bactria, 152
Democracy: Greek, 90, 93-100, 113-114, 119-121, 125-129, 141, 174, 178; Roman, 178
Democritus, 132, 136
Demos [*dee'mos*], 90, 99
Demosthenes [*di-mos'thuh-neez*], 122, 455
Diaz [*dee'az*], Bartholomew, 440
Dicasteries [*dy-kas'tur-eez*], 99
Dictatorship, Roman, 171, 175, 190, 207, 210, 213, 215, 248
Didius, 231, 239
Diocletian, 231, 241-242
Dodona [*doh-doh'nuh*], 141
Doge [*dohj*] of Venice, 432
Domestic system, 442
Dominic, St., 394, 489; Order of, 394, 397
Domitian, 231, 235, 237
Donatello, 453, 460
Dorians, 83, 86, 90, 163
Doris, 84
Draco [*dray'koh*], 93
Drusus, 232
Dürer, Albrecht, 453, 468

EBRO [*ee'broh*] River, 185, 187
Ecbatana [*ek-bat'a-na*] (Hamadan), 56, 73
Ecclesia [*i-klee'zee-uh*], 93, 123, 125
Edward I, 326, 339, 382, 386, 388-389, 411, 419-422, 424
Edward II, 414-415, 421-422
Edward III, 413-415, 422-423, 426-428
Edward VI, 432, 487
Egypt, 15, 18-31, 48, 51, 53, 56-60, 62, 64, 66-68, 71-72, 74, 77-78, 99, 101, 147, 151, 156, 182, 194, 212, 218-219, 221, 240, 250-251, 256, 275, 288, 391; Middle Kingdom, 24; New Kingdom, 24-27; Old Kingdom, 21-24

Einhard, 292
El Cid, 341
Elagabalus, 231, 239
Eleanor of Aquitaine, 322-323
Elijah, 65
Elizabeth I, 255, 432, 492-494, 496
England, 312, 321, 323-327, 331, 338, 377-379, 382-383, 386-389, 401, 413-424, 426, 435-436, 438-440, 465, 469, 473, 476-477, 485-488, 496
Engrossing, 364
Epaminondas [*i-pam-uh-non'dus*], 121-122
Ephesus, 84
Ephors [*ef'urs*], 92-93
Epictetus [*ep-ik-tee'tus*], 240, 251
Epicurus, 152-153; Epicureanism, 152-153, 251
Epidaurus, 143
Epirus [*i-py'rus*], 167
Equestrians, Roman, 173-174, 196-197, 202-203, 207, 209-210, 213, 226-227, 237-238
Erasmus, Desiderius, 453, 465-467, 473, 477, 482-483, 489, 497, 499
Eratosthenes [*air-uh-tos'thuh-neez*], 154, 158
Erechtheum [*air-ek-thee'um*] 118
Eretria [*air-eh'tree-uh*], 105, 146
Erfurt, University of, 478
Esarhaddon, 51
Escheat, 311
Estates-General, 274, 386, 415-416, 424
Ethiopia, 150
Etruria [*i-trur'ee-uh*], 163, 165, 167, 190, 209
Etruscans, 163-164, 166-168, 171, 208, 252, 254
Euboea [*yoo-bee'uh*], 84
Euclid, 154
Euphrates River, 14, 24-25
Euripides, 140-141, 143; *Bacchae*, 143
Evans, Sir Arthur, 77-78, 81
Everyman, 399
Evesham, battle of, 388
Exodus from Egypt, 60
Ezekiel, 66, 69

FABIUS MAXIMUS, Quintus, 190, 221
Fabliaux, 371, 399
Fairs, 363-364, 442
Falerii, 161
Faubourg, 364
Fayum [*fy-yoom'*], 12, 24
Fealty, 310, 333
Federate allies of Rome, 169-170, 201, 206
Ferdinand and Isabella, of Spain, 432, 438, 489
Fertile Crescent, 15, 48, 59
Festivals, Greek, 141-142
Feudalism, 308-314, 321, 324, 331-332, 371, 378, 417
Fief [*feef*], 310, 312, 372
Flanders, 310, 333, 366, 377, 383, 386-387, 414, 434, 442, 473
Flavians, 231, 235-237
Florence, 369, 372, 377, 413, 432, 434, 445, 452, 455, 460-463, 469, 475;
University of, 452, 454

Florida, 440
Forestalling, 364
Forfeiture, 311
Formariage, 319
Fountains Abbey, 393
France, 280, 287, 294, 296, 302, 306-307, 312-313, 321-324, 327, 331, 335, 340, 347, 355, 362, 366, 377-378, 382-386, 388, 391, 394, 401, 403-404, 412-416, 422-424, 435-439, 468-469, 473, 477, 485, 492, 495-496
Francis of Assisi, St., 394, 407, 489;
Rule of, 394, 407-408; Order of, 394, 396, 413, 488
Francis I, 432, 438, 469
Franks, 245, 277, 279-282, 288-296, 299-300, 310, 327
Frederick the Wise, of Saxony, 478, 481
Frederick I, Barbarossa, 321, 329-330, 384, 389, 433
Frederick II, 382, 384, 389-391, 395
Friars, 394
Frisians, 305
Fuggers, 479

GABELLE, 436
Gaiseric [*gy'suh-rik*], 244
Galatia [*guh-lay'shuh*], 194
Galba, 231, 235
Galen [*gay'lun*], 254, 348
Gallipoli peninsula, 439
Gama, Vasco da, 440
Ganges River, 32
Gascony, 388-389, 414, 473
Gaspé peninsula, 440
Gaugamela [*gau-guh-mee'luh*], battle of, 145, 147
Gaul, 166, 186, 207, 209, 211-212, 215, 235, 244-245, 251, 280; Gauls, 166-167, 180, 185, 190, 208
Geneva, 484-485, 497
Genghiz Khan [*jen'giz kawn*], 374-375
Genoa [*jen'oh-uh*], 343, 366, 372
Geometry, 19, 136, 154
Gerbert [*jair-bair'*] (Pope Sylvester II), 346-347, 353
German Empire, 321, 327, 329-331, 383-384, 389-391, 414, 416, 424, 437-439, 478, 481-482, 492
Germanic barbarians, 238-239, 241-246, 249-251, 277-285, 309, 317, 324.
See also Law.
Germanicus, 232
Germany, 235, 237, 280, 294, 307, 329-331, 382, 390-391, 401, 410, 416, 424, 431, 468-469, 473, 477-483, 485, 490, 492, 496
Gerousia [*jair-oo'zee-uh*], 92
Ghent, 387
Ghibellines [*gib'uhl-eenz*], 329
Ghiberti [*gee-bair'tee*], Lorenzo, 453, 460-463
Gibbon, Edward, 247, 262
Gibraltar, 87, 244, 287, 306
Gilgamesh [*gil'guh-mesh*], *Epic of*, 18
Giotto [*jaw'toh*], 453, 456
Gizeh [*gee'zuh*], pyramids of, 22-23
Gladiators, Roman, 208, 236-238, 251, 266

Goa, 440
Godric, 362-363
Golden Bull (1356), 416
Gracchus [gra'kus]: Gaius [gy'us], 198, 201-203, 207, 211, 214, 221; Tiberius Sempronius, 198-203, 207, 221
Granada, 438, 489
Granicus [gruh-ny'kus], battle of, 145
Gravelines, battle of, 494
Gravitas, 171
Great Schism [sism], 412-413, 473-474
Greece, 50, 57, 77, 82, 161-162, 167, 198, 208, 212, 217, 219, 222, 235, 260, 285, 335, 345, 353, 426, 467; Greeks, 31, 56, 72, 82-159, 251, 254, 262, 264, 266; Western Greeks, 108, 164, 167, 184, 190-191; Greek colonies, 87, 100
Greek Orthodox Church, 274, 283, 338
Greenland, 306, 338, 440
Gregory, Bishop of Tours, 282, 299-300, 344
Gregory I, the Great, Pope, 277, 286; Gregorian chant, 277
Gregory III, Pope, 291
Gregory VII (Hildebrand), Pope, 336-338, 353-355, 382
Gregory IX, Pope, 382
Gregory XI, Pope, 412-413, 474
Gregory XII, Pope, 474
Gregory XIII, Pope, 215
Grosseteste [grohs'test], Robert, 395-397, 408
Guelfs, 329-330, 389-390
Guilds, 377, 442; merchant, 365; craft, 365-366, 379-380
Gunpowder, invention of, 373
Gutenberg [goo'tun-burg], Johann, 443

HABIRU [haw-bee'roo] (Hapiru), 59
Habsburg dynasty, 391
Hadrian, 231, 238
Hagia Sophia [hag'yuh soh-fee'uh], 253, 285-286
Hamilcar Barca, 184-185
Hammurabi [haw-mur-aw'bee], 16, 18-19; Code of, 17, 44, 65, 72
Han [hawn] dynasty, 373
Hannibal, 185-193, 195, 198, 221-222
Hanseatic League, 366-367
Harappa, 32
Harold, Saxon king of England, 324
Hasdrubal [haz'dru-bul], 185, 191
Hastings, battle of, 324
Head tax, serf's, 319
Hebrews, 19, 43, 48, 50-51, 57, 59-71, 261, 262
Hegira [hi-jy'ruh], 286
Heliaea [hee-luh-ee'a], 95, 99
Heliopolis [hee-lee-op'uh-lis], 25
Hellas [hel'us], 85
Hellenes, 85, 103, 120
Hellenic League, 122, 145
Hellenistic Age, 151-156; Hellenistic culture, in Rome, 197-198
Hellespont, 105, 107, 120
Helots [hee'luts], 90-91, 94
Henry the Navigator, Prince, 440, 447-448

Henry II, of England, 321, 323, 325-327, 329, 388, 395, 398, 418
Henry III, of England, 382, 388, 418, 419, 421
Henry III, of Germany, 321, 327, 336
Henry IV, of England, 414
Henry IV, of France, 432, 495
Henry IV, of Germany, 321, 329, 337-338, 354-355
Henry V, of England, 435
Henry VI, of England, 432, 436
Henry VI, of Germany, 389
Henry VII, of England, 431-432, 438, 469
Henry VII, of Germany, 413-414, 416
Henry VIII, of England, 432, 438, 485-488, 492
Heraclides [hair-a-kly'deez], 154
Heraclitus [hair-uh-kly'tus], 132
Herculaneum [hur-kyuh-lay'nee-um], 236
Heresy, 382-383, 413-414, 477
Heriot, 319
Hermits, early Christian, 275
Hero of Alexander, 154
Herodotus [hi-rod'uh-tus], 85, 106-107
Hesiod [hee'see-ud], Works and Days, 87
Hieroglyphs, 31, 50
Hildebrand, see Gregory VII
Himalayas, 31-32
Hinduism, 34-37
Hipparchus [hi-par'kus], 96-97
Hippias, 96-98, 105
Hippo, 344
Hippocrates [hi-pok'ruh-teez], 137, 348
Histiaea [his-ti-ee'uh], 120
Hittites, 16, 19, 26, 48-49
Hohenstaufen dynasty, 391
Holbein, Hans the Younger, 453, 467-468, 478
Holy Roman Empire, 327. See German Empire.
Homage, 310, 312, 333
Homer, 77-78, 84, 86, 140; Iliad, 84, 86, 140, 180; Odyssey [od'uh-see], 84, 86, 140
Honan, 40
Honorius III, Pope, 407
Hoplites, 92, 104-106, 108
Horace, 255; Odes, 255
Horse collar, 318
Horseshoes, 318
Hosea, 66-68
Huguenots, 485, 495
Humanism, 454-456, 465-467, 472, 477, 482-483, 488-489
Hundred Years' War, 414-415, 422-424, 438
Hungary, 307, 438, 473, 490, 492
Huns, 242-243, 245, 279
Huss, John, 474, 481
Hydaspes [hi-das'peez], battle of, 145, 148
Hyksos [hik'sohs], 24, 29, 60

ICE AGES, 4-5, 11
Iceland, 150, 306, 338
Iconoclasm, 290-291
Idealism, Platonic, 134-135
Ile de France, 322

Imperium, 165, 171
Incidents, feudal, 311
Index of Prohibited Books, 490
India, 31-37, 43, 57, 71, 147-148, 151-152, 155, 250, 287, 440, 470, 490
Indo-Europeans, 16, 24, 27, 34, 48, 56, 82, 162
Indulgences, 343, 476, 479-480, 499-500
Indus River, 31-35
Industry: Athenian, 142; Roman, 250
Innocent III, Pope, 343, 382-383, 388, 394, 407, 413, 475
Innocent VII, Pope, 474
Innocent VIII, Pope, 471
Inquisition, 382-383, 438, 446
Investiture Controversy, 336-338, 342, 353-355, 398
Ionia, 84, 87, 89, 103, 105, 121
Iran, 12-13, 32, 48, 56, 73
Iraq, 13, 48
Ireland, 325, 344
Iron Age, 30-31, 48-49, 83, 99, 162
Isaiah, 48, 66, 68; Second Isaiah, 66, 69, 71
Islam, 286-288, 300
Israel, 12-13, 48, 60-61, 64-69, 71, 74
Issus [*is'us*], battle of, 145-147
Italy, 161-165, 284-285, 291-292, 294, 307, 310, 327, 330, 345, 350, 366, 391, 397, 416, 424, 432-435, 437-438, 450-464, 469, 473, 476. *See also* Rome.

JAMESTOWN, 441
Japan, 373, 375-376
Jarmo, 12-13
Java, 6-7
Jeanne d'Arc, 436, 437, 446-447
Jeremiah, 48, 66, 69, 74, 398
Jericho, 12-13, 60
Jeroboam I [*jair-uh-boh'um*], 61
Jeroboam II, 67
Jerome, St., 275, 344-345, 455
Jerusalem, 51, 61, 63-64, 67, 68-69, 72, 74, 235, 255, 257-260, 276, 339-340, 342, 349, 355, 390
Jesuit Order, 489-490, 498
Jesus of Nazareth, 69, 234, 257-261, 264, 267, 398
Jezebel, 65
John, Gospel according to, 257
John, of England, 321, 323, 325, 327, 379, 382-383, 386-388
John I, of Portugal, 447
John II, of France, 414
John XXIII, Pope (1410-15), 474
John XXIII, Pope (1958-63), 491
Joint-stock company, 372
Jordan, 12-13, 48
Joseph, 60
Joshua, 60, 65
Journeyman, 365-366, 442
Judaea, 63, 227, 234-235, 258, 260
Judah, 61, 67-68, 74
Judaism, 256, 286, 288
Judgment of God, 278-279
Jugurtha [*joo-gur'thuh*], 203
Julian calendar, 215
Julio-Claudians, 231-235
Julius II, Pope, 451, 475, 477
Jupiter, 256
Jury system, 325-326
Justin Martyr, 263, 269-270; *First Apology*, 269-270
Justinian, 284-285, 296, 350; *Corpus Juris Civilis*, 284-285, 350, 388-389; *Code*, 284; *Digest*, 284
Jutes, 324

KARNAK, 27-28; Temple of Amon at, 27-28
Kassites, 16, 48
Kells, Book of, 337
Kish, 14
Knighton, Henry, 380
Knights, 310, 312-314, 350, 361, 371, 415, 418-423, 427-429
Knossos [*nos'us*], 78-79, 81
Koine [*koy-nay'*], 156
Koran, 286, 300-301, 347
Krak des Chevaliers, 339
Krypteia [*krip-tee'uh*], 92
Kublai Khan [*koo-bly kawn*], 372, 375-377

LACONIA, 90
Lancastrians, 438
Langland, William, 334, 453, 464
Language: Arabic, 300, 347, 349; Aramaic, 50, 57, 260; Babylonian, 19, 50; Etruscan, 164; French, 307; German, 307; Greek, 78-79, 156, 452, 455-456, 465-467, 472; Latin, 264, 277, 292, 325, 344-345, 348-349, 398-399, 450, 452, 454, 456, 465-467, 472; Minoan, 78-79; Oscan, 165
Laocoön [*lay-ok'oh-won*], 153
Lascaux cave murals, 9
Lateran Council (1215), 382
Latifundia, 195, 314-315
Latin allies of Rome, 164-165, 169-170, 201-203, 206; League, 165-166
Latin Kingdom of Jerusalem, 276, 339, 343
Latium [*lay'shum*], 163-166, 190
Laurium, 95
Law: English common, 325-327, 426; Germanic, 278-279, 285; Mosaic, 64-66, 72; Roman, 238, 255-256, 283-285, 296, 350, 398
Lebanon, 48, 57
Lefèvre d'Etaples, Jacques, 453, 465, 489, 497
Leo I, Pope, 245
Leo III (the Isaurian), Byzantine Emperor, 287, 290
Leo III, Pope, 295
Leo X, Pope, 475-476, 479-481, 485
Leon [*lay'on*], kingdom of, 341
Lepidus, Marcus Aemilius, 217-218
Lesbos, 84
Letter of credit, 372
Leuctra [*look'truh*], battle of, 121
Lewes [*loo'is*], battle of, 418
Libya, 24, 183
Life rent, 362
Lincoln, battle of, 312
Lisbon, 441

Literature: French, 294; Greek, 140-141; Hellenistic, 155; Latin, 254-255; medieval Latin, 349; medieval vernacular, 294, 371, 399
Livia, 227
Livius Andronicus [*an-droh-nee'kus*], 197
Livy, 179-180, 186, 470
Lombards, 285, 291-292, 294
London (England), 369
Long-bow, 415, 427
Long Walls, 113, 120-121
Lords, House of, 422
Lorraine, 322
Lothair, 307, 310, 327
Louis the German, 307
Louis the Pious, 296, 307, 355
Louis IV, of Germany, 414
Louis VI, of France, 321-323
Louis VII, of France, 321, 323
Louis IX, of France, 382, 385-386, 388, 416
Louis X, of France, 414
Louis XI, of France, 432, 436-438
Louis XII, of France, 432, 438
Louvre, 153, 458
Loyang, 40
Loyola, Ignatius, 489, 498
Lübeck, 366-367
Lucania [*loo-kay'nee-uh*], 167, 190
Luke, Gospel according to, 257, 337
Lûll [*lool*], Raymond, 320
Luther, Martin, 473, 477-483, 489, 496; Lutheranism, 482-483, 487
Luxor [*luk'sor*], 27
Lycaonia [*ly-kay-oh'nee-uh*], 194
Lycurgus [*ly-kur'gus*], 91, 100, 467
Lydia, 56, 84, 89, 103

MACEDONIA, 57, 121-122, 141, 144-145, 151, 182, 191, 193-194, 222, 260
Machiavelli, Niccolo, 433-434, 445-446, 453, 456; *The Prince*, 434, 445-446
Macrinus, 231, 239
Maecenas [*mi-see'nus*], Gaius, 255
Magellan, Ferdinand, 440
Magistracies, Roman, 170-176, 178, 195
Magna Carta, 388-389, 416, 421
Magyars, 307
Malacca, 440
Man: Old Stone Age, 4-11, 13; New Stone Age, 11-15
Manorialism, 314-321, 331, 361, 372, 377-378; manorial court, 319
Mantinea [*man-tin-ee'uh*], battle of, 121
Manutius [*muh-noo'shus*], Aldus, 443
Marathon, battle of, 105-106
Marcus Aurelius [*or-ee'lee-us*], 231, 238, 240-241, 251-252, 261, 269, 373, 459; *Meditations*, 251-252
Marduk, 58
Mare, Sir Peter de la, 428
Maritime insurance, 372
Marius, Gaius, 161, 182, 203-204, 206-207, 210, 217, 221, 266
Mark, Gospel according to, 257, 337
Marshal, Carolingian, 303

Marsiglio [*mar-sil'yoh*] of Padua, 413; *Defensor Pacis*, 413
Martin V, Pope, 474
Mary Tudor, 432, 492-493
Mathematics: Chaldean, 56; Egyptian, 28, 31; medieval, 347, 351; Mesopotamian, 19; Moslem, 347
Matthew, Gospel according to, 257, 274, 337, 398
Mayors of the Palace, 288-292
Maximilian I, of Germany, 416, 438
Mecca, 286-287
Medes, 56-57, 106-107
Medici [*may'dee-chee*], 434, 445, 463; Cosimo de', 434; Giuliano [*joo-lee-aw'noh*] de', 434; Lorenzo de', 434, 475
Medicine: Frankish, 356; Greek, 137; Hellenistic, 154; Moslem, 348, 356; Roman, 254
Medina, 286
Megara [*meg'uh-ruh*], 95, 101, 115
Megaron, 82
Melos [*mee'los*], 84, 119-120
Memphis, 25
Menander, 155
Mendicant orders, 394
Menes [*mee'neez*], 21
Mercantilism, 444
Merchet, 319
Merovingian dynasty, 282, 288-292, 296, 346
Mesopotamia, 15-19, 21, 29, 31, 34, 37, 42-43, 59, 77, 99, 146, 237
Messana [*meh-say'nuh*], Straits of, 183-184
Messenia [*meh-see'nee-uh*], 91
Metaurus River, battle of the, 191
Metics, 123
Mexico, 440
Micah, 66, 68
Michelangelo, Buonarroti, 451, 453, 459, 461, 463-464, 469, 491; *David*, 430-431, 463; *Creation and Fall of Man*, 463; *Last Judgment*, 463
Middle class, 371, 378, 407, 442, 444
Middle East, 12, 194, 210, 238, 395, 439, 441, 451
Milan, 262, 369, 432-433, 438; Edict of Toleration at, 262
Miletus [*my-lee'tus*], 84, 87, 95, 105
Miltiades [*mil-ty'uh-deez*], 105-106
Milton, John, 155
Milvian bridge, battle at the, 241-242
Minoan [*mi-noh'un*] civilization, 78-82
Minos [*my'nos*], 78, 81
Minotaur [*min'uh-tor*], 81-82
Missi dominici, 303
Mithraism, 256
Mithridates VI [*mith-ruh-day'teez*], 206, 210
Mohammed, 286-287, 296, 300, 390
Mohammed II, 439
Mohenjo-daro [*moh-hen'joh dah'roh*], 32-33
Monasticism, 275-277, 290, 296-299, 320, 336-337, 344-345, 350, 478, 482, 489

Mongolia, 373-374; Mongols, 242-243, 372, 439
Monotheism, Hebrew, 67-69, 71, 262
Montaigne, Michel de, 453, 467-468
Monte Cassino, 276, 297
Montfort, Simon de, 418-419, 421-422
More, Thomas, 453, 465-466, 477, 486, 489; *Utopia*, 465-466
Mosaics, Byzantine, 285-286
Moses, 60, 62, 64-65, 71, 390
Moslems, 286-289, 291, 294, 300, 305-307, 313, 331, 338-343, 347-350, 353, 372, 375, 386, 389, 394-395, 438, 440, 443, 489
Mummification, 21, 29
Music, medieval, 277, 399
Mycale [*mik'uh-lee*], battle of, 108
Mycenae [*my-see'nee*], 77-79, 81-86, 99
Mylae [*my'lee*], 184
Myron, 138-139; *Discus Thrower,* 138-139
Mytilene [*mit'uh-lee-nee*], 117, 120, 129-130

NANTES, EDICT OF, 495
Naples, Kingdom of, 435
Naramsin, 16
Nationalism, 407, 413, 431, 445
Navarre, 473
Naxos [*nak'sus*], 112
Nebuchadrezzar [*neb-yu-kud-rez'ur*], 53, 56, 62, 69, 74
Nefertiti, 25, 28
Neolithic [*nee-oh-lith'ik*] (New Stone Age), 5, 11-13, 38, 40, 77-79, 162
Neptune, 256
Nero, 208, 231, 233-236, 260
Nerva, 231, 237
New Amsterdam, 441
New Babylonia, *see* Chaldeans
New Carthage, 187
New England, 441, 485
New France, 490
New Testament, 64, 195, 257, 286, 455, 465-466, 477, 481, 489
Newfoundland, 306, 440
Nicaea [*ny-see'uh*], 279; Council of, 274; Nicene Creed, 279
Nile Valley, 19-20, 22-23, 28
Nîmes, 252
Nineveh, 3, 51, 53
Nirvana [*nur-van'uh*], 35-36
Noah, 15
Nomes, Egyptian, 29
Normandy, 307, 312, 322, 324, 383, 387, 409, 417
North Africa, 183-184, 192, 202-204, 279, 284, 287, 305, 307, 344, 386
Numidia, 203

OCTAVIA, 219
Octavian; Octavius: *see* Augustus
Odoacer [*oh-doh-ay'sur*], 246, 273, 279
Old Oligarch, 77, 99; *On the Constitution of Athens,* 125, 128-129
Old Testament, 41, 64, 66, 71, 259, 460-461, 465, 481, 489
Olduvai, 4-6

Oligarchy, Greek, 90-93, 100, 113, 121, 125, 174
Olympus, Mt., 86
Open field system, 315-316
Opis, 150
Optimates, 203, 206, 210, 213
Oracle bones, 38-40
Oratory of Divine Love, 489
Ordeal, 278-279, 326, 340
Ordinances of 1311, 421
Orléans, 436-437
Osiris [*oh-sy'ris*], 27, 44
Ostia, 215, 233
Ostracism, 98, 106, 112
Ostrogoths, 242, 277, 279, 284, 432
Otho, 231, 235
Otto I, 321, 327; Ottonian Empire, 327, 330
Otto IV, 383-384, 389-390
Ovid, 255; *Metamorphoses,* 255
Oxen, 317-318
Oxford, Provisions of, 418, 421
Oxford University, 351

PALEOLITHIC [*pay-lee-oh-lith'ik*] (Old Stone Age), 5-13, 42, 162
Palestine, 12, 24, 51, 53, 59, 64, 276, 287, 338
Panini, Giovanni Paola, 253
Pantheon, 253
Papacy, 263, 274, 277, 290-292, 295-297, 327, 329, 336-338, 343, 382-383, 407, 412-414, 424, 438, 451
Papal states, 292, 412, 434-435
Papyrus, 28-29, 64
Paris, 306-307, 322, 369; University of, 351
Parlement of Paris, 386
Parliament, English, 389, 416-424, 426-429, 438, 444, 485
Parmenides [*par-men'uh-deez*], 132
Parrhasius [*par-ay'zee-us*], 140
Parthenon, 18, 118-119, 138-139, 153
Parthia, 237-238, 373; Parthians, 212, 219, 238
Patricians, 172, 174-175, 180
Paul, St., 255, 259-260, 262-263
Paul III, Pope, 475, 490, 499
Pazzi [*pat'see*], conspiracy of, 434
Peace of God, 338
Peking, 376
Peloponnesus [*pel-uh-puh-nee'sus*], 77, 84, 90, 93, 109, 121-122, 133, 137; Peloponnesian League, 93, 115, 121, 126; Peloponnesian War, 77, 114-121, 127, 134, 136-137, 141-142, 156, 191
Pentecontor, 96-97
Pepin [*pep'in*] the Short, 290-292, 297
Pergamum [*pur'guh-mum*], 115, 194, 201, 254
Pericles [*pair'uh-kleez*], 112-116, 119, 122-123, 125-127, 138-139, 141-142; *Funeral Oration of,* 116, 126-128, 142
Perioeci [*pair-i-ee'sy*], 90-91
Persecutions, of Christians, 260-262, 267-268
Persepolis [*pur-sep'uh-lis*], 58
Perseus of Macedon, 222

Persian Gulf, 13-15, 56
Persian Empire, 48, 56-59, 71, 97, 103, 108, 112, 121, 145, 151, 256, 287-288;
   Persians, 43, 50, 56-59, 71, 93, 126, 132, 161, 239, 241, 284-285, 355;
   Wars of, 105-109
Pertinax, 231, 239
Peter, St., 258-259, 262-263, 273-274
Peter the Hermit, 338-339
Petrarch [*pee'trark*], Francesco, 431, 452-454
Phaestos [*fes'tus*], 78
Phalanx [*fal'angks*], 104-105, 121-122, 145, 168
Pharsalus [*far-say'lus*], 212
Pheidippides [*fy-dip'uh-deez*], 105
Phidias [*fi'dee-us*], 118, 139
Philip II, Augustus, of France, 302, 321, 323-325, 343, 382-385
Philip II, of Macedon, 105, 121-122, 126, 144-145
Philip II, of Spain, 432, 492, 494-496
Philip IV, the Fair, of France, 382, 386, 389, 398, 412, 424
Philip V, of France, 414
Philip V, of Macedon, 191, 193-194
Philip VI, of France, 414
Philippi, 217-218; battle of, 255
Philistines, 60-61, 65
Philosophy: Greek, 131-136; Hellenistic, 251
Philoxenus, 146
Phoenicia [*fuh-nee'shuh*], 61, 146-147; Phoenicians, 49-50, 64, 183
Phrygia [*frij'ee-uh*], 194; Phrygians, 48
Pico della Mirandola, Giovanni [*pee'koh del'lah mee-rahn'doh-lah*], 453, 456, 465, 470-471, 475
Pierce the Ploughman's Crede, 334
Pietas, 171
Pilate, Pontius, 234, 258
Pineda, Alvarez de, 440
Pinzon, Vincente Yañez, 440
Piraeus [*py-ree'us*], 112-113, 120
Pisa, Council of, 474
Pisistratus [*pi-sis'truh-tus*], 93, 95-96, 101-102, 105
Pithecanthropus [*pith-uh-kan'throh-pus*] erectus, 6
Pius II, Pope, 431, 475
Plataea [*pluh-tee'uh*], 105, 108, 115-116
Plato, 87, 125, 134-136, 434, 455, 467; *Apology*, 156, 158; *Republic*, 134-135, 455
Plautus [*plah'tus*], 197-198
Plebeians (Plebs), 172, 174-175, 179-180, 213
Pliny [*pli'nee*] the Younger, 261, 267-268
Plough, heavy wheeled, 317, 369
Plutarch, 91-92, 100-101, 158-159, 183, 202, 219, 223; *Lives*, 100-101, 158, 223
Po River, 161-162
Poitiers, battle of, 415
Polemarch, 93
Polis [*paw-lis*], 90, 120
Polo, Marco, 372, 375-377
Polybius, 188, 194

Pompeii [*pom-pay'ee*], 214, 236
Pompey [*pom'pee*] the Great, 182, 209-212, 218, 221
Pontifex Maximus, 213
Populares, 203-204, 206, 210
Population: Athenian in 5th century B.C., 123; Chinese under T'angs, 373; England in 14th century, 415; Europe in 13th and 14th centuries, 377, 378; European towns in 14th century, 368-369; France in 14th century, 415; Roman Empire in 3rd century, 249; Rome (city of) in 1st century, 252
Portland Vase, 154
Porus, king of the Pauravas, 148
Poseidon, 256
Posidonius [*poh-si-doh'nee-us*], 155
Potestas, 170-171
Potidaea [*pot-uh-dee'uh*], 114-115
Pottery, Greek, 91, 94; Hellenistic, 154
Praetorian Guard, 228, 232, 235, 239
Praetorship [*pree'tur-ship*], 171-172, 175, 195, 203, 207, 210, 213
Praxiteles [*prak-sit'uh-leez*], 139-140
Prima Porta, 227, 252
Printing, 443-445, 456, 480
Promissory notes, 442
Prophets, Hebrew, 41, 62, 65-69, 71, 257
Propylaea [*prop-i-lee'uh*], 118-119
Proscriptions: Sullan, 207, 210, 223; by Second Triumvirate, 217, 224
Protagoras, 133
Protestantism, origin of name, 483. See also Reformation.
Provost, 315
Prytany [*pri'ta-nee*], 98, 123, 125
Ptolemy [*tol'uh-mee*] of Alexandria, 254
Ptolemy I, 151; Ptolemaic dynasty, 194, 218
Publicans, 195
Punic Wars, 168, 183-195. See also Carthage.
Purveyance, 311
Pylos [*py'lus*], 79
Pyramids, 22-23, 27
Pyrrhus [*pir'us*] of Epirus, 167-168, 179
Pythagoras [*pi-thag'ur-us*], 136

QUADRIVIUM, 351
Quaestorship [*kwees'tur-ship*], 171-172, 174-176, 195, 198, 200, 207, 210, 213
Quinqueremes, 183-184

RAMADAN, 287
Ramses I [*ram'seez*], 26
Ramses III, 27
Raphael [*raf'ee-ul*], 453, 459; *School of Athens*, 459
Ratherius of Liège, 333-334
Re [*ray*], 25, 27, 58
Red Sea, 56, 60
Reeve, 315
Reformation, 466, 477-488
Regrating, 364
Rehoboam [*ree-oh-boh'um*], 61
Relief, 311

Religion: Assyrian, 53; Egyptian, 25-27; Greek, 86, 141-142; Hellenistic, 155; Mesopotamian, 17-18; Paleolithic, 9; Persian, 57, 59; Roman, 256
Renaissance: Carolingian, 346, 348, 350, 353; Italian, 39, 389, 435, 439, 449-464, 468-471, 473-474; Northern, 464-469, 473; Twelfth-century, 348-354, 397-398
Reuchlin [*roikh'leen*], Johann, 453, 465, 489
Revolution: agricultural in Middle Ages, 317-318; commercial, 369-372, 377-379, 407, 439-444, 469; food-producing, 11-13, 15; French (1789), 308, 436; price, 442; scientific, 452, 457; urban, 13-15
Rhetoric, 250, 254, 351
Rhodes, 84
Richard I, Lionheart, 321, 323, 325, 327, 388, 418
Richard II, 414-415, 417, 423, 428
Richard III, 432, 438
Robert the Monk, 355
Roman de la Rose, 399
Romance of Reynard, 371
Rome, 151, 154, 156, 161-271, 335, 382, 391, 412, 426, 431, 435, 439, 455, 459, 463, 467, 470, 473-474, 476, 478-479, 485, 487, 489, 491; Empire of, 32, 53, 57, 82, 193, 210, 226-227, 231-266, 273, 294, 305, 308, 314, 361, 373, 378, 398, 450, 455-456; roads of, 3, 234-235, 253; taxation by, 195, 202, 211, 215, 224, 235, 238, 241, 250, 282
Romulus Augustulus, 246
Roses, Wars of the, 438
Rubicon River, 212
Rudolf of Habsburg, 382, 391
Runnymede, 416

SABINES, 166
Sacraments, Christian, 269-270, 482, 484, 488, 490, 498-500
Saguntum [*sa-gun'tum*], 185
St. Denis, abbey church of, 401
St. Peter's basilica, 253, 295, 451, 459, 463-464, 479-480, 491
Salamis [*sal'uh-mis*], battle of, 107
Salisbury Oath (1086), 417
Samaria, 61
Samnium, 166, 190; Samnites, 166-168, 208
Samos [*say'moss*], 84
Santa Sabina, Church of, 263
Saracens, 339, 466
Sardinia, 183, 185, 191, 194, 218
Sargon I, 15-16, 31
Sargon II, 51
Saronic Gulf, 86
Satrap [*sat'rap*], 57
Saul, 60-61
Savonarola [*sav-uh-nuh-roh'luh*], Girolamo, 475
Saxony, 289, 294, 305, 324, 327, 331, 338, 478-479
Scandinavia, 306, 483, 492
Schism between Rome and Constantinople, 338
Schliemann, Heinrich, 77-78, 83

Scholasticism, 369, 397-398
Science: Assyrian, 53; Chaldean, 56; Egyptian, 28-29; Greek, 132, 136-137; Hellenistic, 150, 153-155; medieval, 350, 395-397; Mesopotamian, 19; Moslem, 347, 394-395; Roman, 253-254
Scipio [*sip'ee-oh*], Publius Cornelius, 191-193, 198
Sculpture: Egyptian, 27-28; Hellenistic, 137; medieval, 400-401, 405; Mesopotamian, 18; Paleolithic, 8; Renaissance, 460-461, 463, 469; Roman, 252-253
Seleucus [*si-loo'kus*], 151
Senate, Roman, 165, 170-172, 175-176, 178-179, 183-184, 193-196, 199-204, 206-210, 212-213, 216-217, 222-223, 225-227, 230, 232, 235, 237-238, 246, 266, 452
Seneca the Younger, 233, 251
Seneschals, 303, 386
Sennacherib [*si-nak'ur-ib*], 51, 68, 72
Septimius Severus, 231, 239
Serfdom, 289, 304, 309-310, 314-315, 318, 321, 331, 334, 343, 361, 364, 369, 371, 377-378
Severus Alexander, 231, 239
Sforza [*sfor'tsah*], General Francisco, 433
Shakespeare, William, 453, 467-468
Shang dynasty, 38-40; bronzes of, 38, 41
Sheriffs, 386, 418
Shires, 418-419, 421
Shiva [*shee'vuh*], 34
Shots (of open field system), 315-316
Sicilies, Kingdom of the Two, 432, 435
Sicily, 80, 108, 112, 116, 120, 164-165, 178, 183-184, 186, 194, 209-210, 218, 224, 305, 307, 338, 384, 389, 435
Simeon Stylites, St., 275
Simony, 499
Sinai [*sy'ny*], Mt., 64-65
Sinuhe, Tale of, 27
Sistine Chapel, 463
Sixtus IV, Pope, 434
Slavery: Athenian, 128-129; Germanic, 278; medieval, 364-365, 369; Roman, 174, 195, 221, 229, 249, 261
Slavs, 284-285, 305
Socrates, 133-134, 136, 138, 141, 156, 158, 257
Sohan valley (India), 32
Solomon, 61, 66, 285, 340
Solon [*soh'lun*], 93-95, 99, 101, 174
Sophists [*sof'ists*], 133
Sophocles [*sof'uh-kleez*], 140; *Oedipus* [*ee'du-pus*] *Tyrannus*, 140-141
Sorbonne, Robert de, 353
Spain, 183, 185, 187, 191, 194, 198, 200, 203, 209-210, 212, 217-218, 244, 251, 254, 279, 284, 287-288, 294, 307, 313, 338, 341, 347, 401, 424, 438-440, 443, 470, 489, 492, 494-496, 499
Sparta, 90-93, 97, 100-101, 105-109, 112-116, 118-119, 121, 126, 156, 194
Spartacus, 209
Spartiates, 90-91
Speech, earliest man's, 9-10
Stained glass, medieval, 401, 405, 410

Stephen, of England, 324-325
Steward, 315
Stirrups, introduction of, 289
Stock exchanges, 442
Stoicism, 153, 159, 251-252
Strasbourg Oaths, 307
Strategoi [*stra-tee'goy*], 98
Suetonius, 216
Suger [*soo'zhair*], abbot of St. Denis, 401
Suit to court, serf's, 319; vassal's, 311
Sulla, Lucius Cornelius, 182, 206-209, 215, 217, 221, 223, 225
Sumer [*soo'mur*], 15-16, 31
Susa, 17, 57
Switzerland, 483-485, 492
Synagogue, 62
Syracuse, 120, 154, 183, 224
Syria, 12-13, 21, 24, 30, 48, 50, 53, 68, 146, 151, 182, 193-194, 210, 238, 251, 256, 287-288, 339, 387

TACITUS [*tas'uh-tus*], 230, 235, 237, 254, 261, 268, 454; *Germania,* 278
Taille, 436
Tallage, 319
Tandem harness, 318
T'ang [*tong*] dynasty, 373-374
Tartars, 374-375
Technology: medieval, 317-318, 324; Renaissance, 439; Roman, 249-250, 253
Temples, Greek, 77, 138, 451
Tenedos [*ten'uh-daws*], 84
Terramare folk, 162
Tetzel, John, 479
Thales [*thay'leez*], 132, 136
Thasos [*thay'sos*], 112
Theatre, Greek, 142-143
Thebes, 24-26, 28, 51, 97, 115, 119, 121-122, 126
Themistocles [*thi-mis'tuh-kleez*], 98-99, 106-107, 109, 112-113
Theocritus [*thee-ok'ruh-tus*], 155
Theodora, 284
Theodoric [*thee-od'uh-rik*], 279
Theodosius the Great, 244, 262
Theophilus (Roger), 410
Thera [*thee'ruh*], 84
Thermopylae [*thur-mop'uh-lee*], battle of, 107
Theseus [*thee'see-us*], 81-82
Thessaly, 84, 107, 122
Thetes [*theets*], 94, 99
Thirty Tyrants, 121
Thrace, 244
Thucydides [*thoo-sid'uh-deez*], 114, 116-117, 119-120, 122, 125-129, 137, 144; *Funeral Oration,* 122-123, 126-128
Thutmose III [*tut'mohz*], 24-26
Tiberius, 230-232, 234, 237, 253
Tiglath-pileser I [*tig'lath pil-ee'zur*], 51
Tiglath-pileser III, 51
Tigris-Euphrates Rivers, 12-13, 15, 37, 42, 48, 150
Time Charts: Ancient East, 70; Early Modern Times, 493; Egyptian Dynasties, 21; Explorers, 440; Feudal Monarchs, 321, 382, 414; Great Schism, 474; Greece and Rome, 220; Man's Life on Earth, 5, 13; Middle Ages, 274, 425; Minoan History, 78; Pleb victories, 175; Renaissance and Reformation Monarchs, 432; Renaissance and Reformation Popes, 475; Roman Emperors, 231; Roman Empire, 265
Tithe, 320
Titian [*tish'un*], 453, 459
Titus, 231, 235-236
Torah, 64
Tournaments, 383, 399
Tours, battle of, 287, 289
Towns: first sizeable, 5, 13; charters of, 343, 378; population of medieval, 368-369; rise of in Middle Ages, 364-369, 378
Trajan, 161, 218, 231, 237-238, 254, 261, 267-268
Trasimene, Lake, battle of, 188-190
Treasons Act (1534), 486
Trebia River, battle of, 188-189
Trent, Council of, 490-492, 499-500
Tribal Assembly, 172-176, 178, 200, 203-204
Tribunate, 172, 174-176, 178, 180, 200-203, 207, 209, 213, 215, 226, 266
Triremes, 88, 96-97, 183, 306
Triumvirates: First, 210-212; Second, 217-218
Trivium, 351
Troas, 260
Trojan War, 82, 84
Troyes, 245, 312
Truce of God, 338
Tudors, 438, 485-488, 492-494
Turkey, 13, 266; Turks, 273, 373, 439, 455, 466
Tutankhamon [*toot-awngk-aw'mun*], 26
Twelve Tables, 175, 180-181
Two- and three-field systems, 315, 318, 321
Tyranny, Greek, 90, 100, 103, 105
Tyre, 49, 147, 183

UNAM SANCTAM, 383, 398
Underwriting, 372
Universities, medieval, 350-353, 358, 426
Upanishads [*oo-pan'uh-shadz*], 35
Ur, 15-18
Urban II, Pope, 338, 355
Urban VI, Pope, 412, 473-474
Uruk [*oor'ook*] (Warka), 14

VALENS [*vay'lenz*], 243-244
Valla, Lorenzo, 453, 455, 466
Valley of the Kings, 27
Van Eyck [*van yk*], Jan, 453, 468
Vandals, 244-245, 277, 279, 305
Vassalage, 289, 309-312, 315, 331
Vatican Council II, 491
Vaulting, cathedral, 400-404
Vedas [*vay'duz*], 35
Veii [*vay'aye*], 166
Venice, 366, 369, 432-433, 439, 441
Venus, 256
Verdun, Treaty of, 307, **327**
Verres [*veh'rayz*], Gaius, 133, 209-210, 224

Verrocchio, Andrea del, 433
Vespasian, 231, 235-236
Vespucci, Amerigo, 441
Vesuvius, Mt., 236, 268-269
Vikings (Norsemen, Northmen), 306-307, 324, 331
Villanovan folk, 162-163
Villas: Roman, 234, 314; Charlemagne's Decree concerning, 305, 332
Villon [*vee-yohn'*], François, 453, 465
Vincent of Beauvais, 381
Vinci, Leonardo da, 453, 457-460, 469; *Last Supper*, 458-459; *Mona Lisa*, 458
Virgil, 155, 255; *Aeneid* [*i-nee'id*], 255
Virtus, 171
Visigoths, 242-244, 277, 279, 284
Vitellius, 231, 235
Vizier, 29-30
Volgograd (Stalingrad), 374
Volscians, 166
Vulgate, 344, 455, 466, 490

WARDSHIP, 311
Wartburg [*vart'burg*], 481
Water-wheel, 369
Week-work, 319
Wenceslas IV, 414
Wessex, 306
Westminster Abbey, 469
William, Duke of Aquitaine, 366
William I, the Conqueror, 321, 324-325, 414, 417

William II, Rufus, of England, 417
Windmill, 369
Winged Victory of Samothrace, 153
Witan, 417, 421
Wittenberg [*vit'n-berkh*], 479, 482; University of, 478-479
Wool, 377, 391, 414, 419-420, 434, 442
Worms [*vorms*], Concordat of, 338; Diet of, 481-482, 489
Writing: Chinese, 39; Egyptian, 31; Minoan, 78-79, 84; Sumerian, 18
Wycliffe, John, 413, 474, 476

XERXES [*zurk'seez*], 106-108
Ximines de Cisernos, Cardinal, 488-489

YAHWEH [*yaw'way*], 60, 63-69, 71
Yang-shao [*yang shoo*], 38
Yellow River, 37
Yorkists, 438

ZACHARIAS, Pope, 292
Zama [*zay'muh*], battle of, 192-193
Zeno [*zee'noh*], 153, 159
Zeus [*zoos*], 139, 141, 256
Zeuxis [*zook'sis*], 140
Ziggurat, 17-18
Zinjanthropus, 6
Zoroaster [*zoh-roh-as'tur*], 59, 256; Zoroastrianism, 57, 59
Zwingli, Ulrich, 483-484

PUBLISHER'S ACKNOWLEDGMENTS TO THE FIRST EDITION

Grateful acknowledgment is made to the following for permission to quote from copyrighted material:

GEORGE ALLEN & UNWIN LTD., *Mohammed and Charlemagne* by Henri Pirenne; and *Mediaeval Universities* by N. Schachner.

APPLETON-CENTURY-CROFTS, INC., *The World of the Middle Ages* by John L. LaMonte, copyright 1949, Appleton-Century-Crofts, Inc.; *Western Europe in the Middle Ages* by Joseph R. Strayer, copyright 1955, Appleton-Century-Crofts, Inc.

UNIVERSITY PRESS, CAMBRIDGE, *Liberties and Communities in Mediaeval England* by Helen Cam.

UNIVERSITY PRESS, CAMBRIDGE, *Life on the English Manor*, by H. S. Bennett and *Life in the Middle Ages* by G. G. Coulton.

THE CLARENDON PRESS, *The Greek Commonwealth* by A. Zimmern; *The European Inheritance; From Alexander to Constantine*.

COLUMBIA UNIVERSITY PRESS, *Roman Civilization* by Lewis and Reinhold and *Introduction to Contemporary Civilization in the West*.

DOUBLEDAY & COMPANY INC., *A History of Rome* by Moses Hadas, copyright 1956, by Moses Hadas, reprinted by permission of Doubleday & Co. Inc.

HARVARD UNIVERSITY PRESS, *Plutarch's Lives* translated by B. Perrin, Loeb Classical Library.

LONGMANS, GREEN & CO. LIMITED, *The Scientific Revolution* by A. R. Hall; *The Constitutional History of England* by B. Wilkinson; *A History of Medieval Europe* by R. H. C. Davis.

MCCLELLAND & STEWART LIMITED, *The Birth of Britain* by Sir Winston Churchill.

MESSRS. MAX PARRISH & CO. LTD., charts prepared by the Isotype Institute.

PENGUIN BOOKS LTD., *Thucydides: History of the Peloponnesian War*, translated by Rex Warner.

SIR ISAAC PITMAN & SONS LIMITED, *Cicero and the Roman Republic* by F. R. Cowell.

RINEHART & COMPANY INCORPORATED, *The Medieval World* by L. C. MacKinney.

CHARLES SCRIBNER'S SONS, *The Book of the Courtier* by Count Castiglione, translated by Leonard E. Opdycke.

The Regents of the University of Wisconsin, copyright owners of *Classics in Translation*, Volume I by Paul L. MacKendrick and Herbert M. Howe published by The University of Wisconsin Press.

PUBLISHER'S ACKNOWLEDGMENTS TO THE REVISED EDITION

QUOTATIONS

ABBEY PRESS, *The Rule of St. Benedict*, translated by D. O. H. Blair, copyright 1948.

ALLEN AND UNWIN, *Conquest and Discoveries of Henry the Navigator* by V. de Castro E. Almieda, copyright 1936.

APPLETON-CENTURY-CROFTS, *Medieval Europe* by J. O'Sullivan and J. F. Burns, copyright 1943; *Readings in Medieval History* by J. F. Scott, A. Hyman and A. H. Noyes, copyright 1933; *The Middle Ages, 395 to 1500* by J. R. Strayer and D. C. Munro, copyright 1959.

BASIL BLACKWELL, *The Origins of Modern Germany* by G. Barraclough, copyright 1947; *They Saw It Happen: An Anthology of Eye-Witnesses' Accounts of Events in British History, 55 B.C. to A.D. 1485*, edited by W. O. Hassall, copyright 1957.

GEORGE BRAZILLER, *Islam*, edited by J. A. Williams, copyright 1961.

CAMBRIDGE UNIVERSITY PRESS, *The Italian Renaissance in its Historical Background* by D. Hay, copyright 1961; *Saints and Scholars: Twenty-five Medieval Portraits* by D. Knowles, copyright 1962; *The Greeks* by D. E. Limebeer, copyright 1952; *The Shorter Cambridge Medieval History, Vol. 2* by C. W. Previté-Orton, copyright 1952.

CLARENDON PRESS, OXFORD, *The Legacy of Islam* by T. W. Arnold and A. Guillaume, copyright 1931.

COLUMBIA UNIVERSITY PRESS, *Introduction to Contemporary Civilization in the West, Vol. 1*, copyright 1954; *Hellenic Civilization* edited by G. W. Botsford and E. G. Sihler, copyright 1915; *Introduction to Oriental Civilizations: Sources of Chinese Tradition*, edited by W. T. deBary, copyright 1960; *Introduction to Oriental Civilizations: Sources of Indian Tradition*, edited by W. T. deBary, copyright 1958; *Roman Civilization: Selected Readings Vol. 1, The Republic* by N. Lewis and M. Reinhold, copyright 1951; *Roman Civilization:*

Selected Readings, Vol 2, The Empire by N. Lewis and M. Reinhold, copyright 1955; *Chaucer's World*, edited by E. Rickert, C. C. Olson and M. M. Crow, copyright 1948; *University Records and Life in the Middle Ages* by Lynn Thorndike, copyright 1944.

CORNELL UNIVERSITY PRESS, *The Decline of Rome and the Rise of Medieval Europe* by S. Katz, copyright 1955; *Heirs of the Roman Empire* by R. E. Sullivan, copyright 1960.

DOUBLEDAY AND COMPANY, *A History of Rome* by M. Hadas, copyright 1956.

ELEK, *Pompeii and Herculaneum, The Glory and the Grief* by M. Brion, copyright 1960.

EYRE & SPOTTISWOODE, *King John* by W. L. Warren, copyright 1961.

EYRE & SPOTTISWOODE AND THE ESTATE OF H. A. L. FISHER, *A History of Europe* by H. A. L. Fisher, copyright 1936.

HARCOURT BRACE AND WORLD, *Art Through the Ages* by H. Gardner, copyright 1948.

HARPER AND ROW, *The Western Intellectual Tradition From Leonardo to Hegel* by J. Bronowski and B. Mazlish, copyright 1960.

HARVARD UNIVERSITY PRESS, *Plutarch's Antony*, translated by B. Perrin, copyright 1920, Loeb Classical Library Edition; *Plutarch's Lives*, translated by B. Perrin, copyright 1915, Loeb Classical Library Edition.

HOLT, RINEHART AND WINSTON, *The Tenth Century: How Dark the Dark Ages?* by R. S. Lopez, copyright 1959.

HOUGHTON MIFFLIN COMPANY, *Europe in Transition, 1300 to 1520* by W. K. Ferguson, copyright 1962; *The Making of the Modern Mind*, revised, by J. H. Randall Jr., copyright 1940.

LAWRENCE AND WISHART, *Studies in Ancient Greek Society, Vol. 2*, by G. D. Thomson, copyright 1955.

J. B. LIPPINCOTT COMPANY, *Readings in Western Civilization* by Knoles and Snyder.

THE MACMILLAN COMPANY (NEW YORK), *Ancient History* by C. A. Robinson, copyright 1951.

METHUEN & CO., *A History of Rome During the Later Republic and Early Principate, Vol. 1*, by A. H. J. Greenidge, copyright 1904; *The Archaeology of Crete* by Pendlebury, copyright 1939.

ASHLEY MONTAGU, *Man: His First Million Years*, World Publishing Co.

THOMAS NELSON AND SONS (CANADA), *Revised Standard Version of the Bible*, copyrighted 1946 and 1952.

MAX PARRISH, *Handbook to the Palace of Minos, Knossos* by Pendlebury, copyright 1955.

PENGUIN BOOKS, *Sacred Books of the World* by A. C. Bouquet, copyright 1954; *Roman Readings* by M. Grant, copyright 1958; *The Twelve Caesars* by R. Graves; *Thucydides History of the Peloponnesian War*, translated by R. Warner.

PHAIDON PRESS LTD., *Life in Medieval France* by J. Evans, copyright 1957.

PRENTICE-HALL, *Great Problems in European Civilization*, edited by K. M. Setton and H. R. Winkler, copyright 1954.

PRINCETON UNIVERSITY PRESS, *The Consecration of the Church of St. Denis* by E. Panofsky, copyright 1946.

QUADRANGLE BOOKS, *The Face of the Ancient Orient* by S. Moscati, copyright 1960.

RANDOM HOUSE, *Plutarch's Lives* (Dryden translation), copyright 1864.

CHARLES SCRIBNER'S SONS, *Europe, 1450 to 1815* by E. J. Knapton, copyright 1958.

SPECULUM, *Technology and Invention in the Middle Ages* by Lynn White Jr. (Vol. XV, No. 2, April 1940).

RALPH E. TURNER (Yale University), *Great Cultural Traditions, Vol. 1*, published by McGraw-Hill, copyright 1941.

UNIVERSITY OF CHICAGO PRESS, *Petrarch's Letters to Classical Authors*, translated by Mario E. Cosenza, copyright 1910.

UNIVERSITY OF MICHIGAN PRESS, *The Age of Attila: Fifth-Century Byzantium and The Barbarians* by C. D. Gordon, copyright 1960.

UNIVERSITY OF PENNSYLVANIA PRESS, *Translations & Reprints*, Vol. 3, No. 2 and Vol. 6, No. 5; *Opus Majus of Robert Bacon* (Burke translation); *Paganism to Christianity in the Roman Empire* by W .W. Hyde, copyright 1946.

UNIVERSITY OF WISCONSIN PRESS AND PROF. J. P. HEIRONIMUS, *Classics in Translation, Vol. II, Roman Literature* by MacKendrick and Howe, copyright 1958.

D. VAN NOSTRAND, *The Near East in History* by P. K. Hitti, copyright 1961; *The Age of the Reformation* by R. H. Bainton, copyright 1956.

WEIDENFELD & NICOLSON, *The Medieval World: Europe, 1100 to 1350* by F. Heer, copyright 1960.

PHOTOGRAPHS

Aerofilms and Aero Pictorial Limited: 339
American School of Classical Studies, Athens: 85, 115
Amsterdam Museum of Asiatic Art: 36
Art Reference Bureau: ii, iii, 137, 146-7, 161, 192, 211, 227, 236, 263, 431, 433, 461
Bettman Archive: 327, 478
British Museum: 2, 154, 273, 361
British Travel and Holidays Association: 417
Clarendon Press, Oxford: 28
Ewing Galloway: 387
Alison Frantz, Princeton: 79
Franciscan Fathers of La Custodie de Terre Sainte, Jerusalem: 276
French Tourist Office, Montreal: 468
The Louvre: 17, 467
Metropolitan Museum, New York: 163, 171, 462
Monkmeyer: 218
Montreal Museum: 91, 249
National Gallery, Washington: 253
The National Geographic Society: 6
National Tourist Office, Greece: 143
New York Public Library: 484
Oriental Institute, University of Chicago: 18, 58
Roma's Press Service: 491
Royal Library, Windsor: 457
Royal Ontario Museum: 41, 67
Thames and Hudson: 451
University Museum, Philadelphia: 18
Yale University Art Library: 77

Every reasonable care has been taken to trace ownership of copyrighted material in this edition. The author and publisher will welcome information which will enable them to rectify any reference or credit in the next printing of this book.

Map of Western Europe

- KINGDOM OF NORWAY
- SCOTLAND
- NORTH SEA
- KINGDOM OF DENMARK — Copenhagen
- IRELAND — Dublin
- ENGLAND — London
- THE NETHERLANDS — Amsterdam
- FLANDERS — Brussels
- THE GERMAN EMPIRE
- SAXONY
- Frankfurt
- Elbe R.
- Rhine R.
- BAVARIA
- ATLANTIC OCEAN
- FRANCE — Paris
- Seine R.
- Loire R.
- Rhone R.
- SWISS CONFEDERATION
- Geneva
- SAVOY
- Avignon
- Milan
- Venice
- VENE-
- Genoa
- PAPAL STATES
- Rome
- PORTUGAL — Lisbon
- NAVARRE
- ARAGON
- SPAIN
- CASTILE — Madrid
- GRANADA
- Cadiz
- BALEARIC ISLANDS
- CORSICA (Genoese)
- KINGDOM OF SARDINIA
- FEZ AND MOROCCO
- BARBARY STATES
- MEDITERRANEAN

Scale of Miles: 0 — 100 — 200